T0214793

Algebraic geometry is one of the most diverse fields of research in mathematics. It has had an incredible evolution over the past century, with new subfields constantly branching off and spectacular progress in certain directions and, at the same time, with many fundamental unsolved problems still to be tackled.

In the spring of 2009 the first main workshop of the MSRI algebraic geometry program served as an introductory panorama of current progress in the field, addressed to both beginners and experts. This volume reflects that spirit, offering expository overviews of the state of the art in many areas of algebraic geometry. Prerequisites are kept to a minimum, making the book accessible to a broad range of mathematicians. Many chapters present approaches to long-standing open problems by means of modern techniques currently under development and contain questions and conjectures to help spur future research.

Mathematical Sciences Research Institute
Publications

59

Current Developments in Algebraic Geometry

Mathematical Sciences Research Institute Publications

Volumes 1–4, 6–8, and 10–27 are published by Springer-Verlag

Current Developments in Algebraic Geometry

Edited by

Lucia Caporaso

Università Roma Tre

James McKernan

Massachusetts Institute of Technology

Mircea Mustaţă

University of Michigan, Ann Arbor

Mihnea Popa

University of Illinois at Chicago

Lucia Caporaso
Dipartimento di Matematica
Università Roma Tre
caporaso@mat.uniroma3.it

Mircea Mustaţă
Department of Mathematics
University of Michigan, Ann Arbor
mmustata@umich.edu

James McKernan
Department of Mathematics
Massachusetts Institute of Technology
mckernan@math.mit.edu

Mihnea Popa
Department of Mathematics
University of Illinois at Chicago
mpopa@math.uic.edu

Silvio Levy (*Series Editor*)
Mathematical Sciences Research Institute
Berkeley, CA 94720
levy@msri.org

The Mathematical Sciences Research Institute wishes to acknowledge support by the National Science Foundation and the *Pacific Journal of Mathematics* for the publication of this series.

CAMBRIDGE
UNIVERSITY PRESS

32 Avenue of the Americas, New York NY 10013-2473, USA

Cambridge University Press is part of the University of Cambridge.

It furthers the University's mission by disseminating knowledge in the pursuit of education, learning and research at the highest international levels of excellence.

www.cambridge.org
Information on this title: www.cambridge.org/9781107459465

© Mathematical Sciences Research Institute 2012

First published 2012
First paperback edition 2014

A catalogue record for this publication is available from the British Library

ISBN 978-0-521-76825-2 Hardback
ISBN 978-1-107-45946-5 Paperback

Current Developments in Algebraic Geometry
MSRI Publications
Volume **59**, 2011

Contents

Current Developments in Algebraic Geometry
MSRI Publications
Volume 59, 2011

Preface

In the spring of 2009, the entire MSRI semester was devoted to Algebraic Geometry. The first main workshop, held during the week of January 26–30 and entitled "Classical Algebraic Geometry Today", served as an introduction to the whole program. It was designed to give a panorama of current progress in the field, of interest to experts, but presented in an introductory manner in order to make it accessible to those with only a general background. As a result, all talks contained a substantial expository and didactic component for the younger audience for whom the workshop was primarily intended.

One of the goals of the workshop was to reflect the diversity of Algebraic Geometry and its most fascinating recent developments. As such, it covered a wide range of topics such as birational geometry and classification, rational curves on algebraic varieties, moduli spaces of abelian varieties, derived categories and Fourier-Mukai transforms, curve counting invariants, holomorphic symplectic manifolds, and Hodge theory. Topics like modern enumerative geometry, moduli of varieties of general type, and combinatorial geometry, although represented at this event as well, were more strongly emphasized at other workshops organized throughout the semester.

The speakers at the workshop were, in alphabetical order: Mark de Cataldo, Olivier Debarre, David Eisenbud, Sam Grushevsky, Christopher Hacon, Joe Harris, Daniel Huybrechts, Jun-Muk Hwang, János Kollár, Robert Lazarsfeld, Alina Marian, Kieran O'Grady, Martin Olsson, Rita Pardini, Giuseppe Pareschi, Richard Thomas, Burt Totaro, and Yuri Tschinkel.

We hope that this volume closely reflects the spirit of the workshop. Intended to give an overview of the state of the art in the field, it is addressed to both beginners and specialists. As one of the fields in mathematics with the longest history, Algebraic Geometry has had an incredible evolution over the past century, with new subfields constantly branching off, with spectacular progress in certain directions, but at the same time, with many fundamental unsolved problems still to be tackled. The focus of the volume is on approaches to long-standing open problems (in other words, problems with a "classical" flavor) by means of modern techniques currently under development. Many of the surveys contain questions and conjectures which will hopefully lead to a flurry of activity and wonderful new results in the future.

The volume contains fifteen contributions. Most of them are expanded versions of the speakers' lectures and, as mentioned above, cover a wide range of topics:

- birational geometry (T. de Fernex and C. Hacon, B. Totaro)

- rational curves and differential geometry (J.-M. Hwang)

- moduli spaces of abelian varieties in connection with the Schottky problem (S. Grushevsky) and log-geometry (M. Olsson)

- moduli spaces via period maps (O. Debarre)

- derived categories in the study of K3 surfaces (D. Huybrechts) and irregular varieties (G. Pareschi)

- anabelian geometry (F. Bogomolov and Y. Tschinkel)

- holomorphic symplectic manifolds (K. O'Grady)

- classical problems on interpolation (J. Harris) and projections (R. Beheshti and D. Eisenbud)

- classification of surfaces of general type (M. Mendes Lopes and R. Pardini)

- Hodge theory of character varieties (M. de Cataldo)

- quotients of algebraic varieties (J. Kollár, appendix by C. Raicu).

We would like to thank the authors for the effort they put into submitting truly valuable contributions. We are grateful to Silvio Levy, the MSRI editor, for his help in preparing this volume, and to the organizers of the MSRI semester, Brendan Hassett, Sándor Kovács and Ravi Vakil, for suggesting that we organize the original workshop and for logistical help. Special thanks are due to Brendan Hassett, who co-organized the workshop together with us, and who initiated and further supported the idea of these proceedings.

Lucia Caporaso
James McKernan
Mircea Mustaţă
Mihnea Popa

Current Developments in Algebraic Geometry
MSRI Publications
Volume **59**, 2011

Fibers of projections and submodules of deformations

ROYA BEHESHTI AND DAVID EISENBUD

We bound the complexity of the fibers of the generic linear projection of a smooth variety in terms of a new family of invariants. These invariants are closely related to ideas of John Mather, and we give a simple proof of his bound on the Thom–Boardman invariants of a generic projection as an application.

1. Introduction

Let $X \subset \mathbf{P}^r$ be a smooth projective variety of dimension n over an algebraically closed field k of characteristic zero, and let $\pi : X \to \mathbf{P}^{n+c}$ be a general linear projection. In this note we introduce some new ways of bounding the complexity of the fibers of π. Our ideas are closely related to the groundbreaking work of John Mather, and we explain a simple proof of his result [1973] bounding the Thom–Boardman invariants of π as a special case.

This subject was studied classically for small n. In our situation the map π will be finite and generically one-to-one, so we are asking for bounds on the complexity of finite schemes, and the degree of the scheme is the obvious invariant. Consider, for simplicity, the case $c = 1$. It is well-known that the maximal degree of the fiber of a general projection of a curve to the plane is 2, and that the maximal degree of a fiber of a general projection of a smooth surface to three-space is 3. These results were extended to higher dimension and more general ground fields at the expense of strong hypotheses on the structure of the fibers by Kleiman, Roberts, Ran and others.

In characteristic zero, the most striking results are those of John Mather. In the case $c = 1$ and $n \leq 14$ he proved that a general projection π would be a stable map, and as a consequence he was able to show that, in this case, the fibers of π have degree $\leq n + 1$. More generally, in case $n \leq 6c + 7$, or $n \leq 6c + 8$ and $c \leq 3$, he showed that the degree of any fiber of π is bounded by $n/c + 1$. He also proved that for any n and c, the number of distinct points in any fiber is bounded by $n/c + 1$; this is a special case of his result bounding the Thom–Boardman invariants.

An optimist (such as the second author), seeing these facts, might hope that the degree of the fibers of π would be bounded by $n/c + 1$ for any n and c. However, Lazarsfeld [2004, II, Proposition 7.2.17] showed that the singularities of $\pi(X)$ could have very high multiplicity when n is large. His ideas can also be used to prove that for large n and a sufficiently positive embedding of any smooth variety X in a projective space, a general linear projection of X to $\mathbf{P}^{n+c}$ will have fibers of degree exponentially greater than n/c. The first case with $c = 1$ in which his argument gives something interesting is $n = 56$, where it shows that (if the embedding is sufficiently positive) then there will be fibers of degree ≥ 70. For a proof see [Beheshti and Eisenbud 2010, Proposition 2.2].

Although we know no upper bound on the degrees of the fibers of π that depends only on n and c, we showed in [Beheshti and Eisenbud 2010] that there is a natural invariant of the fiber that agrees "often" with the degree and that is always bounded by $n/c + 1$.

In this note, we generalize the construction there and give a general mechanism for producing such invariants. Our approach is closely related to that of John Mather.

Here is a sample of the results we prove. We first ask how "bad" a finite scheme Y can be and still appear inside the fiber of the generic projection of a smooth n-fold to $\mathbf{P}^{n+c}$? Our result is written in terms of the degree of Y and the degree of the *tangent sheaf* to Y, defined as

$$\mathcal{T}_Y = Hom_{\mathcal{O}_Y}(\Omega_{Y/K}, \mathcal{O}_Y).$$

We prove the following in Theorem 4.1.

Theorem 1.1. *Let $X \subset \mathbf{P}^r$ be a smooth projective variety of dimension n, and let Y be a scheme of dimension zero. If for a general linear projection $\pi_\Sigma : X \to \mathbf{P}^{n+c}$, there is a fiber of π_Σ that contains Y as a closed subscheme, then*

$$\deg Y + \frac{1}{c} \deg \mathcal{T}_Y \leq \frac{n}{c} + 1.$$

This result easily implies (the special case for projections) of Mather's result bounding the Thom–Boardman invariants (itself a special case of the transversality theorem he proves.) This is because, as Mather shows, the Thom Boardman invariant of a germ at a point is determined by knowing whether certain sub-schemes are or are not contained in the fiber. By way of example, we carry out the proof of the following useful special case:

Corollary 1.2 (Mather). *Let $X \subset \mathbf{P}^r$ be a smooth subvariety of dimension n, and let $\pi : X \to \mathbf{P}^{n+c}$ be a general projection with $c \geq 1$. Let p be a point in $\mathbf{P}^{n+c}$, and assume $\pi^{-1}(p)$ consists of r distinct points $q_1, \ldots, q_r$. Denote by d_i the corank of π at q_i. Then*

$$\sum_{1 \leq i \leq r} \left(\frac{d_i^2}{c} + d_i + 1 \right) \leq \frac{n}{c} + 1.$$

In particular, the number of distinct points in every fiber is bounded by $n/c + 1$.

Mather's approach to this theorem works because the subschemes involved in defining the Thom–Boardman singularities have no moduli—there is a discrete family of "test schemes". In other situations it is much more common for a certain "type" of subscheme to appear in a fiber, although the subschemes themselves have nontrivial moduli. We can prove a result (Theorem 4.2) taking the dimension of the moduli space into account that sometimes gives sharper results. Suppose, for example, that you know that a generic projection from the smooth n-fold X to $\mathbf{P}^{n+1}$ always has a fiber isomorphic to one of the schemes $Y_F := \operatorname{Spec} k[x, y, z]/F + (x, y, z)^5$, where F varies over all nonsingular cubic forms. This "truncated cone over an elliptic curve", which has degree 31, varies with one parameter of moduli. The only obvious subscheme common to all the Y_F is $\operatorname{Spec} k[x, y, z]/(x, y, z)^3$. With Theorem 1.1 we get the bound $n \geq 36$. But if we apply Theorem 4.2 to the 1-dimensional moduli family of Y_F, we get the much stronger bound $n \geq 69$.

One motivation for the study of the complexity of the fibers of general projections comes from the Eisenbud–Goto conjecture [1984], which states that the regularity of a projective subvariety of $\mathbf{P}^r$ is $\leq \deg(X) - \operatorname{codim}(X) + 1$. An approach to this conjecture, which has been used to prove the conjecture for smooth surfaces and to prove a slightly weaker bound for smooth varieties of dimension at most 5 (see [Lazarsfeld 1987] and [Kwak 2000]), is to bound the regularity of the fibers of general projections.

Conjecture 1.3. *Let $X \subset \mathbf{P}^r$ be a smooth projective variety of dimension n, and let $\pi : X \to \mathbf{P}^{n+c}$ be a general linear projection. If $Z \subset X$ is any fiber, then the Castelnuovo–Mumford regularity of Z as a subscheme of $\mathbf{P}^r$ is at most $n/c + 1$.*

The truth of the conjecture would imply that the Eisenbud–Goto conjecture holds up to a constant that depends only on n and r and is given explicitly in [Beheshti and Eisenbud 2010]. If true, the conjecture is sharp in some cases: The

"Reye Construction" gives an Enriques surface in $\mathbf{P}^5$ whose projections to $\mathbf{P}^3$ all have 3 colinear points in some fibers; and an argument of Lazarsfeld shows that if X is a Cohen–Macaulay variety of codimension 2 in $\mathbf{P}^{n+2}$, and if X is not contained in a hypersurface of degree $\leq n$, then any projection of X into $\mathbf{P}^{n+1}$ has fibers of length $n+1$. In this case any fiber is colinear. Since a scheme consisting of $n+1$ colinear points has regularity $n+1$, we get fibers of regularity $= n+1$ in these examples (see [Beheshti and Eisenbud 2010] for proofs.)

2. Notation

We will work over an algebraically closed field k of characteristic zero. If T is a coherent sheaf of finite support on some scheme, we identify T with its module of global sections and write $\deg T$ for the vector space dimension of this module over k.

We fix $r \geq 2$, and we denote by G_k the Grassmannian of linear subvarieties of codimension k in $\mathbf{P}^r$. Let X be a smooth projective variety of dimension n, and let $c \geq 1$. A linear projection $X \to \mathbf{P}^{n+c}$ is determined by a sequence $l_1, \ldots, l_{n+c+1}$ of $n+c+1$ independent linear forms on $\mathbf{P}^r$ that do not simultaneously vanish at any point of X. Associated to such a projection is the projection center Σ, the linear space of codimension $n+c+1$ defined by the vanishing of the l_i. The map taking a linear projection to the associated projection center makes this set of projections into a $\mathrm{PGL}(n+c)$-bundle over the subset $U \subset G_{n+c+1}$ of planes Σ that do not meet X.

We denote by π_Σ the linear projection $X \to \mathbf{P}^{n+c}$ with center Σ. The morphism π_Σ is birational, and its fibers are all zero-dimensional. The fibers of π have the form $X \cap \Lambda$, where $\Lambda \in G_{n+c}$ contains Σ.

We will keep this notation throughout this paper.

3. Measuring the complexity of the fibers

Let $X \subset \mathbf{P}^r$ be a smooth subvariety of dimension n, and let H be a subscheme of G_{n+c}. For $[\Lambda] \in H$, set $Z = \Lambda \cap X$, and assume $\dim Z = 0$. Consider the restriction map

$$\rho : T_{G_{n+c},[\Lambda]} = H^0(N_{\Lambda/\mathbf{P}^r}) \to N_{\Lambda/\mathbf{P}^r}|_Z$$

Let $V_G \subset N_{\Lambda/\mathbf{P}^r}|_Z$ be the image of ρ and let $V_H = \rho(T_{H,[\Lambda]}) \subset V_G$. Denote by $\mathbb{O}_Z V_H$ the $\mathbb{O}_Z$-submodule of $N_{\Lambda/\mathbf{P}^r}|_Z$ generated by V_H, and let Q be the quotient module:

$$0 \to \mathbb{O}_Z V_H \to N_{\Lambda/\mathbf{P}^r}|_Z \to Q \to 0.$$

Here is our main technical result:

Theorem 3.1. *Let $X \subset \mathbf{P}^r$ be a smooth subvariety of dimension n, and let H be a locally closed irreducible subvariety of $\mathbf{G}_{n+c}$, $c \geq 1$. Assume that, for a general $[\Sigma]$ in $\mathbf{G}_{n+c+1}$, there is $[\Lambda] \in H$ such that $\Sigma \subset \Lambda$. Then for a general $[\Lambda] \in H$, either $\Lambda \cap X$ is empty, or*

$$\deg Q \leq n + c.$$

The proof uses the following result, which will also be used in the proof of Theorem 5.2.

Lemma 3.2. *Let X be a smooth variety of dimension n in $\mathbf{P}^r$, and let H be a smooth locally closed subvariety of $\mathbf{G}_{n+c}$. Assume that for a general $[\Sigma]$ in $\mathbf{G}_{n+c+1}$, there is $[\Lambda] \in H$ such that $\Sigma \subset \Lambda$. Let $[\Sigma]$ be a general point of $\mathbf{G}_{n+c+1}$ and let $[\Lambda]$ be a point of H such that $\Sigma \subset \Lambda$. If Q is as in Theorem 3.1, then the map*

$$H^0(N_{\Lambda/\mathbf{P}^r} \otimes \mathcal{O}_\Lambda(-1)) \to Q$$

is surjective.

Proof. The restriction map $N_{\Lambda/\mathbf{P}^r} \to N_{\Lambda/\mathbf{P}^r}|_Z$ followed by the surjective map $N_{\Lambda/\mathbf{P}^r}|_Z \to Q$ gives a surjective map of $\mathcal{O}_\Lambda$-modules $N_{\Lambda/\mathbf{P}^r} \to Q$. We denote the kernel by F:

$$0 \to F \to N_{\Lambda/\mathbf{P}^r} \to Q \to 0.$$

We first show that the restriction map $H^0(F) \to H^0(F|_\Sigma)$ is surjective. Consider the incidence correspondence

$$J = \{([\Sigma], [\Lambda]) : \Sigma \subset \Lambda, [\Lambda] \in H\} \subset \mathbf{G}_{n+c+1} \times H,$$

and assume that $[\Sigma]$ is a general point of $\mathbf{G}_{n+c+1}$. By our assumption the projection map $\pi_1 : J \to \mathbf{G}_{n+c+1}$ is dominant. Since H is smooth, and since the projection map $\pi_2 : J \to H$ makes J a Grassmannian bundle over H, the space J is smooth as well. Thus by generic smoothness, π_1 is smooth at $([\Sigma], [\Lambda])$, and so the map on Zariski tangent spaces $T_{J,([\Sigma],[\Lambda])} \to T_{\mathbf{G}_{n+c+1},[\Sigma]}$ is surjective.

The short exact sequence of $\mathcal{O}_\Sigma$-modules

$$0 \to N_{\Sigma/\Lambda} \to N_{\Sigma/\mathbf{P}^r} \to N_{\Lambda/\mathbf{P}^r}|_\Sigma \to 0$$

gives a surjective map $H^0(N_{\Sigma/\mathbf{P}^r}) \to H^0(N_{\Lambda/\mathbf{P}^r}|_\Sigma)$. Note that since Σ is general, $\Sigma \cap X = \varnothing$, and since Q is supported on $\Lambda \cap X$, $F|_\Sigma = N_{\Lambda/\mathbf{P}^r}|_\Sigma$. It follows from the commutative diagram

$$
\begin{array}{ccccc}
T_{J,([\Sigma],[\Lambda])} & \twoheadrightarrow & T_{\mathbf{G}_{n+c+1},[\Sigma]} = H^0(N_{\Sigma/\mathbf{P}^r}) & \longrightarrow & H^0(N_{\Lambda/\mathbf{P}^r}|_\Sigma) \\
\downarrow & & & & \| \\
T_{H,[\Lambda]} & \longrightarrow & H^0(F) & \longrightarrow & H^0(F|_\Sigma)
\end{array}
$$

then that $H^0(F) \to H^0(F|_\Sigma)$ is surjective.

Consider now the short exact sequence

$$0 \to F \otimes \mathcal{O}_\Lambda(-1) \to F \to F|_\Sigma \to 0.$$

Since $F|_\Sigma = N_{\Lambda/\mathbf{P}^r}|_\Sigma$, we have $H^1(F|_\Sigma) = H^1(N_{\Lambda/\mathbf{P}^r}|_\Sigma) = 0$. Since

$$H^0(F) \to H^0(F|_\Sigma)$$

is surjective, we get $H^1(F \otimes \mathcal{O}_\Lambda(-1)) \cong H^1(F)$. Therefore, the image of the map $H^0(N_{\Lambda/\mathbf{P}^r}) \to Q$ is the same as that of the map $H^0(N_{\Lambda/\mathbf{P}^r} \otimes \mathcal{O}_\Lambda(-1)) \to Q$, and thus by Proposition 3.3 both of these maps are surjective.   □

Proposition 3.3 [Beheshti and Eisenbud 2010, Proposition 3.1]. *Suppose that* $\delta : A \to B$ *is an epimorphism of coherent sheaves on* $\mathbf{P}^r$, *and suppose that A is generated by global sections. If* $\delta(H^0(A)) \subset H^0(B)$ *has the same dimension as* $\delta(H^0(A(1))) \subset H^0(B(1))$, *then* $\dim B = 0$ *and* $\delta(H^0(A(m))) = H^0(B(m)) \cong H^0(B)$ *for all* $m \geq 0$.   □

Proof of Theorem 3.1. Assume that for a general $[\Lambda]$ in H, $\Lambda \cap X$ is nonempty. It follows from Lemma 3.2, applied to the smooth locus in H, that the map $H^0(N_{\Lambda/\mathbf{P}^r} \otimes \mathcal{O}_\Lambda(-1)) \to Q$ is surjective. Therefore,

$$\deg Q \leq \dim H^0(N_{\Lambda/\mathbf{P}^r} \otimes \mathcal{O}_\Lambda(-1)) = \dim H^0(\mathcal{O}_\Lambda^{n+c}) = n + c.$$

□

Since we have

$$(n+c) \deg Z = \deg N_{\Lambda/\mathbf{P}^r}|_Z = \deg \mathcal{O}_Z V_H + \deg Q,$$

where $Z = \Lambda \cap X$, it follows from the above theorem that any upper bound on the degree of $\mathcal{O}_Z V_H$ puts some restrictions on the fibers of π_Σ for general Σ.

Corollary 3.4. *Let* $X \subset \mathbf{P}^r$ *be a smooth projective variety of dimension n, and let* $c \geq 1$ *be an integer. Let H be a locally closed irreducible subvariety of* G_{n+c}, *and assume that for a general projection* π_Σ, *there is* $[\Lambda] \in H$ *that contains* Σ. *Then for a general* $[\Lambda] \in H$

$$\deg Z \leq \frac{\deg \mathcal{O}_Z V_H}{n+c} + 1.$$

where $Z = \Lambda \cap X$.

For example, if we apply Theorem 3.1 to the scheme $H \subset G_{n+c}$ whose points correspond to planes that intersect X in schemes of length $\geq l$, for some integer $l \geq 1$, we recover the central result of [Beheshti and Eisenbud 2010]. Recall that for varieties $X, Y \subset \mathbf{P}^r$ that meet in a scheme $Z = X \cap Y$ of dimension zero, with codim $Y - \dim X > 0$ we there defined $q(X, Y)$ to be

$$q(X, Y) = \frac{\deg \operatorname{coker} \operatorname{Hom}(\mathscr{I}_{Z/X}/\mathscr{I}^2_{Z/X}, \mathscr{O}_Z) \to \operatorname{Hom}(\mathscr{I}_{Y/P}/\mathscr{I}^2_{Y/P}, \mathscr{O}_Z)}{\operatorname{codim} Y - \dim X}.$$

For example, if X, Y are smooth and Z is a locally complete intersection scheme then $q(X, Y) = \deg X \cap Y$, and more generally q is a measure of the difficulty of flatly deforming Y in such a way that $Z = X \cap Y$ deforms flatly as well. Using this notation, and applying Theorem 3.1 to the fiber of a general projection, we derive:

Theorem 3.5. *If X is a smooth projective variety of dimension n in $\mathbf{P}^r$, and if $\pi : X \to \mathbf{P}^{n+c}$ is a general projection, then every fiber $X \cap \Lambda$, where Λ is a linear subspace containing the projection center in codimension 1, satisfies:*

$$q(X, \Lambda) \leq \frac{n}{c} + 1.$$

$\square$

In [Beheshti and Eisenbud 2010] we derived explicit bounds on the lengths of fibers from this result.

3.1. *A Problem.* Fix positive integers l and m, and let H be the reduced subscheme of $\mathbf{G}_{n+c}$ consisting of those planes Λ such that $\deg \Lambda \cap X = l$ and $\deg \Omega_{\Lambda \cap X} = m$. Assume that for a general projection $\pi : \Sigma \to \mathbf{P}^{n+c}$, there is a fiber $\Lambda \cap X$ such that $[\Lambda]$ is in H. We would like to use Corollary 3.4 in this case to get a bound on the fibers of general projections stronger than, say, that of Corollary 1.2 in [Beheshti and Eisenbud 2010].

Assume that $[\Lambda] \in H$ is a general point, and let V_H be the image of $T_{H, [\Lambda]}$ in $N_{\Lambda/\mathbf{P}^r}|_Z$. Then since we assume that the length of the intersection with X is fixed for all points of H, V_H is a subspace of $N_{Z/X}$. If we denote by $V' \subset N_{Z/X}$ the tangent space to the space of first order deformations of Z in X that keep the degree of Ω_Z fixed, then $V_H \subset V'$.

If Z is an arbitrary zero-dimensional subscheme of a smooth variety X, then V' is not necessarily a submodule of $N_{Z/X}$. For example, if we let Z be the subscheme of $\mathbf{A}^2$ defined by the ideal $I = \langle x^4 + y^4, xy(x-y)(x+y)(x-2y) \rangle$, then Z is supported at the origin and is of degree 20. Using Macaulay 2, we find that the space of deformations of Z in $\mathbf{A}^2$ that fix the degree of Ω_Z is a vector space of dimension 17, but the $\mathscr{O}_Z$-module generated by this space is a vector space of dimension 18.

Is there an upper bound on the dimension of the submodule generated by V_H that is stronger than $\deg N_{Z/X}$? In the special case, when Z is curvilinear of degree m (so deg $\Omega_Z = m-1$), V' is a submodule of degree m dim $X-(m-1) <$ deg $N_{Z/X}$. But if Z is of arbitrary degree and not curvilinear, we currently know no bound that could improve Corollary 1.2 in [Beheshti and Eisenbud 2010].

4. General projections whose fibers contain given subschemes

Fix a zero-dimensional scheme Y. We wish to give a bound on invariants of Y that must hold if Y appears as a subscheme of a fiber of the general projection of X to $\mathbf{P}^{n+c}$. Our result generalizes a key part of the proof of Mather's theorem bounding the Thom–Boardman invariants of a general projection.

In the following we write $\mathcal{T}_Y$ for the tangent sheaf $\mathcal{T}_Y := Hom_{\mathcal{O}_Y}(\Omega_Y, \mathcal{O}_Y)$ of Y, and similarly for X.

Theorem 4.1. *Let $X \subset \mathbf{P}^r$ be a smooth projective variety of dimension n, and let Y be a scheme of dimension zero. If for a general linear projection $\pi_\Sigma : X \to \mathbf{P}^{n+c}$, there is a fiber of π_Σ that contains Y as a closed subscheme, then*

$$\deg Y + \frac{1}{c} \deg \mathcal{T}_Y \le \frac{n}{c} + 1.$$

Proof. Let $Hom(Y, X)$ be the space of morphisms from Y to X, and let $I \subset Hom(Y, X) \times G_{n+c}$ be the incidence correspondence parametrizing the pairs $([i], [\Lambda])$ such that i is a closed immersion from Y to $\Lambda \cap X$. Denote by H the image of I under the projection map $Hom(Y, X) \times G_{n+c} \to G_{n+c}$. We give H the reduced induced scheme structure.

Let $([i], [\Lambda])$ be a general point of I, and set $Z := \Lambda \cap X$. We consider Y as a closed subscheme of X, and we let $N_{Y/X} = Hom(I_{Y/X}, \mathcal{O}_Y)$ denote the normal sheaf of Y in X. The Zariski tangent space to $Hom(Y, X)$ at $[i]$ is isomorphic to $H^0(\mathcal{T}_X|_Y)$; see [Kollár 1996, Theorem I.2.16].

Denote by M' the $\mathcal{O}_Y$-submodule of $N_{\Lambda/\mathbf{P}^r}|_Y$ generated by the image of the restriction map from the Zariski tangent space:

$$\rho_Y : T_{H,[\Lambda]} \hookrightarrow H^0(N_{\Lambda/\mathbf{P}^r}) \longrightarrow N_{\Lambda/\mathbf{P}^r}|_Y.$$

We first claim that deg $M' \ge (n+c)$ deg $Y - (n+c)$. Let Q' be the quotient of $N_{\Lambda/\mathbf{P}^r}|_Y$ by M':

$$0 \to M' \to N_{\Lambda/\mathbf{P}^r}|_Y \to Q' \to 0.$$

Denote by M the submodule of $N_{\Lambda/\mathbf{P}^r}|_Z$ generated by the image of

$$\rho_Z : T_{H,[\Lambda]} \hookrightarrow H^0(N_{\Lambda/\mathbf{P}^r}) \longrightarrow N_{\Lambda/\mathbf{P}^r}|_Z,$$

and let Q be the cokernel:

$$0 \to M \to N_{\Lambda/\mathbf{P}^r}|_Z \to Q \to 0.$$

The surjective map $N_{\Lambda/\mathbf{P}^r}|_Z \to N_{\Lambda/\mathbf{P}^r}|_Y$ carries M into M', and thus induces a surjective map $Q \to Q'$. Since by Theorem 3.1, $\deg Q \le n + c$, we have $\deg Q' \le n + c$, and so

$$\deg M' = \deg N_{\Lambda/\mathbf{P}^r}|_Y - \deg Q' \ge (n + c) \deg Y - (n + c).$$

This establishes the desired lower bound.

We next give an upper bound on $\deg M'$. Since X is smooth, dualizing the surjective map $\Omega_X|_Y \to \Omega_Y$ into $\mathcal{O}_Y$, we get an injective map $\mathcal{T}_Y \to \mathcal{T}_X|_Y$ and an exact sequence

$$0 \longrightarrow \mathcal{T}_Y \longrightarrow \mathcal{T}_X|_Y \overset{\phi}{\longrightarrow} N_{Y/X}.$$

Let π_1 and π_2 denote the projections maps from I to $\mathrm{Hom}(X, Y)$ and H respectively. We have a diagram

$$
\begin{array}{ccccc}
T_{I,([i],[\Lambda])} & \overset{d\pi_2}{\longrightarrow} & T_{H,[\Lambda]} & \hookrightarrow & T_{G_{n+c},[\Lambda]} = H^0(N_{\Lambda/\mathbf{P}^r}) \\
\downarrow{\scriptstyle d\pi_1} & & & & \\
T_{\mathrm{Hom}(X,Y),[i]} = \mathcal{T}_X|_Y & & & & \downarrow{\scriptstyle \rho_Y} \\
\downarrow{\scriptstyle \phi} & & & & \\
N_{Y/X} & \overset{\psi}{\longrightarrow} & & & N_{\Lambda/\mathbf{P}^r}|_Y
\end{array}
$$

where ψ is obtained by dualizing the map $I_{\Lambda/\mathbf{P}^r} \otimes \mathcal{O}_X \to I_{Y/X}$ into $\mathcal{O}_Y$.

It follows from the diagram that $\rho_Y(T_{H,[\Lambda]})$ is contained in the image of $\psi \circ \phi$. Since the image of $\psi \circ \phi : \mathcal{T}_X|_Y \to N_{\Lambda/\mathbf{P}^r}|_Y$ is a submodule of $N_{\Lambda/\mathbf{P}^r}|_Y$, M' is contained in the image of $\psi \circ \phi$ as well. Thus the degree of M' is less than or equal to the degree of the image of ϕ, that is

$$\deg M' \le \deg \mathcal{T}_X|_Y - \deg \mathcal{T}_Y = n \deg Y - \deg \mathcal{T}_Y.$$

Comparing this upper bound with the lower bound we got earlier completes the proof. $\qquad\square$

Theorem 4.1 was inspired by the results of Mather on the Thom–Boardman invariants. Mather shows that the Thom–Boardman symbol of a germ of a map is determined by which of a certain discrete set of different subschemes the fiber contains. These are the schemes of the form $\mathrm{Spec}\, k[x_1, \ldots, x_n]/(x_1)^{t_1} +$

$(x_1, x_2)^{t_2} + \cdots + (x_1, \ldots, x_n)^{t_n}$. His result that the general projection is transverse to the Thom–Boardman strata is closely related (see also Section 5.) We illustrate by proving the special case announced in the introduction.

Proof of Corollary 1.2. For $d \geq 1$, let $A_d = \frac{k[x_1, \ldots, x_d]}{m^2}$ where m is the ideal generated by $x_1, \ldots, x_d$. If $q \in X$ is a point of corank d for the projection π, then there is a surjective map $\mathbb{O}_{\pi^{-1}(\pi(q)), q} \to A_d$.

Fix an integer $r \geq 1$, and fix a sequence of coranks $d_1 \geq \cdots \geq d_r \geq 0$. If we denote by Y the disjoint union of the schemes Spec A_{d_i}, $1 \leq i \leq r$, then we have $\deg Y = \sum_{1 \leq i \leq r} (d_i + 1)$ and $\deg \mathcal{T}_Y = \sum_{1 \leq i \leq r} d_i^2$.

Assume now that for a general linear projection $\pi : X \to \mathbf{P}^{n+c}$, there is a fiber consisting of at least r points $q_1, \ldots, q_r$ such that the corank of π at q_i is at least d_i for $1 \leq i \leq r$. Then for a general linear projection with center Σ, there is a fiber $X \cap \Lambda$ and a closed immersion $i : Y \to \Lambda \cap X$. It follows from the previous theorem that

$$\sum_{1 \leq i \leq r} (\frac{d_i^2}{c} + d_i + 1) = \deg Y + \frac{1}{c} \deg \mathcal{T}_Y \leq \frac{n}{c} + 1.$$

$\square$

Except in a few situations, such as the Thom–Boardman computation above, it is more likely that for a general projection the fiber might be "of a certain type", or contain one of a given family of special subschemes. In the next theorem, we generalize Theorem 4.1 to such a family of zero-dimensional schemes. We have separated the proofs because this version involves considerably more technique. But we do not repeat the final part of the argument, since it is the same as before.

Theorem 4.2. *Let $X \subset \mathbf{P}^r$ be a smooth projective variety of dimension n. Suppose that B is an integral scheme of dimension m and that $p : U \to B$ is a flat family of zero dimensional schemes over B. For a point $b \in B$, let $\mathcal{T}_{U_b} = \mathrm{Hom}(\Omega_{U_b}, \mathbb{O}_{U_b})$. If for a general projection $\pi_\Sigma : X \to \mathbf{P}^{n+c}, c \geq 1$, there is a fiber U_b of $p : U \to B$ such that U_b can be embedded in one of the fibers of π_Σ, then*

$$\left(1 - \frac{m}{c}\right) \deg U_b + \min_{b \in B} \left\{ \frac{1}{c} \deg \mathcal{T}_{U_b} \right\} \leq \frac{n}{c} + 1.$$

Proof. Passing to a desingularization, we can assume that B is smooth. Let $\mathrm{Hom}_B(U, B \times X)$ be the functor

$$\mathrm{Hom}_B(U, X \times B)(S) = \{B\text{-morphisms} : U \times_B S \to (X \times B) \times_B S\}.$$

By [Kollár 1996, I.1.10], this functor is represented by a scheme $\mathrm{Hom}_B(U, X \times B)$ over B that is isomorphic to an open subscheme of $\mathrm{Hilb}(U \times X/B)$. The closed points of $\operatorname{Hom}_B(U, X \times B)$ parametrize morphisms from fibers of $p : U \to B$ to X.

Denote by $I \subset \operatorname{Hom}_B(U, X \times B) \times G_{n+c}$ the incidence correspondence consisting of the points $([i], [\Lambda])$ such that i is a closed immersion to $\Lambda \cap X$, and let H be the image of the projection map $I \to G_{n+c}$. We give H the reduced structure as a subscheme of G_{n+c}.

Assume that $([i], [\Lambda])$ is a general point of I, and assume that $[i]$ represents the closed immersion $i : U_b \to \Lambda \cap X$, $b \in B$. Set $Y = U_b$ and $Z = \Lambda \cap X$. Denote by M' the $\mathcal{O}_Y$-submodule of $N_{\Lambda/\mathbf{P}^r}|_Y$ generated by the image of the restriction map

$$\rho_Y : T_{H,[\Lambda]} \longhookrightarrow H^0(N_{\Lambda/\mathbf{P}^r}) \longrightarrow N_{\Lambda/\mathbf{P}^r}|_Y.$$

As in the proof of Theorem 4.1 we need an upper bound on the degree of M'. Since $i : Y \to X$ is a closed immersion, there is a natural map on the Zariski tangent spaces:

$$\phi : T_{\operatorname{Hom}_B(U, X \times B), [i]} \to T_{\operatorname{Hilb}(X), [Y]} = N_{Y/X}.$$

(If V is a flat family over $D := \operatorname{Spec} k[\epsilon]/\epsilon^2$, and if $f : V \to X \times D$ is such that $f_0 : V_0 \to X$ is a closed immersion, then so is f. Thus a morphism $D \to \operatorname{Hom}_B(U, X \times B)$ gives a natural morphism $D \to \operatorname{Hilb}(X)$.) As a first step toward bounding the degree of M' we will show that the $\mathcal{O}_Y$-submodule of N_Y generated by the image of ϕ has degree at most $(n + m) \deg Y - \dim \mathcal{T}_Y$.

Note that the fiber of the map $\operatorname{Hom}_B(U, X \times B) \to B$ over b is $\operatorname{Hom}(Y, X)$. Therefore, the vertical Zariski tangent space to $\operatorname{Hom}_B(U, X \times B)$ at $[i]$ is isomorphic to $H^0(\mathcal{T}_X|_Y)$. Since Y is zero-dimensional, we may identify $H^0(\mathcal{T}_X|_Y)$ with $\mathcal{T}_X|_Y$.

Let $(Q, \mathfrak{m}_Q)$ be the local ring of $\operatorname{Hom}_B(U, X \times B)$ at $[i]$, and let $\mathfrak{m}_b$ be the maximal ideal of the local ring of B at b. There is an exact sequence of k-vector spaces:

$$(\mathfrak{m}_b Q + \mathfrak{m}_Q^2)/\mathfrak{m}_Q^2 \to \mathfrak{m}_Q/\mathfrak{m}_Q^2 \to \mathfrak{m}_Q/(\mathfrak{m}_b Q + \mathfrak{m}_Q^2) \to 0.$$

Since $(\mathfrak{m}_Q/(\mathfrak{m}_b Q + \mathfrak{m}_Q^2))^*$ is the vertical Zariski tangent space at $[i]$, it is equal to $H^0(\mathcal{T}_X|_Y)$. So dualizing the above sequence, we get an exact sequence

$$0 \to H^0(\mathcal{T}_X|_Y) \to T_{\operatorname{Hom}_B(U, X \times B), [i]} \to V.$$

Since B is smooth of dimension m we see that

$$V = \operatorname{Hom}((\mathfrak{m}_b Q + \mathfrak{m}_Q^2)/\mathfrak{m}_Q^2, Q/\mathfrak{m}_b)$$

is a vector space of dimension $\leq m$.

Denote by N the quotient of $\mathscr{T}_X|_Y$ by $\mathscr{T}_Y$, and consider the diagram

$$
\begin{array}{ccccccc}
 & & \mathscr{T}_Y & & & & \\
 & & \downarrow & & & & \\
0 & \longrightarrow & \mathscr{T}_X|_Y & \longrightarrow & T_{\mathrm{Hom}_B(U,X\times B),[i]} & \longrightarrow & V \\
 & & \downarrow & & \downarrow \phi & & \\
 & & N & \lhook\joinrel\longrightarrow & N_{Y/X} & &
\end{array}
$$

The image of $\mathscr{T}_X|_Y$ under ϕ is contained in the image of $N \to N_{Y/X}$, which is a $\mathcal{O}_Y$-submodule of degree $\leq \deg \mathscr{T}_X|_Y - \deg \mathscr{T}_Y = n \deg Y - \deg \mathscr{T}_Y$. The degree of the $\mathcal{O}_Y$-submodule generated by the image of ϕ in $N_{Y/X}$ is therefore

$$\leq n \deg Y - \deg \mathscr{T}_Y + \dim V \cdot \deg Y = (n + \dim B) \deg Y - \dim \mathscr{T}_Y.$$

This establishes the desired upper bound on the degree of the $\mathcal{O}_Y$-submodule of N_Y generated by the image of ϕ.

Let π_1 and π_2 denote the projections maps from I to $\mathrm{Hom}_B(U, B \times X)$ and H respectively. We have a diagram

$$
\begin{array}{ccccc}
T_{I,([i],[\Lambda])} & \xrightarrow{d\pi_2} & T_{H,[\Lambda]} & \lhook\joinrel\longrightarrow & T_{G_{n+c},[\Lambda]} = H^0(N_{\Lambda/\mathbf{P}^r}) \\
\downarrow d\pi_1 & & & & \downarrow \rho_Y \\
T_{\mathrm{Hom}_B(U,B\times X),[i]} & & & & \\
\downarrow \phi & & & & \\
N_{Y/X} & & \xrightarrow{\psi} & & N_{\Lambda/\mathbf{P}^r}|_Y
\end{array}
$$

The diagram shows that M' is contained in the submodule generated by the image of the composition of the maps $T_{\mathrm{Hom}_B(U,B\times X),[i]} \to N_{Y/X} \to N_{\Lambda/\mathbf{P}^r}|_Y$, so by the upper bound established before we have

$$\deg M' \leq (n + m) \deg Y - \dim \mathscr{T}_Y.$$

On the other hand, the same argument as in the proof of Theorem 4.1 shows that

$$\deg M' \geq (n + c) \deg Y - (n + c).$$

Thus, we get

$$(c - m) \deg Y + \deg \mathscr{T}_Y \leq n + c.$$

Since $\deg \mathscr{T}_{U_b}$ is an upper semicontinuous function on B, we get the desired result. $\qquad\square$

5. A transversality theorem

Mather defines a property of a smooth subvariety of a jet bundle that he calls *modularity*, and proves that a general projection has jets that are transverse to any modular subvariety. In this section we study a related definition that we hope will lead to bounds stronger than the one in Theorem 3.5.

Let H be a locally closed subscheme of $G := G_{n+c}$. Let $[\Lambda] \in H$ be such that $Z := \Lambda \cap X$ is zero-dimensional. Consider the restriction map

$$\rho : T_{G,[\Lambda]} = H^0(N_{\Lambda/\mathbf{P}^r}) \to N_{\Lambda/\mathbf{P}^r}|_Z,$$

and set $V_G = \rho(T_{G,[\Lambda]})$ and $V_H = \rho(T_{H,[\Lambda]}) \subset V_G$. Denote by $\mathcal{O}_Z V_H$ the $\mathcal{O}_Z$-submodule of $N_{\Lambda/\mathbf{P}^r}|_Z$ generated by V_H.

We call H *semimodular with respect to* X if whenever $Z = \Lambda \cap X$ is zero-dimensional, we have

$$V_G \cap \mathcal{O}_Z V_H = V_H.$$

Example 5.1. We describe examples of semimodular and nonsemimodular sub-schemes of $G = G_{n+c}$. First, let $l \geq 0$ be an integer. Let $U \subset G_{n+c} \times \mathbf{P}^r$ be the universal family over G_{n+c}, and let

$$U_X = U \cap (G_{n+c} \times X) \subset G_{n+c} \times \mathbf{P}^r$$

be the scheme theoretic intersection. By [Kollár 1996, I, Theorem 1.6], there is a locally closed subscheme H of G_{n+c} with the following property: For any morphism $S \to G_{n+c}$, the pullback of U_X to S is flat of relative dimension zero and relative degree l if and only if $S \to G_{n+c}$ factors through H. The closed points of H parametrize those linear subvarieties whose degree of intersection with X is l.

Denote the normal sheaf of Z in X by $N_{Z/X} = \mathrm{Hom}(I_{Z/X}, \mathcal{O}_Z)$. If I and J are the ideal sheaves of X and Λ in $\mathbf{P}^r$ respectively, then there is a surjective map of $\mathcal{O}_Z$-modules

$$\frac{J}{J^2 + IJ} \to \frac{J+I}{J^2 + I}.$$

Dualizing the above map into $\mathcal{O}_Z$, we get an injective map

$$N_{Z/X} \to N_{\Lambda/\mathbf{P}^r}|_Z.$$

If M is the image of this map, then M is an $\mathcal{O}_Z$-module, and $V_G \cap M = V_H$, as one sees by considering morphisms from $\mathrm{Spec}\, k[t]/(t^2)$. Therefore H is semimodular with respect to X.

For the next example, fix a point $p \in X$, and let $q \notin X$ be a point on a tangent line to X at p. Let H be the subvariety of G that consists of those linear subvarieties of codimension $n + c$ that pass through q. We claim that H is not a semimodular subvariety. Pick $[\Lambda] \in H$ so that it passes through p and q, and set $Z = \Lambda \cap X$. Choose a system of homogenous coordinates $x_0, \ldots, x_{r-n-c}$ for Λ such that $p = (1:0:\cdots:0)$, $q = (0:1:0:\cdots:0)$, and $Z \subset U := \{x \in \Lambda \mid x_0 \neq 0\}$. We have $\mathcal{O}_U(U) = k[x_1, \ldots, x_{r-n-c}]$, and since q is on the tangent plane to X, for any linear polynomial that vanishes on Z, the coefficient of x_1 is zero.

If we identify $T_{G,[\Lambda]}$ with the global sections of $N_{\Lambda/\mathbf{P}^r} \simeq \mathcal{O}_\Lambda(1)^{n+c}$, then $T_{H,[\Lambda]}$ is identified with the $(n+c)$-tuples of linear forms $(L_1, \ldots, L_{n+c})$ in $x_0, \ldots, x_{r-n-c}$ vanishing at q. The image of

$$\rho : H^0(N_{\Lambda/\mathbf{P}^r}) \to N_{\Lambda/\mathbf{P}^r}|_Z \simeq \mathcal{O}_Z^{n+c}$$

contains $(1, 0, \ldots, 0)$ and $(x_1|_Z, 0, \ldots, 0)$. But $(1, 0, \ldots, 0) \in \rho(T_{H,[\Lambda]})$ and $(x_1|_Z, 0, \ldots, 0) \notin \rho(T_{H,[\Lambda]})$. Thus H is not semimodular.

We now turn to the transversality result. If $f : Y_1 \to Y_2$ is a regular morphism between smooth varieties, and if H is a smooth subvariety of Y_2, then f is called *transverse* to H if for every y in Y_1, either $f(y) \notin H$ or

$$T_{H, f(y)} + df(T_{Y_1, y}) = T_{Y_2, f(y)}$$

where $df : T_{Y_1, y} \to T_{Y_2, f(y)}$ is the map induced by f on the Zariski tangent spaces.

Theorem 5.2. *Let $X \subset \mathbf{P}^r$ be a smooth projective variety of dimension n, fix $c \geq 1$, and let H be a smooth subvariety of $\mathbf{G}_{n+c}$ that is semimodular with respect to X. For a general linear projection $\pi_\Sigma : X \to \mathbf{P}^{n+c}$, the map*

$$\phi_\Sigma : \mathbf{P}^{n+c} \to \mathbf{G}_{n+c}$$

that sends $y \in \mathbf{P}^{n+c}$ to the corresponding linear subvariety in $\mathbf{P}^r$ is transverse to H.

Proof. Let $[\Sigma]$ be a general point of $\mathbf{G}_{n+c+1}$. For $y \in \mathbf{P}^{n+c}$, let $\Lambda \subset \mathbf{P}^r$ be the corresponding linear subvariety, the preimage of y under the projection map from $\mathbf{P}^{n+c}$, and set $Z = \Lambda \cap X$.

Assume $[\Lambda] \in H$, and let V_H be the image of $T_{H,[\Lambda]}$ under the restriction map

$$T_{G,[\Lambda]} = H^0(N_{\Lambda/\mathbf{P}^r}) \to N_{\Lambda/\mathbf{P}^r}|_Z.$$

Denote by Q the quotient of $N_{\Lambda/\mathbf{P}^r}|_Z$ by $\mathcal{O}_Z V_H$:

$$0 \to \mathcal{O}_Z V_H \to N_{\Lambda/\mathbf{P}^r}|_Z \to Q \to 0.$$

Then we can consider Q as a sheaf of $\mathcal{O}_\Lambda$-modules that is supported on Z. Let $F = \ker(N_{\Lambda/\mathbf{P}^r} \to Q)$, so that $\rho(T_{H,\Lambda}) \subset H^0(F)$. Since H is semimodular, $\mathcal{O}_Z V_H \cap V_G = V_H$, and hence $T_{H,[\Lambda]} = H^0(F)$.

To prove the statement, note that if for a general Σ, there is no $y \in \mathbf{P}^{n+c}$ with $\phi([\Sigma], y) \in H$, then there is nothing to prove. Otherwise, by Lemma 3.2, the map $H^0(N_{\Lambda/\mathbf{P}^r} \otimes \mathcal{O}_\Lambda(-1)) \to Q$ is surjective. Hence if we consider $H^0(F)$ and $H^0(N_{\Lambda/\mathbf{P}^r} \otimes \mathcal{O}_\Lambda(-1))$ as subspaces of $H^0(N_{\Lambda/\mathbf{P}^r})$, then we get

$$H^0(F) + H^0(N_{\Lambda/\mathbf{P}^r} \otimes \mathcal{O}_\Lambda(-1)) = H^0(N_{\Lambda/\mathbf{P}^r}).$$

If we identify $T_{G_{n+c},[\Lambda]}$ with the space of global sections of $N_{\Lambda/\mathbf{P}^r}$, then $d\phi_\Sigma(T_{\mathbf{P}^{n+c},y})$ is identified with $H^0(N_{\Lambda/\mathbf{P}^r} \otimes \mathcal{O}_\Lambda(-1))$, and thus

$$T_{H,\phi_\Sigma(y)} + d\phi_\Sigma(T_{\mathbf{P}^{n+c},y}) = H^0(N_{\Lambda/\mathbf{P}^r}) = T_{G_{n+c},\phi_\Sigma(y)}. \qquad \square$$

References

[Beheshti and Eisenbud 2010] R. Beheshti and D. Eisenbud, "Fibers of generic projections", *Compos. Math.* **146**:2 (2010), 435–456. MR 2011i:14087

[Eisenbud and Goto 1984] D. Eisenbud and S. Goto, "Linear free resolutions and minimal multiplicity", *J. Algebra* **88**:1 (1984), 89–133. MR 85f:13023

[Kollár 1996] J. Kollár, *Rational curves on algebraic varieties*, Ergebnisse der Math. (3) **32**, Springer, Berlin, 1996. MR 98c:14001

[Kwak 2000] S. Kwak, "Generic projections, the equations defining projective varieties and Castelnuovo regularity", *Math. Z.* **234**:3 (2000), 413–434. MR 2001e:14042

[Lazarsfeld 1987] R. Lazarsfeld, "A sharp Castelnuovo bound for smooth surfaces", *Duke Math. J.* **55**:2 (1987), 423–429. MR 89d:14007

[Lazarsfeld 2004] R. Lazarsfeld, *Positivity in algebraic geometry, I: Classical setting: line bundles and linear series*, Ergebnisse der Math. (3) **48**, Springer, Berlin, 2004. MR 2005k:14001a

[Mather 1973] J. N. Mather, "On Thom–Boardman singularities", pp. 233–248 in *Dynamical systems* (Salvador, Brazil, 1971), Academic Press, New York, 1973. MR 50 #5843

beheshti@math.wustl.edu   *Department of Mathematics, Washington University, St. Louis, MO 63130, United States*

eisenbud@math.berkeley.edu   *Department of Mathematics, University of California, Berkeley, CA 94720, United States*

Current Developments in Algebraic Geometry
MSRI Publications
Volume 59, 2011

Introduction to birational anabelian geometry

FEDOR BOGOMOLOV AND YURI TSCHINKEL

We survey recent developments in the Birational Anabelian Geometry program aimed at the reconstruction of function fields of algebraic varieties over
algebraically closed fields from pieces of their absolute Galois groups.

Introduction

The essence of Galois theory is to *lose* information, by passing from a field k,
an algebraic structure with two compatible operations, to a (profinite) group, its
absolute Galois group G_k or some of its quotients. The original goal of testing
solvability in radicals of polynomial equations in one variable over the rationals
was superseded by the study of deeper connections between the arithmetic in
k, its ring of integers, and its completions with respect to various valuations on
the one hand, and (continuous) representations of G_k on the other hand. The
discovered structures turned out to be extremely rich, and the effort led to the
development of deep and fruitful theories: class field theory (the study of abelian
extensions of k) and its nonabelian generalizations, the Langlands program. In
fact, techniques from class field theory (Brauer groups) allowed one to deduce
that Galois groups of global fields encode the field:

Keywords: Galois groups, function fields.

17

Theorem 1 [Neukirch 1969; Uchida 1977]. *Let K and L be number fields or function fields of curves over finite fields with isomorphic Galois groups*

$$G_{K^{\mathrm{solv}}/K} \simeq G_{L^{\mathrm{solv}}/L}$$

of their maximal solvable extensions. Then

$$L \simeq K.$$

In another, more geometric direction, Galois theory was subsumed in the theory of the étale fundamental group. Let X be an algebraic variety over a field k. Fix an algebraic closure $\bar{k}/k$ and let $K = k(X)$ be the function field of X. We have an associated exact sequence

$$1 \to \pi_1(X_{\bar{k}}) \to \pi_1(X) \xrightarrow{\mathrm{pr}_X} G_k \to 1 \qquad (\Psi_X)$$

of étale fundamental groups, exhibiting an action of the Galois group of the ground field k on the *geometric fundamental group* $\pi_1(X_{\bar{k}})$. Similarly, we have an exact sequence of Galois groups

$$1 \to G_{\bar{k}(X)} \to G_K \xrightarrow{\mathrm{pr}_K} G_k \to 1. \qquad (\Psi_K)$$

Each k-rational point on X gives rise to a section of pr_X and pr_K.

When X is a smooth projective curve of genus $\mathrm{g} \geq 2$, its geometric fundamental group $\pi_1(X_{\bar{k}})$ is a profinite group in 2g generators subject to one relation. Over fields of characteristic zero, these groups depend only on g but not on the curve. However, the sequence (Ψ_X) gives rise to a plethora of representations of G_k and the resulting configuration is so strongly rigid[1] that it is natural to expect that it encodes much of the geometry and arithmetic of X over k.

For example, let k be a *finite field* and X an abelian variety over k of dimension g. Then G_k is the procyclic group $\hat{\mathbb{Z}}$, generated by the Frobenius, which acts on the Tate module

$$T_\ell(X) = \pi_{1,\ell}^a(X_{\bar{k}}) \simeq \mathbb{Z}_\ell^{2\mathrm{g}},$$

where $\pi_{1,\ell}^a(X_{\bar{k}})$ is the ℓ-adic quotient of the abelianization $\pi_1^a(X_{\bar{k}})$ of the étale fundamental group. By a theorem of Tate [1966], the characteristic polynomial of the Frobenius determines X, up to isogeny. Moreover, if X and Y are abelian varieties over k then

$$\mathrm{Hom}_{G_k}(T_\ell(X), T_\ell(Y)) \simeq \mathrm{Hom}_k(X, Y) \otimes \mathbb{Z}_\ell.$$

Similarly, if k is a *number field* and X, Y abelian varieties over k then

$$\mathrm{Hom}_{G_k}(\pi_1^a(X), \pi_1^a(Y)) \simeq \mathrm{Hom}_k(X, Y) \otimes \hat{\mathbb{Z}},$$

by a theorem of Faltings [1983].

---

[1]"ausserordentlich stark" [Grothendieck 1997].

With these results at hand, Grothendieck [1997] conjectured that there is a certain class of *anabelian* varieties, defined over a field k (which is finitely generated over its prime field), which are *characterized* by their fundamental groups. Main candidates are hyperbolic curves and varieties which can be successively fibered by hyperbolic curves. There are three related conjectures:

Isom: An anabelian variety X is determined by (Ψ_X), i.e., by the profinite group $\pi_1(X)$ together with the action of G_k.

Hom: If X and Y are anabelian, then there is a bijection

$$\mathrm{Hom}_k(X, Y) = \mathrm{Hom}_{G_k}(\pi_1(X), \pi_1(Y))/\sim$$

between the set of dominant k-morphisms and G_k-equivariant open homomorphisms of fundamental groups, modulo conjugacy (inner automorphisms by the geometric fundamental group of Y).

Sections: If X is anabelian then there is a bijection between the set of rational points $X(k)$ and the set of sections of pr_X (modulo conjugacy).

Similar conjectures can be made for nonproper varieties. Excising points from curves makes them "more" hyperbolic. Thus, one may reduce to the generic point of X, replacing the fundamental group by the Galois group of the function field $K = k(X)$. In the resulting *birational* version of Grothendieck's conjectures, the exact sequence (Ψ_X) is replaced by (Ψ_K) and the projection pr_X by pr_K.

These conjectures have generated wide interest and stimulated intense research. Here are some of the highlights of these efforts:

- proof of the birational **Isom**-conjecture for function fields over k, where k is finitely generated over its prime field, by Pop [1994];

- proof of the birational **Hom**-conjecture over sub-p-adic fields k, i.e., k which are contained in a finitely generated extension of $\mathbb{Q}_p$, by Mochizuki [1999];

- proof of the birational **Section**-conjecture for local fields of characteristic zero, by Königsmann [2005].

Here is an incomplete list of other significant result in this area: [Nakamura 1990; Voevodsky 1991a; 1991b; Tamagawa 1997]. In all cases, the proofs relied on *nonabelian* properties in the structure of the Galois group G_K, respectively, the relative Galois group. Some of these developments were surveyed in [Ihara and Nakamura 1997; Faltings 1998; Nakamura et al. 1998; Pop 1997; 2000; Mochizuki 2003].

After the work of Iwasawa the study of representations of the maximal pro-ℓ-quotient $\mathcal{G}_K$ of the absolute Galois group G_K developed into a major branch of number theory and geometry. So it was natural to turn to pro-ℓ-versions of the hyperbolic anabelian conjectures, replacing the fundamental groups by their maximal pro-ℓ-quotients and the absolute Galois group G_K by $\mathscr{G}_K$. Several results in this direction were obtained in [Corry and Pop 2009; Saïdi and Tamagawa 2009b].

A very different intuition evolved from higher-dimensional birational algebraic geometry. One of the basic questions in this area is the characterization of fields isomorphic to purely transcendental extensions of the ground field, i.e., varieties birational to projective space. Interesting examples of function fields arise from faithful representations of *finite* groups

$$G \to \mathrm{Aut}(V),$$

where $V = \mathbb{A}^n_k$ is the standard affine space over k. The corresponding variety

$$X = V/G$$

is clearly unirational. When $n \leq 2$ and k is algebraically closed the quotient is rational (even though there exist unirational but nonrational surfaces in positive characteristic). The quotient is also rational when G is abelian and k algebraically closed.

Noether's problem (inspired by invariant theory and the inverse problem in Galois theory) asks whether or not $X = V/G$ is rational for nonabelian groups. The first counterexamples were constructed in [Saltman 1984]. Geometrically, they are quotients of products of projective spaces by projective actions of finite *abelian groups*. The first obstruction to (retract) rationality was described in terms of Azumaya algebras and the *unramified* Brauer group

$$\mathrm{Br}_{nr}(k(X)) = \mathrm{H}^2_{nr}(X),$$

(see Section 7). A group cohomological interpretation of these examples was given in [Bogomolov 1987]; it allowed one to generate many other examples and elucidated the key structural properties of the obstruction group. This obstruction can be computed in terms of G, in particular, it does not depend on the chosen representation V of G:

$$B_0(G) := \mathrm{Ker}\left(\mathrm{H}^2(G, \mathbb{Q}/\mathbb{Z}) \to \prod_B \mathrm{H}^2(B, \mathbb{Q}/\mathbb{Z}) \right),$$

where the product ranges over the set of subgroups $B \subset G$ which are generated by *two commuting* elements. A key fact is that, for $X = V/G$,

$$B_0(G) = \mathrm{Br}_{nr}(k(X)) = \mathrm{H}^2_{nr}(X),$$

see Section 7 and Theorem 22.

One has a decomposition into primary components

$$B_0(G) = \oplus_\ell B_{0,\ell}(G), \qquad (0\text{-}1)$$

and computation of each piece reduces to computations of cohomology of the ℓ-Sylow subgroups of G, with coefficients in $\mathbb{Q}_\ell/\mathbb{Z}_\ell$.

We now restrict to this case, i.e., finite ℓ-groups G and $\mathbb{Q}_\ell/\mathbb{Z}_\ell$-coefficients. Consider the exact sequence

$$1 \to Z \to G^c \to G^a \to 1,$$

where

$$G^c = G/[[G, G], G]$$

is the canonical central extension of the abelianization

$$G^a = G/[G, G].$$

We have

$$B_0(G^c) \hookrightarrow B_0(G) \qquad (0\text{-}2)$$

(see Section 7); in general, the image is a *proper* subgroup. The computation of $B_0(G^c)$ is a problem in linear algebra: We have a well-defined map (from "skew-symmetric matrices" on G^a, considered as a linear space over $\mathbb{Z}/\ell$) to the center of G^c:

$$\wedge^2(G^a) \xrightarrow{\lambda} Z, \qquad (\gamma_1, \gamma_2) \mapsto [\tilde{\gamma}_1, \tilde{\gamma}_2],$$

where $\tilde{\gamma}$ is some lift of $\gamma \in G^a$ to G^c. Let

$$R(G^c) := \mathrm{Ker}(\lambda)$$

be the subgroup of relations in $\wedge^2(G^a)$ (the subgroup generated by "matrices" of rank one). We say that γ_1, γ_2 form a *commuting pair* if

$$[\tilde{\gamma}_1, \tilde{\gamma}_2] = 1 \in Z.$$

Let

$$R_\wedge(G^c) := \langle \gamma_1 \wedge \gamma_2 \rangle \subset R(G^c)$$

be the subgroup generated by commuting pairs. It is proved in [Bogomolov 1987] that

$$B_0(G^c) = \left(R(G^c)/R_\wedge(G^c)\right)^\vee.$$

Using this representation it is easy to produce examples with nonvanishing $B_0(G)$, thus nonrational fields of G-invariants, already for central extensions of $(\mathbb{Z}/\ell)^4$ by $(\mathbb{Z}/\ell)^3$ [Bogomolov 1987].

Note that for $K = k(V)^G$ the group G is naturally a quotient of the absolute Galois group G_K. The sketched arguments from group cohomology suggested to focus on $\mathcal{G}_K$, the pro-ℓ-quotient of G_K and the pro-ℓ-cohomology groups introduced above. The theory of *commuting pairs* explained in Section 4 implies that the groups $\mathcal{G}_K$ are very special: for any function field K over an algebraically closed field one has

$$B_{0,\ell}(G_K) = B_0(\mathcal{G}_K) = B_0(\mathcal{G}_K^c).$$

This lead to a *dismantling* of nonabelian aspects of anabelian geometry. For example, from this point of view it is unnecessary to assume that the Galois group of the ground field k is large. On the contrary, it is preferable if k is algebraically closed, or at least contains all ℓ^n-th roots of 1. More significantly, while the *hyperbolic anabelian geometry* has dealt primarily with curves C, the corresponding $B_0(\mathcal{G}_{k(C)})$, and hence $B_0(\mathcal{G}_{k(C)}^c)$, are *trivial*, since the ℓ-Sylow subgroups of $G_{k(C)}$ are free. Thus we need to consider function fields K of transcendence degree at least 2 over k. It became apparent that in these cases, at least over $k = \bar{\mathbb{F}}_p$,

$$B_0(\mathcal{G}_K^c) = H_{nr}^2(k(X))$$

encodes a wealth of information about $k(X)$. In particular, it determines all higher unramified cohomological invariants of X (see Section 3).

Let p and ℓ be distinct primes and $k = \bar{\mathbb{F}}_p$ an algebraic closure of $\mathbb{F}_p$. Let X be an algebraic variety over k and $K = k(X)$ its function field (X will be called a *model* of K). In this situation, $\mathcal{G}_K^a$ is a torsion-free $\mathbb{Z}_\ell$-module. Let Σ_K be the set of not procyclic subgroups of $\mathcal{G}_K^a$ which lift to abelian subgroups in the canonical central extension

$$\mathcal{G}_K^c = \mathcal{G}_K/[[\mathcal{G}_K, \mathcal{G}_K], \mathcal{G}_K] \to \mathcal{G}_K^a.$$

The set Σ_K is canonically encoded in

$$R_\wedge(\mathcal{G}_K^c) \subset \bigwedge\nolimits^2(\mathcal{G}_K^a),$$

a group that carries *less* information than $\mathcal{G}_K^c$ (see Section 6).

The main goal of this survey is to explain the background of this result:

Theorem 2 [Bogomolov and Tschinkel 2008b; 2009b]. *Let K and L be function fields over algebraic closures of finite fields k and l, of characteristic $\neq \ell$. Assume that the transcendence degree of K over k is at least two and that there exists an isomorphism*

$$\Psi = \Psi_{K,L} : \mathcal{G}_K^a \xrightarrow{\sim} \mathcal{G}_L^a \tag{0-3}$$

of abelian pro-ℓ-groups inducing a bijection of sets

$$\Sigma_K = \Sigma_L.$$

Then $k = l$ and there exists a constant $\epsilon \in \mathbb{Z}_\ell^\times$ such that $\epsilon^{-1} \cdot \Psi$ is induced from a unique isomorphism of perfect closures

$$\bar{\Psi}^* : \bar{L} \xrightarrow{\sim} \bar{K}.$$

The intuition behind Theorem 2 is that the arithmetic and geometry of varieties of transcendence degree ≥ 2 over algebraically closed ground fields is governed by *abelian* or *almost abelian* phenomena. One of the consequences is that central extensions of abelian groups provide *universal* counterexamples to Noether's problem, and more generally, provide *all* finite cohomological obstructions to rationality, at least over $\bar{\mathbb{F}}_p$ (see Section 3).

Conceptually, the proof of Theorem 2 explores a *skew-symmetric* incarnation of the field, which is a *symmetric object*, with two symmetric operations. Indeed, by Kummer theory, we can identify

$$\mathcal{G}_K^a = \mathrm{Hom}(K^\times / k^\times, \mathbb{Z}_\ell).$$

Dualizing again, we obtain

$$\mathrm{Hom}(\mathcal{G}_K^a, \mathbb{Z}_\ell) = \hat{K}^\times,$$

the pro-ℓ-completion of the multiplicative group of K. Recall that

$$K^\times = \mathrm{K}_1^M(K),$$

the first Milnor K-group of the field. The elements of $\bigwedge^2(\mathcal{G}_K^a)$ are matched with symbols in Milnor's K-group $\mathrm{K}_2^M(K)$. The symbol (f, g) is infinitely divisible in $\mathrm{K}_2^M(K)$ if and only if f, g are algebraically dependent, i.e., $f, g \in E = k(C)$ for some curve C (in particular, we get no information when $\mathrm{tr\,deg}_k(K) = 1$). In Section 2 we describe how to reconstruct homomorphisms of fields from compatible homomorphisms

$$\mathrm{K}_1^M(L) \xrightarrow{\psi_1} \mathrm{K}_1^M(K),$$
$$\mathrm{K}_2^M(L) \xrightarrow{\psi_2} \mathrm{K}_2^M(K).$$

Indeed, the multiplicative group of the ground field k is characterized as the subgroup of infinitely divisible elements of $K^\times$, thus

$$\psi_1 : \mathbb{P}(L) = L^\times / l^\times \to \mathbb{P}(K) = K^\times / k^\times,$$

a homomorphism of multiplicative groups (which we assume to be injective).

Compatibility with ψ_2 means that infinitely divisible symbols are mapped to infinitely divisible symbols, i.e., ψ_1 maps multiplicative groups $F^\times$ of 1-dimensional subfields $F \subset L$ to $E^\times \subset K^\times$, for 1-dimensional $E \subset K$. This implies that already each $\mathbb{P}^1 \subset \mathbb{P}(L)$ is mapped to a $\mathbb{P}^1 \subset \mathbb{P}(K)$. The Fundamental theorem of projective geometry (see Theorem 5) shows that (some rational power of) ψ_1 is a restriction of a homomorphisms of fields $L \to K$.

Theorem 2 is a pro-ℓ-version of this result. Kummer theory provides the isomorphism

$$\Psi^* : \hat{L}^\times \to \hat{K}^\times$$

The main difficulty is to recover the *lattice*

$$K^\times / k^\times \otimes \mathbb{Z}_{(\ell)}^\times \subset \hat{K}^\times.$$

This is done in several stages. First, the theory of *commuting pairs* [Bogomolov and Tschinkel 2002a] allows one to reconstruct abelianized inertia and decomposition groups of valuations

$$\mathcal{I}_\nu^a \subset \mathcal{D}_\nu^a \subset \mathcal{G}_K^a.$$

Note that for divisorial valuations ν we have $\mathcal{I}_\nu^a \simeq \mathbb{Z}_\ell$, and the set

$$\mathcal{I}^a = \{\mathcal{I}_\nu^a\}$$

resembles a $\mathbb{Z}_\ell$-*fan* in $\mathcal{G}_K^a \simeq \mathbb{Z}_\ell^\infty$. The key issue is to pin down, canonically, a topological generator for each of these $\mathcal{I}_\nu^a$. The next step is to show that

$$\Psi^*(F^\times / l^\times) \subset \hat{E}^\times \subset \hat{K}^\times$$

for some 1-dimensional $E \subset K$. This occupies most of [Bogomolov and Tschinkel 2008b], for function fields of surfaces. The higher-dimensional case, treated in [2009b], proceeds by induction on dimension. The last step, i.e., matching of projective structures on multiplicative groups, is then identical to the arguments used above in the context of Milnor K-groups.

The Bloch–Kato conjecture says that $\mathcal{G}_K^c$ contains all information about the cohomology of G_K, with finite constant coefficients (see Section 3 for a detailed discussion). Thus we can consider Theorem 2 as a *homotopic* version of the Bloch–Kato conjecture, i.e., $\mathcal{G}_K^c$ determines the field K itself, modulo purely inseparable extension.

Almost abelian anabelian geometry evolved from the Galois-theoretic interpretation of Saltman's counterexamples described above and the Bloch–Kato conjecture. These ideas, and the "recognition" technique used in the proof of Theorem 2, were put forward in [Bogomolov 1987; 1991a; 1991b; 1992; 1995b], and developed in [Bogomolov and Tschinkel 2002a; 2008b; 2009a; 2009b]. In

recent years, this approach has attracted the attention of several experts, including
F. Pop; see [Pop 2003], as well as his webpage, for other preprints on this topic,
which contain his version of the recognition procedure of K from $\mathcal{G}_K^c$, for the
same class of fields K. Other notable contributions are found in [Chebolu et al.
2009; Chebolu and Mináč 2009].

Several ingredients of the proof of Theorem 2 sketched above appeared already
in Grothendieck's anabelian geometry, relating the *full* absolute Galois group
of function fields to the geometry of projective models. Specifically, even
before Grothendieck's insight, it was understood by Uchida and Neukirch (in
the context of number fields and function fields of curves over finite fields)
that the identification of decomposition groups of valuations can be obtained
in purely group-theoretic terms as, roughly speaking, subgroups with nontrivial
center. Similarly, it was clear that Kummer theory essentially captures the
multiplicative structure of the field and that the projective structure on $\mathbb{P}_k(K)$
encodes the additive structure. The main difference between our approach and the
techniques of, e.g., Mochizuki [1999] and Pop [2003] is the theory of commuting
pairs which is based on an unexpected coincidence: the *minimal* necessary
condition for the commutation of two elements of the absolute Galois group of a
function field K is also sufficient and already implies that these elements belong
to the same decomposition group. It suffices to check this condition on $\mathcal{G}_K^c$,
which *linearizes* the commutation relation. Another important ingredient in our
approach is the correspondence between *large* free quotients of $\mathcal{G}_K^c$ and integrally
closed 1-dimensional subfields of K. Unfortunately, in full generality, this
conjectural equivalence remains open (see the discussion in Section 6). However,
by exploiting geometric properties of projective models of K we succeed in
proving it in many important cases, which suffices for solving the recognition
problem and for several other applications.

Finally, in Section 9 we discuss almost abelian phenomena in Galois groups
of *curves* that occur for completely different reasons. An application of a recent
result of Corvaja–Zannier concerning the divisibility of values of recurrence sequences leads to a Galois-theoretic Torelli-type result for curves over finite fields.

1. Abstract projective geometry

Definition 3. A *projective structure* is a pair $(S, \mathcal{L})$ where S is a set (of points)
and $\mathcal{L}$ a collection of subsets $\mathfrak{l} \subset S$ (lines) such that

P1 there exist an $s \in S$ and an $\mathfrak{l} \in \mathcal{L}$ such that $s \notin \mathfrak{l}$;

P2 for every $\mathfrak{l} \in \mathcal{L}$ there exist at least three distinct $s, s', s'' \in \mathfrak{l}$;

P3 for every pair of distinct $s, s' \in S$ there exists exactly one

$$\mathfrak{l} = \mathfrak{l}(s, s') \in \mathfrak{L}$$

such that $s, s' \in \mathfrak{l}$;

P4 for every quadruple of pairwise distinct $s, s', t, t' \in S$ one has

$$\mathfrak{l}(s, s') \cap \mathfrak{l}(t, t') \neq \varnothing \;\Rightarrow\; \mathfrak{l}(s, t) \cap \mathfrak{l}(s', t') \neq \varnothing.$$

In this context, one can define (inductively) the dimension of a projective space: a two-dimensional projective space, i.e., a projective plane, is the set of points on lines passing through a line and a point outside this line; a three-dimensional space is the set of points on lines passing through a plane and a point outside this plane, etc.

A *morphism* of projective structures $\rho : (S, \mathfrak{L}) \to (S', \mathfrak{L}')$ is a map of sets $\rho : S \to S'$ preserving lines, i.e., $\rho(\mathfrak{l}) \in \mathfrak{L}'$, for all $\mathfrak{l} \in \mathfrak{L}$.

A projective structure $(S, \mathfrak{L})$ satisfies *Pappus' axiom* if

PA for all 2-dimensional subspaces and every configuration of six points and lines in these subspaces as in the figure, the intersections are collinear.

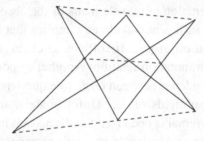

The following Fundamental theorem of abstract projective geometry goes back at least to Schur and Hessenberg, but there were many researchers before and after exploring the various interconnections between different sets of axioms (Poncelet, Steiner, von Staudt, Klein, Pasch, Pieri, Hilbert, and others).[2]

Theorem 4 (Reconstruction). *Let $(S, \mathfrak{L})$ be a projective structure of dimension $n \geq 2$ which satisfies Pappus' axiom. Then there exists a vector space V over a field k and an isomorphism*

$$\sigma : \mathbb{P}_k(V) \xrightarrow{\sim} S.$$

For any two such triples (V, k, σ) and (V', k', σ') there is an isomorphism

---

[2]"But there is one group of deductions which cannot be ignored in any consideration of the principles of Projective Geometry. I refer to the theorems, by which it is proved that numerical coordinates, with the usual properties, can be defined without the introduction of distance as a fundamental idea. The establishment of this result is one of the triumphs of modern mathematical thought." A. N. Whitehead, *The axioms of projective geometry*, 1906, p. v.

$$V/k \xrightarrow{\sim} V'/k'$$

compatible with σ, σ' *and unique up to homothety* $v \mapsto \lambda v, \lambda \in k^{\times}$.

Main examples are of course the sets of k-rational points of the usual projective $\mathbb{P}^n$ space over k of dimension $n \geq 2$. Then $\mathbb{P}^n(k)$ carries a projective structure: lines are the usual projective lines $\mathbb{P}^1(k) \subset \mathbb{P}^n(k)$.

A related example arises as follows: Let K/k be an extension of fields. Then

$$S := \mathbb{P}_k(K) = (K \setminus 0)/k^{\times}$$

carries a natural (possibly, infinite-dimensional) projective structure. Moreover, the *multiplication* in $K^{\times}/k^{\times}$ preserves this structure. In this setup we have the following reconstruction theorem:

Theorem 5 (Reconstructing fields [Bogomolov and Tschinkel 2008b, Theorem 3.6]). *Let* K/k *and* K'/k' *be field extensions of degree* ≥ 3 *and*

$$\bar{\psi} : S = \mathbb{P}_k(K) \to \mathbb{P}_{k'}(K') = S'$$

an injective homomorphism of abelian groups compatible with projective struc-tures. Then $k \simeq k'$ *and* K *is isomorphic to a subfield of* K'.

The following strengthening is due to M. Rovinsky.

Theorem 6. *Let* S *be an abelian group equipped with a compatible structure of a projective space. Then there exist fields* k *and* K *such that* $S = \mathbb{P}_k(K)$.

Proof. There is an embedding of $S = \mathbb{P}(V)$ as a projective subspace into $\mathrm{PGL}(V)$. Its preimage in $\mathrm{GL}(V)$ is a linear subspace minus a point. Since V is invariant under products (because $\mathbb{P}(V)$ is) we obtain that V is a commutative subalgebra of $\mathrm{Mat}(V)$ and every element is invertible — hence it is a field. $\square$

Related *reconstruction theorems* of "large" fields have recently emerged in model theory. The setup there is as follows: A *combinatorial pregeometry* (*finitary matroid*) is a pair $(\mathcal{P}, cl)$ where $\mathcal{P}$ is a set and

$$cl : \text{Subsets}(\mathcal{P}) \to \text{Subsets}(\mathcal{P}),$$

such that for all $a, b \in \mathcal{P}$ and all $Y, Z \subseteq \mathcal{P}$ one has:

- $Y \subseteq cl(Y)$,
- if $Y \subseteq Z$, then $cl(Y) \subseteq cl(Z)$,
- $cl(cl(Y)) = cl(Y)$,
- if $a \in cl(Y)$, then there is a finite subset $Y' \subset Y$ such that $a \in cl(Y')$ (finite character),
- (exchange condition) if $a \in cl(Y \cup \{b\}) \setminus cl(Y)$, then $b \in cl(Y \cup \{a\})$.

A *geometry* is a pregeometry such that $cl(a) = a$, for all $a \in \mathcal{P}$, and $cl(\varnothing) = \varnothing$. Standard examples are provided by:

(1) $\mathcal{P} = V/k$, a vector space over a field k and $cl(Y)$ the k-span of $Y \subset \mathcal{P}$;

(2) $\mathcal{P} = \mathbb{P}_k(V)$, the usual *projective* space over a field k;

(3) $\mathcal{P} = \mathcal{P}_k(K)$, a field K containing an algebraically closed subfield k and $cl(Y)$ — the normal closure of $k(Y)$ in K, note that a *geometry* is obtained after factoring by $x \sim y$ if and only if $cl(x) = cl(y)$.

It turns out that a sufficiently large field can reconstructed from the *geometry* of its 1-dimensional subfields.

Theorem 7 [Evans and Hrushovski 1991; 1995; Gismatullin 2008]. *Let k and k' be algebraically closed fields, K/k and K'/k' field extensions of transcendence degree ≥ 5 over k, resp. k'. Then, every isomorphism of combinatorial geometries*

$$\mathcal{P}_k(K) \to \mathcal{P}_{k'}(K')$$

is induced by an isomorphism of purely inseparable closures

$$\bar{K} \to \overline{K'}.$$

In the next section, we show how to reconstruct a field of transcendence degree ≥ 2 from its *projectivized* multiplicative group and the "geometry" of multiplicative groups of 1-dimensional subfields.

2. K-theory

Let $K_i^M(K)$ be i-th Milnor K-group of a field K. Recall that

$$K_1^M(K) = K^\times$$

and that there is a canonical surjective homomorphism

$$\sigma_K : K_1^M(K) \otimes K_1^M(K) \to K_2^M(K);$$

we write (x, y) for the image of $x \otimes y$. The kernel of σ_K is generated by symbols $x \otimes (1-x)$, for $x \in K^\times \setminus 1$. Put

$$\bar{K}_i^M(K) := K_i^M(K)/\text{infinitely divisible elements}, \quad i = 1, 2.$$

Theorem 8 [Bogomolov and Tschinkel 2009a]. *Let K and L be function fields of transcendence degree ≥ 2 over an algebraically closed field k, resp. l. Let*

$$\bar{\psi}_1 : \bar{K}_1^M(K) \to \bar{K}_1^M(L)$$

be an injective homomorphism.

Assume that there is a commutative diagram

$$
\begin{array}{ccc}
\bar{K}_1^M(K) \otimes \bar{K}_1^M(K) & \xrightarrow{\ \bar{\psi}_1 \otimes \bar{\psi}_1\ } & \bar{K}_1^M(L) \otimes \bar{K}_1^M(L) \\
\ \downarrow{\scriptstyle \sigma_K} & & \ \downarrow{\scriptstyle \sigma_L} \\
\bar{K}_2^M(K) & \xrightarrow[\ \bar{\psi}_2\ ]{} & \bar{K}_2^M(L).
\end{array}
$$

Assume that $\bar{\psi}_1(K^\times/k^\times) \not\subseteq E^\times/l^\times$, *for a* 1-*dimensional field* $E \subset L$ (*i.e., a field of transcendence degree* 1 *over* l).

Then there exist an $m \in \mathbb{Q}$ *and a homomorphism of fields*

$$\psi : K \to L$$

such that the induced map on $K^\times/k^\times$ *coincides with* $\bar{\psi}_1^m$.

Sketch of proof. First we reconstruct the multiplicative group of the ground field as the subgroup of infinitely divisible elements: An element $f \in K^\times = K_1^M(K)$ is infinitely divisible if and only if $f \in k^\times$. In particular,

$$\bar{K}_1^M(K) = K^\times/k^\times.$$

Next, we characterize multiplicative groups of 1-dimensional subfields: Given a nonconstant $f_1 \in K^\times/k^\times$, we have

$$\mathrm{Ker}_2(f_1) = E^\times/k^\times,$$

where $E = \overline{k(f_1)}^K$ is the normal closure in K of the 1-dimensional field generated by f_1 and

$$\mathrm{Ker}_2(f) := \{\, g \in K^\times/k^\times = \bar{K}_1^M(K) \mid (f, g) = 0 \in \bar{K}_2^M(K) \,\}.$$

At this stage we know the infinite-dimensional projective subspaces $\mathbb{P}(E) \subset \mathbb{P}(K)$. To apply Theorem 5 we need to show that projective *lines* $\mathbb{P}^1 \subset \mathbb{P}(K)$ are mapped to projective lines in $\mathbb{P}(L)$. It turns out that lines can be characterized as intersections of (shifted) $\mathbb{P}(E)$, for appropriate 1-dimensional $E \subset K$. The following technical result lies at the heart of the proof. □

Proposition 9 [Bogomolov and Tschinkel 2009a, Theorem 22]. *Let k be an algebraically closed field, K be an algebraically closed field extension of k, $x, y \in K$ algebraically independent over k, and take $p \in \overline{k(x)}^\times \setminus k \cdot x^{\mathbb{Q}}$ and $q \in \overline{k(y)}^\times \setminus k \cdot y^{\mathbb{Q}}$. Suppose that*

$$\overline{k(x/y)}^\times \cdot y \cap \overline{k(p/q)}^\times \cdot q \neq \varnothing.$$

Then there exist $a \in \mathbb{Q}$ and $c_1, c_2 \in k^\times$ such that

$$p \in k^\times \cdot (x^a - c_1)^{1/a}, \quad q \in k^\times \cdot (y^a - c_2)^{1/a}$$

and

$$\overline{k(x/y)}^\times \cdot y \cap \overline{k(p/q)}^\times \cdot q = k^\times \cdot (x^a - cy^a)^{1/a},$$

where $c = c_1/c_2$.

Proof. The following proof, which works in characteristic zero, has been suggested by M. Rovinsky (the general case in [Bogomolov and Tschinkel 2009a] is more involved).

Assume that there is a nontrivial

$$I \in \overline{k(x/y)}^\times \cdot y \cap \overline{k(p/q)}^\times \cdot q.$$

We obtain equalities in $\Omega_{K/k}$:

$$\frac{d(I/y)}{I/y} = r \cdot \frac{d(x/y)}{x/y} \quad \text{and} \quad \frac{d(I/q)}{I/q} = s \cdot \frac{d(p/q)}{p/q}, \tag{2-1}$$

for some

$$r \in \overline{k(x/y)}^\times \quad \text{and} \quad s \in \overline{k(p/q)}^\times.$$

Using the first equation, rewrite the second as

$$r \cdot \frac{d(x/y)}{x/y} + \frac{d(y/q)}{y/q} = s \cdot \frac{d(p/q)}{p/q},$$

or

$$r\frac{dx}{x} - s\frac{dp}{p} = r \cdot \frac{dy}{y} + \frac{d(q/y)}{q/y} - s\frac{dq}{q}.$$

The differentials on the left and on the right are linearly independent, thus both are zero, i.e., $r = sf = sg - g + 1$, where

$$f = xp'/p \in \overline{k(x)}^\times \quad \text{and} \quad g = yq'/q \in \overline{k(y)}^\times,$$

and p' is derivative with respect to x, q' the derivative In particular, $s = \frac{1-g}{f-g}$. Applying d log to both sides, we get

$$\frac{ds}{s} = \frac{g'dy}{g-1} + \frac{g'dy - f'dx}{f-g} = \frac{f'}{g-f}dx + \frac{g'(1-f)}{(1-g)(f-g)}dy.$$

As ds/s is proportional to

$$\frac{d(p/q)}{p/q} = \frac{p'}{p}dx - \frac{q'}{q}dy = f\frac{dx}{x} - g\frac{dy}{y}dy,$$

we get

$$x\frac{f'}{f} = y\frac{g'(1-f)}{(1-g)g}, \quad x\frac{f'}{(1-f)f} = y\frac{g'}{(1-g)g}.$$

Note that the left side is in $\overline{k(x)}^{\times}$, while the right hand side is in $\overline{k(y)}^{\times}$. It follows that

$$x\frac{f'}{(1-f)f} = y\frac{g'}{(1-g)g} = a \in k.$$

Solving the ordinary differential equation(s), we get

$$\frac{f}{f-1} = c_1^{-1}x^a \quad \text{and} \quad \frac{g}{g-1} = c_2^{-1}y^a$$

for some $c_1, c_2 \in k^{\times}$ and $a \in \mathbb{Q}$, so

$$f = (1 - c_1x^{-a})^{-1} = x\frac{\mathrm{d}}{\mathrm{d}x}\log(x^a - c_1)^{1/a},$$

$$g = (1 - c_2y^{-a})^{-1} = y\frac{\mathrm{d}}{\mathrm{d}y}\log(y^a - c_2)^{1/a}.$$

Thus finally,

$$p = b_1 \cdot (x^a - c_1)^{1/a} \quad \text{and} \quad q = b_2 \cdot (y^a - c_2)^{1/a}.$$

We can now find

$$s = \frac{(1 - c_1x^{-a})^{-1}c_2y^{-a}}{c_2y^{-a} - c_1x^{-a}} = \frac{c_2(x^a - c_1)}{c_2x^a - c_1y^a}$$

and then

$$r = sf = \frac{c_2x^a}{c_2x^a - c_1y^a} = (1 - c(x/y)^{-a})^{-1},$$

where $c = c_1/c_2$. From (2-1) we find

$$\mathrm{d}\log(I/y) = -\frac{1}{a}\frac{\mathrm{d}T}{T(1-T)},$$

where $T = c(x/y)^{-a}$, and thus,

$$I = y \cdot b_3(1 - c^{-1}(x/y)^a)^{1/a} = b_0(x^a - cy^a)^{1/a}. \qquad \square$$

This functional equation has the following projective interpretation: If $E = k(x)$ then the image of each $\mathbb{P}^1 \subset \mathbb{P}(E)$ under Ψ lies in a rational normal curve given by the conclusion of Proposition 9, where a may *a priori* depend on x. However, a simple lemma shows that it is actually independent of x (in characteristic zero), thus $\Psi^{1/a}$ extends to a field homomorphism. (In general, it is well-defined modulo powers of p, this brings up purely inseparable extensions, which are handled by an independent argument.)

3. Bloch-Kato conjecture

Let K be a field and ℓ a prime distinct from the characteristic of K. Let

$$\boldsymbol{\mu}_{\ell^n} := \{\sqrt[\ell^n]{1}\} \quad \text{and} \quad \mathbb{Z}_\ell(1) = \varprojlim \boldsymbol{\mu}_{\ell^n}.$$

We will assume that K contains all ℓ^n-th roots of unity and identify $\mathbb{Z}_\ell$ and $\mathbb{Z}_\ell(1)$. Let $\mathscr{G}_K^a$ be the abelianization of the maximal pro-ℓ-quotient of the absolute Galois group G_K.

Theorem 10 (Kummer theory). *There is a canonical isomorphism*

$$\mathrm{H}^1(G_K, \mathbb{Z}/\ell^n) = \mathrm{H}^1(\mathscr{G}_K^a, \mathbb{Z}/\ell^n) = K^\times/\ell^n. \tag{3-1}$$

More precisely, the discrete group $K^\times/(K^\times)^{\ell^n}$ and the compact profinite group $\mathscr{G}_K^a/\ell^n$ are Pontryagin dual to each other, for a $\boldsymbol{\mu}_{\ell^n}$-duality, i.e., there is a perfect pairing

$$K^\times/(K^\times)^{\ell^n} \times \mathscr{G}_K^a/\ell^n \to \boldsymbol{\mu}_{\ell^n}.$$

Explicitly, this is given by

$$(f, \gamma) \mapsto \gamma(\sqrt[\ell^n]{f})/\sqrt[\ell^n]{f} \in \boldsymbol{\mu}_{\ell^n}.$$

For $K = k(X)$, with k algebraically closed of characteristic $\neq \ell$, we have

- $K^\times/k^\times$ is a free $\mathbb{Z}$-module and

$$K^\times/(K^\times)^{\ell^n} = (K^\times/k^\times)/\ell^n \quad \text{for all} \quad n \in \mathbb{N};$$

- identifying $K^\times/k^\times \xrightarrow{\sim} \mathbb{Z}^{(\mathrm{I})}$, one has $K^\times/(K^\times)^{\ell^n} \xrightarrow{\sim} (\mathbb{Z}/\ell^n)^{(\mathrm{I})}$ and

$$\mathscr{G}_K^a/\ell^n \xrightarrow{\sim} (\mathbb{Z}/\ell^n(1))^{\mathrm{I}};$$

in particular, the duality between $\hat{K}^\times = \widehat{K^\times/k^\times}$ and $\mathscr{G}_K^a$ is modeled on that between

$$\{\text{functions } \mathrm{I} \to \mathbb{Z}_\ell \text{ tending to } 0 \text{ at } \infty\} \text{ and } \mathbb{Z}_\ell^{\mathrm{I}}.$$

Since the index set I is not finite, taking double-duals increases the space of *functions with finite support* to the space of *functions with support converging to zero*, i.e., the support modulo ℓ^n is finite, for all $n \in \mathbb{N}$. For function fields, the index set is essentially the set of irreducible divisors on a projective model of the field. This description is a key ingredient in the reconstruction of function fields from their Galois groups.

In particular, an isomorphism of Galois groups

$$\Psi_{K,L} : \mathscr{G}_K^a \xrightarrow{\sim} \mathscr{G}_L^a$$

as in Theorem 2 implies a canonical isomorphism

$$\Psi^*: \hat{K}^\times \simeq \hat{L}^\times.$$

The Bloch–Kato conjecture, now a theorem established by Voevodsky [2003; 2010], with crucial contributions by Rost and Weibel [2009; 2009], describes the cohomology of the absolute Galois group G_K through Milnor K-theory for all n:

$$K_n^M(K)/\ell^n = H^n(G_K, \mathbb{Z}/\ell^n). \tag{3-2}$$

There is an alternative formulation. Let $\mathcal{G}_K^c$ be the canonical central extension of $\mathcal{G}_K^a$ as in the Introduction. We have the diagram

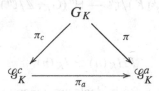

Theorem 11. *The Bloch–Kato conjecture (3-2) is equivalent to:*

(1) *The map*

$$\pi^*: H^*(\mathcal{G}_K^a, \mathbb{Z}/\ell^n) \to H^*(G_K, \mathbb{Z}/\ell^n)$$

   is surjective and

(2) $\mathrm{Ker}(\pi_a^*) = \mathrm{Ker}(\pi^*).$

Proof. The proof uses the first two cases of the Bloch–Kato conjecture. The first is (3-1), i.e., Kummer theory. Recall that the cohomology ring of a torsion-free abelian group is the exterior algebra on H^1. We apply this to $\mathcal{G}_K^a$; combining with (3-1) we obtain:

$$H^*(\mathcal{G}_K^a, \mathbb{Z}/\ell^n) = \bigwedge{}^*(K^\times/\ell^n).$$

Since $\mathcal{G}^c$ is a central extension of the torsion-free abelian group $\mathcal{G}_K^a$, the kernel of the ring homomorphism

$$\pi_a^*: H^*(\mathcal{G}_K^a, \mathbb{Z}/\ell^n) \to H^*(\mathcal{G}_K^c, \mathbb{Z}/\ell^n)$$

is an ideal $IH_K(n)$ generated by

$$\mathrm{Ker}\left(H^2(\mathcal{G}_K^a, \mathbb{Z}/\ell^n) \to H^2(\mathcal{G}_K^c, \mathbb{Z}/\ell^n)\right)$$

(as follows from the standard spectral sequence argument). We have an exact sequence

$$0 \to IH_K(n) \to \bigwedge{}^*(K^\times/\ell^n) \to H^*(\mathcal{G}^c, \mathbb{Z}/\ell^n).$$

On the other hand, we have a diagram for the Milnor K-functor:

$$
\begin{array}{ccccccccc}
1 & \longrightarrow & \tilde{I}_K(n) & \longrightarrow & \otimes^*(K^\times/\ell^n) & \longrightarrow & K_*^M(K)/\ell^n & \longrightarrow & 1 \\
 & & \downarrow & & \downarrow & & \| & & \\
1 & \longrightarrow & I_K(n) & \longrightarrow & \bigwedge^*(K^\times/\ell^n) & \longrightarrow & K_*^M(K)/\ell^n & \longrightarrow & 1 \\
 & & & & \| & & & & \\
 & & & & H^*(\mathcal{G}_K^a, \mathbb{Z}/\ell^n) & & & &
\end{array}
$$

Thus the surjectivity of π^* is equivalent to the surjectivity of

$$
K_n^M(K)/\ell^n \to H^n(G_K, \mathbb{Z}/\ell^n).
$$

Part (2) is equivalent to

$$
I H_K(n) \simeq I_K(n),
$$

under the isomorphism above. Both ideals are generated by degree 2 components. In degree 2, the claimed isomorphism follows from the Merkurjev–Suslin theorem

$$
H^2(G_K, \mathbb{Z}/\ell^n) = K_2^M(K)/\ell^n. \qquad \square
$$

Thus the Bloch–Kato conjecture implies that $\mathcal{G}_K^c$ completely captures the ℓ-part of the cohomology of G_K. This led the first author to conjecture in [Bogomolov 1991a] that the "homotopy" structure of G_K is also captured by $\mathcal{G}_K^c$ and that morphisms between function fields $L \to K$ should be captured (up to purely inseparable extensions) by morphisms $\mathcal{G}_K^c \to \mathcal{G}_L^c$. This motivated the development of the *almost abelian anabelian geometry*.

We now describe a recent related result in Galois cohomology, which could be considered as one of the incarnations of the general principle formulated above. Let G be a group and ℓ a prime number. The descending ℓ^n-central series of G is given by

$$
G^{(1,n)} = G, \quad G^{(i+1,n)} := (G^{(i,n)})^{\ell^n} [G^{(i,n)}, G], \quad i = 1, \ldots.
$$

We write

$$
G^{c,n} = G/G^{(3,n)}, \quad G^{a,n} = G/G^{(2,n)},
$$

so that

$$
G^c = G^{c,0}, \quad G^a = G^{a,0}.
$$

Theorem 12 [Chebolu et al. 2009]. *Let K and L be fields containing ℓ^n-th roots of 1 and*

$$
\Psi : \mathcal{G}_K \to \mathcal{G}_L
$$

a continuous homomorphism. The following are equivalent:

(i) *the induced homomorphism*

$$\Psi^c : \mathcal{G}_K^{c,n} \to \mathcal{G}_L^{c,n}$$

   is an isomorphism;

(ii) *the induced homomorphism*

$$\Psi^* : H^*(\mathcal{G}_L, \mathbb{Z}/\ell^n) \to H^*(\mathcal{G}_K, \mathbb{Z}/\ell^n)$$

   is an isomorphism.

4. Commuting pairs and valuations

A *value group*, Γ, is a totally ordered (torsion-free) abelian group. A (nonarchimedean) *valuation* on a field K is a pair $v = (v, \Gamma_v)$ consisting of a value group Γ_v and a map

$$v : K \to \Gamma_{v,\infty} = \Gamma_v \cup \infty$$

such that

- $v : K^\times \to \Gamma_v$ is a surjective homomorphism;
- $v(\kappa + \kappa') \geq \min(v(\kappa), v(\kappa'))$ for all $\kappa, \kappa' \in K$;
- $v(0) = \infty$.

The set of all valuations of K is denoted by $\mathcal{V}_K$.

Note that $\bar{\mathbb{F}}_p$ admits only the trivial valuation; we will be mostly interested in function fields $K = k(X)$ over $k = \bar{\mathbb{F}}_p$. A valuation is a *flag map* on K: every finite-dimensional $\bar{\mathbb{F}}_p$-subspace, and also $\mathbb{F}_p$-subspace, $V \subset K$ has a flag $V = V_1 \supset V_2 \cdots$ such that v is constant on $V_j \setminus V_{j+1}$. Conversely, every flag map gives rise to a valuation.

Let K_v, $\mathfrak{o}_v$, $\mathfrak{m}_v$, and $\mathbf{K}_v := \mathfrak{o}_v/\mathfrak{m}_v$ be the completion of K with respect to v, the valuation ring of v, the maximal ideal of $\mathfrak{o}_v$, and the residue field, respectively. A valuation of $K = \bar{\mathbb{F}}_p(X)$, is called *divisorial* if the residue field is the function field of a divisor on X; the set of such valuations is denoted by $\mathcal{DV}_K$. We have exact sequences:

$$1 \to \mathfrak{o}_v^\times \to K^\times \to \Gamma_v \to 1,$$
$$1 \to (1 + \mathfrak{m}_v) \to \mathfrak{o}_v^\times \to \mathbf{K}_v^\times \to 1.$$

A homomorphism $\chi : \Gamma_v \to \mathbb{Z}_\ell(1)$ gives rise to a homomorphism

$$\chi \circ v : K^\times \to \mathbb{Z}_\ell(1),$$

thus to an element of $\mathcal{G}_K^a$, an *inertia element* of v. These form the *inertia* subgroup $\mathcal{I}_v^a \subset \mathcal{G}_K^a$. The *decomposition group* $\mathcal{D}_v^a$ is the image of $\mathcal{G}_{K_v}^a$ in $\mathcal{G}_K^a$. We have an embedding $\mathcal{G}_{K_v}^a \hookrightarrow \mathcal{G}_K^a$ and an isomorphism

$$\mathcal{D}_v^a / \mathcal{I}_v^a \simeq \mathcal{G}_{K_v}^a.$$

We have a dictionary (for $K = k(X)$ and $k = \bar{\mathbb{F}}_p$):

$$\mathcal{G}_K^a = \{\text{homomorphisms } \gamma \, : \, K^\times / k^\times \to \mathbb{Z}_\ell(1)\},$$
$$\mathcal{D}_v^a = \qquad \{\mu \in \mathcal{G}_K^a \mid \mu \text{ trivial on } (1 + \mathfrak{m}_v)\},$$
$$\mathcal{I}_v^a = \qquad\qquad \{\iota \in \mathcal{G}_K^a \mid \iota \text{ trivial on } \mathfrak{o}_v^\times\}.$$

In this language, inertia elements define flag maps on K. If $E \subset K$ is a subfield, the corresponding homomorphism of Galois groups $\mathcal{G}_K \to \mathcal{G}_E$ is simply the restriction of special $\mathbb{Z}_\ell(1)$-valued functions on the space $\mathbb{P}_k(K)$ to the projective subspace $\mathbb{P}_k(E)$.

The following result is fundamental in our approach to anabelian geometry.

Theorem 13 [Bogomolov and Tschinkel 2002a; 2008b, Section 4]. *Let K be any field containing a subfield k with $\#k \geq 11$. Assume that there exist nonproportional homomorphisms*

$$\gamma, \gamma' : K^\times \to R$$

where R is either $\mathbb{Z}$, $\mathbb{Z}_\ell$ or $\mathbb{Z}/\ell$, such that

(1) *γ, γ' are trivial on $k^\times$;*

(2) *the restrictions of the R-module $\langle \gamma, \gamma', 1 \rangle$ to every projective line $\mathbb{P}^1 \subset \mathbb{P}_k(K) = K^\times / k^\times$ has R-rank ≤ 2.*

Then there exists a valuation v of K with value group Γ_v, a homomorphism $\iota : \Gamma_v \to R$, and an element ι_v in the R-span of γ, γ' such that

$$\iota_v = \iota \circ v.$$

In (2), γ, γ', and 1 are viewed as functions on a projective line and the condition states simply that these functions are linearly dependent.

This general theorem can be applied in the following contexts: K is a function field over k, where k contains all ℓ-th roots of its elements and $R = \mathbb{Z}/\ell$, or $k = \bar{\mathbb{F}}_p$ with $\ell \neq p$ and $R = \mathbb{Z}_\ell$. In these situations, a homomorphism $\gamma : K^\times \to R$ (satisfying the first condition) corresponds via Kummer theory to an element in $\mathcal{G}_K^a / \ell$, resp. $\mathcal{G}_K^a$. Nonproportional elements $\gamma, \gamma' \in \mathcal{G}_K^a$ lifting to commuting elements in $\mathcal{G}_K^c$ satisfy condition (2). Indeed, for 1-dimensional function fields $E \subset K$ the group $\mathcal{G}_E^c$ is a free central extension of $\mathcal{G}_E^a$. This holds in particular for $k(x) \subset K$. Hence γ, γ' are proportional on any $\mathbb{P}^1$ containing 1; the restriction of $\sigma = \langle \gamma, \gamma' \rangle$ to such $\mathbb{P}^1$ is isomorphic to $\mathbb{Z}_\ell$. Property (2) follows since every $\mathbb{P}^1 \subset P_k(K)$ is a translate, with respect to multiplication in $P_k(K) = K^\times/k^\times$, of the "standard" $\mathbb{P}^1 = P_k(k \oplus kx)$, $x \in K^\times$. Finally, the element ι_ν obtained in the theorem is an inertia element for ν, by the dictionary above.

Corollary 14. *Let K be a function field of an algebraic variety X over an algebraically closed field k of dimension n. Let $\sigma \in \Sigma_K$ be a liftable subgroup. Then*

- $\mathrm{rk}_{\mathbb{Z}_\ell}(\sigma) \leq n$;
- *there exists a valuation $\nu \in \mathcal{V}_K$ and a subgroup $\sigma' \subseteq \sigma$ such that $\sigma' \subseteq \mathcal{I}_\nu^a$, $\sigma \subset \mathcal{D}_\nu^a$, and σ/σ' is topologically cyclic.*

Theorem 13 and its Corollary 14 allow to *recover* inertia and decomposition groups of valuations from $(\mathcal{G}_K^a, \Sigma_K)$. In reconstructions of function fields we need only divisorial valuations; these can be characterized as follows:

Corollary 15. *Let K be a function field of an algebraic variety X over $k = \bar{\mathbb{F}}_p$ of dimension n. If $\sigma_1, \sigma_2 \subset \mathcal{G}_K^a$ are maximal liftable subgroups of $\mathbb{Z}_\ell$-rank n such that $\mathcal{I}^a := \sigma_1 \cap \sigma_2$ is topologically cyclic then there exists a divisorial valuation $\nu \in \mathcal{DV}_K$ such that $\mathcal{I}^a = \mathcal{I}_\nu^a$.*

Here we restricted to $k = \bar{\mathbb{F}}_p$ to avoid a discussion of mixed characteristic phenomena. For example, the obtained valuation may be a divisorial valuation of a *reduction* of the field, and not of the field itself.

This implies that an isomorphism of Galois groups

$$\Psi : \mathcal{G}_K^a \to \mathcal{G}_L^a$$

inducing a bijection of the sets of liftable subgroups

$$\Sigma_K = \Sigma_L$$

induces a bijection of the sets of inertial and decomposition subgroups of valuations

$$\{\mathcal{I}_\nu^a\}_{\nu \in \mathcal{DV}_K} = \{\mathcal{I}_\nu^a\}_{\nu \in \mathcal{DV}_L}, \quad \{\mathcal{D}_\nu^a\}_{\nu \in \mathcal{DV}_K} = \{\mathcal{D}_\nu^a\}_{\nu \in \mathcal{DV}_L}.$$

Moreover, Ψ maps topological generators $\delta_{\nu, K}$ of procyclic subgroups $\mathcal{I}_\nu^a \subset \mathcal{G}_K^a$, for $\nu \in \mathcal{DV}_K$, to generators $\delta_{\nu, L}$ of corresponding inertia subgroups in $\mathcal{G}_L^a$, which pins down a generator up to the action of $\mathbb{Z}_\ell^\times$.

Here are two related results concerning the reconstruction of valuations.

Theorem 16 [Efrat 1999]. *Assume that $\mathrm{char}(K) \neq \ell$, $-1 \in (K^\times)^\ell$, and that*

$$\textstyle\bigwedge^2 (K^\times/(K^\times)^\ell) \xrightarrow{\sim} K_2^M(K)/\ell.$$

Then there exists a valuation ν on K such that

- char$(K_\nu) \neq \ell$;
- $\dim_{\mathbb{F}_\ell}(\Gamma_\nu/\ell) \geq \dim_{\mathbb{F}_\ell}(K^\times/(K^\times)^\ell) - 1$;
- *either* $\dim_{\mathbb{F}_\ell}(\Gamma_\nu/\ell) = \dim_{\mathbb{F}_\ell}(K^\times/(K^\times)^\ell)$ *or* $K_\nu \neq K_\nu^\ell$.

In our terminology, under the assumption that K contains an algebraically closed subfield k and $\ell \neq 2$, the conditions mean that G_K^a modulo ℓ is liftable, i.e., $G_K^c = G_K^a$. Thus there exists a valuation with abelianized inertia subgroup (modulo ℓ) of corank at most one, by Corollary 14. The third assumption distinguishes the two cases, when the corank is zero versus one. In the latter case, the residue field K_ν has nontrivial ℓ-extensions, hence satisfies $K_\nu^\times \neq (K_\nu^\times)^\ell$.

Theorem 17 [Engler and Koenigsmann 1998; Engler and Nogueira 1994]. *Let K be a field of characteristic $\neq \ell$ containing the roots of unity of order ℓ. Then K admits an ℓ-Henselian valuation ν (i.e., ν extends uniquely to the maximal Galois ℓ-extension of K) with* char$(K_\nu) \neq \ell$ *and non-ℓ-divisible Γ_ν if and only if $\mathcal{G}_K$ is noncyclic and contains a nontrivial normal abelian subgroup.*

Again, under the assumption that K contains an algebraically closed field k, of characteristic $\neq \ell$, we can directly relate this result to our Theorem 13 and Corollary 14 as follows: The presence of an abelian normal subgroup in $\mathcal{G}_K$ means that modulo ℓ^n there is a nontrivial center. Thus there is a valuation ν such that $\mathcal{G}_K = \mathcal{D}_\nu$, the corresponding decomposition group. Note that the inertia subgroup $\mathcal{I}_\nu \subset \mathcal{G}_K$ maps injectively into $\mathcal{I}_\nu^a$.

We now sketch the proof of Theorem 13. Reformulating the claim, we see that the goal is to produce a *flag map* on $\mathbb{P}_k(K)$. Such a map ι jumps only on projective subspaces of $\mathbb{P}_k(K)$, i.e., every finite dimensional projective space $\mathbb{P}^n \subset \mathbb{P}_k(K)$ should admit a flag by projective subspaces

$$\mathbb{P}^n \supset \mathbb{P}^{n-1} \supset \cdots$$

such that ι is constant on $\mathbb{P}^r(k) \setminus \mathbb{P}^{r-1}(k)$, for all r. Indeed, a flag map defines a partial order on $K^\times$ which is preserved under shifts by multiplication in $K^\times/k^\times$, hence a scale of k-subspaces parametrized by some ordered abelian group Γ.

We proceed by contradiction. Assuming that the R-span $\sigma := \langle \gamma, \gamma' \rangle$ does not contain a flag map we find a distinguished $\mathbb{P}^2 \subset \mathbb{P}_k(K)$ such that σ contains no maps which would be flag maps on this $\mathbb{P}^2$ (this uses that $\#k \geq 11$). To simplify the exposition, assume now that $k = \mathbb{F}_p$.

Step 1. If $p > 3$ then $\alpha : \mathbb{P}^2(\mathbb{F}_p) \to R$ is a flag map if and only if the restriction to *every* $\mathbb{P}^1(\mathbb{F}_p) \subset \mathbb{P}^2(\mathbb{F}_p)$ is a flag map, i.e., constant on the complement of one point.

A counterexample for $p = 2$ and $R = \mathbb{Z}/2$ is provided by the Fano plane:

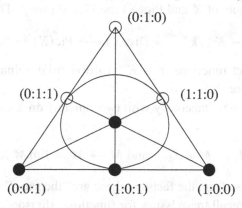

Step 2. On the other hand, assumptions (1) and (2) imply that the map

$$K^\times / k^\times = \mathbb{P}_k(K) \xrightarrow{\varphi} \mathbb{A}^2(R), \qquad f \mapsto (\gamma(f), \gamma'(f)),$$

maps every *projective line* into an *affine* line, a collineation. This imposes strong conditions on $\varphi = \varphi_{\gamma, \gamma'}$ and both γ, γ'. For example, for all $\mathbb{P}^2 \subset \mathbb{P}_k(K)$ the image $\varphi(\mathbb{P}^2)$ is contained in a union of an affine line and at most one extra point in $\mathbb{A}^2(R)$.

Step 3. At this stage we are working with maps

$$\mathbb{P}^2(\mathbb{F}_p) \to \mathbb{A}^2(R),$$

preserving the geometries as above. Using Step 2 we may even reduce to considerations of maps with image consisting of 3 points:

$$\mathbb{P}^2(\mathbb{F}_p) \to \{\bullet, \circ, \star\}$$

and such that every line $\mathbb{P}^1(\mathbb{F}_p) \subset \mathbb{P}^2(\mathbb{F}_p)$ is mapped to exactly two points. Projective/affine geometry considerations produce a *flag map* in the R-linear span of γ, γ', contradicting the assumption.

The case where K is of characteristic 0 is more complicated; see [Bogomolov and Tschinkel 2002a].

5. Pro-ℓ-geometry

One of the main advantages in working with function fields K as opposed to arbitrary fields is the existence of normal *models*, i.e., algebraic varieties X with $K = k(X)$, and a *divisor* theory on X. Divisors on these models give rise to a rich supply of valuations of K, and we can employ geometric considerations in the study of relations between them.

We now assume that $k = \bar{\mathbb{F}}_p$, with $p \neq \ell$. Let $\mathrm{Div}(X)$ be the group of (locally principal) Weil divisors of X and $\mathrm{Pic}(X)$ the Picard group. The exact sequence

$$0 \to K^{\times}/k^{\times} \xrightarrow{\mathrm{div}_X} \mathrm{Div}(X) \xrightarrow{\varphi} \mathrm{Pic}(X) \to 0, \qquad (5\text{-}1)$$

allows us to connect functions $f \in K^{\times}$ to divisorial valuations, realized by irreducible divisors on X.

We need to work simultaneously with two functors on $\mathbb{Z}$-modules of possibly infinite rank:

$$M \mapsto M_{\ell} := M \otimes \mathbb{Z}_{\ell} \quad \text{and} \quad M \mapsto \hat{M} := \varprojlim M \otimes \mathbb{Z}/\ell^n.$$

Some difficulties arise from the fact that these are "the same" at each finite level, (mod ℓ^n). We now recall these issues for functions, divisors, and Picard groups of normal projective models of function fields (see [Bogomolov and Tschinkel 2008b, Section 11] for more details).

Equation (5-1) gives rise to an exact sequence

$$0 \to K^{\times}/k^{\times} \otimes \mathbb{Z}_{\ell} \xrightarrow{\mathrm{div}_X} \mathrm{Div}^0(X)_{\ell} \xrightarrow{\varphi_{\ell}} \mathrm{Pic}^0(X)\{\ell\} \to 0. \qquad (5\text{-}2)$$

where

$$\mathrm{Pic}^0(X)\{\ell\} = \mathrm{Pic}^0(X) \otimes \mathbb{Z}_{\ell}$$

is the ℓ-primary component of the torsion group of $k = \bar{\mathbb{F}}_p$-points of $\mathrm{Pic}^0(X)$, the algebraic group parametrizing classes of algebraically equivalent divisors modulo rational equivalence. Put

$$\mathcal{T}_{\ell}(X) := \varprojlim \mathrm{Tor}_1(\mathbb{Z}/\ell^n, \mathrm{Pic}^0(X)\{\ell\}).$$

We have $\mathcal{T}_{\ell}(X) \simeq \mathbb{Z}_{\ell}^{2g}$, where g is the dimension of $\mathrm{Pic}^0(X)$. In fact, $\mathcal{T}_{\ell}$ is a contravariant functor, which stabilizes on some normal projective model X, i.e., $\mathcal{T}_{\ell}(\tilde{X}) = \mathcal{T}_{\ell}(X)$ for all $\tilde{X}$ surjecting onto X. In the sequel, we will implicitly work with such X and we write $\mathcal{T}_{\ell}(K)$.

Passing to pro-ℓ-completions in (5-2) we obtain an exact sequence:

$$0 \to \mathcal{T}_{\ell}(K) \to \hat{K}^{\times} \xrightarrow{\mathrm{div}_X} \widehat{\mathrm{Div}^0}(X) \longrightarrow 0, \qquad (5\text{-}3)$$

since $\mathrm{Pic}^0(X)$ is an ℓ-divisible group. Note that all groups in this sequence are torsion-free. We have a diagram

$$
\begin{array}{ccccccccc}
0 & \longrightarrow & K^{\times}/k^{\times} \otimes \mathbb{Z}_{\ell} & \xrightarrow{\mathrm{div}_X} & \mathrm{Div}^0(X)_{\ell} & \xrightarrow{\varphi_{\ell}} & \mathrm{Pic}^0(X)\{\ell\} & \longrightarrow & 0 \\
& & \downarrow & & \downarrow & & \downarrow & & \\
0 & \longrightarrow & \mathcal{T}_{\ell}(K) & \longrightarrow & \hat{K}^{\times} & \xrightarrow{\mathrm{div}_X} & \widehat{\mathrm{Div}^0}(X) & \longrightarrow & 0
\end{array}
$$

Galois theory allows one to "reconstruct" the second row of this diagram. The reconstruction of fields requires the first row. The passage from the second to the first employs the theory of valuations. Every $v \in \mathcal{DV}_K$ gives rise to a homomorphism

$$v : \hat{K}^\times \to \mathbb{Z}_\ell.$$

On a normal model X, where $v = v_D$ for some divisor $D \subset X$, $v(\hat{f})$ is the ℓ-adic coefficient at D of $\mathrm{div}_X(\hat{f})$. "Functions", i.e., elements $f \in K^\times$, have *finite support* on models X of K, i.e., only finitely many coefficients $v(f)$ are nonzero. However, the passage to blowups of X introduces more and more divisors (divisorial valuations) in the support of f. The strategy in [Bogomolov and Tschinkel 2008b], specific to dimension two, was to extract elements of $\hat{K}^\times$ with *intrinsically* finite support, using the interplay between one-dimensional subfields $E \subset K$, i.e., projections of X onto curves, and divisors of X, i.e., curves $C \subset X$. For example, Galois theory allows one to distinguish valuations v corresponding to *rational* and *nonrational* curves on X. If X had only *finitely many* rational curves, then every blowup $\tilde{X} \to X$ would have the same property. Thus elements $\hat{f} \in \hat{K}^\times$ with finite *nonrational* support, i.e., $v(f) = 0$ for all but finitely many nonrational v, have necessarily finite support on every model X of K, and thus have a chance of being functions. A different geometric argument applies when X admits a fibration over a curve of genus ≥ 1, with rational generic fiber. The most difficult case to treat, surprisingly, is the case of rational surfaces. See Section 12 of [Bogomolov and Tschinkel 2008b] for more details.

The proof of Theorem 2 in [Bogomolov and Tschinkel 2009b] reduces to dimension two, via Lefschetz pencils.

6. Pro-ℓ-K-theory

Let k be an algebraically closed field of characteristic $\neq \ell$ and X a smooth projective variety over k, with function field $K = k(X)$. A natural generalization of (5-1) is the Gersten sequence (see, e.g., [Suslin 1984]):

$$0 \to K_2(X) \to K_2(K) \to \bigoplus_{x \in X_1} K_1(k(x)) \to \bigoplus_{x \in X_2} \mathbb{Z} \to \mathrm{CH}^2(X) \to 0,$$

where X_d is the set of points of X of codimension d and $\mathrm{CH}^2(X)$ is the second Chow group of X. Applying the functor

$$M \mapsto M^\vee := \mathrm{Hom}(M, \mathbb{Z}_\ell)$$

and using the duality

$$\mathcal{G}_K^a = \mathrm{Hom}(K^\times, \mathbb{Z}_\ell)$$

we obtain a sequence

$$K_2(X)^\vee \longleftarrow K_2(K)^\vee \longleftarrow \prod_{D \subset X} \mathcal{G}^a_{k(D)}.$$

Dualizing the sequence

$$0 \to I_K \to \wedge^2(K^\times) \to K_2(K) \to 0$$

we obtain

$$I_K^\vee \leftarrow \wedge^2(\mathcal{G}^a_K) \leftarrow K_2(K)^\vee \leftarrow 0$$

On the other hand, we have the exact sequence

$$0 \to Z_K \to \mathcal{G}^c_K \to \mathcal{G}^a_K \to 0$$

and the resolution of $Z_K = [\mathcal{G}^c_K, \mathcal{G}^c_K]$

$$0 \to R(K) \to \wedge^2(\mathcal{G}^a_K) \to Z_K \to 0.$$

Recall that $\mathcal{G}^a_K = \mathrm{Hom}(K^\times/k^\times, \mathbb{Z}_\ell)$ is a torsion-free $\mathbb{Z}_\ell$-module, with topology induced from the discrete topology on $K^\times/k^\times$. Thus any primitive finitely generated subgroup $A \subset K^\times/k^\times$ is a direct summand and defines a continuous surjection $\mathcal{G}^a_K \to \mathrm{Hom}(A, \mathbb{Z}_\ell)$. The above topology on $\mathcal{G}^a_K$ defines a natural topology on $\wedge^2(\mathcal{G}^a_K)$. On the other hand, we have a topological profinite group $\mathcal{G}^c_K$ with topology induced by finite ℓ-extensions of K, which contains a closed abelian subgroup $Z_K = [\mathcal{G}^c_K, \mathcal{G}^c_K]$.

Proposition 18 [Bogomolov 1991a]. *We have*

$$R(K) = (\mathrm{Hom}(K_2(K)/\mathrm{Image}(k^\times \otimes K^\times), \mathbb{Z}_\ell) = K_2(K)^\vee.$$

Proof. There is *continuous* surjective homomorphism

$$\begin{aligned} \wedge^2(\mathcal{G}^a_K) &\to Z_K \\ \gamma \wedge \gamma' &\mapsto [\gamma, \gamma'] \end{aligned}$$

The kernel $R(K)$ is a profinite group with the induced topology. Any $r \in R(K)$ is trivial on symbols $(x, 1-x) \in \wedge^2(K^\times/k^\times)$ (since the corresponding elements are trivial in $\mathrm{H}^2(\mathcal{G}^a_K, \mathbb{Z}/\ell^n)$, for all $n \in \mathbb{N}$). Thus $R(K) \subseteq K_2(K)^\vee$.

Conversely, let $\alpha \in K_2(K)^\vee \setminus R(K)$; so that it projects nontrivially to Z_K, i.e., to a nontrivial element modulo ℓ^n, for some $n \in \mathbb{N}$. Finite quotient groups of $\mathcal{G}^c_K$ with $Z(G^c_i) = [G^c_i, G^c_i]$ form a basis of topology on $\mathcal{G}^c_K$. The induced surjective homomorphisms $\mathcal{G}^a_K \to G^a_i$ define surjections $\wedge^2(\mathcal{G}^a_K) \to [G_i, G_i]$ and

$$R(K) \to R_i := \mathrm{Ker}(\wedge^2(G^a_i) \to [G_i, G_i]).$$

Fix a G_i such that α is nontrivial of G_i^c. Then the element α is nonzero in the image of $H^2(G_i^a, \mathbb{Z}/\ell^n) \to H^2(G_i^c, \mathbb{Z}/\ell^n)$. But this is incompatible with relations in $K_2(K)$, modulo ℓ^n. □

It follows that $R(K)$ contains a distinguished $\mathbb{Z}_\ell$-submodule

$$R_\wedge(K) = \text{Image of } \prod_{D \subset X} \mathcal{G}^a_{k(D)} \tag{6-1}$$

and that

$$K_2(X)^\vee \supseteq R(K)/R_\wedge(K).$$

In general, let

$$K_{2,nr}(K) = \text{Ker}(K_2(K) \to \bigoplus_{v \in \mathcal{D}\mathcal{V}_K} K_v^\times)$$

be the *unramified* K_2-group. Combining Proposition 18 and (6-1), we find that

$$\widehat{K_{2,nr}}(K) \subseteq \text{Hom}(R(K)/R_\wedge(K), \mathbb{Z}_\ell).$$

This sheds light on the connection between relations in $\mathcal{G}^c_K$ and the K-theory of the field, more precisely, the unramified Brauer group of K. This in turn helps to reconstruct multiplicative groups of 1-dimensional subfields of K.

We now sketch a closely related, alternative strategy for the reconstruction of these subgroups of $\hat{K}^\times$ from Galois-theoretic data. We have a diagram

$$
\begin{array}{ccccccc}
0 & \longrightarrow & \mathcal{G}^c_K & \longrightarrow & \prod_E \mathcal{G}^c_K & \xrightarrow{\rho^c_E} & \mathcal{G}^c_E \\
 & & \downarrow & & \downarrow & & \downarrow \\
0 & \longrightarrow & \mathcal{G}^a_K & \longrightarrow & \prod_E \mathcal{G}^a_K & \xrightarrow{\rho^a_E} & \mathcal{G}^a_E
\end{array}
$$

where the product is taken over all normally closed 1-dimensional subfields $E \subset K$, equipped with the direct product topology, and the horizontal maps are closed embeddings. Note that $\mathcal{G}^a_K$ is a primitive subgroup given by equations

$$\mathcal{G}^a_K = \{\gamma \mid (xy)(\gamma) - (x)(\gamma) - (y)(\gamma) = 0\} \subset \prod_E \mathcal{G}^a_E$$

where x, y are algebraically independent in K and $xy, x, y \in K^\times$ are considered as functionals on $\mathcal{G}^a_{k(xy)}, \mathcal{G}^a_{k(x)}, \mathcal{G}^a_{k(y)}$, respectively. The central subgroup

$$Z_K \subset \mathcal{G}^c_K \subset \prod_E \wedge^2(\mathcal{G}^a_E)$$

is the image of $\wedge^2(\mathcal{G}^a_K)$ in $\prod_E \wedge^2(\mathcal{G}^a_E)$. Thus for any finite quotient ℓ-group G of $\mathcal{G}^c_K$ there is an intermediate quotient which is a subgroup of finite index in the product of free central extensions. The following fundamental conjecture lies at the core of our approach.

Conjecture 19. *Let K be a function field over $\bar{\mathbb{F}}_p$, with $p \neq \ell$, F^a a torsion-free topological $\mathbb{Z}_\ell$-module of infinite rank. Assume that*

$$\Psi_F^a : \mathcal{G}_K^a \to F^a$$

is a continuous surjective homomorphism such that

$$\text{rk}_{\mathbb{Z}_\ell}(\Psi_F^a(\sigma)) \leq 1$$

for all liftable subgroups $\sigma \in \Sigma_K$. Then there exist a 1-dimensional subfield $E \subset K$, a subgroup $\tilde{F}^a \subset F^a$ of finite corank, and a diagram

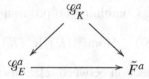

We expect that $\tilde{F}_a = F_a$, when $\pi_1(X)$ is finite. Note that there can exist at most one normally closed subfield $E \subset F$ satisfying this property.

The intuition behind this conjecture is that such maps should arise from surjective homomorphisms onto free central extensions, i.e., we should be able to factor as follows:

$$\Psi_F^c = \mathcal{G}_K^c \xrightarrow{\rho_F^c} \mathcal{G}_F^c \to F^c$$

where F^c is a free central extension of F^a:

$$0 \to \wedge^2(F^a) \to F^c \to F^a \to 0.$$

We can prove the conjecture under some additional geometric assumptions. Assuming the conjecture, the proofs in [Bogomolov and Tschinkel 2008b; 2009b] would become much more straightforward. Indeed, consider the diagram

$$\mathcal{G}_K^a \xrightarrow{\sim} \mathcal{G}_L^a$$
$$\downarrow$$
$$\mathcal{G}_F^a$$

Applying Conjecture 19 we find a unique normally closed subfield $E \subset K$ and a canonical isomorphism

$$\Psi : \mathcal{G}_E^a \to \mathcal{G}_F^a, \quad F \subset L,$$

Moreover, this map gives a bijection between the set of inertia subgroups of divisorial valuations on E and of F; these are the images of inertia subgroups of divisorial valuations on K and L. At this stage, the simple rationality argument (see [Bogomolov and Tschinkel 2008b, Proposition 13.1 and Corollary 15.6]) implies that

$$\Psi^* \colon \hat{L}^\times \xrightarrow{\sim} \hat{K}^\times$$

induces an isomorphism

$$L^\times / l^\times \otimes \mathbb{Z}_{(\ell)} \xrightarrow{\sim} \epsilon \left(K^\times / k^\times \otimes \mathbb{Z}_{(\ell)} \right),$$

for some $\epsilon \in \mathbb{Z}_\ell^\times$, respecting multiplicative subgroups of 1-dimensional subfields. Moreover, for each 1-dimensional rational subfield $l(y) \subset L$ we obtain

$$\Psi^*(l(y)^\times / l^\times) = \epsilon \cdot \epsilon_y \cdot \left(k(x)^\times / k^\times \right)$$

for some $\epsilon_y \in \mathbb{Q}$. Proposition 2.13 in [Bogomolov and Tschinkel 2008b] shows that this implies the existence of subfields $\bar{L}$ and $\bar{K}$ such that $L/\bar{L}$ and $K/\bar{K}$ are purely inseparable extensions and such that $\epsilon^{-1} \cdot \Psi^*$ induces an *isomorphism* of multiplicative groups

$$\mathbb{P}(\bar{L}) = \bar{L}^\times / l^\times \xrightarrow{\sim} \mathbb{P}(\bar{K}) = \bar{K}^\times / k^\times.$$

Moreover, this isomorphism maps lines $\mathbb{P}^1 \subset \mathbb{P}(l(y))$ to lines $\mathbb{P}^1 \subset \mathbb{P}(k(x))$. Arguments similar to those in Section 2 allow us to show that Ψ^* induces an bijection of the sets of all *projective lines* of the projective structures. The Fundamental theorem of projective geometry (Theorem 5) allows to match the additive structures and leads to an isomorphism of fields.

The proof of Theorem 2 in [Bogomolov and Tschinkel 2008b] is given for the case of the fields of transcendence degree two. However, the general case immediately follows by applying Theorem 5 from Section 1 (or [Bogomolov and Tschinkel 2009b]). Indeed, it suffices to show that for all $x, y \in L^\times / l^\times$

$$\Psi^*(l(x,y)^\times / l^\times) \subset \overline{k(x,y)}^\times / k^\times \otimes \mathbb{Z}_{(\ell)} \subset K^\times / k^\times \otimes \mathbb{Z}_{(\ell)}.$$

Note that the groups $\overline{l(x)}^\times / l^\times$ map into subgroups $\overline{k(x)}^\times / k^\times \times \mathbb{Z}_{(\ell)}$ since Ψ^* satisfies the conditions of [Bogomolov and Tschinkel 2009b, Lemma 26], i.e., the symbol

$$(\Psi^*(y), \Psi^*(z)) \in K_2^M(K) \otimes \mathbb{Z}_\ell$$

is infinitely ℓ-divisible, for any $y, z \in \overline{l(x)}^\times / l^\times$. Thus

$$\Psi^*(\overline{l(x/y)}^\times) \in \overline{k(x,y)}^\times / k^\times \otimes \mathbb{Z}_{(\ell)}$$

and similarly for $\Psi^*(\overline{l(x+by)}^\times)/l^\times$, $b \in l$, since by multiplicativity

$$\Psi^*(\overline{l(x+y)}^\times/l^\times) \subset \cup_n (y^n \cdot \Psi^*(\overline{l(x/y+b)}^\times/l^\times) = \cup_n (y^n \cdot \Psi^*(\overline{l(x/y)}^\times/l^\times)).$$

Thus

$$\Psi^*(x/y)/l^\times, \Psi^*(x+y)/l^\times \in \overline{k(x,y)}^\times/k^\times \otimes \mathbb{Z}_{(\ell)},$$

so that Theorem 2, for fields of arbitrary transcendence degree, follows from the result for transcendence degree two.

7. Group theory

Our intuition in Galois theory and Galois cohomology is based on the study of *finite* covers and *finite* groups. Our goal is to recover fields or some of their invariants from invariants of their absolute Galois groups and their quotients.

In this section, we study some group-theoretic constructions which appear, in disguise, in the study of function fields. Let G be a finite group. We have

$$G^c = G/[[G,G],G], \quad G^a = G/[G,G].$$

Let

$$B_0(G) := \mathrm{Ker}\left(H^2(G,\mathbb{Q}/\mathbb{Z}) \to \prod_B H^2(B,\mathbb{Q}/\mathbb{Z})\right)$$

be the subgroup of those Schur multipliers which restrict trivially to all bicyclic subgroups $B \subset G$. The first author conjectured in [Bogomolov 1992] that

$$B_0(G) = 0$$

for all finite simple groups. Some special cases were proved in [Bogomolov et al. 2004], and the general case was settled [Kunyavski 2010].

In computations of this group it is useful to keep in mind the following diagram

$$
\begin{array}{ccccc}
B_0(G^c) & & H^2(G^a) & & B_0(G) \\
\downarrow & & \downarrow & & \downarrow \\
H^2(G^c) & =\!=\!= & H^2(G^c) & \longrightarrow & H^2(G) \\
\downarrow & & \downarrow & & \downarrow \\
\prod_{B \subset G^c} H^2(B) & =\!=\!= & \prod_{B \subset G^c} H^2(B) & \twoheadrightarrow & \prod_{B \subset G} H^2(B).
\end{array}
$$

Thus we have a homomorphism

$$B_0(G^c) \to B_0(G).$$

We also have an isomorphism

$$\mathrm{Ker}\left(H^2(G^a, \mathbb{Q}/\mathbb{Z}) \to H^2(G,\mathbb{Q}/\mathbb{Z})\right) = \mathrm{Ker}\left(H^2(G^a,\mathbb{Q}/\mathbb{Z}) \to H^2(G^c,\mathbb{Q}/\mathbb{Z})\right)$$

Combining with the fact that $B_0(G^c)$ is in the image of

$$\pi_a^*: H^2(G^a, \mathbb{Q}/\mathbb{Z}) \to H^2(G, \mathbb{Q}/\mathbb{Z})$$

this implies that

$$B_0(G^c) \hookrightarrow B_0(G). \tag{7-1}$$

Let ℓ be a prime number. We write G_ℓ for the maximal ℓ-quotient of G and fix an ℓ-Sylow subgroup $\mathrm{Syl}_\ell(G) \subset G$, all considerations below are independent of the conjugacy class. We have a diagram

$$
\begin{array}{ccccc}
G & \longrightarrow & G^c & \longrightarrow & G^a \\
\downarrow & & \downarrow & & \downarrow \\
\mathrm{Syl}_\ell(G) & \longrightarrow & G_\ell & \longrightarrow & G_\ell^c & \longrightarrow & G_\ell^a
\end{array}
$$

Note that

$$G_\ell^c = \mathrm{Syl}_\ell(G^c) \quad \text{and} \quad G_\ell^a = \mathrm{Syl}_\ell(G^a),$$

but that, in general, $\mathrm{Syl}_\ell(G)$ is much bigger than G_ℓ.

We keep the same notation when working with pro-ℓ-groups.

Proposition 20 [Bogomolov 1995b]. *Let X be a projective algebraic variety of dimension n over a field k. Assume that $X(k)$ contains a smooth point. Then*

$$\mathrm{Syl}_\ell(G_{k(X)}) = \mathrm{Syl}_\ell(G_{k(\mathbb{P}^n)}).$$

Proof. First of all, let X and Y be algebraic varieties over a field k with function fields $K = k(X)$, resp. $L = k(Y)$. Let $X \to Y$ be a map of degree d and ℓ a prime not dividing d and $\mathrm{char}(k)$. Then

$$\mathrm{Syl}_\ell(G_K) = \mathrm{Syl}_\ell(G_L).$$

Let $X \to \mathbb{P}^{n+1}$ be a birational embedding as a (singular) hypersurface of degree d'. Consider two projections onto $\mathbb{P}^n$: the first, π_x from a smooth point x in the image of X and the second, π_y, from a point y in the complement of X in $\mathbb{P}^{n+1}$. We have $\deg(\pi_y) = d'$ and $\deg(\pi_y) - \deg(\pi_x) = 1$, in particular, one of these degrees is coprime to ℓ. The proposition follows from the first step. $\square$

Remark 21. This shows that the full Galois group G_K is, in some sense, *too large*: the isomorphism classes of its ℓ-Sylow subgroups depend only on the dimension and the ground field. We may write

$$\mathrm{Syl}_\ell(G_K) = \mathrm{Syl}_{\ell,n,k}.$$

In particular, they *do not* determine the function field. However, the maximal pro-ℓ-quotients do [Mochizuki 1999; Pop 1994]. Thus we have a surjection from a *universal* group, depending only on the dimension and ground field k, onto a *highly individual* group $\mathscr{G}_K^c$, which by Theorem 2 determines the field K, for $k = \bar{\mathbb{F}}_p$, $\ell \neq p$, and $n \geq 2$.

The argument shows in particular that the group $\mathrm{Syl}_{\ell,k,n}$ belongs to the class of self-similar groups. Namely any open subgroup of finite index in $\mathrm{Syl}_{\ell,k,n}$ is isomorphic to $\mathrm{Syl}_{\ell,k,n}$. The above construction provides with isomorphisms parametrized by smooth k-points of n-dimensional algebraic varieties. Note that the absence of smooth k-points in K may lead to a nonisomorphic group $\mathrm{Syl}_{\ell,k,n}$, as seen already in the example of a conic C over $k = \mathbb{R}$ with $C(\mathbb{R}) = \varnothing$ [Bogomolov 1995b].

Theorem 22 [Bogomolov 1991a, Theorem 13.2]. *Let G_K be the Galois group of a function field $K = k(X)$ over an algebraically closed ground field k. Then, for all $\ell \neq \mathrm{char}(k)$ we have*

$$\mathrm{B}_{0,\ell}(G_K) = \mathrm{B}_0(\mathscr{G}_K^c).$$

Here is a sample of known facts:

- If X is stably rational over k, then

$$\mathrm{B}_0(G_K) = 0.$$

- If $X = V/G$, where V is a faithful representation of G over an algebraically closed field of characteristic coprime to the order of G, and $K = k(X)$, then

$$\mathrm{B}_0(G) = \mathrm{B}_0(G_K),$$

thus nonzero in many cases.

Already this shows that the groups G_K are quite special. The following "Freeness conjecture" is related to the Bloch–Kato conjecture discussed in Section 3; it would imply that all cohomology of G_K is induced from metabelian finite ℓ-groups.

Conjecture 23 (Bogomolov). *For $K = k(X)$, with k algebraically closed of characteristic $\neq \ell$, let*

$$\mathrm{Syl}_{\ell,n,k}^{(2)} = [\mathrm{Syl}_{\ell,n,k}, \mathrm{Syl}_{\ell,n,k}],$$

and let M be a finite $\mathrm{Syl}_{\ell,n,k}^{(2)}$-module. Then

$$\mathrm{H}^i(\mathrm{Syl}_{\ell,n,k}^{(2)}, M) = 0, \quad \text{for all} \quad i \geq 2.$$

Further discussions in this direction, in particular, concerning the connections between the Bloch–Kato conjecture, "Freeness", and the Koszul property of the algebra $\mathrm{K}_*^M(K)/\ell$, can be found in [Positselski 2005] and [Positselski and Vishik 1995].

8. Stabilization

The varieties V/G considered in the Introduction seem very special. On the other hand, let X be any variety over a field k and let

$$G_{k(X)} \to G$$

be a continuous homomorphism from its Galois group onto some *finite* group. Let V be a faithful representation of G. Then we have two homomorphisms (for cohomology with finite coefficients and trivial action)

$$\kappa_X : H^*(G) \to H^*(G_{k(X)})$$

and

$$\kappa_{V/G} : H^*(G) \to H^*(G_{k(V/G)}).$$

These satisfy

- $\mathrm{Ker}(\kappa_{V/G}) \subseteq H^*(G)$ is independent of V, and the quotient

  $$H_s^*(G) := H^*(G)/\mathrm{Ker}(\kappa_{V/G})$$

  is well-defined;
- $\mathrm{Ker}(\kappa_{V/G}) \subseteq \mathrm{Ker}(\kappa_X)$.

The groups $H_s^i(G)$ are called *stable* cohomology groups of G. They were introduced and studied in [Bogomolov 1992]. *A priori*, these groups depend on the ground field k. We get a surjective homomorphism

$$H_s^*(G) \to H^*(G)/\mathrm{Ker}(\kappa_X).$$

This explains the interest in stable cohomology—all group-cohomological invariants arising from finite quotients of $G_{k(X)}$ arise from similar invariants of V/G. On the other hand, there is no effective procedure for the computation of stable cohomology, except in special cases. For example, for abelian groups the stabilization can be described already on the group level:

Proposition 24 (see, e.g., [Bogomolov 1992]). *Let G be a finite abelian group and $\sigma : \mathbb{Z}^m \to G$ a surjective homomorphism. Then $\kappa^* : H^*(G) \to H^*(\mathbb{Z}^m)$ coincides with the stabilization map, i.e.,*

$$\mathrm{Ker}(\kappa^*) = \mathrm{Ker}(\kappa_{V/G})$$

for any faithful representation V of G, for arbitrary ground fields k with $\mathrm{char}(k)$ coprime to the order of G.

Geometrically, stabilization is achieved on the variety $T/G \subset V/G$, where G acts faithfully on V by diagonal matrices and $T \subset V$ is a G-invariant subtorus in V (see, e.g., [Bogomolov 1995a]).

Similar actions exist for any finite group G: there is faithful representation V and a torus $T \subset \mathrm{Aut}(V)$, with normalizer $N = N(T)$ such that $G \subset N \subset \mathrm{Aut}(V)$, and such that G acts *freely* on T. We have an exact sequence

$$1 \to \pi_1(T) \to \pi_1(T/G) \to G \to 1$$

of topological fundamental groups. Note that $\pi_1(T)$ decomposes as a sum of G-permutation modules and that $\pi_1(T/G)$ is torsion-free of cohomological dimension $\dim(T) = \dim(V)$. Torus actions were considered in [Saltman 1987], and the special case of actions coming from restrictions to open tori in linear representations in [Bogomolov 1995a].

The following proposition, a consequence of the Bloch–Kato conjecture, describes a partial stabilization for central extensions of abelian groups.

Proposition 25. *Let G^c be a finite ℓ-group which is a central extension of an abelian group*

$$0 \to Z \to G^c \to G^a \to 0, \quad Z = [G^c, G^c], \tag{8-1}$$

and $K = k(V/G^c)$. Let

$$\phi_a \colon \mathbb{Z}_\ell^m \to G^a$$

be a surjection and

$$0 \to Z \to D^c \to \mathbb{Z}_\ell^m \to 0$$

the central extension induced from (8-1). Then

$$\mathrm{Ker}(\mathrm{H}^*(G^a) \to \mathrm{H}^*(D^c)) = \mathrm{Ker}(\mathrm{H}^*(G^a) \to \mathrm{H}^*(G_K)),$$

for cohomology with $\mathbb{Z}/\ell^n$-coefficients, $n \in \mathbb{N}$.

Proof. Since $\mathscr{G}_K^a$ is a torsion-free $\mathbb{Z}_\ell$-module we have a diagram

$$
\begin{array}{ccccccccc}
G_K & \twoheadrightarrow & \mathscr{G}_K^c & \longrightarrow & \mathscr{G}_K^a & \longrightarrow & 0 \\
& & \downarrow & & \downarrow & & \\
0 & \longrightarrow & Z & \longrightarrow & D^c & \longrightarrow & \mathbb{Z}_\ell^m & \longrightarrow & 0 \\
& & \| & & \downarrow & & \downarrow{\scriptstyle \phi_a} & & \\
0 & \longrightarrow & Z & \longrightarrow & G^c & \longrightarrow & G^a & \longrightarrow & 0
\end{array}
$$

By Theorem 11,

$$\mathrm{Ker}\left(\mathrm{H}^*(G^a) \to \mathrm{H}^*(G_K)\right) = \mathrm{Ker}\left(\mathrm{H}^*(G^a) \to \mathrm{H}^*(\mathscr{G}_K^c)\right).$$

Note that

$$I := \mathrm{Ker}\left(\mathrm{H}^*(G^a) \to \mathrm{H}^*(D^c)\right)$$

is an ideal generated by its degree-two elements I_2 and that

$$I_2 = \mathrm{Ker}\left(\mathrm{H}^2(G^a) \to \mathrm{H}^2(G^c)\right) \oplus \delta(\mathrm{H}^1(G^a)).$$

Similarly, for all intermediate D^c

$$\mathrm{Ker}\left(\mathrm{H}^*(G^a) \to \mathrm{H}^*(D^c)\right)$$

is also generated by I_2, and hence equals I.   □

Corollary 26. *Let G^c be a finite ℓ-group as above, $\mathrm{R} \subseteq \bigwedge^2(G^a)$ the subgroup of relations defining D^c, and let*

$$\Sigma = \{\sigma_i \subset G^a\}$$

be the set of subgroups of G^a liftable to abelian subgroups of G^c. Then the image of $\mathrm{H}^*(G^a, \mathbb{Z}/\ell^n)$ in $\mathrm{H}_s^*(G^c, \mathbb{Z}/\ell^n)$ coincides with $\bigwedge^*(G^a)^*/I_2$, where $I_2 \subseteq \bigwedge^2(G^a)$ are the elements orthogonal to R (with respect to the natural pairing).

Lemma 27. *For any finite group G^c there is a torsion-free group $\mathscr{G}^c$ with $\mathscr{G}^a = \mathbb{Z}_\ell^n$ and $[\mathscr{G}^c, \mathscr{G}^c] = \mathbb{Z}_\ell^m$ with a natural surjection $\mathscr{G}^c \to G^c$ and a natural embedding*

$$\mathrm{Ker}(\mathrm{H}^2(G^a) \to \mathrm{H}^2(G^c)) = \mathrm{Ker}(\mathrm{H}^2(\mathscr{G}^a) \to \mathrm{H}^2(\mathscr{G}^c)),$$

for cohomology with $\mathbb{Q}_\ell/\mathbb{Z}_\ell$-coefficients.

Proof. Assume that we have a diagram of central extensions

$$
\begin{array}{ccccccccc}
0 & \longrightarrow & Z_{\mathscr{G}} & \longrightarrow & \mathscr{G}^c & \xrightarrow{\ \pi_{a,\mathscr{G}}\ } & \mathscr{G}^a & \longrightarrow & 0 \\
& & \downarrow & & \downarrow{\scriptstyle \pi_c} & & \| & & \\
0 & \longrightarrow & Z_{\mathscr{H}} & \longrightarrow & H^c & \xrightarrow{\ \pi_{a,\mathscr{H}}\ } & \mathscr{H}^a & \longrightarrow & 0
\end{array}
$$

with $\mathscr{G}^a = \mathscr{H}^a$, $Z_{\mathscr{G}}$, and $Z_{\mathscr{H}}$ finite rank torsion-free $\mathbb{Z}_\ell$-modules. Assume that

$$\mathrm{Ker}(\pi_{a,\mathscr{H}}^*) := \mathrm{Ker}\left(\mathrm{H}^2(\mathscr{H}^u, \mathbb{Z}_\ell) \to \mathrm{H}^?(\mathscr{H}^c, \mathbb{Z}_\ell)\right)$$

coincides with

$$\mathrm{Ker}(\pi_{a,\mathscr{G}}^*) := \mathrm{Ker}\left(\mathrm{H}^2(\mathscr{G}^a, \mathbb{Z}_\ell) \to \mathrm{H}^2(\mathscr{G}^c, \mathbb{Z}_\ell)\right).$$

Then there is a section

$$s : \mathscr{H}^c \to \mathscr{G}^c, \quad \pi^c \circ s = id.$$

Indeed, since $\mathcal{H}^a$, $\mathcal{G}^a$ are torsion-free $\mathbb{Z}_\ell$-modules we have

$$\mathrm{H}^2(\mathcal{H}^a, \mathbb{Z}_\ell)) = \mathrm{H}^2(\mathcal{H}^a, \mathbb{Z}_\ell)) \pmod{\ell^n}, \quad \text{for all } n \in \mathbb{N},$$

and $\mathrm{H}^2(\mathcal{H}^a, \mathbb{Z}_\ell))$ is a free $\mathbb{Z}_\ell$-module. The groups $\mathcal{G}^c$, $\mathcal{H}^c$ are determined by the surjective homomorphisms

$$\wedge^2(\mathcal{H}^a) \to Z_{\mathcal{H}} = [\mathcal{H}^c, \mathcal{H}^c], \quad \wedge^2(\mathcal{G}^a) \to Z_{\mathcal{G}} = [\mathcal{G}^c, \mathcal{G}^c].$$

Since $Z_{\mathcal{H}}$, $Z_{\mathcal{G}}$ are free $\mathbb{Z}_\ell$-modules, $\mathrm{Ker}(Z_{\mathcal{G}} \to Z_{\mathcal{H}})$ is also a free $\mathbb{Z}_\ell$-module. $\square$

Let G be a finite group, V a faithful representation of G over k and $K = k(V/G)$. We have a natural homomorphism $G_K \to G$. Every valuation $v \in \mathcal{V}_K$ defines a *residue* homomorphism

$$\mathrm{H}_s^*(G, \mathbb{Z}/\ell^n) \hookrightarrow \mathrm{H}^*(G_K, \mathbb{Z}/\ell^n) \xrightarrow{\delta_v} \mathrm{H}^*(G_{K_v}, \mathbb{Z}/\ell^n),$$

and we define the stable *unramified* cohomology as the kernel of this homomorphism, over all divisorial valuations v:

$$\mathrm{H}_{s,nr}^*(G, \mathbb{Z}/\ell^n) = \{\alpha \in \mathrm{H}_s^*(G, \mathbb{Z}/\ell^n) \mid \delta_v(\alpha) = 0 \quad \text{for all } v \in \mathcal{DV}_K\}.$$

Again, this is independent of the choice of V and is functorial in G. Fix an element $g \in G$. We say that $\alpha \in \mathrm{H}_s^*(G, \mathbb{Z}/\ell^n)$ is g-unramified if the restriction of α to the centralizer $Z(g)$ of g in G is unramified (see [Bogomolov 1992] for more details).

Lemma 28. *Let G be a finite group of order coprime to $p = \mathrm{char}(k)$. Then*

$$\mathrm{H}_{s,nr}^*(G, \mathbb{Z}/\ell^n) \subseteq \mathrm{H}_s^*(G, \mathbb{Z}/\ell^n)$$

is the subring of elements which are g-unramified for all $g \in G$.

Proof. We may assume that G is an ℓ-group, with ℓ coprime to $\mathrm{char}(k)$. By functoriality, a class $\alpha \in \mathrm{H}_{s,nr}^*(G, \mathbb{Z}/\ell^n)$ is also g-unramified.

Conversely, let $v \in \mathcal{DV}_K$ be a divisorial valuation and X a normal projective model of $K = k(V/G)$ such that v is realized by a divisor $D \subset X$ and both D, X are smooth at the generic point of D. Let D^* be a formal neighborhood of this point. The map $V \to V/G$ defines a G-extension of the completion K_v. Geometrically, this corresponds to a union of finite coverings of formal neighborhoods of D^*, since G has order coprime to p: the preimage of D^* in $\bar{V}$ is a finite union of smooth formal neighborhoods D_i^* of irreducible divisors $D_i \subset \bar{V}$. If the covering $\pi_i : D_i^* \to D$ is unramified at the generic point of D_i then $\delta_v(\alpha) = 0$. On the other hand, if there is ramification, then there is a $g \in G$ which acts trivially on some D_i, and we may assume that g is a generator of a cyclic subgroup acting trivially on D_i. Consider the subgroup of G which preserves D_i and acts linearly on the normal bundle of D_i. This group is a subgroup of

$Z(g)$; hence there is a $Z(g)$-equivariant map $D_i^* \to V$ for some faithful linear representation of $Z(g)$ such that α on $D_i^*/Z(g)$ is induced from $V/Z(g)$. In particular, if $\alpha \in H_{s,nr}^*(Z(g), \mathbb{Z}/\ell^n)$ then $\delta_v(\alpha) = 0$. Thus an element which is unramified for any $g \in G$ in $H_s^*(G, \mathbb{Z}/\ell^n)$ is unramified. $\qquad\square$

The considerations above allow to *linearize* the construction of *all* finite cohomological obstructions to rationality.

Corollary 29. *Let*

$$1 \to Z \to G^c \to G^a \to 1$$

be a central extension, $g \in G^a$ a nontrivial element, and $\tilde{g}$ a lift of g to G^c. Then $Z(\tilde{g})$ is a sum of liftable abelian subgroups σ_i containing g.

Lemma 30. *An element in the image of* $H^*(G^a, \mathbb{Z}/\ell^n) \subset H_{s,nr}^*(G^c, \mathbb{Z}/\ell^n)$ *is $\tilde{g}$-unramified for a primitive element g if and only if its restriction to $Z(\tilde{g})$ is induced from $Z(\tilde{g})/\langle g \rangle$.*

Proof. One direction is clear. Conversely, $Z(\tilde{g})$ is a central extension of its abelian quotient. Hence the stabilization homomorphism coincides with the quotient by the ideal $I H_K(n)$ (see the proof of Theorem 11). $\qquad\square$

Corollary 31. *The subring* $H_{s,nr}^*(G^a, \mathbb{Z}/\ell^n) \subset H_s^*(G^a, \mathbb{Z}/\ell^n)$ *is defined by Σ, i.e., by the configuration of liftable subgroups σ_i.*

Such cohomological obstructions were considered in [Colliot-Thélène and Ojanguren 1989], where they showed that unramified cohomology is an invariant under stable birational equivalence. In addition, they produced explicit examples of nontrivial obstructions in dimension 3. Subsequently, Peyre [1993; 2008] gave further examples with $n = 3$ and $n = 4$; see also [Saltman 1995; 1997]. Similarly to the examples with nontrivial $H_{nr}^2(G)$ in [Bogomolov 1987], one can construct examples with nontrivial higher cohomology using as the only input the combinatorics of the set of liftable subgroups $\Sigma = \Sigma(G^c)$ for suitable *central extensions* G^c. Since we are interested in function fields $K = k(V/G^c)$ with trivial $H_{nr}^2(K)$, we are looking for groups G^c with $R(G) = R_\wedge(G)$. Such examples can be found by working with analogs of *quaternionic* structures on linear spaces $G^a = \mathbb{F}_\ell^{4n}$, for $n \in \mathbb{N}$.

9. What about curves?

In this section we focus on anabelian geometry of curves over finite fields. By Uchida's theorem (see Theorem 1), a curve over $k = \mathbb{F}_q$ is uniquely determined by its absolute Galois group. Recently, Saïdi–Tamagawa proved the **Isom**-version of Grothendieck's conjecture for the prime-to-characteristic geometric fundamental (and absolute Galois) groups of hyperbolic curves [Saïdi and Tamagawa 2009b] (generalizing results of Tamagawa and Mochizuki which dealt with the full groups). A **Hom**-form appears in [Saïdi and Tamagawa 2009a]. The authors are interested in *rigid* homomorphisms of *full* and prime-to-characteristic Galois groups of function fields of curves. Modulo passage to open subgroups, a homomorphism

$$\Psi : G_K \to G_L$$

is called rigid if it preserves the *decomposition* subgroups, i.e., if for all $v \in \mathscr{D}\mathscr{V}_K$

$$\Psi(D_v) = D_{v'},$$

for some $v' \in \mathscr{D}\mathscr{V}_L$. The main result is that there is a bijection between *admissible* homomorphisms of fields and rigid homomorphisms of Galois groups

$$\mathrm{Hom}^{\mathrm{adm}}(L, K) \xrightarrow{\sim} \mathrm{Hom}^{\mathrm{rig}}(G_K, G_L)/\sim,$$

modulo conjugation (here *admissible* essentially means that the extension of function fields K/L is finite of degree coprime to the characteristic, see [Saïdi and Tamagawa 2009a, p. 3] for a complete description of this notion).

Our work on higher-dimensional anabelian geometry led us to consider homomorphisms of Galois groups preserving *inertia* subgroups.

Theorem 32 [Bogomolov et al. 2010]. *Let $K = k(X)$ and $L = l(Y)$ be function fields of curves over algebraic closures of finite fields. Assume that $\mathrm{g}(X) > 2$ and that*

$$\Psi : G_K^a \to G_L^a$$

is an isomorphism of abelianized absolute Galois groups such that for all $v \in \mathscr{D}\mathscr{V}_K$ there exists a $v' \in \mathscr{D}\mathscr{V}_L$ with

$$\Psi(I_v^a) = I_{v'}^a.$$

Then $k = l$ and the corresponding Jacobians are isogenous.

This theorem is a Galois-theoretic incarnation of a finite field version of the "Torelli" theorem for curves. Classically, the setup is as follows: let k be any field and C/k a smooth curve over k of genus $\mathrm{g}(C) \geq 2$, with $C(k) \neq \varnothing$. For each $n \in \mathbb{N}$, let J^n be Jacobian of rational equivalence classes of degree n zero-cycles on C. Put $J^0 = J$. We have

$$C^n \longrightarrow \mathrm{Sym}^n(C) \xrightarrow{\lambda_n} J^n$$

Choosing a point $c_0 \in C(k)$, we may identify $J^n = J$. The image $\mathrm{Image}(\lambda_{\mathrm{g}-1}) = \Theta \subset J$ is called the theta divisor. The Torelli theorem asserts that the pair (J, Θ) determines C, up to isomorphism.

Theorem 33 [Bogomolov et al. 2010]. *Let C, $\tilde{C}$ be smooth projective curves of genus $g \geq 2$ over closures of finite fields k and $\tilde{k}$. Let*

$$\Psi: J(k) \xrightarrow{\sim} \tilde{J}(\tilde{k})$$

be an isomorphism of abelian groups inducing a bijection of sets

$$C(k) \leftrightarrow \tilde{C}(\tilde{k}).$$

Then $k = \tilde{k}$ and J is isogenous to $\tilde{J}$.

We expect that the curves C and $\tilde{C}$ are isomorphic over $\bar{k}$.

Recall that

$$J(\bar{\mathbb{F}}_p) = p\text{-part} \oplus \bigoplus_{\ell \neq p} (\mathbb{Q}_\ell/\mathbb{Z}_\ell)^{2g}.$$

The main point of Theorem 33 is that the set $C(\bar{\mathbb{F}}_p) \subset J(\bar{\mathbb{F}}_p)$ *rigidifies* this very large torsion abelian group. Moreover, we have

Theorem 34 [Bogomolov et al. 2010]. *There exists an N, bounded effectively in terms of g, such that*

$$\Psi(\mathrm{Fr})^N \quad and \quad \tilde{\mathrm{Fr}}^N$$

(the respective Frobenius) commute, as automorphisms of $\tilde{J}(\tilde{k})$.

In some cases, we can prove that the curves C and $\tilde{C}$ are actually isomorphic, as algebraic curves. Could Theorem 33 hold with k and $\tilde{k}$ replaced by $\mathbb{C}$? Such an isomorphism Ψ matches all "special" points and linear systems of the curves. Thus the problem may be amenable to techniques developed in [Hrushovski and Zilber 1996], where an algebraic curve is reconstructed from an abstract "Zariski geometry" (*ibid.*, Proposition 1.1), analogously to the reconstruction of projective spaces from an "abstract projective geometry" in Section 1.

The proof of Theorem 33 has as its starting point the following sufficient condition for the existence of an isogeny:

Theorem 35 [Bogomolov et al. 2010; Bogomolov and Tschinkel 2008a]. *Let A and $\tilde{A}$ be abelian varieties of dimension g over finite fields k_1, resp. $\tilde{k}_1$ (of sufficiently divisible cardinality). Let k_n/k_1, resp. $\tilde{k}_n/\tilde{k}_1$, be the unique extensions of degree n. Assume that*

$$\#A(k_n) \mid \#\tilde{A}(\tilde{k}_n)$$

for infinitely many $n \in \mathbb{N}$. Then $\mathrm{char}(k) = \mathrm{char}(\tilde{k})$ and A and $\tilde{A}$ are isogenous over $\bar{k}$.

The proof of this result is based on the theorem of Tate:

$$\mathrm{Hom}(A, \tilde{A}) \otimes \mathbb{Z}_\ell = \mathrm{Hom}_{\mathbb{Z}_\ell[\mathrm{Fr}]}(T_\ell(A), T_\ell(\tilde{A}))$$

and the following, seemingly unrelated, theorem concerning divisibilities of values of *recurrence sequences*.

Recall that a *linear recurrence* is a map $R : \mathbb{N} \to \mathbb{C}$ such that

$$R(n+r) = \sum_{i=0}^{r-1} a_i R(n+i),$$

for some $a_i \in \mathbb{C}$ and all $n \in \mathbb{N}$. Equivalently,

$$R(n) = \sum_{\gamma \in \Gamma^0} c_\gamma(n)\gamma^n, \qquad (9\text{-}1)$$

where $c_\gamma \in \mathbb{C}[x]$ and $\Gamma^0 \subset \mathbb{C}^\times$ is a finite set of *roots* of R. Throughout, we need only *simple* recurrences, i.e., those where the *characteristic polynomial* of R has no multiple roots so that $c_\gamma \in \mathbb{C}^\times$, for all $\gamma \in \Gamma^0$. Let $\Gamma \subset \mathbb{C}^\times$ be the group generated by Γ^0. In our applications we may assume that it is torsion-free. Then there is an isomorphism of rings

$$\{\text{Simple recurrences with roots in } \Gamma\} \Leftrightarrow \mathbb{C}[\Gamma],$$

where $\mathbb{C}[\Gamma]$ is the ring of Laurent polynomials with exponents in the finite-rank $\mathbb{Z}$-module Γ. The map

$$R \mapsto F_R \in \mathbb{C}[\Gamma]$$

is given by

$$R \mapsto F_R := \sum_{\gamma \in \Gamma^0} c_\gamma x^\gamma.$$

Theorem 36 [Corvaja and Zannier 2002]. *Let R and $\tilde{R}$ be simple linear recurrences such that*

(1) $R(n), \tilde{R}(\tilde{n}) \neq 0$, *for all* $n, \tilde{n} \gg 0$;

(2) *the subgroup $\Gamma \subset \mathbb{C}^\times$ generated by the roots of R and $\tilde{R}$ is torsion-free;*

(3) *there is a finitely generated subring $\mathfrak{A} \subset \mathbb{C}$ with $R(n)/\tilde{R}(n) \in \mathfrak{A}$, for infinitely many $n \in \mathbb{N}$.*

Then

$$Q : \mathbb{N} \to \mathbb{C}$$
$$n \mapsto R(n)/\tilde{R}(n)$$

is a simple linear recurrence. In particular, $F_Q \in \mathbb{C}[\Gamma]$ and

$$F_Q \cdot F_{\tilde{R}} = F_R.$$

This very useful theorem concerning divisibilities is actually an application of a known case of the Lang–Vojta conjecture concerning nondensity of integral points on "hyperbolic" varieties, i.e., quasi-projective varieties of log-general type. In this case, one is interested in subvarieties of algebraic tori and the needed result is *Schmidt's subspace theorem*. Other applications of this result to integral points and diophantine approximation are discussed in [Bilu 2008], and connections to Vojta's conjecture in [Silverman 2005; 2007].

A rich source of interesting simple linear recurrences is geometry over finite fields. Let X be a smooth projective variety over $k_1 = \mathbb{F}_q$ of dimension d, $\bar{X} = X \times_{k_1} \bar{k}_1$, and let k_n/k_1 be the unique extension of degree n. Then

$$\#X(k_n) := \operatorname{tr}(\operatorname{Fr}^n) = \sum_{i=0}^{2d} (-1)^i c_{ij} \alpha_{ij}^n,$$

where Fr is Frobenius acting on étale cohomology $\mathrm{H}_{et}^*(\bar{X}, \mathbb{Q}_\ell)$, with $\ell \nmid q$, and $c_{ij} \in \mathbb{C}^\times$. Let $\Gamma^0 := \{\alpha_{ij}\}$ be the set of corresponding eigenvalues. and $\Gamma_X \subset \mathbb{C}^\times$ the multiplicative group generated by α_{ij}. It is torsion-free provided the cardinality of k_1 is sufficiently divisible.

For example, let A be an abelian variety over k_1, $\{\alpha_j\}_{j=1,\dots,2g}$ the set of eigenvalues of the Frobenius on $\mathrm{H}_{et}^1(\bar{A}, \mathbb{Q}_\ell)$, for $\ell \neq p$, and $\Gamma_A \subset \mathbb{C}^\times$ the multiplicative subgroup spanned by the α_j. Then

$$R(n) := \#A(k_n) = \prod_{j=1}^{2g} (\alpha_j^n - 1). \tag{9-2}$$

is a simple linear recurrence with roots in Γ_A. Theorem 35 follows by applying Theorem 36 to this recurrence and exploiting the special shape of the Laurent polynomial associated to (9-2).

We now sketch a proof of Theorem 33, assuming for simplicity that C be a nonhyperelliptic curve of genus $g(C) \geq 3$.

Step 1. For all finite fields k_1 with sufficiently many elements ($\geq cg^2$) the group $J(k_1)$ is generated by $C(k_1)$, by [Bogomolov et al. 2010, Corollary 5.3]. Let

$$k_1 \subset k_2 \subset \cdots \subset k_n \subset \cdots$$

be the tower of degree 2 extensions. To *characterize* $J(k_n)$ it suffices to characterize $C(k_n)$.

Step 2. For each $n \in \mathbb{N}$, the abelian group $J(k_n)$ is generated by $c \in C(k)$ such that there exists a point $c' \in C(k)$ with

$$c + c' \in J(k_{n-1}).$$

Step 3. Choose $k_1, \tilde{k}_1$ (sufficiently large) such that

$$\Psi(J(k_1)) \subset \tilde{J}(\tilde{k}_1)$$

Define $C(k_n)$, resp. $\tilde{C}(\tilde{k}_n)$, intrinsically, using only the group- and set-theoretic information as above. Then

$$\Psi(J(k_n)) \subset \tilde{J}(\tilde{k}_n), \quad \text{for all} \quad n \in \mathbb{N}.$$

and

$$\#J(k_n) \mid \#\tilde{J}(\tilde{k}_n).$$

To conclude the proof of Theorem 33 it suffices to apply Theorem 36 and Theorem 35 about divisibility of recurrence sequences.

One of the strongest and somewhat counter-intuitive results in this area is a theorem of Tamagawa:

Theorem 37 [Tamagawa 2004]. *There are at most finitely many (isomorphism classes of) curves of genus g over $k = \bar{\mathbb{F}}_p$ with given (profinite) geometric fundamental group.*

On the other hand, in 2002 we proved:

Theorem 38 [Bogomolov and Tschinkel 2002b]. *Let C be a hyperelliptic curve of genus ≥ 2 over $k = \bar{\mathbb{F}}_p$, with $p \geq 5$. Then for every curve C' over k there exists an étale cover $\pi : \tilde{C} \to C$ and surjective map $\tilde{C} \to C'$.*

This shows that the geometric fundamental groups of hyperbolic curves are "almost" independent of the curve: every such $\pi_1(C)$ has a subgroup of small index and such that the quotient by this subgroup is almost abelian, surjecting onto the fundamental group of another curve C'.

This relates to the problem of couniformization for hyperbolic curves (see [Bogomolov and Tschinkel 2002b]). The Riemann theorem says that the unit disc in the complex plane serves as a universal covering for all complex projective curves of genus ≥ 2, simultaneously. This provides a canonical embedding of the fundamental group of a curve into the group of complex automorphisms of the disc, which is isomorphic to $\mathrm{PGL}_2(\mathbb{R})$. In particular, it defines a natural embedding of the field of rational functions on the curve into the field of meromorphic functions on the disc. The latter is unfortunately too large to be of any help in solving concrete problems.

However, in some cases there is an algebraic substitute. For example, in the class of modular curves there is a natural pro-algebraic object *Mod* (introduced by Shafarevich) which is given by a tower of modular curves; the corresponding pro-algebraic field, which is an inductive union M of the fields of rational functions on modular curves. Similarly to the case of a disc the space *Mod* has a wealth of symmetries which contains a product $\prod_p \mathrm{SL}_2(\mathbb{Z}_p)$ and the absolute Galois group $\mathrm{G}(\bar{\mathbb{Q}}/\mathbb{Q})$.

The above result alludes to the existence of a similar disc-type algebraic object for all hyperbolic curves defined over $\bar{\mathbb{F}}_p$ (or even for arithmetic hyperbolic curves).

For example consider C_6 given by $y^6 = x(x-1)$ over $\mathbb{F}_p$, with $p \neq 2, 3$, and define $\tilde{C}_6$ as a pro-algebraic universal covering of C_6. Thus $\bar{\mathbb{F}}_p(\tilde{C}_6) = \bigcup \bar{\mathbb{F}}_p(C_i)$, where C_i range over all finite geometrically unramified coverings of C_6. Then $\bar{\mathbb{F}}_p(\tilde{C}_6)$ contains all other fields $\bar{\mathbb{F}}_p(C)$, where C is an arbitrary curve defined over some $\mathbb{F}_q \subset \bar{\mathbb{F}}_p$. Note that it also implies that étale fundamental group $\pi_1(C_6)$ contains a subgroup of finite index which surjects onto $\pi_1(C)$ with the action of $\hat{\mathbb{Z}} = \mathrm{G}(\bar{\mathbb{F}}_p/\mathbb{F}_q)$.

The corresponding results in the case of curves over number fields $K \subset \bar{\mathbb{Q}}$ are weaker, but even in the weak form they are quite intriguing.

Acknowledgments

We have benefited from conversations with J.-L. Colliot-Thélène, B. Hassett, and M. Rovinsky. We are grateful to the referee for helpful remarks and suggestions. Bogomolov was partially supported by NSF grant DMS-0701578. Tschinkel was partially supported by NSF grants DMS-0739380 and 0901777.

References

[Bilu 2008] Y. F. Bilu, "The many faces of the subspace theorem [after Adamczewski, Bugeaud, Corvaja, Zannier. . .]", Exp. 967, pp. 1–38 in *Séminaire Bourbaki* 2006/2007, Astérisque **317**, 2008. MR 2010a:11141

[Bogomolov 1987] F. Bogomolov, "The Brauer group of quotient spaces of linear representations", *Izv. Akad. Nauk SSSR Ser. Mat.* **51**:3 (1987), 485–516, 688. MR 88m:16006

[Bogomolov 1991a] F. Bogomolov, "Abelian subgroups of Galois groups", *Izv. Akad. Nauk SSSR Ser. Mat.* **55**:1 (1991), 32–67. MR 93b:12007

[Bogomolov 1991b] F. Bogomolov, "On two conjectures in birational algebraic geometry", pp. 26–52 in *Algebraic geometry and analytic geometry* (Tokyo, 1990), ICM-90 Satell. Conf. Proc., Springer, Tokyo, 1991. MR 94k:14013

[Bogomolov 1992] F. Bogomolov, "Stable cohomology of groups and algebraic varieties", *Mat. Sb.* **183**:5 (1992), 3–28. MR 94c:12008

[Bogomolov 1995a] F. Bogomolov, "Linear tori with an action of finite groups", *Mat. Zametki* **57**:5 (1995), 643–652, 796. MR 96e:20081

[Bogomolov 1995b] F. Bogomolov, "On the structure of Galois groups of the fields of rational functions", pp. 83–88 in *K-theory and algebraic geometry: connections with quadratic forms and division algebras* (Santa Barbara, CA, 1992), Proc. Sympos. Pure Math. **58**, Amer. Math. Soc., Providence, RI, 1995. MR 1 327 291

[Bogomolov and Tschinkel 2002a] F. Bogomolov and Y. Tschinkel, "Commuting elements in Galois groups of function fields", pp. 75–120 in *Motives, Polylogarithms and Hodge theory*, International Press, 2002.

[Bogomolov and Tschinkel 2002b] F. Bogomolov and Y. Tschinkel, "Unramified correspondences", pp. 17–25 in *Algebraic number theory and algebraic geometry*, Contemp. Math. **300**, Amer. Math. Soc., Providence, RI, 2002. MR 2003k:14032

[Bogomolov and Tschinkel 2008a] F. Bogomolov and Y. Tschinkel, "On a theorem of Tate", *Cent. Eur. J. Math.* **6**:3 (2008), 343–350. MR 2009g:14054

[Bogomolov and Tschinkel 2008b] F. Bogomolov and Y. Tschinkel, "Reconstruction of function fields", *Geom. Funct. Anal.* **18**:2 (2008), 400–462. MR 2009g:11155

[Bogomolov and Tschinkel 2009a] F. Bogomolov and Y. Tschinkel, "Milnor K_2 and field homomorphisms", pp. 223–244 in *Surveys in Differential Geometry XIII*, International Press, 2009.

[Bogomolov and Tschinkel 2009b] F. Bogomolov and Y. Tschinkel, "Reconstruction of higher-dimensional function fields", 2009. arXiv 0912.4923

[Bogomolov et al. 2004] F. Bogomolov, J. Maciel, and T. Petrov, "Unramified Brauer groups of finite simple groups of Lie type A_l", *Amer. J. Math.* **126**:4 (2004), 935–949. MR 2005k:14035

[Bogomolov et al. 2010] F. Bogomolov, M. Korotiaev, and Y. Tschinkel, "A Torelli theorem for curves over finite fields", *Pure and Applied Math Quarterly, Tate Festschrift* **6**:1 (2010), 245–294.

[Chebolu and Mináč 2009] S. K. Chebolu and J. Mináč, "Absolute Galois groups viewed from small quotients and the Bloch–Kato conjecture", pp. 31–47 in *New topological contexts for Galois theory and algebraic geometry*, Geom. Topol. Monogr. **16**, Geom. Topol. Publ., Coventry, Banff, 2008, 2009. MR 2544385

[Chebolu et al. 2009] S. K. Chebolu, I. Efrat, and J. Mináč, "Quotients of absolute Galois groups which determine the entire Galois cohomology", 2009. arXiv 0905.1364

[Colliot-Thélène and Ojanguren 1989] J.-L. Colliot-Thélène and M. Ojanguren, "Variétés unirationnelles non rationnelles: au-delà de l'exemple d'Artin et Mumford", *Invent. Math.* **97**:1 (1989), 141–158. MR 90m:14012

[Corry and Pop 2009] S. Corry and F. Pop, "The pro-p Hom-form of the birational anabelian conjecture", *J. Reine Angew. Math.* **628** (2009), 121–127. MR 2010b:14045

[Corvaja and Zannier 2002] P. Corvaja and U. Zannier, "Finiteness of integral values for the ratio of two linear recurrences", *Invent. Math.* **149**:2 (2002), 431–451. MR 2003g:11015

[Efrat 1999] I. Efrat, "Construction of valuations from K-theory", *Math. Res. Lett.* **6**:3-4 (1999), 335–343. MR 2001i:12011

[Engler and Koenigsmann 1998] A. J. Engler and J. Koenigsmann, "Abelian subgroups of pro-p Galois groups", *Trans. Amer. Math. Soc.* **350**:6 (1998), 2473–2485. MR 98h:12004

[Engler and Nogueira 1994] A. J. Engler and J. B. Nogueira, "Maximal abelian normal subgroups of Galois pro-2-groups", *J. Algebra* **166**:3 (1994), 481–505. MR 95h:12004

[Evans and Hrushovski 1991] D. M. Evans and E. Hrushovski, "Projective planes in algebraically closed fields", *Proc. London Math. Soc.* (3) **62**:1 (1991), 1–24. MR 92a:05031

[Evans and Hrushovski 1995] D. M. Evans and E. Hrushovski, "The automorphism group of the combinatorial geometry of an algebraically closed field", *J. London Math. Soc.* (2) **52**:2 (1995), 209–225. MR 97f:51012

[Faltings 1983] G. Faltings, "Endlichkeitssätze für abelsche Varietäten über Zahlkörpern", *Invent. Math.* **73**:3 (1983), 349–366. MR 85g:11026a

[Faltings 1998] G. Faltings, "Curves and their fundamental groups (following Grothendieck, Tamagawa and Mochizuki)", Exp. 840, pp. 131–150 in *Séminaire Bourbaki* 1997/98, Astérisque **252**, 1998. MR 2000j:14036

[Gismatullin 2008] J. Gismatullin, "Combinatorial geometries of field extensions", *Bull. Lond. Math. Soc.* **40**:5 (2008), 789–800. MR 2009h:03049

[Grothendieck 1997] A. Grothendieck, "Brief an G. Faltings", pp. 49–58 in *Geometric Galois actions*, vol. 1, London Math. Soc. Lecture Note Ser. **242**, Cambridge Univ. Press, Cambridge, 1997. MR 99c:14023

[Haesemeyer and Weibel 2009] C. Haesemeyer and C. Weibel, "Norm varieties and the chain lemma (after Markus Rost)", pp. 95–130 in *Algebraic topology*, Abel Symp. **4**, Springer, Berlin, 2009. MR 2597737

[Hrushovski and Zilber 1996] E. Hrushovski and B. Zilber, "Zariski geometries", *J. Amer. Math. Soc.* **9**:1 (1996), 1–56. MR 96c:03077

[Ihara and Nakamura 1997] Y. Ihara and H. Nakamura, "Some illustrative examples for anabelian geometry in high dimensions", pp. 127–138 in *Geometric Galois actions*, vol. 1, London Math. Soc. Lecture Note Ser. **242**, Cambridge Univ. Press, Cambridge, 1997. MR 99b:14021

[Koenigsmann 2005] J. Koenigsmann, "On the 'section conjecture' in anabelian geometry", *J. Reine Angew. Math.* **588** (2005), 221–235. MR 2006h:14032

[Kunyavski 2010] B. Kunyavski, "The Bogomolov multiplier of finite simple groups", pp. 209–217 in *Cohomological and geometric approaches to rationality problems*, Progr. in Math. **282**, Birkhäuser, Basel, 2010.

[Mochizuki 1999] S. Mochizuki, "The local pro-p anabelian geometry of curves", *Invent. Math.* **138**:2 (1999), 319–423. MR 2000j:14037

[Mochizuki 2003] S. Mochizuki, "Topics surrounding the anabelian geometry of hyperbolic curves", pp. 119–165 in *Galois groups and fundamental groups*, Math. Sci. Res. Inst. Publ. **41**, Cambridge Univ. Press, Cambridge, 2003. MR 2004m:14052

[Nakamura 1990] H. Nakamura, "Galois rigidity of the étale fundamental groups of punctured projective lines", *J. Reine Angew. Math.* **411** (1990), 205–216. MR 91m:14021

[Nakamura et al. 1998] H. Nakamura, A. Tamagawa, and S. Mochizuki, "Grothendieck's conjectures concerning fundamental groups of algebraic curves", *Sūgaku* **50**:2 (1998), 113–129. MR 2000e:14038

[Neukirch 1969] J. Neukirch, "Kennzeichnung der p-adischen und der endlichen algebraischen Zahlkörper", *Invent. Math.* **6** (1969), 296–314. MR 39 #5528

[Peyre 1993] E. Peyre, "Unramified cohomology and rationality problems", *Math. Ann.* **296**:2 (1993), 247–268. MR 94e:14015

[Peyre 2008] E. Peyre, "Unramified cohomology of degree 3 and Noether's problem", *Invent. Math.* **171**:1 (2008), 191–225. MR 2008m:12011

[Pop 1994] F. Pop, "On Grothendieck's conjecture of birational anabelian geometry", *Ann. of Math.* (2) **139**:1 (1994), 145–182. MR 94m:12007

[Pop 1997] F. Pop, "Glimpses of Grothendieck's anabelian geometry", pp. 113–126 in *Geometric Galois actions, 1*, London Math. Soc. Lecture Note Ser. **242**, Cambridge Univ. Press, Cambridge, 1997. MR 99f:14026

[Pop 2000] F. Pop, "Alterations and birational anabelian geometry", pp. 519–532 in *Resolution of singularities* (Obergurgl, 1997), Progr. Math. **181**, Birkhäuser, Basel, 2000. MR 2001g:11171

[Pop 2003] F. Pop, "Pro-ℓ birational anabelian geometry over algebraically closed fields I", 2003. arXiv math/0307076

[Positselski 2005] L. Positselski, "Koszul property and Bogomolov's conjecture", *Int. Math. Res. Not.* **2005**:31 (2005), 1901–1936. MR MR2171198 (2006h:19002)

[Positselski and Vishik 1995] L. Positselski and A. Vishik, "Koszul duality and Galois cohomology", *Math. Res. Lett.* **2**:6 (1995), 771–781. MR 97b:12008

[Saïdi and Tamagawa 2009a] M. Saïdi and A. Tamagawa, "On the Hom-form Grothendieck's birational anabelian conjecture in characteristic $p > 0$", 2009. arXiv 0912.1972

[Saïdi and Tamagawa 2009b] M. Saïdi and A. Tamagawa, "A prime-to-p version of Grothendieck's anabelian conjecture for hyperbolic curves over finite fields of characteristic $p > 0$", *Publ. Res. Inst. Math. Sci.* **45**:1 (2009), 135–186. MR 2512780

[Saltman 1984] D. J. Saltman, "Noether's problem over an algebraically closed field", *Invent. Math.* **77**:1 (1984), 71–84. MR 85m:13006

[Saltman 1987] D. J. Saltman, "Multiplicative field invariants", *J. Algebra* **106**:1 (1987), 221–238. MR 88f:12007

[Saltman 1995] D. J. Saltman, "Brauer groups of invariant fields, geometrically negligible classes, an equivariant Chow group, and unramified H^3", pp. 189–246 in *K-theory and algebraic geometry: connections with quadratic forms and division algebras* (Santa Barbara, CA, 1992), Proc. Sympos. Pure Math. **58**, Amer. Math. Soc., Providence, RI, 1995. MR 96c:12008

[Saltman 1997] D. J. Saltman, "H^3 and generic matrices", *J. Algebra* **195**:2 (1997), 387–422. MR 99b:12003

[Silverman 2005] J. H. Silverman, "Generalized greatest common divisors, divisibility sequences, and Vojta's conjecture for blowups", *Monatsh. Math.* **145**:4 (2005), 333–350. MR 2006e:11087

[Silverman 2007] J. H. Silverman, "Greatest common divisors and algebraic geometry", pp. 297–308 in *Diophantine geometry*, CRM Series **4**, Ed. Norm., Pisa, 2007. MR 2008k:11069

[Suslin 1984] A. A. Suslin, "Algebraic K-theory and the norm residue homomorphism", pp. 115–207 in *Current problems in mathematics, Vol. 25*, Itogi Nauki i Tekhniki, Akad. Nauk SSSR Vsesoyuz. Inst. Nauchn. i Tekhn. Inform., Moscow, 1984. MR 86j:11121

[Tamagawa 1997] A. Tamagawa, "The Grothendieck conjecture for affine curves", *Compositio Math.* **109**:2 (1997), 135–194. MR 99a:14035

[Tamagawa 2004] A. Tamagawa, "Finiteness of isomorphism classes of curves in positive characteristic with prescribed fundamental groups", *J. Algebraic Geom.* **13**:4 (2004), 675–724. MR 2005c:14032

[Tate 1966] J. Tate, "Endomorphisms of abelian varieties over finite fields", *Invent. Math.* **2** (1966), 134–144. MR 34 #5829

[Uchida 1977] K. Uchida, "Isomorphisms of Galois groups of algebraic function fields", *Ann. Math.* (2) **106**:3 (1977), 589–598. MR 57 #273

[Voevodsky 1991a] V. A. Voevodskiĭ, "Galois groups of function fields over fields of finite type over **Q**", *Uspekhi Mat. Nauk* **46**:5(281) (1991), 163–164. MR 93c:11104

[Voevodsky 1991b] V. A. Voevodskiĭ, "Galois representations connected with hyperbolic curves", *Izv. Akad. Nauk SSSR Ser. Mat.* **55**:6 (1991), 1331–1342. MR 93h:14019

[Voevodsky 2003] V. Voevodsky, "Reduced power operations in motivic cohomology", *Publ. Math. Inst. Hautes Études Sci.* **98** (2003), 1–57. MR 2005b:14038a

[Voevodsky 2010] V. Voevodsky, "On motivic cohomology with $\mathbb{Z}/\ell$-coefficients", 2010. *Annals of Math.*, to appear.

[Weibel 2009] C. Weibel, "The norm residue isomorphism theorem", *J. Topol.* **2**:2 (2009), 346–372. MR 2529300

bogomolo@cims.nyu.edu   *Courant Institute of Mathematical Sciences, N.Y.U.,*
251 Mercer Street, New York, NY 10012, United States

tschinkel@cims.nyu.edu   *Courant Institute of Mathematical Sciences, N.Y.U.,*
251 Mercer Street, New York, NY 10012, United States

Current Developments in Algebraic Geometry
MSRI Publications
Volume **59**, 2011

Periods and moduli

OLIVIER DEBARRE

This text is an introduction, without proofs and by means of many examples, to some elementary aspects of the theory of period maps, period domains, and their relationship with moduli spaces. We start with the definitions of Jacobians of curves, Prym varieties, and intermediate Jacobians, then move on to Griffiths' construction of period domains and period maps. We review some instances of the Torelli problem and discuss some recent results of Allcock, Carlson, Laza, Looijenga, Swierstra, and Toledo, expressing some moduli spaces as ball quotients.

It has been known since the nineteenth century that there is a group structure on the points of the smooth cubic complex plane curve (called an *elliptic curve*) and that it is isomorphic to the quotient of $\mathbf{C}$ by a lattice. Conversely, any such quotient is an elliptic curve.

The higher-dimensional analogs are *complex tori* V/Γ, where Γ is a lattice in a (finite-dimensional) complex vector space V. The group structure and the analytic structure are obvious, but not all tori are algebraic. For that, we need an additional condition, which was formulated by Riemann: the existence of a positive definite Hermitian form on V whose (skew-symmetric) imaginary part is integral on Γ. An algebraic complex torus is called an *abelian variety*. When this skew-symmetric form is in addition unimodular on Γ, we say that the abelian variety is *principally polarized*. It contains a hypersurface uniquely determined up to translation ("the" *theta divisor*).

The combination of the algebraic and group structures makes the geometry of abelian varieties very rich. This is one of the reasons why it is useful to associate, whenever possible, an abelian variety (if possible principally polarized) to a given geometric situation. This can be done only in a few specific cases, and the theory of periods, mainly developed by Griffiths, constitutes a far-reaching extension.

This is a slightly expanded version of a talk given in January 2009 for the workshop "Classical Algebraic Geometry Today" while I was a member of MSRI. I would like to thank the organizers of the workshop, Lucia Caporaso, Brendan Hassett, James McKernan, Mircea Mustață, and Mihnea Popa, for the invitation, and MSRI for support.

Our aim is to present an elementary introduction to this theory. We show many examples to illustrate its diversity, with no pretense at exhaustivity, and no proofs. For those interested in pursuing this very rich subject, we refer to [Carlson et al. 2003] and its bibliography.

Here is a short description of the contents of this text. In section 1, we review some of the classical cases where one can attach an abelian variety to a geometric situation; they are very special, because they correspond to varieties with a Hodge decomposition of level one. In section 2, we show, following Griffiths, how to extend drastically this construction by defining period maps and period domains. In section 3, we review some instances of the Torelli problem: when does the abelian variety associated to a given situation (as in section 1) characterize it? In the framework of section 2, this translates into the question of the injectivity of the period map. There is no general principle here, and we give examples where the answer is yes, and other examples where the period map is not injective, and even has positive-dimensional fibers. In the last section, we briefly discuss various questions relative to moduli spaces; the period map can sometimes be used to relate them to more concrete geometrical objects. There is also the important matter of the construction of compactifications for these moduli spaces and of the extension of the period map to these compactifications.

We work over the field of complex numbers.

1. Attaching an abelian variety to an algebraic object

1.1. *Curves.* Given a smooth projective curve C of genus g, we have the Hodge decomposition

$$H^1(C, \mathbf{Z}) \subset H^1(C, \mathbf{C}) = H^{0,1}(C) \oplus H^{1,0}(C),$$

where the right side is a $2g$-dimensional complex vector space and $H^{1,0}(C) = \overline{H^{0,1}(C)}$. The g-dimensional complex torus

$$J(C) = H^{0,1}(C)/H^1(C, \mathbf{Z})$$

is a principally polarized abelian variety (the polarization corresponds to the unimodular intersection form on $H^1(C, \mathbf{Z})$). We therefore have an additional geometric object: the theta divisor $\Theta \subset J(C)$, uniquely defined up to translation.

The geometry of the theta divisor of the Jacobian of a curve has been intensively studied since Riemann. One may say that it is well-known; see [Arbarello et al. 1985].

1.2. *Prym varieties.* Given a double étale cover $\pi : \tilde{C} \to C$ between smooth projective curves, one can endow the abelian variety $P(\tilde{C}/C) = J(\tilde{C})/\pi^* J(C)$

with a natural principal polarization. The dimension of $P(\tilde{C}/C)$ is

$$g(\tilde{C}) - g(C) = g(C) - 1;$$

it is called the *Prym variety* attached to π. By results from [Mumford 1974; Tjurin 1975a; Beauville 1977a], we have a rather good understanding of the geometry of the theta divisor of $P(\tilde{C}/C)$.

1.3. Threefolds. If X is a smooth projective threefold, the Hodge decomposition is

$$H^3(X, \mathbf{Z}) \subset H^3(X, \mathbf{C}) = \left(H^{0,3}(X) \oplus H^{1,2}(X)\right) \oplus (\text{complex conjugate})$$

and we may again define the *intermediate Jacobian* of X as the complex torus

$$J(X) = \left(H^{0,3}(X) \oplus H^{1,2}(X)\right)/H^3(X, \mathbf{Z}).$$

It is in general not algebraic. In case $H^{0,3}(X)$ vanishes, however, we have again a principally polarized abelian variety.

In some situations, the intermediate Jacobian is a Prym. For example if X has a conic bundle structure $X \to \mathbf{P}^2$ (i.e., a morphism with fibers isomorphic to conics), define the discriminant curve $C \subset \mathbf{P}^2$ as the locus of points whose fibers are reducible conics, i.e., unions of two lines. The choice of one of these lines defines a double cover $\tilde{C} \to C$. Although C may have singular points, we can still define a Prym variety $P(\tilde{C}/C)$. We have $H^{0,3}(X) = 0$, and

$$J(X) \simeq P(\tilde{C}/C)$$

as principally polarized abelian varieties. This isomorphism is a powerful tool for proving nonrationality of some threefolds: the intermediate Jacobian of a rational threefold must have a very singular theta divisor and the theory of Prym varieties can sometimes tell that this does not happen.

Example 1.1 (Cubic threefolds). If $X \subset \mathbf{P}^4$ is a smooth cubic hypersurface, we have $h^{0,3}(X) = 0$ and $h^{1,2}(X) = 5$, so that $J(X)$ is a 5-dimensional principally polarized abelian variety.

Any such X contains a line ℓ. Projecting from this line induces a conic bundle structure $\tilde{X} \to \mathbf{P}^2$ on the blow-up $\tilde{X}$ of ℓ in X. The discriminant curve $C \subset \mathbf{P}^2$ is a quintic and $J(X) \simeq P(\tilde{C}/C)$ (this agrees with the fact that $J(X)$ has dimension $g(C) - 1 = \binom{4}{2} - 1 = 5$). This isomorphism can be used to prove that the theta divisor $\Theta \subset J(X)$ has a unique singular point, which has multiplicity 3 [Beauville 1982]. In particular, it is not "singular enough," and X is not rational [Clemens and Griffiths 1972].

Example 1.2 (Quartic double solids). If $p : X \to \mathbf{P}^3$ is a double cover branched along a quartic surface $B \subset \mathbf{P}^3$ (a *quartic double solid*), we have $h^{0,3}(X) = 0$

and $h^{1,2}(X) = 10$, so that $J(X)$ has dimension 10. There is in general no conic bundle structure on X. However, when B acquires an ordinary double point s, the variety X becomes singular, and there is a (rational) conic bundle structure $X \dashrightarrow \mathbf{P}^2$ obtained by composing p with the projection $\mathbf{P}^3 \dashrightarrow \mathbf{P}^2$ from s. The discriminant curve is a sextic. This degeneration can be used to prove that for X general, the singular locus of the theta divisor $\Theta \subset J(X)$ has dimension 5 [Voisin 1988] and has a unique component of that dimension [Debarre 1990]. Again, this implies that X is not rational.

Example 1.3 (Fano threefolds of degree 10). If $X \subset \mathbf{P}^9$ is the smooth complete intersection of the Grassmannian $G(2, 5)$ in its Plücker embedding, two hyperplanes, and a smooth quadric, we have $h^{0,3}(X) = 0$ and $h^{1,2}(X) = 10$, so that $J(X)$ has dimension 10.

1.4. Odd-dimensional varieties. Let X be a smooth projective variety of dimension $2n + 1$ whose Hodge decomposition is of the form

$$H^{2n+1}(X, \mathbf{C}) = H^{n,n+1}(X) \oplus H^{n+1,n}(X).$$

We may define the *intermediate Jacobian* of X as

$$J(X) = H^{n,n+1}(X)/H^{2n+1}(X, \mathbf{Z}).$$

This is again a principally polarized abelian variety.

Example 1.4 (Intersections of two quadrics). If $X \subset \mathbf{P}^{2n+3}$ is the smooth baselocus of a pencil Λ of quadrics, its Hodge decomposition satisfies the conditions above, so we can form the principally polarized abelian variety $J(X)$.

The choice of one of the two components of the family of $\mathbf{P}^{n+1}$ contained in a member of Λ defines a double cover $C \to \Lambda$ ramified exactly over the $2n + 4$ points corresponding to the singular members of the pencil. The curve C is smooth, hyperelliptic, of genus $n + 1$, and its Jacobian is isomorphic to $J(X)$ [Reid 1972; Donagi 1980].

Example 1.5 (Intersections of three quadrics). If $X \subset \mathbf{P}^{2n+4}$ is the smooth baselocus of a net of quadrics $\Pi = \langle Q_1, Q_2, Q_3 \rangle$, its Hodge decomposition satisfies the conditions above, so we can form the principally polarized abelian variety $J(X)$.

When $n \geq 1$, the variety X contains a line ℓ. The map $X \dashrightarrow \Pi$ defined by sending a point $x \in X$ to the unique quadric in Π that contains the 2-plane $\langle \ell, x \rangle$, is a (rational) *quadric* bundle structure on X. The discriminant curve $C \subset \Pi$ parametrizing singular quadrics has equation

$$\det(\lambda_1 Q_1 + \lambda_2 Q_2 + \lambda_3 Q_3) = 0,$$

hence degree $2n + 5$. The choice of a component of the set of $\mathbf{P}^{n+1}$ contained in a singular quadric of the net Π defines a double étale cover $\tilde{C} \to C$, and $J(X) \simeq P(\tilde{C}/C)$ [Beauville 1977b; Tjurin 1975b].

2. Periods and period maps

Assume now that we have a family $\mathcal{X} \to S$ of smooth projective threefolds, whose fibers X_s all satisfy $H^{0,3}(X_s) = 0$. We can construct for each $s \in S$ the intermediate Jacobian $J(X_s)$. Let us look at this from a slightly different point of view. Assume that the base S is simply connected and fix a point 0 in S, with fiber X_0; we can then identify each $H^3(X_s, \mathbf{Z})$ with the fixed rank-$2g$ lattice $H_{\mathbf{Z}} = H^3(X_0, \mathbf{Z})$ and define an algebraic *period map* with values in a Grassmannian:

$$ \wp : S \longrightarrow G(g, H_{\mathbf{C}}), \quad s \longmapsto H^{2,1}(X_s), $$

where $H_{\mathbf{C}} = H_{\mathbf{Z}} \otimes_{\mathbf{Z}} \mathbf{C}$. Letting Q be the skew-symmetric intersection form on $H_{\mathbf{C}}$, the following properties hold:

- $H^{2,1}(X_s)$ is totally isotropic for Q,
- the Hermitian form $i Q(\cdot, \bar{\cdot})$ is positive definite on $H^{2,1}(X_s)$,

so that $\wp$ takes its values into a dense open subset of an isotropic Grassmannian which is isomorphic to the Siegel upper half-space $\mathcal{H}_g = \mathrm{Sp}(2g, \mathbf{R})/\mathrm{U}(g)$. If $\wp(s)$ correspond to $\tau(s) \in \mathcal{H}_g$, we have

$$ J(X_s) = H_{\mathbf{C}}/(H_{\mathbf{Z}} \oplus \tau(s)H_{\mathbf{Z}}). $$

Back to the case where the base S is general, with universal cover $\tilde{S} \to S$, we obtain a diagram

$$
\begin{array}{ccc}
\tilde{S} & \xrightarrow{\tilde{\wp}} & \mathcal{H}_g \\
\downarrow & & \downarrow \\
S & \xrightarrow{\wp} & \mathcal{H}_g / \mathrm{Sp}(2g, \mathbf{Z}) = \mathcal{A}_g
\end{array}
$$

where $\tilde{\wp}$ is *holomorphic*, $\wp$ is algebraic if S is algebraic, and

$$ \mathcal{A}_g = \{\text{ppavs of dimension } g\}/\text{isomorphism} $$

is the *moduli space* of principally polarized abelian varieties of dimension g. It has a natural structure of a quasiprojective variety of dimension $g(g+1)/2$.

We want to generalize this construction to any smooth projective variety X of dimension n. Even if the Hodge decomposition

$$H^k(X, \mathbf{C}) = \bigoplus_{p=0}^{k} H^{p,k-p}(X)$$

does not have level one (i.e., only two pieces), we can still use it to define a period map as follows [Griffiths 1969; 1970]. Choose an ample class $h \in H^2(X, \mathbf{Z}) \cap H^{1,1}(X)$ and define the *primitive cohomology* by

$$H^k(X, \mathbf{C})_{\mathrm{prim}} = \mathrm{Ker}\left(H^k(X, \mathbf{C}) \xrightarrow{\smile h^{n-k+1}} H^{2n-k+2}(X, \mathbf{C})\right).$$

Set $H^{p,q}(X)_{\mathrm{prim}} = H^{p,q}(X) \cap H^k(X, \mathbf{C})_{\mathrm{prim}}$ and

$$F^r = \bigoplus_{p \geq r} H^{p,k-p}(X)_{\mathrm{prim}}.$$

Define a bilinear form on $H^k(X, \mathbf{C})_{\mathrm{prim}}$ by

$$Q(\alpha, \beta) = \alpha \smile \beta \smile h^{n-k}.$$

The associated *period domain* $\mathcal{D}$ is then the set of flags

$$0 = F^{k+1} \subset F^k \subset \cdots \subset F^1 \subset F^0 = H^k(X, \mathbf{C})$$

satisfying the following conditions:

- $F^r \oplus \overline{F^{k-r+1}} = F^0$,
- $F^r = (F^{k-r+1})^{\perp_Q}$,
- for each p and k, the Hermitian form $i^{2p-k}Q(\cdot, \bar{\cdot})$ is positive definite on $H^{p,k-p}(X)_{\mathrm{prim}}$.

It is a homogeneous complex manifold, quotient of a real Lie group by a compact subgroup. We already encountered the period domain $\mathcal{H}_g = \mathrm{Sp}(2g, \mathbf{R})/\mathrm{U}(g)$; however, the subgroup may be not maximal, so that $\mathcal{D}$ is not in general Hermitian symmetric (see Examples 2.5 and 3.3).

Given a family $\mathcal{X} \to S$ of polarized varieties, we obtain as above a *holomorphic* map

$$\tilde{\wp} : \tilde{S} \longrightarrow \mathcal{D}.$$

Note that the lattice $H^k(X, \mathbf{Z})$ has played no role here yet. It will however if we want to define a period map on S: one needs to quotient by the action of $\pi_1(S, 0)$ and this group acts via the monodromy representation

$$\pi_1(S, 0) \longrightarrow \mathrm{Aut}(H^k(X_0, \mathbf{Z})).$$

The discrete group $\Gamma = \text{Aut}(H^k(X_0, \mathbf{Z}))$ acts on $H^k(X_0, \mathbf{C})_{\text{prim}}$, and properly on $\mathcal{D}$, hence we obtain a diagram

$$
\begin{array}{ccc}
\tilde{S} & \xrightarrow{\tilde{\wp}} & \mathcal{D} \\
\downarrow & & \downarrow \\
S & \xrightarrow{\wp} & \mathcal{D}/\Gamma
\end{array}
$$

where $\mathcal{D}/\Gamma$ is an analytic space (not algebraic in general; see Examples 2.5 and 2.6) and $\wp$ is holomorphic.

Example 2.1 (Quartic surfaces). If $B \subset \mathbf{P}^3$ is a quartic (hence K3) surface, we have

$$
\begin{array}{ccccccc}
H^2(B, \mathbf{C}) & = & H^{0,2}(B) & \oplus & H^{1,1}(B) & \oplus & H^{2,0}(B) \\
\text{dimensions:} & & 1 & & 20 & & 1
\end{array}
$$

$$
\begin{array}{ccccccc}
H^2(B, \mathbf{C})_{\text{prim}} & = & H^{0,2}(B) & \oplus & H^{1,1}(B)_{\text{prim}} & \oplus & H^{2,0}(B) \\
\text{dimensions:} & & 1 & & 19 & & 1
\end{array}
$$

Because of its properties, explained above, relative to the intersection form Q, this decomposition is completely determined by the point of $\mathbf{P}(H^2(B, \mathbf{C})_{\text{prim}})$ defined by the line $H^{2,0}(B)$. The period map takes values in the 19-dimensional period domain

$$
\begin{aligned}
\mathcal{D}^{19} &= \{[\omega] \in \mathbf{P}^{20} \mid Q(\omega, \omega) = 0, \ Q(\omega, \bar{\omega}) > 0\} \\
&\simeq \text{SO}(19, 2)^0 / \text{SO}(19) \times \text{SO}(2),
\end{aligned}
$$

where Q is a quadratic form, integral on a lattice $H_{\mathbf{Z}}$, with signature $(19, 2)$ on $H_{\mathbf{R}}$. It is a bounded symmetric domain of type IV and the discrete group Γ^{19} can be explicitly described [Beauville et al. 1985].

Note that quartic surfaces are in one-to-one correspondence with quartic double solids (Example 1.2), so we may also associate to B the 5-dimensional intermediate Jacobian $J(X)$ of the double solid $X \to \mathbf{P}^3$ branched along B and get another kind of period map with values in $\mathcal{A}_5$.

Example 2.2 (Cubic fourfolds). If $X \subset \mathbf{P}^5$ is a smooth cubic fourfold, the situation is completely analogous: the decomposition

$$
\begin{array}{ccccccc}
H^4(X, \mathbf{C})_{\text{prim}} & = & H^{1,3}(X) & \oplus & H^{2,2}(X)_{\text{prim}} & \oplus & H^{3,1}(X) \\
\text{dimensions:} & & 1 & & 20 & & 1
\end{array}
$$

is completely determined by the point $[H^{3,1}(X)]$ of $\mathbf{P}(H^4(X, \mathbf{C})_{\text{prim}})$, and

$$
\begin{aligned}
\mathcal{D}^{20} &= \{[\omega] \in \mathbf{P}^{21} \mid Q(\omega, \omega) = 0, \ Q(\omega, \bar{\omega}) > 0\} \\
&\simeq \text{SO}(20, 2)^0 / \text{SO}(20) \times \text{SO}(2).
\end{aligned}
$$

Here the quadratic form Q, integral on a lattice $H_{\mathbf{Z}}$, has signature $(20, 2)$ on $H_{\mathbf{R}}$. The domain $\mathscr{D}^{20}$ is again is a bounded symmetric domain of type IV. The discrete group Γ^{20} can be explicitly described [Laza 2009].

Example 2.3 (Cubic surfaces). If $X \subset \mathbf{P}^3$ is a smooth cubic surface, with equation $F(x_0, x_1, x_2, x_3) = 0$, we have $H^2(X, \mathbf{C}) = H^{1,1}(X)$, so the period map is trivial.

Proceeding as in Example 2.1, where we associated to a quartic surface in $\mathbf{P}^3$ the double cover of $\mathbf{P}^3$ branched along this surface, we may consider the cyclic triple cover $\tilde{X} \to \mathbf{P}^3$ branched along X. It is isomorphic to the cubic threefold with equation $F(x_0, x_1, x_2, x_3) + x_4^3 = 0$ in $\mathbf{P}^4$, so its Hodge structure is as in Example 1.1 and the period domain is $\mathscr{H}_5$. On the other hand, the Hodge structure carries an action of the group μ_3 of cubic roots of unity. The eigenspace $H_\omega^3(\tilde{X})$ for the eigenvalue $e^{2i\pi/3}$ splits as

$$H_\omega^3(\tilde{X}) = H_\omega^{1,2}(\tilde{X}) \oplus H_\omega^{2,1}(\tilde{X}).$$
$$\text{dimensions:} \qquad\quad 1 \qquad\qquad\quad 4$$

Following Allcock, Carlson, and Toledo [2002], one can then define a period map with values in the 4-dimensional space

$$\{[\omega] \in \mathbf{P}^4 \mid Q(\omega, \bar{\omega}) < 0\},$$

where the quadratic form Q is integral on a lattice $H_{\mathbf{Z}}$, with signature $(4, 1)$ on $H_{\mathbf{R}}$. It is isomorphic to the complex hyperbolic space $\mathbf{B}^4$, which is much smaller than $\mathscr{H}_5$! The discrete group Γ^4 can be explicitly described.

Example 2.4 (Cubic threefolds, II). Similarly, if $X \subset \mathbf{P}^4$ is a smooth cubic threefold, we consider the cyclic triple cover $\tilde{X} \to \mathbf{P}^4$ branched along X. It is a cubic fourfold in $\mathbf{P}^5$, so its Hodge structure is as in Example 2.2, with an extra symmetry of order three. With analogous notation as above, $H_\omega^{3,1}(\tilde{X})$ has dimension 1 and $H_\omega^{2,2}(\tilde{X})$ has dimension 10. Allcock, Carlson, and Toledo [2011] then define a period map with values in the 10-dimensional space

$$\{\omega \in \mathbf{P}^{10} \mid Q(\omega, \bar{\omega}) < 0\} \simeq \mathbf{B}^{10},$$

where again the quadratic form Q is integral on a lattice $H_{\mathbf{Z}}$, with signature $(10, 1)$ on $H_{\mathbf{R}}$.

Example 2.5 (Calabi–Yau threefolds of mirror quintic type). Consider the quintic hypersurface $Q_\lambda \subset \mathbf{P}^4$ with equation

$$x_0^5 + \cdots + x_4^5 + \lambda x_0 \cdots x_4 = 0,$$

its quotient Q_λ / G by the diagonal action of the finite group

$$G = \{(\alpha_0, \ldots, \alpha_4) \in \mathbf{C}^5 \mid \alpha_0^5 = \cdots = \alpha_4^5 = \alpha_0 \cdots \alpha_4 = 1\},$$

and a minimal desingularization $\widetilde{Q_\lambda/G} \to Q_\lambda/G$. Its Hodge numbers are

$$h^{0,3}(\widetilde{Q_\lambda/G}) = h^{1,2}(\widetilde{Q_\lambda/G}) = h^{2,1}(\widetilde{Q_\lambda/G}) = h^{3,0}(\widetilde{Q_\lambda/G}) = 1.$$

The corresponding period domain $\mathcal{D}^4$ has dimension 4. It is the first instance that we meet of what is called the "nonclassical" situation, where the analytic space $\mathcal{D}^4/\Gamma^4$ is not quasiprojective in any way compatible with its analytic structure.

Example 2.6 (Hypersurfaces in the projective space). If X is a smooth hypersurface of degree d in $\mathbf{P}^{n+1}$, the only interesting Hodge structure is that of $H^n(X, \mathbf{C})$. If $F(x_0, \ldots, x_{n+1}) = 0$ is an equation for X, and

$$R(F) := \mathbf{C}[x_0, \ldots, x_{n+1}] \Big/ \left\langle \frac{\partial F}{\partial x_0}, \ldots, \frac{\partial F}{\partial x_{n+1}} \right\rangle$$

is the (graded) Jacobian quotient ring, we have, by [Griffiths 1969],

$$H^{p,n-p}(X)_{\mathrm{prim}} \simeq R(F)^{(n+1-p)d-n-2},$$

where $R(F)^e$ is the graded piece of degree e in $R(F)$. So again, except for small d and n (as in Examples 2.1, 2.2, 2.3, and 2.4), we are most of the times in a nonclassical situation.

3. Is the period map injective?

3.1. *Curves*. This is the famous *Torelli theorem* [Torelli 1913; Arbarello et al. 1985]: a smooth projective curve C is determined (up to isomorphism) by the pair $(J(C), \Theta)$. In fancy terms, the period map

$$\mathcal{M}_g = \{\text{smooth projective curves of genus } g\}/\text{isomorphism}$$

$$\wp_g \downarrow$$

$$\mathcal{A}_g = \{\text{ppavs of dimension } g\}/\text{isomorphism}$$

is injective.

More generally, it is customary to call *Torelli problem* the question of deciding whether an algebraic object is determined by a polarized abelian variety attached to it.

3.2. *Prym varieties*. The period map

$$\mathcal{R}_g = \{\text{double étale covers of genus-}g \text{ curves}\}/\text{isom.}$$

$$\downarrow \wp_g$$

$$\mathcal{A}_{g-1}$$

cannot be injective in low genera for dimensional reasons. The following table sums up the situation (see [Donagi and Smith 1981; Friedman and Smith 1982; Welters 1987]):

g	$\dim(\mathcal{R}_g)$	$\dim(\mathcal{A}_{g-1})$	$\wp_g$
2	3	1	dominant, not injective
3	6	3	dominant, not injective
4	9	6	dominant, not injective
5	12	10	dominant, not injective
6	15	15	dominant, generically 27:1
$g \geq 7$	$3g - 3$	$g(g-1)/2$	generically injective, *not* injective

The injectivity defect for $g \geq 7$ is not yet entirely understood (see [Donagi 1981; Debarre 1989b; Verra 2004; Izadi and Lange 2010]).

3.3. *Hypersurfaces in the projective space.* Donagi [1983] proved that the period map

$$\mathcal{M}_{d,n} = \left\{ \begin{array}{c} \text{smooth hypersurfaces} \\ \text{of degree } d \text{ in } \mathbf{P}^n \end{array} \right\} \Big/ \text{isom.}$$

$$\downarrow \wp_{d,n}$$

$$\mathcal{D}_{d,n} / \Gamma_{d,n}$$

for hypersurfaces is *generically injective,* except perhaps in the following cases (see also [Cox and Green 1990]):

- $n = 2$ and $d = 3$, i.e., cubic surfaces, where this is obviously false (see Example 2.3);

- d divides $n + 2$ (the answer in these cases is unknown, except for $d = 5$ and $n = 3$; see [Voisin 1999]);

- $d = 4$ and $4 \mid n$.

The proof relies on Griffiths' theory of infinitesimal variation of Hodge structures and a clever argument in commutative algebra. Using analogous techniques, this result was extended later to hypersurfaces of more general homogeneous spaces [Konno 1989].

The period domain is in general much too big for the period map to be dominant (not to mention the fact that it is in general not even algebraic!). But for some small d and n, this can happen, as shown in the next examples.

3.4. *Cubic threefolds.* The period map

$$\mathcal{M}_{ct}^{10} = \left\{ \begin{array}{c} \text{smooth cubic} \\ \text{threefolds} \end{array} \right\} \Big/ \text{isom.}$$

$$\downarrow \wp_{ct}$$

$$\mathcal{A}_5$$

for cubic threefolds is injective. This can be seen as follows [Beauville 1982]: if $X \subset \mathbf{P}^4$ is a cubic, we explained in Example 1.1 that the theta divisor $\Theta \subset J(X)$ has a unique singular point s, which has multiplicity 3. It turns out that the projectified tangent cone

$$\mathbf{P}(TC_{\Theta,s}) \subset \mathbf{P}(T_{J(X),s}) = \mathbf{P}^4$$

is isomorphic to X.

Of course, $\wp_{ct}$ is not dominant, since it maps a 10-dimensional space to a 15-dimensional space. Its image was characterized geometrically in [Casalaina-Martin and Friedman 2005]: it is essentially the set of elements of $\mathcal{A}_5$ whose theta divisor has a point of multiplicity 3.

Recall (Example 2.4) that Allcock, Carlson, and Toledo defined [2011] another period map

$$\wp'_{ct} : \mathcal{M}_{ct} \longrightarrow \mathbf{B}^{10}/\Gamma^{10}.$$

They prove (among other things) that $\wp'_{ct}$ induces an isomorphism onto an open subset whose complement is explicitly described.

3.5. *Quartic double solids and quartic surfaces.* The period map

$$\mathcal{M}_{qds}^{19} = \left\{ \begin{array}{c} \text{smooth quartic} \\ \text{double solids} \end{array} \right\} \Big/ \text{isom.}$$

$$\downarrow \wp_{qds}$$

$$\mathcal{A}_{10}$$

for quartic double solids is injective. This can be seen as follows: as mentioned in Example 1.2, if $X \to \mathbf{P}^3$ is a smooth quartic double solid, the singular locus of the theta divisor $\Theta \subset J(X)$ has a unique 5-dimensional component S. General points s of S are double points on Θ, and the projectified tangent cones $\mathbf{P}(TC_{\Theta,s})$ are, after translation, quadrics in $\mathbf{P}(T_{J(X),0}) = \mathbf{P}^9$. The intersection of these quadrics is isomorphic to the image of the branch quartic surface $B \subset \mathbf{P}^3$ by the Veronese morphism $v_2 : \mathbf{P}^3 \to \mathbf{P}^9$ [Clemens 1983].

Again, $\wp_{qds}$ maps a 19-dimensional space to a 45-dimensional space, so it cannot be dominant. However, X is determined by the quartic surface $B \subset \mathbf{P}^3$, and we have another period map (Example 2.1)

$$\wp'_{qds} : \mathcal{M}^{19}_{qds} \longrightarrow \mathcal{D}^{19}/\Gamma^{19},$$

which is an isomorphism onto an explicitly described open subset [Piatetski-Shapiro and Shafarevich 1971].

3.6. Intersections of two quadrics. The period map

$$\mathcal{M}^{2n+1}_{i2q} = \left\{ \begin{array}{c} \text{smooth intersections of} \\ \text{two quadrics in } \mathbf{P}^{2n+3} \end{array} \right\} \Big/ \text{isom.}$$

$$\downarrow \wp_{i2q}$$

$$\mathcal{A}_{n+1}$$

for intersections of two quadrics is injective. This is because, by the Torelli theorem for curves (§3.1), one can reconstruct from the intermediate Jacobian $J(X)$ the hyperelliptic curve C (see Example 1.4), hence its Weierstrass points, hence the pencil of quadrics that defines X. The image of $\wp_{i2q}$ is the set of hyperelliptic Jacobians, hence it is not dominant for $n \geq 2$.

3.7. Intersections of three quadrics. The period map

$$\mathcal{M}^{2n^2+13n+12}_{i3q} = \left\{ \begin{array}{c} \text{smooth intersections of} \\ \text{three quadrics in } \mathbf{P}^{2n+4} \end{array} \right\} \Big/ \text{isom.}$$

$$\downarrow \wp_{i3q}$$

$$\mathcal{A}_{(n+1)(2n+5)}$$

for intersections of three quadrics is injective: using *ad hoc* geometric constructions, we showed in [Debarre 1989a] how to recover, from the theta divisor $\Theta \subset J(X)$, the double cover $\tilde{C}$ of the discriminant curve C and from there, it was classically known how to reconstruct X.

Again, for dimensional reasons, $\wp_{i3q}$ is not dominant.

3.8. Cubic surfaces. Allcock, Carlson, and Toledo [2002] proved that the modified period map

$$\mathcal{M}^4_{cs} = \left\{ \begin{array}{c} \text{smooth cubic} \\ \text{surfaces} \end{array} \right\} \Big/ \text{isom.}$$

$$\downarrow \wp'_{cs}$$

$$\mathbf{B}^4/\Gamma^4$$

constructed in Example 2.3 induces an isomorphism with an explicit open subset of $\mathbf{B}^4/\Gamma^4$.

3.9. *Cubic fourfolds, II.* The period map (Example 2.2)

$$\mathcal{M}_{\mathrm{cf}}^{20} = \left\{ \begin{array}{c} \text{smooth cubic} \\ \text{fourfolds} \end{array} \right\} \Big/ \text{isom.}$$

$$\downarrow \wp_{\mathrm{cf}}$$

$$\mathcal{D}^{20} / \Gamma^{20}$$

for cubic fourfolds is injective [Voisin 1986; Looijenga 2009] and induces an isomorphism with an explicitly described open subset of $\mathcal{D}^{20} / \Gamma^{20}$ [Laza 2010].

3.10. *Calabi–Yau threefolds of mirror quintic type, II.* In the situation considered in Example 2.5, we have a period map $\wp : U \to \mathcal{D}^4 / \Gamma^4$, where U is the open set of those $\lambda \in \mathbf{C}$ for which the quintic Q_λ is smooth. Using techniques from log-geometry, Usui [2008] showed that $\wp$ is generically injective.

3.11. *Fano threefolds of degree* 10. I am referring here to the threefolds $X \subset \mathbf{P}^7$ considered in Example 1.3. Their moduli space $\mathcal{M}^{22}$ has dimension 22, so the period map

$$\wp : \mathcal{M}^{22} \longrightarrow \mathcal{A}_{10}$$

can certainly not be dominant. Furthermore, Debarre, Iliev, and Manivel proved that its fibers have everywhere dimension 2 [Debarre et al. 2011].

Here is a sketch of the construction. Conics $c \subset X$ are parametrized by a smooth connected projective surface $F(X)$ which is the blow-up at one point of a smooth minimal surface $F_m(X)$ of general type. Given such a smooth conic c, one can construct another smooth Fano threefold X_c of degree 10 and a birational map $X \dashrightarrow X_c$ which is an isomorphism in codimension 1. In particular, it induces an isomorphism $J(X_c) \simeq J(X)$. However, one shows that the surface $F(X_c)$ is isomorphic to the blow-up of $F_m(X)$ at the point corresponding to c. In particular, it is (in general) *not* isomorphic to $F(X)$, so X_c is also *not* isomorphic to X. We actually prove that this construction (and a variant thereof) produces *two smooth proper 2-dimensional connected components* of each general fiber.

4. Moduli spaces

Up to now, little care was taken to define the exact structure of the various "moduli spaces" we encountered. There are two main methods for constructing quasiprojective moduli spaces:

- Geometric invariant theory [Mumford et al. 1994]: roughly speaking, one "naturally" embeds the objects one wants to classify into some fixed projective space, then quotients the corresponding subset of the Hilbert scheme by the action of the special linear group using GIT.

- One constructs directly an ample line bundle "on the functor;" roughly speaking, one needs to construct, for every family $X \to S$ of objects, a "functorial" ample line bundle on the base S.

The advantage of the GIT method is that it also produces automatically a compactification of the moduli space. Its drawback is that it is difficult to apply in practice. The second method, pioneered by Kollár and Viehweg, is more general, but technically more difficult. It can also produce compactifications, but there, one needs to decide what kind of singular objects one needs to add to make the moduli space compact. This approach now seems to have had complete success for varieties with ample canonical bundle.

Once a compactification is constructed, one may then try to extend the various period maps constructed above to compactifications of the period domain $\mathcal{D}/\Gamma$. In the "classical" situation, i.e., when the period domain is an arithmetic quotient of a bounded symmetric domain, one can use the Baily–Borel theory [1966]. In general, this is much more difficult (see [Usui 2008]). These extensions turn out to be very useful, in some cases, in order to characterize the image of the original period map, or to prove that it is birational.

Here are a few examples.

4.1. *Curves*. It has been known for a long time that the moduli space $\mathcal{M}_g$ of smooth projective curves of genus g and the moduli space $\mathcal{A}_g$ of principally polarized abelian varieties of dimension g are quasiprojective varieties (over **Z**, this was established in [Mumford et al. 1994], Theorem 5.11 and Theorem 7.10, using GIT).

A compactification $\overline{\mathcal{M}_g}$ of $\mathcal{M}_g$ is obtained by adding certain singular curves called *stable curves* and the resulting moduli space was proved by Mumford, Knudsen, and Gieseker to be projective (see the discussion in [Mumford et al. 1994], Appendix D).

As explained in §2, $\mathcal{A}_g$ is an arithmetic quotient of the Siegel upper half-space, a bounded symmetric domain. A first compactification was constructed by Satake (this was the starting point of the Baily–Borel theory!). Set theoretically, it is simply the disjoint union of $\mathcal{A}_g, \mathcal{A}_{g-1}, \ldots, \mathcal{A}_0$, but it is very singular. *Toroidal* compactifications were later constructed by Ash, Mumford, Rapoport, and Tai (see the references in [Mumford et al. 1994], Appendix E) and some of them are smooth. More recently, Alexeev [2004] constructed a compactification which is a moduli space.

The period map $\wp_g : \mathcal{M}_g \to \mathcal{A}_g$ defined in §3.1 does extend to a morphism from $\overline{\mathcal{M}_g}$ to the Satake compactification by sending a curve to the product of the Jacobians of the components of its normalization. It also extends to a morphism to some toroidal compactifications and to the Alexeev compactification. However, none of these extensions remain injective: points of $\overline{\mathcal{M}}_g$ which correspond to unions of two curves meeting in one point are sent to the product of the Jacobians of their components, regardless of the gluing points. The fibers of the extended period map are precisely described in [Caporaso and Viviani 2011].

4.2. Hypersurfaces in the projective space. Hypersurfaces of degree d in $\mathbf{P}^{n+1}$ are parametrized by the projective space

$$|dH| = \mathbf{P}(H^0(\mathbf{P}^{n+1}, \mathcal{O}_{\mathbf{P}^{n+1}}(d))).$$

Let $|dH|^0$ be the dense open subset corresponding to *smooth* hypersurfaces. The complement $|dH| - |dH|^0$ is a hypersurface, because the condition that the equation F and its partial derivatives $\partial F / \partial x_0, \ldots, \partial F / \partial x_{n+1}$ have a common zero is given by the vanishing of a single (homogeneous) polynomial in the coefficients of F. It follows that $|dH|^0$ is an affine open set, invariant by the action of the reductive group $\mathrm{SL}(n+2)$.

For $d \geq 3$, this action is *regular* in the sense of GIT (the dimensions of the stabilizers are locally constant) hence closed (the orbits are closed). Since $\mathcal{O}_{\mathbf{P}^{n+1}}(1)$ admits a $\mathrm{SL}(n+2)$-linearization, $|dH|^0$ is contained in the set $|dH|^s$ of stable points associated with these data. The GIT theory implies that the quotient $|dH|^0 / \mathrm{SL}(n+2)$, which is the moduli space of smooth hypersurfaces of degree d in $\mathbf{P}^{n+1}$, can be realized as an open set in the GIT quotient $|dH|^{ss} // \mathrm{SL}(n+2)$, which is a projective variety.

The precise description of the semistable points, i.e., of the kind of singularities one needs to add to obtain the GIT compactification of the moduli space, is a difficult task, impossible to achieve in general. Some cases are known: plane curves of degree ≤ 6 and cubic surfaces (Hilbert; [Mumford 1977; Shah 1976]), quartic surfaces [Shah 1976], cubic threefolds [Allcock 2003], cubic fourfolds [Laza 2009],…

Example 4.1 (Cubic surfaces, II). Stable points correspond to cubic surfaces that have at most ordinary double points ("type A_1"). Semistable points correspond to cubic surfaces whose singular points are all of type A_1 or A_2. GIT theory yields a compactification $\overline{\mathcal{M}}_{cs}^4$ of the moduli space of smooth cubic surfaces (§3.8) and the modified period map extends to an isomorphism

$$\overline{\mathcal{M}}_{cs}^4 \xrightarrow{\sim} \overline{\mathbf{B}^4 / \Gamma^4},$$

where the right side is the Baily–Borel compactification of $\mathbf{B}^4 / \Gamma^4$ [Allcock et al. 2002; Doran 2004b].

Example 4.2 (Quartic surfaces, II). There is a list of all allowed singularities on quartic surfaces corresponding to semistable points [Shah 1976]. Again, the period map (§3.5) induces an isomorphism [Kulikov 1977]

$$\overline{\mathcal{M}^{19}} \xrightarrow{\sim} \overline{\mathcal{D}^{19}/\Gamma^{19}},$$

where the left side is the GIT-compactification and the right side is the Baily–
Borel compactification.

Example 4.3 (Cubic threefolds, III). Stable points correspond to cubic threefolds
whose singular points are of type A_n, with $1 \leq n \leq 4$. There is also a list of all
possible singularities of cubic threefolds that correspond to semistable points
[Allcock 2003]. The modified period map (§3.4) induces a morphism

$$\overline{\mathcal{M}_{\mathrm{ct}}^{10}} \longrightarrow \overline{\mathbf{B}^{10}/\Gamma^{10}}$$

which contracts a rational curve, where $\overline{\mathcal{M}_{\mathrm{ct}}^{10}}$ is an explicit blow-up of the GIT-
compactification and $\overline{\mathbf{B}^{10}/\Gamma^{10}}$ is the Baily–Borel compactification [Allcock et al.
2011; Looijenga and Swierstra 2007].

Example 4.4 (Cubic fourfolds, II). There are complete lists of all possible
singularities of cubic fourfolds that correspond to stable and semistable points
[Laza 2009]. The period map (§3.9) induces an isomorphism

$$\overline{\mathcal{M}_{\mathrm{cf}}^{20}} \longrightarrow \overline{\mathcal{D}^{20}/\Gamma^{20}},$$

where $\overline{\mathcal{M}_{\mathrm{cf}}^{20}}$ is an explicit blow-up of the GIT-compactification and $\overline{\mathcal{D}^{20}/\Gamma^{20}}$ is the
Looijenga compactification, a modification of the Baily–Borel compactification
[Laza 2010; Looijenga 2009].

4.3. Complete intersections. Along the same lines, some complete intersections
have also been considered.

Example 4.5 (Intersections of two quadrics, II). The moduli space of smooth
intersections of two quadrics in $\mathbf{P}^n$ can be constructed as the GIT quotient of an
affine dense open set of the Grassmannian $G(2, H^0(\mathbf{P}^n, \mathcal{O}_{\mathbf{P}^n}(2)))$ by the reductive
group $\mathrm{SL}(n+1)$. Using a slightly different presentation, Avritzer and Miranda
[1999] proved that smooth intersections correspond exactly to stable points.
Therefore, the moduli space is a quasiprojective variety.

Example 4.6 (Fano threefolds of degree 10, II). These threefolds $X \subset \mathbf{P}^7$ were
considered in Example 1.3 and §3.11: they are obtained as intersections, in $\mathbf{P}^9$,
of the Grassmannian $G(2, 5)$ in its Plücker embedding, two hyperplanes, and a
smooth quadric, and their moduli space $\mathcal{M}^{22}$ can be seen as follows.

Let $G = G(8, \wedge^2 \mathbf{C}^5)$ be the 16-dimensional Grassmannian parametrizing
pencils of skew-symmetric forms on 5, and let $\mathcal{T}$ be the tautological rank-8

vector bundle on G. The composition

$$\textstyle\bigwedge^4 V_5^\vee \hookrightarrow \mathrm{Sym}^2\left(\bigwedge^2 V_5^\vee\right) \to \mathrm{Sym}^2 \mathcal{T}^\vee$$

is everywhere injective and its cokernel $\mathcal{E}$ is a vector bundle of rank 31 on G. To each point λ of $\mathbf{P}(\mathcal{E})$, one can associate a codimension-2 linear subspace of $\mathbf{P}\left(\bigwedge^2 V_5\right)$ and a quadric in that subspace, well-defined up to the space of quadrics that contain $G(2, \mathbf{C}^5) \subset \mathbf{P}\left(\bigwedge^2 V_5\right)$, hence a threefold $X_\lambda \subset \mathbf{P}^7$ of degree 10, which is in general smooth.

The group $\mathrm{SL}(5)$ acts on $\mathbf{P}(\mathcal{E})$ and one checks that the stabilizers, which correspond to the automorphisms group of X_λ, are finite when X_λ is smooth. So we expect that the moduli space should be an open subset of the GIT quotient $\mathbf{P}(\mathcal{E}) /\!/ \mathrm{SL}(5)$. However, the relationship between the smoothness of X_λ and the stability of λ (which of course involves the choice of a polarization, since $\mathbf{P}(\mathcal{E})$ has Picard number 2) is not clear at the moment.

Remark 4.7. These examples show that several GIT-moduli spaces admit, as ball quotients, complex hyperbolic structures. Another large class of examples of moduli spaces as ball quotients is due to Deligne and Mostow, in their exploration of moduli spaces of points on $\mathbf{P}^1$ and hypergeometric functions [Deligne and Mostow 1986]. "Our" examples are not directly of Deligne–Mostow type, since the corresponding discrete groups do not appear on the various lists in [Deligne and Mostow 1986; Thurston 1998]. However, Doran found, by taking a view of hypergeometric functions based on intersection cohomology valued in local systems, links between these two types of examples [Doran 2004a].

References

[Alexeev 2004] V. Alexeev, "Compactified Jacobians and Torelli map", *Publ. Res. Inst. Math. Sci.* **40**:4 (2004), 1241–1265. MR 2006a:14016

[Allcock 2003] D. Allcock, "The moduli space of cubic threefolds", *J. Algebraic Geom.* **12**:2 (2003), 201–223. MR 2003k:14043

[Allcock et al. 2002] D. Allcock, J. A. Carlson, and D. Toledo, "The complex hyperbolic geometry of the moduli space of cubic surfaces", *J. Algebraic Geom.* **11**:4 (2002), 659–724. MR 2003m:32011

[Allcock et al. 2011] D. Allcock, J. A. Carlson, and D. Toledo, "The moduli space of cubic threefolds as a ball quotient", *Mem. Amer. Math. Soc.* **209**.985 (2011), xii+70. MR 2789835

[Arbarello et al. 1985] E. Arbarello, M. Cornalba, P. A. Griffiths, and J. Harris, *Geometry of algebraic curves*, vol. I, Grundlehren der Math. Wissenschaften **267**, Springer, New York, 1985. MR 86h:14019

[Avritzer and Miranda 1999] D. Avritzer and R. Miranda, "Stability of pencils of quadrics in $\mathbf{P}^4$", *Bol. Soc. Mat. Mexicana* (3) **5**:2 (1999), 281–300. MR 2000j:14014

[Baily and Borel 1966] W. L. Baily, Jr. and A. Borel, "Compactification of arithmetic quotients of bounded symmetric domains", *Ann. of Math.* (2) **84** (1966), 442–528. MR 35 #6870

[Beauville 1977a] A. Beauville, "Prym varieties and the Schottky problem", *Invent. Math.* **41**:2 (1977), 149–196. MR 58 #27995

[Beauville 1977b] A. Beauville, "Variétés de Prym et jacobiennes intermédiaires", *Ann. Sci. École Norm. Sup.* (4) **10**:3 (1977), 309–391. MR 57 #12532

[Beauville 1982] A. Beauville, "Les singularités du diviseur Θ de la jacobienne intermédiaire de l'hypersurface cubique dans $\mathbf{P}^4$", pp. 190–208 in *Algebraic threefolds* (Varenna, 1981), Lecture Notes in Math. **947**, Springer, Berlin, 1982. MR 84c:14030

[Beauville et al. 1985] A. Beauville, J.-P. Bourguignon, and M. Demazure (editors), *Géométrie des surfaces K3: modules et périodes* (Palaiseau, 1981/1982), Astérisque **126**, Société Mathématique de France, Paris, 1985. MR 87h:32052

[Caporaso and Viviani 2011] L. Caporaso and F. Viviani, "Torelli theorem for stable curves", *J. Eur. Math. Soc.* **13**:5 (2011), 1289–1329. arXiv 0904.4039

[Carlson et al. 2003] J. Carlson, S. Müller-Stach, and C. Peters, *Period mappings and period domains*, Cambridge Studies in Advanced Mathematics **85**, Cambridge University Press, Cambridge, 2003. MR 2005a:32014

[Casalaina-Martin and Friedman 2005] S. Casalaina-Martin and R. Friedman, "Cubic threefolds and abelian varieties of dimension five", *J. Algebraic Geometry* **14**:2 (2005), 295–326. MR 2006g:14071

[Clemens 1983] C. H. Clemens, "Double solids", *Adv. in Mathematics* **47**:2 (1983), 107–230. MR 85e:14058

[Clemens and Griffiths 1972] C. H. Clemens and P. A. Griffiths, "The intermediate Jacobian of the cubic threefold", *Ann. of Math.* (2) **95** (1972), 281–356. MR 46 #1796

[Cox and Green 1990] D. A. Cox and M. L. Green, "Polynomial structures and generic Torelli for projective hypersurfaces", *Compositio Math.* **73**:2 (1990), 121–124. MR 91b:14006

[Debarre 1989a] O. Debarre, "Le théorème de Torelli pour les intersections de trois quadriques", *Invent. Math.* **95**:3 (1989), 507–528. MR 89k:14010

[Debarre 1989b] O. Debarre, "Sur le problème de Torelli pour les variétés de Prym", *Amer. J. Math.* **111**:1 (1989), 111–134. MR 90b:14035

[Debarre 1990] O. Debarre, "Sur le théorème de Torelli pour les solides doubles quartiques", *Compositio Math.* **73**:2 (1990), 161–187. MR 91h:14014

[Debarre et al. 2011] O. Debarre, A. Iliev, and L. Manivel, "On the period map for prime Fano threefolds of degree 10", *J. Algebraic Geom.* (2011). arXiv 0812.3670

[Deligne and Mostow 1986] P. Deligne and G. D. Mostow, "Monodromy of hypergeometric functions and nonlattice integral monodromy", *Inst. Hautes Études Sci. Publ. Math.* **63** (1986), 5–89. MR 88a:22023a

[Donagi 1980] R. Donagi, "Group law on the intersection of two quadrics", *Ann. Scuola Norm. Sup. Pisa Cl. Sci.* (4) **7**:2 (1980), 217–239. MR 82b:14025

[Donagi 1981] R. Donagi, "The tetragonal construction", *Bull. Amer. Math. Soc.* (*N.S.*) **4**:2 (1981), 181–185. MR 82a:14009

[Donagi 1983] R. Donagi, "Generic Torelli for projective hypersurfaces", *Compositio Math.* **50**:2-3 (1983), 325–353. MR 85g:14045

[Donagi and Smith 1981] R. Donagi and R. C. Smith, "The structure of the Prym map", *Acta Math.* **146**:1-2 (1981), 25–102. MR 82k:14030b

[Doran 2004a] B. Doran, "Hurwitz spaces and moduli spaces as ball quotients via pull-back", 2004. arXiv math/0404363

[Doran 2004b] B. Doran, "Moduli space of cubic surfaces as ball quotient via hypergeometric functions", 2004. arXiv math/0404062

[Friedman and Smith 1982] R. Friedman and R. Smith, "The generic Torelli theorem for the Prym map", *Invent. Math.* **67**:3 (1982), 473–490. MR 83i:14017

[Griffiths 1969] P. A. Griffiths, "On the periods of certain rational integrals", *Ann. of Math.* (2) **90** (1969), 460–495 and 496–541. MR 41 #5357

[Griffiths 1970] P. A. Griffiths, "Periods of integrals on algebraic manifolds: Summary of main results and discussion of open problems", *Bull. Amer. Math. Soc.* **76** (1970), 228–296. MR 41 #3470

[Izadi and Lange 2010] E. Izadi and H. Lange, "Counter-examples of high Clifford index to Prym–Torelli", 2010. arXiv 1001.3610

[Konno 1989] K. Konno, "Generic Torelli theorem for hypersurfaces of certain compact homogeneous Kähler manifolds", *Duke Math. J.* **59**:1 (1989), 83–160. MR 90k:32072

[Kulikov 1977] V. S. Kulikov, "Surjectivity of the period mapping for $K3$ surfaces", *Uspehi Mat. Nauk* **32**:4 (1977), 257–258. MR 58 #688

[Laza 2009] R. Laza, "The moduli space of cubic fourfolds", *J. Algebraic Geom.* **18**:3 (2009), 511–545. MR 2010c:14039

[Laza 2010] R. Laza, "The moduli space of cubic fourfolds via the period map", *Ann. of Math.* (2) **172**:1 (2010), 673–711. MR 2680429

[Looijenga 2009] E. Looijenga, "The period map for cubic fourfolds", *Invent. Math.* **177**:1 (2009), 213–233. MR 2010h:32013

[Looijenga and Swierstra 2007] E. Looijenga and R. Swierstra, "The period map for cubic threefolds", *Compos. Math.* **143**:4 (2007), 1037–1049. MR 2008f:32015

[Mumford 1974] D. Mumford, "Prym varieties, I", pp. 325–350 in *Contributions to analysis*, Academic Press, New York, 1974. MR 52 #415

[Mumford 1977] D. Mumford, "Stability of projective varieties", *Enseignement Math.* (2) **23**:1-2 (1977), 39–110. MR 56 #8568

[Mumford et al. 1994] D. Mumford, J. Fogarty, and F. Kirwan, *Geometric invariant theory*, Third ed., Ergebnisse der Math. **34**, Springer, Berlin, 1994. MR 95m:14012

[Piatetski-Shapiro and Shafarevich 1971] I. Piatetski-Shapiro and I. Shafarevich, "A Torelli theorem for algebraic surfaces of type K3", *Izv. Akad. Nauk SSSR Ser. Mat.* **35** (1971), 530–572. In Russian; translated in *Math. USSR Izv.* **5** (1971), 547–588.

[Reid 1972] M. Reid, *The complete intersection of two or more quadrics*, Ph.D. thesis, Cambridge University, 1972.

[Shah 1976] J. Shah, "Surjectivity of the period map in the case of quartic surfaces and sextic double planes", *Bull. Amer. Math. Soc.* **82**:5 (1976), 716–718. MR 54 #5246

[Thurston 1998] W. P. Thurston, "Shapes of polyhedra and triangulations of the sphere", pp. 511–549 in *The Epstein birthday schrift*, Geom. Topol. Monogr. **1**, Geom. Topol. Publ., Coventry, 1998. MR 2000b:57026

[Tjurin 1975a] A. N. Tjurin, "The geometry of the Poincaré divisor of a Prym variety", *Izv. Akad. Nauk SSSR Ser. Mat.* **39**:5 (1975), 1003–1043, 1219. In Russian; errata in **42** (1978), 468; translated in **9** (1975), 951–986, **12** (1978), 438. MR 54 #2664

[Tjurin 1975b] A. N. Tjurin, "The intersection of quadrics", *Uspehi Mat. Nauk* **30**:6(186) (1975), 51–99. in Russian; translated in *Russ. Math. Surveys* **30** (1975), 51–105. MR 54 #12791

[Torelli 1913] R. Torelli, "Sulle varietà di Jacobi", *Atti Accad. Lincei Rend. Cl. Sc. fis. mat. nat.* **22** (1913), 98–103.

[Usui 2008] S. Usui, "Generic Torelli theorem for quintic-mirror family", *Proc. Japan Acad. Ser. A Math. Sci.* **84**:8 (2008), 143–146. MR 2010b:14012

[Verra 2004] A. Verra, "The Prym map has degree two on plane sextics", pp. 735–759 in *The Fano Conference*, Univ. Torino, Turin, 2004. MR 2005k:14057

[Voisin 1986] C. Voisin, "Théorème de Torelli pour les cubiques de $\mathbf{P}^5$", *Invent. Math.* **86**:3 (1986), 577–601. MR 88g:14006

[Voisin 1988] C. Voisin, "Sur la jacobienne intermédiaire du double solide d'indice deux", *Duke Math. J.* **57**:2 (1988), 629–646. MR 90f:14029

[Voisin 1999] C. Voisin, "A generic Torelli theorem for the quintic threefold", pp. 425–463 in *New trends in algebraic geometry* (Warwick, 1996), London Math. Soc. Lecture Note Ser. **264**, Cambridge Univ. Press, Cambridge, 1999. MR 2000i:14012

[Welters 1987] G. E. Welters, "Recovering the curve data from a general Prym variety", *Amer. J. Math.* **109**:1 (1987), 165–182. MR 88c:14041

olivier.debarre@ens.fr   *École Normale Supérieure,*
                                 Département de Mathématiques et Applications,
                                 UMR CNRS 8553, 45 rue d'Ulm, 75230 Paris, France

Current Developments in Algebraic Geometry
MSRI Publications
Volume 59, 2011

The Hodge theory of character varieties

MARK ANDREA A. DE CATALDO

This is a report on joint work with T. Hausel and L. Migliorini, where we prove, for each of the groups $GL_{\mathbb{C}}(2)$, $PGL_{\mathbb{C}}(2)$ and $SL_{\mathbb{C}}(2)$, that the nonabelian Hodge theorem identifies the weight filtration on the cohomology of the character variety with the perverse Leray filtration on the cohomology of the domain of the Hitchin map. We review the decomposition theorem, Ngô's support theorem, the geometric description of the perverse filtration and the subadditivity of the Leray filtration with respect to the cup product.

1. Introduction

This is an expanded version of notes from my talk at the conference "Classical Algebraic Geometry Today", at MSRI in Berkeley, January 25–29, 2009. The talk reported on joint work with T. Hausel at Oxford and L. Migliorini at Bologna, written up in [de Cataldo et al. 2011]. Following the recommendation of the editors, this article is designed to be accessible to nonspecialists and to give a small glimpse into an active area of research. The reader is referred to the introduction of the paper just cited for more details on what follows.

Let C be a nonsingular complex projective curve. We consider the following two moduli spaces associated with C: $\mathcal{M} := \mathcal{M}_{\text{Dolbeault}} :=$ the moduli space of stable holomorphic rank two Higgs bundles on C of degree one (see Section 3) and the character variety $\mathcal{M}' := \mathcal{M}_{\text{Betti}}$, i.e., the moduli space of irreducible complex dimension two representations of $\pi_1(C - p)$ subject to the condition that a loop around the chosen point $p \in C$ is sent to $-\text{Id}$. There is an analogous picture

Partially supported by NSA, NSF, and Simons Foundation summer research funds.

associated with any complex reductive Lie group G and the above corresponds to the case $G = GL_{\mathbb{C}}(2)$. In [de Cataldo et al. 2011] only the cases $G = GL_{\mathbb{C}}(2), PGL_{\mathbb{C}}(2), SL_{\mathbb{C}}(2)$ are dealt with. Both $\mathcal{M}$ and $\mathcal{M}'$ are quasiprojective irreducible and nonsingular of some even dimension $2d$. While $\mathcal{M}$ depends on the complex structure of C, $\mathcal{M}'$ does not. There is a proper flat and surjective map, the Hitchin map, $h : \mathcal{M} \to \mathbb{C}^d$ with general fibers abelian varieties of dimension d; in particular, $\mathcal{M}$ is not affine: it contains complete subvarieties of positive dimension. On the other hand, $\mathcal{M}'$ is easily seen to be affine (it is a GIT quotient of an affine variety).

The nonabelian Hodge theorem states that the two moduli spaces $\mathcal{M}_{\text{Dolbeault}}$ and $\mathcal{M}_{\text{Betti}}$ are naturally diffeomorphic, i.e., that there is a natural diffeomorphism $\varphi : \mathcal{M} \simeq \mathcal{M}'$. Since $\mathcal{M}'$ is affine (resp. Stein) and $\mathcal{M}$ is not affine (resp. not Stein), the map φ is not algebraic (resp. not holomorphic). Of course, we can still deduce that φ^* is a natural isomorphism on the singular cohomology groups.

Let us point out that the mixed Hodge structure on the cohomology groups $H^j(\mathcal{M}, \mathbb{Q})$ is in fact pure, i.e., every class has type (p, q) with $p + q = j$, or equivalently, every class has weight j. This follows easily from the fact that, due to the nonsingularity of $\mathcal{M}$, the weights of $H^j(\mathcal{M}, \mathbb{Q})$ must be $\geq j$. It remains to show that the weights are also $\leq j$: the variety $\mathcal{M}$ admits the fiber $h^{-1}(0)$ of the Hitchin map over the origin $0 \in \mathbb{C}^d$ as a deformation retract; it follows that the restriction map in cohomology, $H^j(\mathcal{M}, \mathbb{Q}) \to H^j(h^{-1}(0), \mathbb{Q})$ is an isomorphism of mixed Hodge structures; since the central fiber is compact, the weights of $H^j(h^{-1}(0), \mathbb{Q})$ are $\leq j$, and we are done.

On the other hand, the mixed Hodge structure on the cohomology groups $H^j(\mathcal{M}', \mathbb{Q})$ is known to be nonpure [Hausel and Rodriguez-Villegas 2008], i.e., there are classes of degree j but weight $> j$.

It follows that the isomorphism φ^* is not compatible with the two weight filtrations $\mathcal{W}$ on $H^*(\mathcal{M}, \mathbb{Q})$ and $\mathcal{W}'$ on $H^*(\mathcal{M}', \mathbb{Q})$. This fact raises the following question: *if we transplant the weight filtration $\mathcal{W}'$ onto $H^*(\mathcal{M}, \mathbb{Q})$ via φ^*, can we interpret the resulting filtration on $H^*(\mathcal{M}, \mathbb{Q})$, still called $\mathcal{W}'$, in terms of the geometry of $\mathcal{M}$?*

The main result in [de Cataldo et al. 2011] is Theorem 5.1 below and it gives a positive answer to the question raised above. In order to state this answer, we need to introduce one more ingredient and to make some trivial renumerations. (In this paper, we only deal with increasing filtrations.) The Hitchin map $h : \mathcal{M} \to \mathbb{C}^d$ gives rise to the perverse Leray filtration $^P\mathcal{L} = {}^P\mathcal{L}_h$ on $H^*(\mathcal{M}, \mathbb{Q})$; this is a suitable variant of the ordinary Leray filtration for h; for a geometric description of the perverse Leray filtration see Theorem 4.5. We re-index the filtration $^P\mathcal{L}$ so that $1 \in H^0(\mathcal{M}, \mathbb{Q})$ is in place zero (see (10)); the resulting re-indexed filtration on $H^*(\mathcal{M}, \mathbb{Q})$ is denoted by P.

All the actual weights appearing in $\mathcal{W}'$ on $H^*(\mathcal{M}', \mathbb{Q})$ turn out to be multiples of four. We renumerate $\mathcal{W}'$ by setting $W'_k := \mathcal{W}'_{2k}$.

Our answer to the question above is: *The nonabelian Hodge theorem isomorphism φ^* identifies the weight filtration W' on $H^*(\mathcal{M}', \mathbb{Q})$ with the perverse Leray filtration P on $H^*(\mathcal{M}, \mathbb{Q})$:*

$$P = W'.$$

The nature of these two filtrations being very different, we find this coincidence intriguing, but at present we cannot explain it.

The proof of Theorem 5.1 uses a few ideas from the topology of algebraic maps. Notably, Ngô's support theorem [2008], the geometric description of the perverse filtration [de Cataldo and Migliorini 2010] and the explicit knowledge of the cohomology ring $H^*(\mathcal{M}_{\text{Betti}}, \mathbb{Q})$ [Hausel and Thaddeus 2003] and of its mixed Hodge structure [Hausel and Rodriguez-Villegas 2008].

One of the crucial ingredients we need is Theorem 5.3, which may be of independent interest: it observes that Ngô's support theorem for the Hitchin fibration, i.e., (4) below, can be refined rather sharply, in the rank two cases we consider, as follows: the intersection complexes appearing in (4) are in fact sheaves (up to a dimensional shift).

What follows is a summary of the contents of this paper. Section 2 is devoted to stating the decomposition theorem for proper maps of algebraic varieties and to defining the associated "supports". Section 3 states Ngô's support theorem [Ngô 2008, §7] and sketches a proof of it in a special case and under a very strong splitting assumption that does not occur in practice; the purpose here is only to explain the main idea behind this beautiful result. Section 4 is a discussion of the main result of [de Cataldo and Migliorini 2010], i.e., a description of the perverse filtration in cohomology with coefficients in a complex via the restriction maps in cohomology obtained by taking hyperplane sections. Section 5 states the main result in [de Cataldo et al. 2011] and discusses some of the other key ingredients in the proof, notably the use of the subadditivity of the ordinary Leray filtration with respect to cup products. Since I could not find a reference in the literature for this well-known fact, I have included a proof of it in the more technical Section 6.

1.1. *Notation.* We work with sheaves of either abelian groups, or of rational vector spaces over complex algebraic varieties. The survey [de Cataldo and Migliorini 2009] is devoted to the decomposition theorem and contains a more detailed discussion of what follows.

A sheaf F on a variety Y is constructible if there is a finite partition $Y = \coprod T_i$ into nonsingular locally closed irreducible subvarieties that is adapted to F, i.e., such that each $F_{|T_i}$ is a local system (a locally constant sheaf) on T_i. A constructible complex K on a variety Y is a bounded complex of sheaves whose cohomology sheaves $\mathcal{H}^i(K)$ are constructible. We denote by D_Y the corresponding full subcategory of the derived category of sheaves on Y. If $K \in D_Y$, then $H^i(Y, K)$ denotes the i-th cohomology group of Y with coefficients in K. Similarly, for $H_c^i(Y, K)$. The complex $K[n]$ has i-th entry K^{n+i} and differential $d_{K[n]}^i = (-1)^n d_K^{n+i}$.

The standard truncation functors are denoted by $\tau_{\leq i}$, the perverse (middle perversity) ones by $^p\tau_{\leq i}$. The perverse cohomology sheaves are denoted $^p\mathcal{H}^i(K)$, $i \in \mathbb{Z}$. We make some use of these notions in Section 6. Recall that if $K \in D_Y$, then $^p\mathcal{H}^i(K) \neq 0$ for finitely many values of $i \in \mathbb{Z}$. In general, the collection of perverse cohomology sheaves $\{^p\mathcal{H}^i(K)\}_{i \in \mathbb{Z}}$ does not determine the isomorphism class of K in D_Y; e.g., if $j : U \to X$ is the open immersion of the complement of a point p in a nonsingular surface X, then the sheaves $j_!\mathbb{Q}_U$ and $\mathbb{Q}_X \oplus \mathbb{Q}_p$, viewed as complexes in D_X, yield the same collection $^p\mathcal{H}^0(-) = \mathbb{Q}_p$, $^p\mathcal{H}^2(-) = \mathbb{Q}_X[2]$. On the other hand, the celebrated decomposition theorem (Theorem 2.4 below) implies that if $f : X \to Y$ is a proper map of algebraic varieties, with X nonsingular for example, then the direct image complex satisfies

$$Rf_*\mathbb{Q}_X \simeq \bigoplus_i {}^p\mathcal{H}^i(Rf_*\mathbb{Q}_X)[-i].$$

This implies that the perverse cohomology sheaves reconstitute, up to an isomorphism, the direct image complex; more is true: each perverse cohomology sheaf splits further into a direct sum of simple intersection complexes (cf. (2)).

We have the following subcategories of D_Y:

$$D_Y^{\leq 0} := \{K \mid \mathcal{H}^i(K) = 0 \text{ for all } i > 0\},$$

$$^pD_Y^{\leq 0} := \{K \mid \dim \text{supp } \mathcal{H}^i(K) \leq -i \text{ for all } i \in \mathbb{Z}\}.$$

More generally, a perversity p gives rise to truncation functors $^p\tau_{\leq i}$, subcategories $^pD_Y^{\leq i}$ and cohomology complexes $^p\mathcal{H}^i(K)$.

Filtrations on abelian groups H are assumed to be finite: if the filtration $F_\bullet$ on H is increasing, then $F_iH = 0$ for $i \ll 0$ and $F_iH = H$ for $i \gg 0$; if $F^\bullet$ is decreasing, then it is the other way around. One can switch type by setting $F_i = F^{-i}$. For $i \in \mathbb{Z}$, the i-th graded objects are defined by setting

$$\text{Gr}_i^F H := F_iH/F_{i-1}H.$$

The increasing standard filtration $\mathcal{S}$ on $H^j(Y, K)$ is defined by setting

$$\mathcal{S}_iH^j(K) := \text{Im}\{H^j(Y, \tau_{\leq i}K) \to H^j(Y, K)\}.$$

Similarly, for $^p\mathscr{S}$ and more generally for $^P\mathscr{S}$. These filtrations are the abutment of corresponding spectral sequences.

Let $f : X \to Y$ be a map of varieties. The symbol f_* ($f_!$, respectively) denotes the derived direct image Rf_* (with proper supports $Rf_!$, respectively). Let $C \in D_X$. The direct image sheaves are denoted $R^j f_* C$. We have $H^j(X, C) = H^j(Y, f_* C)$ and $H_c^j(X, C) = H_c^j(Y, f_! K)$.

The Leray filtration is defined by setting $\mathscr{L}_i H^j(X, C) := \mathscr{S}_i H^j(Y, f_* C)$ and it is the abutment of the Leray spectral sequence. Similarly, for $H_c^j(X, C)$. Given a perversity p, we have the p-Leray spectral sequence abutting to the p-Leray filtration $^P\mathscr{L}$. We reserve the terms perverse Leray spectral sequence and perverse Leray filtration to the case of middle perversity $p = \mathfrak{p}$.

If X is smooth and f is proper, we let $Y_{\mathrm{reg}} \subseteq Y$ be the Zariski open set of regular values of f and we denote by R^i the local system $(R^i f_* \mathbb{Q}_X)_{|Y_{\mathrm{reg}}}$ on Y_{reg}.

2. The decomposition theorem

The purpose of this section is to state the decomposition Theorem 2.4 and to introduce the related notion of supports.

Let $f : X \to Y$ be a map of varieties. The Leray spectral sequence

$$E_2^{pq} = H^p(Y, R^q f_* \mathbb{Q}_X) \Longrightarrow H^{p+q}(X, \mathbb{Q})$$

relates the operation of taking cohomology on Y to the same operation on X. If we have E_2-degeneration, i.e., $E_2 = E_\infty$, then we have an isomorphism

$$H^j(X, \mathbb{Q}) \simeq \bigoplus_{p+q=j} H^p(Y, R^q f_* \mathbb{Q}). \tag{1}$$

Example 2.1. Let $f : X \to Y$ be a resolution of the singularities of the projective variety Y. Let us assume, as it is often the case, that the mixed Hodge structure on $H^j(Y, \mathbb{Q})$ is not pure for some j. Then $f^* : H^j(Y, \mathbb{Q}) \to H^*(X, \mathbb{Q})$ is not injective and E_2-degeneration fails; this is because injectivity would imply the purity of the mixed Hodge structure on $H^j(Y, \mathbb{Q})$.

Example 2.2. Let $f : (\mathbb{C}^2 - \{(0, 0)\})/\mathbb{Z} \to \mathbb{CP}^1$ be a Hopf surface (see [Barth et al. 1984]) together with its natural holomorphic proper submersion onto the projective line. Since the first Betti number of the Hopf surface is one and the one of a fiber is two, E_2-degeneration fails.

These examples show that we cannot expect E_2-degeneration, neither for holomorphic proper submersions of compact complex manifolds, nor for projective maps of complex projective varieties. On the other hand, the following result of P. Deligne [1968] shows that E_2-degeneration is the norm for proper submersions of complex algebraic varieties.

Theorem 2.3. *Let $f : X \to Y$ be a smooth proper map of complex algebraic varieties. Then the Leray spectral sequence for f is E_2-degenerate. More precisely, there is an isomorphism in D_Y*

$$f_* \mathbb{Q}_X \simeq \bigoplus_i R^i f_* \mathbb{Q}_X[-i].$$

The decomposition theorem is a far-reaching generalization of Theorem 2.3 that involves intersection cohomology, a notion that we review briefly next. A complex algebraic variety Y of dimension $\dim_\mathbb{C} Y = n$ carries intersection cohomology groups $IH^*(Y, \mathbb{Q})$ and $IH_c^*(Y, \mathbb{Q})$ satisfying the following conditions.

1. Poincaré duality holds: there is a geometrically defined perfect pairing

$$IH^{n+j}(Y) \times IH_c^{n-j}(Y) \to \mathbb{Q}.$$

2. There is the intersection complex IC_Y; it is a constructible complex of sheaves of rational vector spaces on Y such that

$$IH^j(Y, \mathbb{Q}) = H^{j-n}(Y, IC_Y),$$
$$IH_c^j(Y, \mathbb{Q}) = H_c^{j-n}(Y, IC_Y).$$

3. If Y is nonsingular, then $IH^*(Y, \mathbb{Q}) = H^*(Y, \mathbb{Q})$ and $IC_Y = \mathbb{Q}_Y[n]$ (complex with the one entry $\mathbb{Q}_Y$ in cohomological degree $-n$).

4. If Y^o is a nonempty open subvariety of the nonsingular locus of Y and L is a local system on Y^o, then we have the twisted intersection complex $IC_Y(L)$ on Y and the intersection cohomology groups $IH^j(Y, L) = H^{j-n}(Y, IC_Y(L))$ of Y with coefficients in L.

Theorem 2.4 (Decomposition theorem [Beĭlinson et al. 1982, théorème 6.2.5]). *Let $f : X \to Y$ be a proper map of algebraic varieties. Then*

$$f_* IC_X \simeq \bigoplus_{b \in B} IC_{Z_b}(L_b)[d_b] \tag{2}$$

for an uniquely determined finite collection B of triples (Z_b, L_b, d_b) such that $Z_b \subseteq Y$ is a closed irreducible subvariety, $L_b \neq 0$ is a simple local system on some nonempty and nonsingular Zariski open $Z_b^o \subseteq Z_b$ and $d_b \in \mathbb{Z}$.

If, in Theorem 2.4, we replace "simple" with "semisimple", we obtain a uniquely determined collection B' by grouping together the terms with the same cohomological shift $[d_b]$ and the same irreducible subvariety Z_b.

Definition 2.5. The varieties $Z_b \subseteq Y$ are called the *supports* of the map $f : X \to Y$.

The supports Z_b are *among* the closed irreducible subvarieties $Z \subseteq Y$ with the property that

(1) there exists a nonempty $Z^0 \subseteq Z$ over which all the direct image sheaves $R^i f_* \mathbb{Q}$ are local systems, and

(2) Z is maximal with this property.

The following example shows that a support may appear more than once with distinct cohomological shifts. Of course, that happens already in the situation of Theorem 2.3; the point of the example is that this "repeated support" may be smaller than the image $f(X)$.

Example 2.6. Let $f : X \to Y = \mathbb{C}^3$ be the blowing-up of a point $o \in \mathbb{C}^3$; there is an isomorphism

$$f_* \mathbb{Q}_X[3] \simeq \mathbb{Q}_Y[3] \oplus \mathbb{Q}_o[1] \oplus \mathbb{Q}_o[-1].$$

The next example shows that a variety Z, in this case $Z = v$, that satisfies conditions (1) and (2) above, may fail to be a support.

Example 2.7. Let $f : X \to Y$ be the small resolution of the three-dimensional affine cone $Y \subseteq \mathbb{C}^4$ over a nonsingular quadric surface $\mathbb{P}^1 \times \mathbb{P}^1 \simeq \mathcal{Q} \subseteq \mathbb{P}^3$, given by the contraction to the vertex $v \in Y$ of the zero section in the total space X of the vector bundle $\mathcal{O}_{\mathbb{P}^1}(-1)^2$. In this case, we have

$$R f_* \mathbb{Q}_X[3] = IC_Y.$$

The determination of the supports of a proper map is an important and difficult problem.

3. Ngô's support theorem

B. C. Ngô [2008] proved the "fundamental lemma" in the Langlands program. This is a major advance in geometric representation theory, automorphic representation theory and the arithmetic Langlands program. See [Nadler 2010]. One of the crucial ingredients of the proof is the support Theorem 3.1, whose proof applies the decomposition theorem to the Hitchin map associated with a reductive group and a nonsingular projective curve. The support theorem is a rather general result concerning a certain class of fibrations with general fibers abelian varieties and the Hitchin map is an important example of such a fibration.

In our paper [de Cataldo et al. 2011] (to which I refer the reader for more context and references), we deal with the Hitchin map in the rank-two case, i.e., with the reductive groups $GL_\mathbb{C} 2$, $PGL_\mathbb{C}(2)$ and $SL_\mathbb{C}(2)$. The simpler geometry allows us to refine the conclusion (4) of the support theorem for the Hitchin map in the form of Theorem 5.3, which in turn we use in [de Cataldo et al. 2011] to prove Theorem 5.1.

In this section, we discuss the support theorem in the case of $GL_\mathbb{C}(2)$. This situation is too-simple in the context of the fundamental lemma, but it allows us to concentrate on the main idea underlying the proof of the support theorem, which is to pursue the action of abelian varieties on the fibers of the Hitchin map. In the context of Ngô's work, it is critical to work over finite fields. We ignore this important aspect and, for the sake of exposition, we make the oversimplifying Assumption 3.2 and stick with the situation over $\mathbb{C}$.

Let C be a compact Riemann surface of genus $g \geq 2$. Let $\mathcal{M}$ be the moduli space of stable rank 2 Higgs bundles on C with determinant of degree one. In this context, a point $m \in \mathcal{M}$ parametrizes a stable pair (E, φ), where E is a rank two bundle on C with $\deg(\det E) = 1$ and $\varphi : E \to E \otimes \omega_C$ (where $\omega_C := T_C^*$ denotes the canonical bundle of C) is a map of bundles, i.e., a section of $\text{End}(E) \otimes \omega_C$. Stability is a technical condition on the degrees of the subbundles of E preserved by φ. Only the parity of $\deg(\det E)$ counts here: there are only two isomorphism classes of such moduli spaces; the case of even degree yields a singular moduli space and we do not say anything new in that case.

Let $d := 4g - 3$. The variety $\mathcal{M}$ is nonsingular, quasiprojective and of dimension $2d$. There is a proper and flat map, called the Hitchin map, onto affine space

$$h : \mathcal{M}^{2d} \to \mathbb{A}^d \simeq H^0(C, \omega_C \oplus \omega_C^{\otimes 2}), \tag{3}$$

which is a completely integrable system.

Set-theoretically, the map $h : m = (E, \varphi) \mapsto (\text{trace}(\varphi), \det \varphi)$, where the trace and determinant of the twisted endomorphism φ are viewed as sections of the corresponding powers of ω_C.

A priori, it is far from clear that this map is proper. This fact was first noted and proved by Hitchin. It is a beautiful fact (also due to Hitchin) that each nonsingular fiber $\mathcal{M}_a := h^{-1}(a)$, $a \in \mathbb{A}^d$, is isomorphic to the Jacobian $J(C_a')$ of what is called the spectral curve C_a'. This curve lives on the surface given by the total space of the line bundle ω_C and it is given set-theoretically as the double cover of C given by (and this explains the term "spectral")

$$C \ni \{c\} \longleftrightarrow \{\text{the set of eigenvalues of } \varphi_c\} \in \omega_{C,c}.$$

The genus $g(C_a') = d$ by Riemann–Roch and by the Hurwitz formula.

The singular fibers of the Hitchin map $h : \mathcal{M} \to \mathbb{A}^d$ are, and this is an euphemism, difficult to handle.

Let $V \subseteq \mathbb{A}^d$ be the open locus over which the fibers of h are reduced. The sheaf $R_V^{2d} := (R^{2d} f_* \mathbb{Q})_{|V}$ is the $\mathbb{Q}$-linearization of the sheaf of finite sets given by the sets of irreducible components of the fibers over V. Let $h_V : \mathcal{M}_V := h^{-1}(V) \to V$ be the restriction of the Hitchin map over V.

We can now state Ngô's support theorem in the very special case at hand. Roughly speaking, it states that over V, the highest direct image R_V^{2d} is responsible for all the supports.

Theorem 3.1 (Ngô's support theorem). *A closed and irreducible subvariety $Z \subseteq V$ appears as a support Z_b in the decomposition theorem* (2) *for h_V, if and only if there is a dense open subvariety $Z^o \subseteq Z$ such that the restriction $(R_V^{2d})_{|Z^o}$ is locally constant and Z is maximal with this property.*

If we further restrict to the open set $U \subseteq V$ where the fibers are reduced and irreducible, then the support theorem has the following striking consequence: the only support on U is U itself. The decomposition theorem (2) for h_U takes then the following form (notation as in Section 1.1)

$$h_{U*}\mathbb{Q}_{\mathcal{M}_U}[2d] \simeq \bigoplus_{i=-d}^{d} IC_U(R^{i+d})[-i]. \tag{4}$$

The open U is fairly large: its complement has codimension $\geq 2g - 3$.

The remaining part of this section is devoted to discussing the main idea in the proof of the support theorem.

There is a group-variety $\mathcal{P}_V \to V$ over V acting on the variety $\mathcal{M}_V \to V$ over V, i.e., a commutative diagram

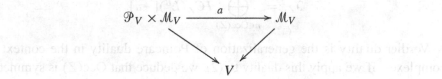

satisfying the axioms of an action.

Let us describe this situation over a point $v \in V$. The fiber $\mathcal{M}_v$ is noncanonically isomorphic to a suitable compactification of the identity component $\mathcal{P}_v$ of the Picard group $\mathrm{Pic}(C_v')$ of the possibly singular spectral curve C_v'. The variety $\mathcal{M}_v$ parametrizes certain torsion free sheaves of rank and degree one on C_v'. The group variety $\mathcal{P}_v$ acts on $\mathcal{M}_v$ via tensor product. There is an exact sequence (Chevalley devissage) of algebraic groups of the indicated dimensions

$$1 \to R_v^{\delta_v} \to \mathcal{P}_v^d \to A_v^{d-\delta_v} \to 1 \tag{5}$$

where A_v is the abelian variety given by the Picard variety of the normalization of the spectral curve C_v' and R_v is an affine algebraic group. The sequence (5) does not split over the complex numbers, but it splits over a finite field. It turns out that this is enough in order to prove the freeness result on which the proof of the support theorem rests. In order to explain the main idea, let us make the following (over)simplifying assumption.

Assumption 3.2. There is a splitting of the Chevalley devissage (5).

A splitting induces an action of A_v on $\mathcal{M}_v$ with finite stabilizers. There is the rational homology algebra $H_*(A_v)$ with product given by the Pontryagin product $H_i(A_v) \otimes H_j(A_v) \to H_{i+j}(A_v)$ induced by the cross product, followed by push-forward via the multiplication map in A_v. We have the following standard ([Ngô 2008], p.134, Proposition 7.4.5)

Fact 3.3. *Let $A \times T \to T$ be an action of an abelian variety A on a variety T such that all stabilizers are finite. Then $H_c^*(T)$ is a free graded $H_*(A)$-module for the action of the rational homology algebra $H_*(A)$ on $H_c^*(T)$.*

Our assumptions imply that

$$\forall v \in V, \quad H^*(\mathcal{M}_v) \text{ is a free graded } H_*(A_v)\text{-module.}$$

Let Z be a support appearing in the decomposition theorem (2) for h_V. Define a finite set of integers as follows

$$\mathrm{Occ}(Z) := \{n \in \mathbb{Z} \mid \exists b \text{ s.t. } Z_b = Z, \ d_b = -n\} \subseteq [-d, d].$$

The integers in $\mathrm{Occ}(Z)$ are in one-to-one correspondence with the summands (2) with support Z. By grouping them, we obtain the graded object

$$\mathfrak{I}_Z := \bigoplus_{n \in \mathrm{Occ}(Z)} IC_Z(L^n)[-n].$$

Verdier duality is the generalization of Poincaré duality in the context of complexes. If we apply this duality to (2), we deduce that $\mathrm{Occ}(Z)$ is symmetric about 0.

Every intersection complex $IC_Y(L)$ on an irreducible variety Y restricts to $L[\dim Y]$ on a suitable nonempty open subvariety $Y^o \subseteq Y$. It follows that there is a nonempty open subvariety $V^o \subseteq V$ such that every $IC_Z(L^n)$ restricts to $L^n[\dim Z]$ on $Z^o := Z \cap V^o$. Let us consider the restriction of $\mathfrak{I}_Z$ to Z^o:

$$\mathfrak{L} := \bigoplus_{n \in \mathrm{Occ}(Z)} L^n[\dim Z][-n].$$

If we set $n^+ := \max \mathrm{Occ}(Z)$, then, by the aforementioned symmetry about the origin, the length $l(\mathfrak{L}) = 2n^+$.

The decomposition theorem (2) over V^o implies that

$$\forall n \in \mathrm{Occ}(Z), \quad L^n \text{ is a direct summand of } (R^{2d+n-\dim Z} h_* \mathbb{Q})_{|Z^o}.$$

Since the fibers of h have dimension d, the higher direct images $R^j h_* \mathbb{Q}$ vanish for every $j > 2d$. It follows that the support theorem is equivalent to the following claim:

Claim. L^{n^+} *is a direct summand of* $(R^{2d} h_* \mathbb{Q})_{|Z^o}$.

This is equivalent to having $n^+ - \dim Z = 0$, and, again by the vanishing for the direct images $R^j h_* \mathbb{Q}$, this is equivalent to having $n^+ - \dim Z \geq 0$.

Let $z \in Z^o$ be any point. By adding and subtracting $\dim A_z = d - \delta_z$, we can reformulate the support theorem as follows:

$$[\operatorname{codim} Z - \delta_z] + [n^+ - (d - \delta_z)] \geq 0.$$

It is thus enough to show that each of the two quantities in square brackets is ≥ 0.

The first inequality $[\operatorname{codim} Z - \delta_z] \geq 0$ follows from the deformation theory of Higgs bundles and Riemann–Roch on the curve C. This point is standard over the complex numbers. At present, in positive characteristic it requires the extra freedom of allowing poles of fixed but arbitrary high order. We do not address this point here.

Since $l(\mathcal{L}) = 2n^+$, in order to prove the second inequality, we need to show that

$$l(\mathcal{L}) = l(\mathcal{L}_z) = 2n^+ \geq 2(d - \delta_z).$$

Since $l(H_*(A_z)) = 2 \dim A_z = 2(d - \delta_z)$, the inequality would follow if we could prove that:

$\mathcal{L}_z$ *is a free graded* $H_*(A_z)$-*module.*

By virtue of the decomposition theorem, the graded vector space $\mathcal{L}_z$ is a graded vector subspace of $H^*(\mathcal{M}_z)$. This is not enough. We need to make sure that it is a free $H_*(A_z)$-submodule. Once it is known that $\mathcal{L}_z$ is a submodule, then its freeness is an immediate consequence of standard results from algebra, notably that a projective module over the local graded commutative algebra $H_*(A_z)$ is free. Showing that $\mathcal{L}_z$ is $H_*(A_z)$-stable is a delicate point, for a priori the contributions from other supports could enter the picture and spoil it. This problem is solved by means of a delicate specialization argument which we do not discuss here.

4. The perverse filtration and the Lefschetz hyperplane theorem

Let us review the classical construction that relates the Leray filtration on the cohomology of the total space a fiber bundle to the filtration by skeleta on the base.

Let $f : X \to Y$ be a topological fiber bundle where Y is a cell complex of real dimension n. Let $Y_* := \{Y_0 \subseteq \ldots \subseteq Y_k \subseteq \ldots \subseteq Y_n = Y\}$ be the filtration by k-skeleta. Let $X_* := \pi^{-1}(Y_*)$ be the corresponding filtration of the total space X.

If $\mathcal{L}$ is the increasing Leray filtration associated with π, then we have (see [Spanier 1966, §9.4])

$$\mathcal{L}_i H^j(X, \mathbb{Z}) = \mathrm{Ker}\,\{H^j(X, \mathbb{Z}) \to H^j(X_{j-i-1}, \mathbb{Z})\}. \qquad (6)$$

The key fact that one needs (see [de Cataldo and Migliorini 2010; de Cataldo 2009]) is the π-cellularity of Y_*, i.e., the fact that

$$H^j(Y_p, Y_{p-1}, R^q f_* \mathbb{Z}_X) = 0 \quad \text{for all } j \neq p \text{ and all } q. \qquad (7)$$

This condition is verified since, for each fixed p, we are really dealing with bouquets of p-spheres.

This classical result can be viewed as a geometric description of the Leray filtration in the sense that the subspaces of the Leray filtration are exhibited as kernels of restrictions maps to the preimages of skeleta. The following result of D. Arapura [2005] gives a geometric description of the Leray filtration for a projective map of quasiprojective varieties: the important point is that the "skeleta" can be taken to be algebraic subvarieties! For generalizations of Arapura's result, see [de Cataldo 2009]. In what follows, for ease of exposition, we concentrate on the case when the target is affine.

Theorem 4.1 (Geometric description of the Leray filtration). *Let $f : X \to Y$ be a proper map of algebraic varieties with Y affine of dimension n. Then there is a filtration Y_* of Y by closed algebraic subvarieties Y_i of dimension i such that (6) holds.*

Remark 4.2. The flag Y_* is constructed inductively as follows. Choose a closed embedding $Y \subseteq \mathbb{A}^N$. Each Y_i is a complete intersection of Y with $n-i$ sufficiently high degree hypersurfaces in special position. Here "special" refers to the fact that in order to achieve the cellularity condition (7), we need to trace, as p decreases, the Y_{p-1} through the positive-codimension strata of a partition of Y_p adapted to the restricted sheaves $(R^q f_* \mathbb{Z}_X)_{|Y_p}$.

Theorem 4.1 affords a simple proof of the following result of M. Saito [1990]. Recall that the integral singular cohomology of complex algebraic varieties carries a canonical and functorial mixed Hodge structure (mHs).

Corollary 4.3 (The Leray filtration is compatible with mHs). *Let $f : X \to Y$ be a proper map of algebraic varieties with Y quasiprojective. Then the subspaces of the Leray filtration $\mathcal{L}$ on $H^q(X, \mathbb{Z})$ are mixed Hodge substructures.*

Let K be a constructible complex of sheaves on an algebraic variety Y. We have the perverse filtration ${}^p\!\mathcal{S}_i H^j(Y, K) := \mathrm{Im}\left\{ H^j\left(Y, {}^p\tau_{\le i} K\right) \to H^j(Y, K)\right\}$. Let $f : X \to Y$ be a map of algebraic varieties and let $C \in D_X$. We have the perverse Leray filtration ${}^p\!\mathcal{L}_i$ on $H^j(X, C)$, i.e the perverse filtration ${}^p\!\mathcal{S}$ on $H^j(Y, f_*C) = H^j(X, C)$. Similarly, for $H^j_c(X, C)$.

Remark 4.4. In the situation of the decomposition Theorem 2.4, if we take X to be nonsingular (if X is singular, then replace cohomology with intersection cohomology in what follows), then the subspace ${}^p\!\mathcal{L}_i H^j(X, C) \subseteq H^j(X, C)$ is given by the images, via the chosen splitting, of the direct sum of the j-th cohomology groups of the terms with $-d_b \le i$. The general theory implies that this image is independent of the chosen splitting. However, different splittings yield different embeddings of each of the direct summands into $H^j(X, C)$.

Let $f : X \to Y$ be a map of varieties where Y is a quasiprojective variety. Let $C \in D_X$ and $K \in D_Y$ (integral coefficients). The main result of [de Cataldo and Migliorini 2010] is a geometric description of the perverse and perverse Leray filtrations. We state a significant special case only.

Theorem 4.5 (Geometric perverse Leray). *Let $f : X \to Y$ be a map of algebraic varieties with Y affine of dimension n. Then there is a filtration Y_* by closed subvarieties Y_i of dimension i such that if we take $X_* := f^{-1}Y_*$, then*

$$ {}^p\!\mathcal{L}_i\, H^j(X, \mathbb{Z}) := {}^p\!\mathcal{S}_i\, H^j(Y, f_*\mathbb{Z}_X) = \mathrm{Ker}\left\{ H^j(X, \mathbb{Z}) \to H^j(X_{n+j-i-1}, \mathbb{Z})\right\}. $$

The main difference with respect to Theorem 4.1 is that Y_* is obtained by choosing general vs. special hypersurfaces (see Remark 4.2). This choice is needed in order to deduce the perverse analogue of the cellularity condition (7), i.e.,

$$ H^j\left(Y_p, Y_{p-1}, {}^p\!\mathcal{H}^q(f_*C)\right) = 0 \text{ for all } j \ne 0 \text{ and all } q. $$

These vanishing conditions are verified by a systematic use of the Lefschetz hyperplane theorem for perverse sheaves. Unlike [Arapura 2005] and [de Cataldo 2009], the proof for compactly supported cohomology is completely analogous to the one for cohomology.

A second difference, is that we do not need the map $f : X \to Y$ to be proper. The choice of general hypersurfaces avoids the usual pitfalls of the failure of the base change theorem (see [de Cataldo 2009]).

The discrepancy "$+n$" between (6) for Theorem 4.1 and Theorem 4.5 boils down to the fact that for the affine variety Y of dimension n, the cohomology groups $H^j(Y, F)$ with coefficients in a sheaf (perverse sheaf, respectively) F are nonzero only in the interval $[0, n]$) ($[-n, 0]$, respectively).

This geometric description of the perverse filtration in terms of the kernels of restriction maps to subvarieties is amenable to applications to the mixed Hodge theory of algebraic varieties. For example, the analogue of Corollary 4.3 holds, with the same proof. For more applications, see [de Cataldo 2010].

5. Character varieties and the Hitchin fibration: $P = W'$

In this section, I report on [de Cataldo et al. 2011], where we prove Theorem 5.1. The main ingredients are the geometric description of the perverse filtration in Theorem 4.5 and the refinement Theorem 5.3 of the support theorem (4) in the case at hand.

We have the Hitchin map (3) for the group $G = \mathrm{GL}_{\mathbb{C}}(2)$. There are analogous maps $\check{h} : \check{\mathcal{M}}^{6g-6} \to \mathbb{A}^{3g-3}$ for $G = \mathrm{SL}_{\mathbb{C}}(2)$ and $\widehat{h} : \widehat{\mathcal{M}}^{6g-6} \to \mathbb{A}^{3g-3}$ for $\mathrm{PGL}_{\mathbb{C}}(2)$.

Though these three geometries are closely related, this is not the place to detail the toing and froing from one group to another. The main point for this discussion is that we have an explicit description of the cohomology algebra $H^*(\mathcal{M}, \mathbb{Q})$ in view of the canonical isomorphism

$$H^*(\mathcal{M}, \mathbb{Q}) \simeq H^*(\widehat{\mathcal{M}}, \mathbb{Q}) \otimes H^*(\mathrm{Jac}(C), \mathbb{Q}) \qquad (8)$$

and of (9) below. In view of (8), the key cohomological considerations towards Theorem 5.1 below can be made in the $\mathrm{PGL}_{\mathbb{C}}(2)$ case, for they will imply easily the ones for $\mathrm{GL}_{\mathbb{C}}(2)$ and, with some extra considerations which we do not address here, the ones for $\mathrm{SL}_{\mathbb{C}}(2)$. For simplicity, ignoring some of the subtle differences between the three groups, let us work with $\widehat{h} : \widehat{\mathcal{M}}^{6g-6} \to \mathbb{A}^{3g-3}$. Though $\widehat{\mathcal{M}}$ is the quotient of a manifold by the action of a finite group, for our purposes we can safely pretend it is a manifold. We set $d := 3g - 3$.

In the context of the nonabelian Hodge theorem ([Simpson 1992]), the quasiprojective variety $\widehat{\mathcal{M}}$ is usually denoted $\widehat{\mathcal{M}}_D$, where D stands for Dolbeault. This is to contrast it with the moduli space $\widehat{\mathcal{M}}_B$ (Betti) of irreducible $\mathrm{PGL}_2(\mathbb{C})$ representations of the fundamental group of C; this is an affine variety.

The nonabelian Hodge theorem states that there is a natural diffeomorphism $\varphi : \widehat{\mathcal{M}}_B \simeq \widehat{\mathcal{M}}_D$. The two varieties are not isomorphic as complex spaces and, a fortiori, neither as algebraic varieties: the latter contains the fibers of the Hitchin map, i.e., d-dimensional abelian varieties, while the former is affine.

The diffeomorphism φ induces an isomorphism of cohomology rings φ^* : $H^*(\widehat{\mathcal{M}}_D, \mathbb{Q}) \simeq H^*(\widehat{\mathcal{M}}_B, \mathbb{Q})$. This isomorphism is not compatible with the mixed Hodge structures. In fact, the mixed Hodge structure on every $H^j(\widehat{\mathcal{M}}_D, \mathbb{Q})$ is known to be pure (see Section 1), while the one on $H^j(\widehat{\mathcal{M}}_B, \mathbb{Q})$ is known to be not pure ([Hausel and Rodriguez-Villegas 2008]).

In particular, the weight filtrations do not correspond to each other via φ^*. Our main result in [de Cataldo et al. 2011] can be stated as follows.

Theorem 5.1 ($P = W'$). *In the cases* $G = \text{GL}_\mathbb{C}(2)$, $\text{PGL}_\mathbb{C}(2)$, $\text{SL}_\mathbb{C}(2)$, *the nonabelian Hodge theorem induces an isomorphism in cohomology that identifies the weight filtration for the mixed Hodge structure on the Betti side with the perverse Leray filtration on the Dolbeault side; more precisely,* (11) *below holds.*

At present, we do not know what happens if the reductive group G has higher rank. Moreover, we do not have a conceptual explanation for the so-far mysterious exchange of structure of Theorem 5.1.

The paper [de Cataldo et al. 2010] deals with a related moduli space, i.e., the Hilbert scheme of n points on the cotangent bundle of an elliptic curve, where a similar exchange takes place.

Let us try and describe some of the ideas that play a role in the proof of Theorem 5.1. We refer to [de Cataldo et al. 2011] for details and attributions.

By the work of several people, the cohomology ring and the mixed Hodge structure of $H^*(\widehat{\mathcal{M}}_B, \mathbb{Q})$ are known. There are tautological classes:

$$\alpha \in H^2, \quad \{\psi_i\}_{i=1}^{2g(C)} \in H^3, \quad \beta \in H^4$$

which generate the cohomology ring. With respect to the mixed Hodge structure, these classes are of weight 4 and pure type $(2, 2)$. Every monomial made of l letters among these tautological classes has weight $4l$ and Hodge type $(2l, 2l)$, i.e., weights are strictly additive for the cup product. In general, weights are only subadditive. There is a graded $\mathbb{Q}$-algebra isomorphism

$$H^*(\widehat{\mathcal{M}}_B, \mathbb{Q}) \simeq \frac{\mathbb{Q}[\alpha, \{\psi_i\}, \beta]}{I}, \tag{9}$$

where I is a certain bihomogeneous ideal with respect to weight and cohomological degree. In particular, we have a canonical splitting for the increasing weight filtration $\mathcal{W}'$ on $H^j(\widehat{\mathcal{M}}_B, \mathbb{Q})$ (the trivial weight filtration $\mathcal{W}$ on the pure $H^j(\widehat{\mathcal{M}}_D, \mathbb{Q})$ plays no role here):

$$H^j(\widehat{\mathcal{M}}_B, \mathbb{Q}) - \bigoplus_{w \geq 0} H^j_w, \qquad \mathcal{W}'_w H^j - \bigoplus_{w' \leq w} H^j_{w'}.$$

The weights occur in the interval $[0, 4d]$ and they are multiples of four: $\mathcal{W}'_{4k-i} = \mathcal{W}'_{4k}$ for every $0 \leq i \leq 3$.

By virtue of the decomposition theorem (2) and of the fact that the Hitchin map $\widehat{h}$ is flat of relative dimension d, the increasing perverse Leray filtration $\mathcal{P}\mathscr{L}$ has type $[-d, d]$.

In order to compare $\mathcal{W}'$ with $^{\mathfrak{p}}\mathscr{L}$, we half the wights, i.e., we set $W_i' := \mathcal{W}_{2i}'$, and we translate $^{\mathfrak{p}}\mathscr{L}$, i.e., we set

$$P := {}^{\mathfrak{p}}\mathscr{L}(-d). \tag{10}$$

We still denote these half-weights by w. We have that both W' and P have nonzero graded groups Gr_i only in the interval $i \in [0, 2d]$. The two modified filtrations could still be completely unrelated. After all, they live on the cohomology of different algebraic varieties! The precise formulation of Theorem 5.1 is

$$P = W'. \tag{11}$$

Let us describe our approach to the proof.

We introduce the notion of perversity and, ultimately, we show that the perversity equals the weight. We say that $0 \neq u \in H^j(\widehat{\mathcal{M}}_D, \mathbb{Q})$ has perversity $p = p(u)$ if $u \in P_p \setminus P_{p-1}$. By definition, $u = 0$ can be given any perversity. Perversities are in the interval $[0, 2d]$. We write monomials in the tautological classes as $\alpha^r \beta^s \psi^t$, where ψ^t is a shorthand for a product of t classes of type ψ. Then (11) can be reformulated as follows:

$$p(\alpha^r \beta^s \psi^t) = w(\alpha^r \beta^s \psi^t) = 2(r + s + t). \tag{12}$$

As it turns out, the harder part is to establish the inequality

$$p(\alpha^r \beta^s \psi^t) \leq 2(r + s + t), \tag{13}$$

for once this is done, the reverse inequality is proved by a kind of simple pigeonhole trick. We thus focus on (13).

Recall that $U \subseteq \mathbb{A}^d$ (see (4)) is the dense open set where the fibers of $\hat{h}$ are irreducible. We have the following sharp estimate

$$\mathrm{codim}\,(\mathbb{A}^d \setminus U) \geq 2g - 3. \tag{14}$$

For every $0 \leq b \leq d$, let $\Lambda^b \subseteq \mathbb{A}^d$ denote a general linear section of dimension b. We have defined the translated perverse Leray filtration P on the cohomology groups of $\widehat{\mathcal{M}}$ for the map $\hat{h}$ that fibers $\widehat{\mathcal{M}}$ over $\mathbb{A}^d$. We can do so, in a compatible way, over U and over the Λ^b so that the restriction maps respect the resulting P filtrations. All these increasing filtrations start at zero and perversities are in the interval $[0, 2d]$.

The test for perversity Theorem 4.5, now reads

Fact 5.2. *Let* $\Lambda^b \subseteq \mathbb{A}^d$ *be a general linear subspace of dimension b. Denote by* $\widehat{\mathcal{M}}_{\Lambda^b} := \hat{h}^{-1}(\Lambda^b)$. *Then*

$$u \in P_{j-b-1} H^j(\widehat{\mathcal{M}}) \iff u_{|\widehat{\mathcal{M}}_{\Lambda^b}} = 0.$$

We need the following strengthening (in the special case we are considering) of the support theorem (4) over U. It is obtained by a study of the local monodromy of the family of spectral curves around the points of U. Let $j : \mathbb{A}^d_{\text{reg}} \to U$ be the open embedding of the set or regular values of $\hat{h}$.

Theorem 5.3. *The intersection complexes $IC_U(R^i)$ are shifted sheaves and we have*

$$\hat{h}_{U*}\mathbb{Q} \simeq \bigoplus j_* R^i [-i].$$

In particular, the translated perverse Leray filtrations P coincides with the Leray filtration $\mathscr{L}$ on $H^*(\widehat{\mathcal{M}}_U, \mathbb{Q})$, and on $H^*(\widehat{\mathcal{M}}_{\Lambda^b}, \mathbb{Q})$ for every $b < 2g - 3$.

The last statement is a consequence of (14): we can trace Λ^b inside U.

We can now discuss the scheme of proof for (13). We start by establishing the perversities of the multiplicative generators, i.e., by proving that

$$p(\alpha) = p(\beta) = p(\psi_i) = 2.$$

By Fact 5.2, we need to show that α vanishes over the empty set, ψ_i over a point, and β over a line. The first requirement is of course automatic. The second one is a result of M. Thaddeus [Thaddeus 1990]. He also proved that β vanishes over a point, but we need more.

Fact 5.4. *The class β is zero over a general line $\mathfrak{l} := \Lambda^1 \subseteq \mathbb{A}^{3g-3}$.*

Idea of proof. By (14), we can choose a general line $\mathfrak{l} = \Lambda^1 \subseteq U$. Let $f : \widehat{\mathcal{M}}_{\mathfrak{l}} := \hat{h}^{-1}(\mathfrak{l}) \to \mathfrak{l}$. In particular, by abuse of notation, we write $\widehat{\mathcal{M}}_{\text{reg}} := \hat{h}^{-1}(\mathbb{A}^d_{\text{reg}})$, where $\mathbb{A}^d_{\text{reg}} \subseteq \mathbb{A}^d$ is the Zariski open and dense set of regular values of the Hitchin map.

By Theorem 5.3 (as it turns out, since we are working over a curve, here (4) is enough to reach the same conclusion) we have

$$Rf_*\mathbb{Q} \simeq \bigoplus \left(j_* R^i \right)_{|\mathfrak{l}} [-i].$$

In particular, there are no skyscraper summands on $\mathfrak{l}$. A simple spectral sequence argument over the affine curve $\mathfrak{l}$, implies that the restriction map $H^4(\widehat{\mathcal{M}}_{\mathfrak{l}}) \to H^4(\widehat{\mathcal{M}}_{\text{reg}})$ is injective. (Note that this last conclusion would be clearly false if we had a skyscraper contribution.) It is enough to show that $\beta_{|\widehat{\mathcal{M}}_{\text{reg}}}$ is zero. The class β is a multiple of $c_2(\mathcal{M})$. On the other hand, since the Hitchin system is a completely integrable system over the affine space, the tangent bundle can be trivialized, in the C^∞-sense, over the open set of regular point $\widehat{\mathcal{M}}_{\text{reg}}$ using the Hamiltonian vector fields. □

Having determined the perversity for the multiplicative generators, we turn to (13) which we can reformulate by saying that perversities are subadditive under cup product.

In general, I do not know if this is the case: see the discussion following the statement of Theorem 6.1 and also Remark 6.8. On the other hand, the analogous subadditivity statement for the Leray filtration $\mathcal{L}$ is well-known to hold; see Theorem 6.1.

Let us outline our procedure to prove the subadditivity of perversity in our case. We want to use the test for perversity Fact 5.2, for the monomials in (13). First we get rid of α^r: in fact, it is a simple general fact that cupping with a class of degree i, raises the perversity by at most i. It follows that we can concentrate on the case $r = 0$.

Here is the outline of the final analysis.

(1) In order to use Theorem 5.3, we need to make sure that we can test the monomials over linear sections Λ^b which can be traced inside U.

(2) Theorem 5.3, combined with the subadditivity of the Leray filtration implies that we have subadditivity over Λ^b.

(3) We deduce that the subadditivity upper bound on the perversity over Λ^b, when compared with the cohomological degree of the monomial, forces the restricted monomial to be zero, so the monomial passes the test and we are done.

The obstacle in Step 1 is the following: the dimension b of the testing Λ^b increases as $s+t$ increases. On the other hand, by (14) we need $b < 2g-3$. There are plenty of monomials for which b exceeds this bound. We use the explicit nature of the relations I (9) to find an upper bound for $s + t$. The corresponding upper bound for b is $b \leq 2g - 3$ (sic!) and the only class that needs to be tested on a Λ^{2g-3} is β^{g-1}. This class turns out to require a separate ad-hoc analysis. Step 2 requires no further comment. Step 3 is standard as it is based on the cohomological dimension of affine varieties with respect to perverse sheaves.

6. Appendix: cup product and Leray filtration

We would like to give a (more or less) self-contained proof of Theorem 6.1, i.e., of the fact that the cup product is compatible with the Leray spectral sequence. We have been unable to locate a suitable reference in the literature. As it is clear from our discussion in Section 5, this fact is used in an essential way in our proof of Theorem 5.1

As it turns out, the same proof shows that the cup product is also compatible with the p-Leray spectral sequence for every nonpositive perversity $p \leq 0$, including $\mathfrak{p}$. However, this statement turns out to be rather weak, unless we are in the standard case when $p \equiv 0$. For example, in the case of middle perversity, it is off the mark by $+d$ with respect to the subadditivity we need in the proof of Theorem 5.1, as it only implies that $p(\beta^2) \le 4 + d$, whereas $p(\beta^2) = 4$. Nevertheless, it seems worthwhile to give a unified proof valid for every $p \le 0$.

The statement involves the cup product operation on the cohomology groups with coefficients in the direct image complex. It is thus natural to state and prove the compatibility result for the p-standard filtration for arbitrary complexes on varieties. The compatibility for the p-Leray filtration is then an immediate consequence. We employ freely the language of derived categories. We work in the context of constructible complexes on algebraic varieties and, just to fix ideas, with integer coefficients. Let us set up the notation necessary to state Theorem 6.1.

Let $p : \mathbb{Z} \to \mathbb{Z}$ be any function; we call it a *perversity*. Given a partition $X = \coprod S_i$ of a variety X into locally closed nonsingular subvarieties S (strata), we set $p(S) := p(\dim S)$. By considering all possible partitions of X, this data gives rise to a t-structure on D_X (see [Beĭlinson et al. 1982], p. 56). The standard t-structure corresponds to $p(S) \equiv 0$ and the middle perversity t-structure corresponds to the perversity $\mathfrak{p}$ defined by setting $\mathfrak{p}(S) := -\dim S$.

For a given perversity p, the subcategories ${}^p D_X^{\le i}$ for the corresponding t-structure are defined as follows:

$$
{}^p D_X^{\le 0} = \left\{ K \in D_X \mid \mathcal{H}^i(K)_{|S} = 0 \quad \text{for all } i > p(S) \right\},
$$

$$
{}^p D_X^{\le i} := {}^p D^{\le 0}[-i].
$$

If $p = 0$, then ${}^p D_X^{\le 0} = D_X^{\le 0}$ is given by the complexes with zero cohomology sheaves in positive degrees. If $p = \mathfrak{p}$ is the middle perversity, then one shows easily that ${}^{\mathfrak{p}} D_X^{\le 0}$ is given by those complexes K such that $\dim \operatorname{supp} \mathcal{H}^i(K) \le -i$. By using the truncation functors ${}^p \tau_{\le i}$, we can define (see Section 1.1) the p-standard ${}^p \mathcal{S}$ and the p-Leray ${}^p \mathcal{L}$ filtrations.

Let $K, L \in D_X$. The tensor product complex $(K \otimes L, d)$ is defined to be

$$
(K \otimes L)^i := \bigoplus_{a+b=i} K^a \otimes L^b, \qquad d(f_a \otimes g_b) = df \otimes g + (-1)^a f \otimes dg. \quad (15)
$$

The left derived tensor product is a bifunctor $\overset{\mathbb{L}}{\otimes} : D_X \times D_X \to D_X$ defined by first taking a flat resolution $L' \to L$ and then by setting $K \overset{\mathbb{L}}{\otimes} L := K \otimes L'$. If we use field coefficients, then the left-derived tensor product coincides with the ordinary tensor product: $\overset{\mathbb{L}}{\otimes} = \otimes$.

Let

$$
H^a(X, K) \otimes H^b(X, L) \to H^{a+b}(X, K \overset{\mathbb{L}}{\otimes} L) \quad (16)
$$

be the cup product ([Kashiwara and Schapira 1990], p. 134).

The following establishes that the filtration in cohomology associated with a nonpositive perversity is compatible with the cup product operation (16).

Theorem 6.1. *Let $p \leq 0$ be a nonpositive perversity. The p-standard filtration and, for a map $f : X \to Y$, the p-Leray filtration are compatible with the cup product*:

$$^p\mathcal{S}_i H^a(X, K) \otimes {}^p\mathcal{S}_j H^b(X, L) \to {}^p\mathcal{S}_{i+j} H^{a+b}\big(X, K \overset{\mathbb{L}}{\otimes} L\big),$$

$$^p\mathcal{L}_i H^a(X, L) \otimes {}^p\mathcal{L}_j H^b(X, L) \to {}^p\mathcal{L}_{i+j} H^{a+b}\big(X, K \overset{\mathbb{L}}{\otimes} L\big).$$

Theorem 6.1 is proved in Section 6.2. Section 6.3 shows how the cup product and its variants for cohomology with compact supports are related to each other; these variants are listed in (26). Moreover, if we specialize (26) to the case of constant coefficients, and also to the case of the dualizing complex, then we get the usual cup products in cohomology (see the left-hand side of (27)) and the usual cap products involving homology and Borel–Moore homology (see the right-hand side of (27)).

Remark 6.2. The obvious variants of the statement of Theorem 6.1 hold also for each of the variants of the cup product mentioned above. The same is true for Theorem 6.7, which is merely a souped-up version of Theorem 6.1. The reader will have no difficulty repeating, for each of these variants, the proof of Theorems 6.1 and 6.7 given in Section 6.2.

Example 6.3. We consider only the two cases $p \equiv 0$ and $p = \mathfrak{p}$; in the former case we drop the index $p = 0$. Let X be a nonsingular variety of dimension d.

(1) Let $K = L = \mathbb{Z}_X$. Then $1 \in \mathcal{S}_0 H^0$ and $1 \cup 1 = 1 \in \mathcal{S}_0 H^0$.

(2) Let $K = L = \mathbb{Z}_X[d]$. Then $K \overset{\mathbb{L}}{\otimes} L = \mathbb{Z}_X[2d]$. While

$$1 = 1 \cup 1 \in {}^{\mathfrak{p}}\mathcal{S}_{-d} H^{-2d}(X, \mathbb{Z}_X[2d]),$$

Theorem 6.1 only predicts

$$1 \cup 1 \in {}^{\mathfrak{p}}\mathcal{S}_0 H^{-2d}(X, \mathbb{Z}_X[2d]).$$

(3) Let $K = L = \mathbb{Z}_p$, where $p \in X$. We have $1_p = 1_p \cup 1_p \in {}^{\mathfrak{p}}\mathcal{S}_0 H^0(X, \mathbb{Z}_p)$ and this agrees with the prediction of Theorem 6.1.

(4) Let $f : X = Y \times F \to Y$ be the projection, and let $K = L = \mathbb{Q}_X$. We have that

$$\mathcal{L}_i H^a(X, \mathbb{Q}) = \bigoplus_{i' \leq i} \big(H^{a-i'}(Y, \mathbb{Q}) \otimes_{\mathbb{Q}} H^{i'}(F, \mathbb{Q})\big).$$

In this case, Theorem 6.1 is a simple consequence of the compatibility of the Künneth formula with the cup product.

(5) Now let us consider ${}^{p}\mathscr{L}$ for the same projection map $f : Y \times F \to Y$ as above. We have $Rf_*\mathbb{Q} \simeq \bigoplus_{i \geq 0} R^i[-i]$, where R^i is the constant local system $R^i f_*\mathbb{Q}$. Let us assume that Y is nonsingular of pure dimension d. Then ${}^{p}\mathscr{L}_i = \mathscr{L}_{i-d}$, where we use the fact that each $R^i[d]$ is a perverse sheaf due to the nonsingularity of Y (which stems from the one of X). We have that $1 \in {}^{p}\mathscr{L}_d H^0(X, \mathbb{Q})$. On the other hand, Theorem 6.1 predicts only that $1 = 1 \cup 1 \in {}^{p}\mathscr{L}_{2d} H^0(X, \mathbb{Q})$.

These examples, which as the reader can verify are not an illusion due to indexing schemes, show that Theorem 6.1 is indeed sharp. However, its conclusions for ${}^{p}\mathscr{S}$ and ${}^{p}\mathscr{L}$ are often off the mark. See also Remark 6.8.

Remark 6.4. I do not know an example of a map $f : X \to Y$, with X and Y nonsingular, f proper and flat of relative dimension d, for which the cup product on $H^*(X, \mathbb{Q})$ does not satisfy

$$ {}^{p}\mathscr{L}_i \otimes {}^{p}\mathscr{L}_j \to {}^{p}\mathscr{L}_{i+j-d} \tag{17} $$

(Theorem 6.1 predicts that the cup product above lands in the bigger ${}^{p}\mathscr{L}_{i+j}$.) In the paper [de Cataldo et al. 2011] we need to establish (17) for the Hitchin map. If (17) were true a priori, the proof of the main result of our paper [de Cataldo et al. 2011] could be somewhat shortened.

Note also that if the shifted perverse Leray filtration $P := {}^{p}\mathscr{L}(-d)$ (see (10)) for the Hitchin map h coincided a priori with the ordinary Leray filtration $\mathscr{L}$ of the map h, then (17) would follow immediately from the case $p = 0$ of Theorem 6.1. At present, we do not know if $P = \mathscr{L}$ for the Hitchin map. In general, i.e., for a map f as above, we have $\mathscr{L} \subseteq P$, but the inclusion can be strict: e.g., the projection to $\mathbb{P}^1$ of the blowing-up of $\mathbb{P}^2$ at a point, where the class of the exceptional divisor is in P_1, but it is not in $\mathscr{L}_1$.

6.1. *A simple lemma relating tensor product and truncation.*

The key simple fact behind Theorem 6.1 in the standard case when $p \equiv 0$ is that if two complexes K, L lie in $D_X^{\leq 0}$, i.e., if they have nonzero cohomology sheaves in nonnegative degrees only, then the same is true for their derived tensor product.

Lemma 6.5 shows that the Künneth spectral sequence implies that the analogous fact is true for any nonpositive perversity p.

Let us recall the Künneth spectral sequence for the derived tensor product of complexes of sheaves. Define the Tor-sheaves, a collection of bifunctor with variables sheaves A and B, by setting

$$ \mathcal{T}or_i(A, B) := \mathcal{H}^{-i}(A \overset{\mathbb{L}}{\otimes} B). $$

We have $\mathcal{T}or_0(A, B) = A \otimes B$ and $\mathcal{T}or_i(A, B) = 0$ for every $i < 0$. Let $K, L \in D_X$. We have the Künneth spectral sequence (see [Grothendieck 1963, III.2., 6.5.4.2] or [Verdier 1996, p. 7])

$$E_2^{st} = \bigoplus_{a+b=t} \mathcal{T}or_{-s}(\mathcal{H}^a(K), \mathcal{H}^b(L)) \Longrightarrow \mathcal{H}^{s+t}(K \overset{\mathbb{L}}{\otimes} L). \tag{18}$$

This sequence lives in the II-III quadrants, i.e., where $s \le 0$. The edge sequence gives a natural map

$$\bigoplus_{a+b=t} \mathcal{H}^a(K) \otimes \mathcal{H}^b(L) \to \mathcal{H}^t(K \overset{\mathbb{L}}{\otimes} L). \tag{19}$$

Lemma 6.5 (Tensor product and truncation). *Let $p \le 0$ be any nonpositive perversity. Then*

$$\overset{\mathbb{L}}{\otimes} : {}^p D_X^{\le i} \times {}^p D_X^{\le j} \to {}^p D_X^{\le i+j}.$$

Proof. We simplify the notation by dropping the decorations X and p. Since

$$D^{\le i} \overset{\mathbb{L}}{\otimes} D^{\le j} = D^{\le 0}[-i] \overset{\mathbb{L}}{\otimes} D^{\le 0}[-j] = D^{\le 0} \overset{\mathbb{L}}{\otimes} D^{\le 0}[-i-j],$$

it is enough to show that $\overset{\mathbb{L}}{\otimes} : D^{\le 0} \times D^{\le 0} \to D^{\le 0}$. We need to verify that the equality

$$\mathcal{H}^q(K \overset{\mathbb{L}}{\otimes} L)_{|S} = 0 \quad \text{for all } q > p(S) \tag{20}$$

holds as soon as the same equality is assumed to hold for K and L.

It is enough to prove the analogous equality for the Tor-sheaves on the left-hand side of (18).

Note that $p \le 0$ implies that if $L \in D^{\le 0}$, then $\mathcal{H}^b(L) = 0$ for every $b > 0$. Let $\sigma \ge 0$ and consider

$$\bigoplus_{a+b=q+\sigma} \mathcal{T}or_\sigma \left(\mathcal{H}^a(K)_{|S}, \mathcal{H}^b(L)_{|S} \right) \quad \text{for all } q > p(S).$$

If $a > p(S)$, then $\mathcal{H}^a(K)_{|S} = 0$.

If $a \le p(S)$, then $b = q - a + \sigma > \sigma \ge 0$, so that $\mathcal{H}^b(L) = 0$. $\qquad\square$

6.2. *Spectral sequences and multiplicativity.*

In this section we prove Theorem 6.1 and we also observe that it is the reflection at the level of the abutted filtrations of the more general statement Theorem 6.7 involving spectral sequences.

The most efficient formulation is perhaps the one involving the filtered derived category $D_X F$. We shall quote freely from [Illusie 1971], pp. 285–288. To fix ideas, we deal with the cup product in cohomology. The formulations for the other products in Section 6.3 are analogous.

Let (K, F_1) and (L, F_2) be two filtered complexes. The filtered derived tensor product

$$(K \overset{\mathbb{L}}{\otimes} L, F_{12})$$

is defined as follows. Let $(L', F_2') \to (L, F_2)$ be a left flat filtered resolution. Define

$$K \overset{\mathbb{L}}{\otimes} L := K \otimes L'$$

and define F_{12} to be the product filtration of F_1 and F_2'. We have natural isomorphisms

$$\bigoplus_{s+s'=\sigma} (\mathrm{Gr}_{F_1}^s K \overset{\mathbb{L}}{\otimes} \mathrm{Gr}_{F_2}^{s'} L) \overset{\simeq}{\to} \mathrm{Gr}_{F_{12}}^\sigma (K \overset{\mathbb{L}}{\otimes} L). \tag{21}$$

We have the filtered version of [Kashiwara and Schapira 1990, p. 134], i.e., a map in $D_{\mathrm{pt}} F$

$$(R\Gamma(X, K), F_1) \overset{\mathbb{L}}{\otimes} (R\Gamma(X, L), F_2) \to (R\Gamma(X, K \overset{\mathbb{L}}{\otimes} L), F_{12}) \tag{22}$$

inducing (see Ex. I.24a of the same reference) the filtered cup product map

$$(H^a(X, K), F_1) \otimes (H^b(X, L), F_2) \to (H^{a+b}(X, K \overset{\mathbb{L}}{\otimes} L), F_{12}). \tag{23}$$

In view of Theorem 6.7, by first recalling the notion of bilinear pairing of spectral sequences [Spanier 1966, p. 491], we have a bilinear pairing of spectral sequences

$$E_1^{st}(K, F_1) \otimes E_1^{s't'}(L, F_2) \to E_1^{s+s', t+t'}(K \overset{\mathbb{L}}{\otimes} L, F_{12}) \tag{24}$$

that on the E_1-term coincides with the cup product map (16) induced by (21), and on the E_∞-term is the graded cup product associated with the filtered cup product (23).

Given $(M, F) \in D_X F$, we have the spectral sequence

$$E_1^{st} = E_1^{st}(M, F) = II^{s+t}(X, \mathrm{Gr}_F^s M) \Longrightarrow II^{s+t}(X, M), \quad E_\infty^{st} = \mathrm{Gr}_F^s II^{s+t}(X, M),$$

with abutment the filtration induced by (M, F) on the cohomology groups $H^*(X, M)$. Clearly, we can always compose with the map of spectral sequences induced by any filtered map $(K \overset{\mathbb{L}}{\otimes} L, F_{12}) \to (M, F)$.

We apply the machinery above to the case when the filtrations F_i are the p-standard decreasing filtrations $^p\mathscr{S}$ induced by the t-structure associated with a nonpositive perversity $p \leq 0$. The construction of $^p\mathcal{S}$ is performed via the use of injective resolutions [de Cataldo and Migliorini 2010, §3.1]. The product filtration F_{12} on the derived tensor product is *not* the p-standard filtration, not even up to isomorphism in the filtered derived category $D_X F$; see Remark 6.8.

The upshot of this discussion is that Lemma 6.5 implies this:

Lemma 6.6. *There is a canonical lift*

$$u : \left(K \overset{\mathbb{L}}{\otimes} L, F_{12}\right) \to \left(K \overset{\mathbb{L}}{\otimes} L, {}^p\mathcal{S}\right)$$

of the identity on $K \overset{\mathbb{L}}{\otimes} L$ to $D_X F$.

Proof. Let N denote the derived tensor product of K with L. It is enough to show that $F_{12}^{\sigma} N \in {}^p D_X^{\leq -\sigma}$, for every $\sigma \in \mathbb{Z}$. We prove this by decreasing induction on σ. The statement is clearly true for $\sigma \gg 0$. We have the short exact sequence

$$0 \to F_{12}^{\sigma+1} N \to F_{12}^{\sigma} N \to \mathrm{Gr}_{F_{12}}^{\sigma} N \to 0.$$

Lemma 6.5 implies that $\mathrm{Gr}_{F_{12}}^{\sigma} N \in {}^p D_X^{\leq -\sigma}$ and the inductive hypothesis gives $F_{12}^{\sigma+1} N \in {}^p D_X^{\leq -\sigma-1} \subseteq {}^p D_X^{\leq -\sigma}$. We have the following simple fact: if

$$A \to B \to C \to A[1]$$

is a distinguished triangle and $A, C \in {}^p D_X^{\leq i}$, then $B \in {}^p D_X^{\leq i}$. We conclude the proof by applying this fact to the distinguished triangle associated with the short exact sequence above. $\qquad\square$

Proof of Theorem 6.1. Apply the construction (23) to the case $F_i = {}^p\mathcal{S}$. Compose the resulting filtered cup product map with the canonical lift u of Lemma 6.6 and obtain the filtered cup product map of Theorem 6.1. $\qquad\square$

As mentioned earlier, Theorem 6.1 is the abutted reflection of the following statement concerning spectral sequences:

Theorem 6.7. *There is a natural bilinear pairing of spectral sequences*

$$E_1^{st}(K, {}^p\mathcal{S}) \otimes E_1^{s't'}(L, {}^p\mathcal{S}) \to E_1^{s+s', t+t'}\left(K \overset{\mathbb{L}}{\otimes} L, {}^p\mathcal{S}\right)$$

such that:

(1) *on the E_1-term it coincides with the cup product map induced by* (21), *which in this case reads*

$$ {}^p\mathcal{H}^{-s}(K)[s] \overset{\mathbb{L}}{\otimes} {}^p\mathcal{H}^{-s'}(K)[s'] \to {}^p\mathcal{H}^{-s-s'}\left(K \overset{\mathbb{L}}{\otimes} L\right)[s+s'], \qquad (25)$$

(2) *on the E_{∞}-term it is the graded cup product associated with the filtered cup product* (23).

Proof. Compose (24) with the map of spectral sequences induced by the canonical map u of Lemma 6.6. $\qquad\qquad\qquad\qquad\qquad\qquad\qquad\qquad\qquad\qquad\qquad\qquad$ □

Remark 6.8. Unless we are in the case $p \equiv 0$, the product filtration F_{12} of the p-standard filtrations is often strictly smaller than the p-standard filtration. As a result, the graded pairing is often trivial. One can see this on the E_1-page in terms of the map (25). Here is an example. Let X be nonsingular of pure dimension d, take middle perversity and perverse complexes $K = L = \mathbb{Q}_X[d]$. The pairing in question is

$$\mathbb{Q}_X[d] \otimes \mathbb{Q}_X[d] \to {}^{p}\mathcal{H}^0(\mathbb{Q}_X)[2d] = 0.$$

6.3. *Cup and cap.* The methods employed in the previous sections are of course susceptible of being applied to the other usual constructions, such as the cup product in cohomology with compact supports and cap products in homology and in Borel–Moore homology.

By taking various flavors of (22) with compact supports, we obtain the commutative diagram of cup product maps

$$
\begin{array}{ccc}
H_c^i(X, K) \otimes H_c^j(X, L) & \longrightarrow & H_c^{i+j}(X, K \overset{\mathbb{L}}{\otimes} L) \\
\downarrow & \searrow & \Big\| = \\
H^i(X, K) \otimes H_c^j(X, L) & \longrightarrow & H_c^{i+j}(X, K \overset{\mathbb{L}}{\otimes} L) \\
\downarrow & \searrow & \downarrow \\
H^i(X, K) \otimes H^j(X, L) & \longrightarrow & H^{i+j}(X, K \overset{\mathbb{L}}{\otimes} L).
\end{array}
\qquad (26)
$$

Theorem 6.7 applies to each row, each vertical arrow is a filtered map for the product filtrations and, as a result, the conclusion of Theorem 6.7 apply to the diagonal products as well.

We have the following important special cases. Take K and L to be either $\mathbb{Z}_X$ and/or ω_X (the Verdier dualizing complex of X). We have

$$\mathbb{Z}_X \overset{\mathbb{L}}{\otimes} \omega_X = \omega_X$$

as well as the following equalities (decorations omitted):

$$H^i(X, \mathbb{Z}) = H^i(X, \mathbb{Z}_X) = H^i \mathbb{Z}_X = H^i,$$
$$H_c^i = H_c^i \mathbb{Z}_X, \qquad H_i = H_c^{-i} \omega_X, \qquad H_i^{BM} = H^{-i} \omega_X.$$

Then we have the following commutative diagrams expressing the well-known compatibilities of the cup and cap products:

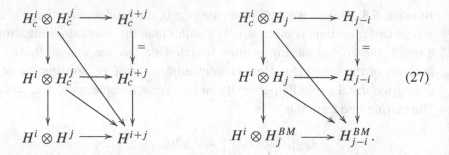

$$(27)$$

One also has the variants in relative cohomology and in relative cohomology with compact supports, the variants with supports on locally closed subvarieties, as well as the variants involving a map $f : X \to Y$ (e.g., $H^*(X)$ as a $H^*(Y)$-module etc). The reader can sort these variants out.

Acknowledgments

I thank D. Nadler for remarks on the support theorem, T. Hausel and L. Migliorini for their very helpful comments, and the referee for excellent suggestions and remarks.

References

[Arapura 2005] D. Arapura, "The Leray spectral sequence is motivic", *Invent. Math.* **160**:3 (2005), 567–589. MR 2006m:14025

[Barth et al. 1984] W. Barth, C. Peters, and A. Van de Ven, *Compact complex surfaces*, Ergebnisse der Math. (3) **4**, Springer, Berlin, 1984. MR 86c:32026

[Beĭlinson et al. 1982] A. A. Beĭlinson, J. Bernstein, and P. Deligne, "Faisceaux pervers", pp. 5–171 in *Analysis and topology on singular spaces* (Luminy, 1981), vol. I, Astérisque **100**, Soc. Math. France, Paris, 1982. MR 86g:32015

[de Cataldo 2009] M. A. A. de Cataldo, "The standard filtration on cohomology with compact supports", pp. 199–220 in *Interactions of classical and numerical algebraic geometry*, Contemp. Math. **496**, Amer. Math. Soc., Providence, RI, 2009. MR 2010j:14036

[de Cataldo 2010] M. A. de Cataldo, "The perverse filtration and the Lefschetz hyperplane theorem, II", 2010. To appear in *J. Alg. Geom.* arXiv 1006.2582

[de Cataldo and Migliorini 2009] M. A. A. de Cataldo and L. Migliorini, "The decomposition theorem, perverse sheaves and the topology of algebraic maps", *Bull. Amer. Math. Soc. (N.S.)* **46**:4 (2009), 535–633. MR 2011a:14012

[de Cataldo and Migliorini 2010] M. A. A. de Cataldo and L. Migliorini, "The perverse filtration and the Lefschetz hyperplane theorem", *Ann. of Math.* (2) **171**:3 (2010), 2089–2113. MR 2011h:14020

[de Cataldo et al. 2010] M. A. de Cataldo, T. Hausel, and L. Migliorini, "Exchange between perverse and weight filtrations for the Hilbert schemes of points of two surfaces", 2010. arXiv 1012:2583

[de Cataldo et al. 2011] M. A. de Cataldo, T. Hausel, and L. Migliorini, "Topology of Hitchin systems and Hodge theory of character varieties", 2011. arXiv 1004.1420v2

[Deligne 1968] P. Deligne, "Théorème de Lefschetz et critères de dégénérescence de suites spectrales", *Inst. Hautes Études Sci. Publ. Math.* **35** (1968), 259–278. MR 39 #5582

[Grothendieck 1963] A. Grothendieck, "Éléments de géométrie algébrique, III: Étude cohomologique des faisceaux cohérents, II", *Inst. Hautes Études Sci. Publ. Math.* **17** (1963), 91. MR 29 #1210

[Hausel and Rodriguez-Villegas 2008] T. Hausel and F. Rodriguez-Villegas, "Mixed Hodge polynomials of character varieties", *Invent. Math.* **174**:3 (2008), 555–624. MR 2010b:14094

[Hausel and Thaddeus 2003] T. Hausel and M. Thaddeus, "Relations in the cohomology ring of the moduli space of rank 2 Higgs bundles", *J. Amer. Math. Soc.* **16**:2 (2003), 303–327. MR 2004b:14055

[Illusie 1971] L. Illusie, *Complexe cotangent et déformations, I*, Lecture Notes in Mathematics **239**, Springer, Berlin, 1971. MR 58 #10886a

[Kashiwara and Schapira 1990] M. Kashiwara and P. Schapira, *Sheaves on manifolds*, Grundlehren der Math. Wissenschaften **292**, Springer, Berlin, 1990. MR 92a:58132

[Nadler 2010] D. Nadler, "The geometric nature of the Fundamental Lemma", preprint, 2010. arXiv 1009.1862

[Ngô 2008] B. C. Ngô, "Le lemme fondamental pour les algèbres de Lie", 2008. arXiv 0801.0446

[Saito 1990] M. Saito, "Mixed Hodge modules", *Publ. Res. Inst. Math. Sci.* **26**:2 (1990), 221–333. MR 91m:14014

[Simpson 1992] C. T. Simpson, "Higgs bundles and local systems", *Inst. Hautes Études Sci. Publ. Math.* **75** (1992), 5–95. MR 94d:32027

[Spanier 1966] E. H. Spanier, *Algebraic topology*, McGraw-Hill, New York, 1966. MR 35 #1007

[Thaddeus 1990] M. Thaddeus, *Topology of the moduli space of stable bundles over a compact Riemann surface*, Diploma Thesis, Oxford University, 1990.

[Verdier 1996] J.-L. Verdier, *Dès catégories dérivées des catégories abéliennes*, Astérisque **239**, Soc. Math. France, Paris, 1996. MR 98c:18007

mde@math.sunysb.edu   *Department of mathematics, Stony Brook University, Stony Brook, NY 11794-3651, United States*

Current Developments in Algebraic Geometry
MSRI Publications
Volume 59, 2011

Rigidity properties of Fano varieties

TOMMASO DE FERNEX AND CHRISTOPHER D. HACON

We overview some recent results on Fano varieties giving evidence of their rigid nature under small deformations.

1. Introduction

From the point of view of the Minimal Model Program, Fano varieties constitute the building blocks of uniruled varieties. Important information on the biregular and birational geometry of a Fano variety is encoded, via Mori theory, in certain combinatorial data corresponding to the Néron–Severi space of the variety. It turns out that, even when there is actual variation in moduli, much of this combinatorial data remains unaltered, provided that the singularities are "mild" in an appropriate sense. One should regard any statement of this sort as a rigidity property of Fano varieties.

This paper gives an overview of Fano varieties, recalling some of their most important properties and discussing their rigid nature under small deformations. We will keep a colloquial tone, referring the reader to the appropriate references for many of the proofs. Our main purpose is indeed to give a broad overview of some of the interesting features of this special class of varieties. Throughout the paper, we work over the complex numbers.

2. General properties of Fano varieties

A *Fano manifold* is a projective manifold X whose anticanonical line bundle $-K_X := \bigwedge^n T_X$ is ample (here $n = \dim X$).

The simplest examples of Fano manifolds are given by the projective spaces $\mathbb{P}^n$. In this case, in fact, even the tangent space is ample. (By [Mori 1979], we know that projective spaces are the only manifolds with this property.)

In dimension two, Fano manifolds are known as del Pezzo surfaces. This class of surfaces has been widely studied in the literature (it suffices to mention

The first author was partially supported by NSF CAREER Grant DMS-0847059. The second author was supported by NSF Grant 0757897.
MSC2000: primary 14J45; secondary 14E22, 14D15.
Keywords: Fano varieties, Mori theory.

that several books have been written just on cubic surfaces), and their geometry is quite well understood. There are ten families of del Pezzo surfaces. The following theorem, obtained as a result of a series of papers [Nadel 1990; 1991; Campana 1991; 1992; Kollár et al. 1992a; 1992b], shows that this is a general phenomenon.

Theorem 2.1. *For every n, there are only finitely many families of Fano manifolds of dimension n.*

This theorem is based on the analysis of rational curves on Fano manifolds. In this direction, we should also mention the following important result:

Theorem 2.2 [Campana 1992; Kollár et al. 1992a]. *Fano manifolds are rationally connected.*

Fano varieties arise naturally in the context of the Minimal Model Program. This however leads us to work with possibly singular varieties. The smallest class of singularities that one has to allow is that of $\mathbb{Q}$-factorial varieties with terminal singularities. However, one can enlarge the class of singularities further, and work with $\mathbb{Q}$-Gorenstein varieties with log terminal (or, in some cases, even log canonical) singularities. In either case, one needs to consider $-K_X$ as a Weil divisor. The hypothesis guarantees that some positive multiple $-mK_X$ is Cartier (i.e., $\mathcal{O}_X(-mK_X)$ is a line bundle), so that one can impose the condition of ampleness.

For us, a *Fano variety* will be a variety with $\mathbb{Q}$-Gorenstein log terminal singularities such that $-K_X$ is ample. We will however be mostly interested in the case where the singularities are $\mathbb{Q}$-factorial and terminal.

The above results are however more delicate in the singular case. By a recent result of Zhang [2006], it is known that Fano varieties are rationally connected (see [Hacon and Mckernan 2007] for a related statement). Boundedness of Fano varieties is instead an open problem. The example of a cone over a rational curve of degree d shows that even for surfaces, we must make some additional assumptions. In this example one has that the minimal log discrepancies are given by $1/d$. One may hope that if we bound the minimal log discrepancies away from 0 the boundedness still holds. More precisely, the BAB conjecture (due to Alexeev, Borisov, Borisov) states that for every $n > 0$ and any $\epsilon > 0$, there are only finitely many families of Fano varieties of dimension n with ϵ-log terminal singularities (in particular, according to this conjecture, for every n there are only finitely many families of Fano varieties with canonical singularities).

Note that, by Theorem 2.2, it follows that Fano manifolds have the same cohomological invariants as rational varieties (namely $h^i(\mathcal{O}_X) = h^0(\Omega_X^q) = 0$ for all $i, q > 0$). On the other hand, by celebrated results of Iskovskikh and Manin [1971] and of Clemens and Griffiths [1972], it is known that there are examples of Fano manifolds that are nonrational. The search for these examples was motivated by the Lüroth Problem. Note that it is still an open problem to find examples of Fano manifolds that are not unirational.

Perhaps the most important result known to hold for Fano varieties (for mild singularities and independently of their dimension), concerns the combinatorial structure associated to the cone of effective curves. The first instance of this was discovered by Mori [1982] in the smooth case. It is a particular case of the Cone Theorem (which holds for all varieties with log terminal singularities).

Theorem 2.3 (Cone theorem for Fano varieties). *The Mori cone of a Fano variety is rational polyhedral, generated by classes of rational curves.*

Naturally one may also ask if there are similar results concerning the structure of other cones of curves. From a dual perspective, one would like to understand the structure of the various cones of divisors on a Fano variety. The strongest result along these lines was conjectured by Hu and Keel [2000] and recently proved by Birkar, Cascini, Hacon, and McKernan:

Theorem 2.4 [Birkar et al. 2010]. *Fano varieties are Mori dream spaces in the sense of Hu and Keel.*

The meaning and impact of these results will be discussed in the next section.

3. Mori-theoretic point of view

Let X be a normal projective variety and consider the dual $\mathbb{R}$-vector spaces

$$N_1(X) := (Z_1(X)/\equiv) \otimes \mathbb{R} \quad \text{and} \quad N^1(X) := (\text{Pic}(X)/\equiv) \otimes \mathbb{R},$$

where $\equiv$ denotes numerical equivalence. The *Mori cone* of X is the closure $\overline{\text{NE}}(X) \subset N_1(X)$ of the cone spanned by classes of effective curves. Its dual cone is the *nef cone* $\text{Nef}(X) \subset N^1(X)$, which by Kleiman's criterion is the closure of the cone spanned by ample classes. The closure of the cone spanned by effective classes in $N^1(X)$ is the *pseudo-effective cone* $\text{PEff}(X)$. Sitting in between the nef cone and the pseudo-effective cone is the *movable cone of divisors* $\text{Mov}(X)$, given by the closure of the cone spanned by classes of divisors moving in a linear system with no fixed components. All of these cones,

$$\text{Nef}(X) \subset \text{Mov}(X) \subset \text{PEff}(X) \subset N^1(X),$$

carry important geometric information about the variety X.

The Cone Theorem says that $\overline{\text{NE}}(X)$ is generated by the set of its K_X positive classes $\overline{\text{NE}}(X)_{K_X \geq 0} = \{\alpha \in \overline{\text{NE}}(X) | K_X \cdot \alpha \geq 0\}$ and at most countably many K_X negative rational curves $C_i \subset X$ of bounded anti-canonical degree $0 < -K_X \cdot C_i \leq 2 \dim(X)$. In particular the only accumulation points for the curve classes $[C_i]$

in $\overline{\mathrm{NE}}(X)$ lie along the hyperplane determined by $K_X \cdot \alpha = 0$. Thus, for a Fano variety, the Mori cone $\overline{\mathrm{NE}}(X)$ is a rational polyhedral cone. By duality, it follows that the nef cone $\mathrm{Nef}(X) = (\overline{\mathrm{NE}}(X))^\vee$ is also a rational polyhedral cone.

The geometry of X is reflected to a large extent in the combinatorial properties of $\overline{\mathrm{NE}}(X)$. Every extremal face F of $\overline{\mathrm{NE}}(X)$ corresponds to a surjective morphism $\mathrm{cont}_F \colon X \to Y$, which is called a *Mori contraction*. The morphism cont_F contracts precisely those curves on X with class in F. Conversely, any morphism with connected fibers onto a normal variety arises in this way.

Remark 3.1. When X is not a Fano variety, $\overline{\mathrm{NE}}(X)_{K_X < 0}$ may fail to be finitely generated, and even in very explicit examples such as blow-ups of $\mathbb{P}^2$, the structure of the K_X positive part of the Mori cone is in general unknown. Consider, for example, the long-standing open conjectures of Nagata and Segre–Harbourne–Gimigliano–Hirschowitz. Also, the fact that the extermal faces of the cone of curves can always be contracted is only known when X is Fano (or, more generally, "log Fano").

A similar behavior, that we will now describe, also occurs for the cone of nef curves. By definition the cone of nef curves $\overline{\mathrm{NM}}(X) \subset N_1(X)$ is the closure of the cone generated by curves belonging to a covering family (a family of curves that dominates the variety X). It is clear that if $\alpha \in \overline{\mathrm{NM}}(X)$ and D is an effective Cartier divisor on X, then $\alpha \cdot D \geq 0$. It follows that $\alpha \cdot D \geq 0$ for any pseudo-effective divisor D on X. We have this remarkable result:

Theorem 3.2 [Boucksom et al. 2004]. *The cone of nef curves is dual to the cone of pseudo-effective divisors, i.e.,* $\overline{\mathrm{NM}}(X) = \mathrm{PEff}(X)^\vee$.

We now turn our attention to the case of $\mathbb{Q}$-factorial Fano varieties. In this case, the cone of nef curves $\overline{\mathrm{NM}}(X)$ is also rational polyhedral and every extremal ray corresponds to a *Mori fiber space* $X' \to Y'$ on a model X' birational to X. More precisely:

Theorem 3.3 [Birkar et al. 2010, 1.3.5]. *R is an extremal ray of* $\overline{\mathrm{NM}}(X)$ *if and only if there exists a $\mathbb{Q}$-divisor D such that (X, D) is Kawamata log terminal, and a $(K_X + D)$ Minimal Model Program $X \dashrightarrow X'$ ending with a Mori fiber space $X' \to Y'$, such that the numerical transform of any curve in the fibers of $X' \to Y'$ (e.g., the proper transform of a general complete intersection curve on a general fiber of $X' \to Y'$) has class in R.*

We will refer to the induced rational map $X \dashrightarrow Y'$ as a *birational Mori fiber structure* on X. We stress that we only consider Mori fiber structures that are the output of a Minimal Model Program. This was first studied by Batyrev [1992] in dimension three. The picture in higher dimensions was recently established by Birkar–Cascini–Hacon–McKernan for Fano varieties and, in a more general context, by Araujo [2010] and Lehman [2008]. As a side note, even if it is known that the fibers of any Mori fibration $X' \to Y'$ are covered by rational curves, it still remains an open question whether the extremal rays of $\overline{\text{NM}}(X)$ are spanned by classes of rational curves. This is related to a delicate question on the rational connectivity of the smooth locus of singular varieties.

The dual point of view (looking at $N^1(X)$ rather than $N_1(X)$), also offers a natural way of refining the above results. As mentioned above, if X is a $\mathbb{Q}$-factorial Fano variety, then it is a *Mori dream space* [Hu and Keel 2000; Birkar et al. 2010]. The movable cone $\text{Mov}(X)$ of a Mori dream space admits a finite decomposition into rational polyhedral cones, called *Mori chambers*. One of these chambers is the nef cone of X. The other chambers are given by nef cones of $\mathbb{Q}$-factorial birational models $X' \sim_{\text{bir}} X$ which are isomorphic to X in codimension one. Note indeed that any such map gives a canonical isomorphism between $N^1(X)$ and $N^1(X')$. Wall-crossings between contiguous Mori chambers correspond to flops (or flips, according to the choice of the log pair structure) between the corresponding birational models. We can therefore view the Mori chamber decomposition of $\text{Mov}(X)$ as encoding information not only on the biregular structure of X but on its birational structure as well.

There is a way of recovering all this information from the total coordinate ring, or Cox ring, of a Mori dream space X, via a GIT construction. For simplicity, we assume that the map $\text{Pic}(X) \to N^1(X)$ is an isomorphism and that the class group of Weil divisors $\text{Cl}(X)$ of X is finitely generated. These properties hold if X is a Fano variety. The property that $\text{Pic}(X) \cong N^1(X)$ simply follows by the vanishing of $H^i(X, \mathcal{O}_X)$ for $i > 0$. The finite generation of $\text{Cl}(X)$ is instead a deeper property; a proof can be found in [Totaro 2009]. Specifically, see Theorem 3.1 there, which implies that the natural map $\text{Cl}(X) \to H_{2n-2}(X, \mathbb{Z})$ is an isomorphism for any n-dimensional Fano variety X (recall that in our definition of Fano variety we assume that the singularities are log terminal).

A *Cox ring* of X is, as defined in [Hu and Keel 2000], a ring of the type

$$R(L_1, \ldots, L_r) := \bigoplus_{m \in \mathbb{Z}^\rho} H^0(X, \mathcal{O}_X(m_1 L_1 + \cdots + m_\rho L_\rho)),$$

for any choice of line bundles $L_1, \ldots, L_\rho$ inducing a basis of $N^1(X)$. Here $\rho = \rho(X)$ is the Picard number of X, and $m = (m_1, \ldots, m_r)$. We will call the *full Cox ring* of X the ring

$$R(X) := \bigoplus_{[D] \in \text{Cl}(X)} H^0(X, \mathcal{O}_X(D)).$$

If X is factorial (that is, if the map $\text{Pic}(X) \to \text{Cl}(X)$ is an isomorphism) and the line bundles L_i induce a basis of the Picard group, then the two rings coincide.

These rings were first systematically studied by Cox [1995] when X is a toric variety. If X is a toric variety and Δ is the fan of X, then the full Cox ring is the polynomial ring

$$R(X) = \mathbb{C}[x_\lambda \mid \lambda \in \Delta(1)],$$

where each x_λ defines a prime toric invariant divisor of X. When X is smooth, this property characterizes toric varieties. More precisely:

Theorem 3.4 [Hu and Keel 2000]. *Assume that X is a smooth Mori dream space. Then $R(X)$ is isomorphic to a polynomial ring if and only if X is a toric variety.*

More generally, Hu and Keel prove that a $\mathbb{Q}$-factorial Mori dream space X can be recovered from any of its Cox rings via a GIT construction. Moreover, the Mori chamber decomposition of X descends to X via this construction from a chamber decomposition associated to variations of linearizations in the GIT setting. From this perspective, the Cox ring of a Fano variety is a very rich invariant, encoding all essential information on the biregular and birational geometry of the variety.

The above discussion shows how the main features of the geometry of a Fano variety X, both from a biregular and a birational standpoint, are encoded in combinatorial data embedded in the spaces $N_1(X)$ and $N^1(X)$. Loosely speaking, we will say that geometric properties of X that are captured by such combinatorial data constitute the *Mori structure* of X.

In the remaining part of the paper, we will discuss to what extent the Mori structure of a Fano variety is preserved under flat deformations. Any positive result in this direction should be thought of as a rigidity statement.

The following result is the first strong evidence that Fano varieties should behave in a somewhat rigid way under deformations:

Theorem 3.5 [Wiśniewski 1991; 2009]. *The nef cone is locally constant in smooth families of Fano varieties.*

First notice that if $f : X \to T$ is a smooth family of Fano varieties, then f is topologically trivial, and thus, if we denote by $X_t := f^{-1}(t)$ the fiber over t, the space $N^1(X_t)$, being naturally isomorphic to $H^2(X_t, \mathbb{R})$, varies in a local system. By the polyhedrality of the nef cone, this local system has finite monodromy. This implies that, after suitable étale base change, one can reduce to a setting where the spaces $N^1(X_t)$ are all naturally isomorphic. The local constancy can therefore be intended in the étale topology.

Wiśniewski's result is the underlying motivation for the results that will be discussed in the following sections.

4. Deformations of the Cox rings

The proof of Theorem 3.5 has three main ingredients: the theory of deformations of embedded rational curves, Ehresmann's Theorem, and the Hard Lefschetz Theorem. All these ingredients use in an essential way the fact that the family is smooth. On the other hand, the very definitions involved in the whole Mori structure of a Fano variety use steps in the Minimal Model Program, which unavoidably generate singularities. With this in mind, we will present a different approach to the general problem of studying the deformation of Mori structures. The main ingredients of this approach will be the use of the Minimal Model Program in families, and an extension theorem for sections of line bundles (and, more generally, of divisorial reflexive sheaves). The first implications of such approach will be on the Cox rings. These applications will be discussed in this section. Further applications will then presented in the following section.

When working with families of singular Fano varieties, one needs to be very cautious. This is evident for instance in the simple example of quadric surfaces degenerating to a quadric cone: in this case allowing even the simplest surface singularity creates critical problems (the Picard number dropping in the central fiber), yielding a setting where the questions themselves cannot be posed.

We will restrict ourselves to the smallest category of singularities which is preserved in the Minimal Model Program, that of $\mathbb{Q}$-factorial terminal singularities. This is the setting considered in [de Fernex and Hacon 2011]. As explained in [Totaro 2009], many of the results presented below hold in fact under weaker assumptions on the singularities.

We consider a small flat deformation $f : X \to T$ of a Fano variety X_0. Here T is a smooth curve with a distinguished point $0 \in T$, and $X_0 = f^{-1}(0)$. We assume that X_0 has terminal $\mathbb{Q}$-factorial singularities. A proof of the following basic result can be found in [de Fernex and Hacon 2011, Corollary 3.2 and Proposition 3.8], where an analogous but less trivial result is also proven to hold for small flat deformations of weak log Fano varieties with terminal $\mathbb{Q}$-factorial singularities.

Proposition 4.1. *For every t in a neighborhood of 0 in T, the fiber X_t is a Fano variety with terminal $\mathbb{Q}$-factorial singularities.*

After shrinking T near 0, we can therefore assume that $f : X \to T$ is a flat family of Fano varieties with terminal $\mathbb{Q}$-factorial singularities. If $t \in T$ is a general point, the monodromy on $N^1(X_t)$ has finite order. This can be seen using the fact that the monodromy action preserves the nef cone of X_t, which is finitely generated and spans the whole space. After a suitable base change, one may always reduce to a setting where the monodromy is trivial.

If f is a smooth family, then it is topologically trivial, and we have already noticed that the spaces $N^1(X_t)$ vary in a local system. We have remarked how in general the dimension of these spaces may jump if f is not smooth. Under our assumptions on singularities the property remains however true. The proof of the following result is given in [de Fernex and Hacon 2011, Proposition 6.5], and builds upon results form [Kollár and Mori 1992, (12.10)].

Theorem 4.2. *The spaces $N^1(X_t)$ and $N_1(X_t)$ form local systems on T with finite monodromy. After suitable base change, for every $t \in T$ there are natural isomorphisms $N^1(X/T) \cong N^1(X_t)$ and $N_1(X_t) \cong N_1(X/T)$ induced, respectively, by pull-back and push-forward.*

A similar property holds for the class group, and is stated next. The proof of this property is given in [de Fernex and Hacon 2011, Lemma 7.2], and uses the previous result in combination with a generalization of the Lefschetz hyperplane theorem of Ravindra and Srinivas [2006] (the statement is only given for toric varieties, but the proof works in general). As shown in [Totaro 2009, Theorem 4.1], the same result holds more generally, only imposing that X is a projective variety with rational singularities and $H^1(X, \mathcal{O}_X) = H^2(X, \mathcal{O}_X) = 0$ (these conditions hold for any Fano variety) and that X_0 is smooth in codimension 2 and $\mathbb{Q}$-factorial in codimension three.

Theorem 4.3. *With the same assumptions as in Theorem 4.2, the class groups $\mathrm{Cl}(X_t)$ form a local system on T with finite monodromy. After suitable base change, for every $t \in T$ there are natural isomorphisms $\mathrm{Cl}(X/T) \cong \mathrm{Cl}(X_t)$ induced by restricting Weil divisors to the fiber (the restriction is well-defined as the fibers are smooth in codimension one and their regular locus is contained in the regular locus of X).*

For simplicity, we henceforth assume that the monodromy is trivial. It follows by the first theorem that one can fix a common grading for Cox rings of the fibers X_t of the type considered in [Hu and Keel 2000]. The second theorem implies that the there is also a common grading, given by $\mathrm{Cl}(X/T)$, for the full Cox rings of the fibers. This is the first step needed to control the Cox rings along the deformation. The second ingredient is the following extension theorem.

Theorem 4.4. *With the above assumptions, let L be any Weil divisor on X that does not contain any fiber of f in its support. Then, after possibly restricting T (and consequently X) to a neighborhood of 0, the restriction map*

$$H^0(X, \mathcal{O}_X(L)) \to H^0(X_0, \mathcal{O}_{X_0}(L|_{X_0}))$$

is surjective (here $L|_{X_0}$ denotes the restriction of L to X_0 as a Weil divisor).

When L is Cartier, this theorem is a small generalization of Siu's invariance of plurigenera for varieties of general type. The formulation for Weil divisors follows by a more general result of [de Fernex and Hacon 2011], which is recalled below in Theorem 5.4.

As a corollary of the above theorems, we obtain the flatness of the Cox rings.

Corollary 4.5. *The full Cox ring $R(X_0)$ of X_0, or any Cox ring $R(L_{0,1}, \ldots, L_{0,\rho})$ of X_0 (provided the line bundles $L_{0,i}$ on X_0 are sufficiently divisible), deforms flatly in the family.*

The flatness of the deformation of the full Cox ring has a very interesting consequence when applied to deformations of toric Fano varieties.

Corollary 4.6. *Simplicial toric Fano varieties with terminal singularities are rigid.*

The proof of this corollary is based on the simple observation that a polynomial ring has no non-isotrivial flat deformations. This theorem appears in [de Fernex and Hacon 2011]. When X is smooth, the result was already known, and follows by a more general result of Bien and Brion on the vanishing of $H^1(X, T_X)$ for any smooth projective variety admitting a toroidal embedding (these are also known as *regular varieties*). The condition that the toric variety is simplicial is the translation, in toric geometry, of the assumption of $\mathbb{Q}$-factoriality. The above rigidity result holds in fact more in general, only assuming that the toric Fano variety is smooth in codimension 2 and $\mathbb{Q}$-factorial in codimension 3. This was proven in [Totaro 2009] using the vanishing theorems of Danilov and Mustaţă $H^i(\tilde{\Omega}^j \otimes \mathcal{O}(D)^{**}) = 0$ for $i > 0$, $j > 0$ and D an ample Weil divisor on a projective toric variety.

The above results can also be used to show that also the Picard group is locally constant.

Corollary 4.7. *With the same assumptions as in Theorem 4.2, the Picard groups $\mathrm{Pic}(X_t)$ form a local system on T with finite monodromy. After suitable base change, for every $t \in T$ there are natural isomorphisms $\mathrm{Pic}(X/T) \cong \mathrm{Pic}(X_t)$ induced by restriction.*

Proof. After suitable étale base change, we can assume that there is no monodromy on $\mathrm{Cl}(X_t)$. Then, as $\mathrm{Pic}(X_t)$ is a subgroup of $\mathrm{Cl}(X_t)$, in view of Theorem 4.3 it suffices to show that every line bundle on X_0 extends, up to isomorphism, to a line bundle on X. Or, equivalently, that given any Cartier divisor D_0 on X_0, there exists a Cartier divisor D on X not containing X_0 in its support and such that $D|_{X_0} \sim D_0$. Since any Cartier divisor can be written as the difference of two very ample divisors, we may assume that D_0 is very ample.

By Theorem 4.3, we can find a Weil divisor D on X not containing X_0 in its support and such that $D|_{X_0} \sim D_0$. We need to show that D is Cartier in a neighborhood of X_0. We can replace D_0 with $D|_{X_0}$. After possibly shrinking T near 0, every section of $\mathcal{O}_{X_0}(D_0)$ extends to a section of $\mathcal{O}_X(D)$ by Theorem 4.4. Since $\mathcal{O}_{X_0}(D_0)$ is generated by its global sections, it follows that the natural homomorphism $\mathcal{O}_X(D) \to \mathcal{O}_{X_0}(D_0)$ is surjective. As $\mathcal{O}_{X_0}(D_0)$ is invertible, this implies that so is $\mathcal{O}_X(D)$, and thus that D is Cartier. $\square$

5. Deformations of the Mori structure

The flatness of Cox rings in flat families of Fano varieties with terminal $\mathbb{Q}$-factorial singularities is already evidence of a strong rigidity property of such varieties. In this section, we consider a flat family $f : X \to T$ of Fano varieties with terminal $\mathbb{Q}$-factorial singularities, parametrized by a smooth curve T.

An immediate corollary of Theorem 4.4 is the following general fact.

Corollary 5.1. *For any flat family $f : X \to T$ of Fano varieties with terminal $\mathbb{Q}$-factorial singularities over a smooth curve T, the pseudo-effective cones* $\mathrm{PEff}(X_t)$ *of the fibers of f are locally constant in the family.*

If one wants to further investigate how the Mori structure varies in the family, it becomes necessary to run the Minimal Model Program. This requires us to step out, for a moment, from the setting of families of Fano varieties.

Suppose for now that $f : X \to T$ is just a flat projective family of normal varieties with $\mathbb{Q}$-factorial singularities. Let X_0 be the fiber over a point $0 \in T$. We assume that the restriction map $N^1(X) \to N^1(X_0)$ is surjective, that X_0 has canonical singularities, and that there is an effective $\mathbb{Q}$-divisor D on X, not containing X_0 in its support, such that $(X_0, D|_{X_0})$ is a Kawamata log terminal pair. Assume furthermore that $D|_{X_0} - aK_{X_0}$ is ample for some $a > -1$. Note that this last condition always holds for Fano varieties.

The following result is crucial for our investigation.

Theorem 5.2. *With the above notation, every step $X^i \dashrightarrow X^{i+1}$ in the Minimal Model Program of (X, D) over T with scaling of $D - aK_X$ is either trivial on the fiber X_0^i of X^i over 0, or it induces a step of the same type (divisorial contraction, flip, or Mori fibration) $X_0^i \dashrightarrow X_0^{i+1}$ in the Minimal Model Program of $(X_0, D|_{X_0})$ with scaling of $D|_{X_0} - aK_{X_0}$. In particular, at each step X_0^i is the proper transform of X_0.*

For a proof of this theorem, we refer the reader to [de Fernex and Hacon 2011] (specifically, see Theorem 4.1 and the proof of Theorem 4.5 there). The key observation is that, by running a Minimal Model Program with scaling of $D - aK_X$, we can ensure that the property that $D|_{X_0} - aK_{X_0}$ is ample for some

$a > -1$ is preserved after each step of the program. By the semicontinuity of fiber dimensions, it is easy to see that $X^i \dashrightarrow X^{i+1}$ is a Mori fiber space if and only if so is $X_0^i \dashrightarrow X_0^{i+1}$. If $X^i \dashrightarrow X^{i+1}$ is birational, then the main issue is to show that if $X^i \dashrightarrow X^{i+1}$ is a flip and $X^i \to Z^i$ is the corresponding flipping contraction, then $X_0^i \to Z_0^i$ is also a flipping contraction. If this were not the case, then $X_0^i \to Z_0^i$ would be a divisorial contraction and hence Z_0^i would be $\mathbb{Q}$-factorial. Since $D^i|_{X_0^i} - aK_{X_0^i}$ is nef over Z_0^i, it follows that $-K_{X_0^i}$ is ample over Z_0^i and hence that Z_0^i is canonical. By [de Fernex and Hacon 2011, Proposition 3.5] it then follows that Z^i is $\mathbb{Q}$-factorial. This is the required contradiction as the target of a flipping contraction is never $\mathbb{Q}$-factorial. Therefore it follows that $X^i \to Z^i$ is a flipping contraction if and only if so is $X_0^i \to Z_0^i$.

Remark 5.3. The theorem implies that $X_0^i \dashrightarrow X_0^{i+1}$ is a divisorial contraction or a Mori fibration if and only if so is $X_t^i \dashrightarrow X_t^{i+1}$ for general $t \in T$. However, there exist flipping contractions $X^i \dashrightarrow X^{i+1}$ which are the identity on a general fiber X_t^i. This follows from the examples of Totaro that we will briefly sketch at the end of the section.

One of the main applications of this result is the following extension theorem, which in particular implies the statement of Theorem 4.4 in the case of families of Fano varieties.

Theorem 5.4 [de Fernex and Hacon 2011, Theorem 4.5]. *With the same notation as in Theorem 5.2, assume that $(X_0, D|_{X_0})$ is canonical and, moreover, that either $D|_{X_0}$ or $K_{X_0} + D|_{X_0}$ is big. Let L be any Weil divisor whose support does not contain X_0 and such that $L|_{X_0} \equiv k(K_X + D)|_{X_0}$ for some rational number $k > 1$. Then the restriction map*

$$H^0(X, \mathcal{O}_X(L)) \to H^0(X_0, \mathcal{O}_{X_0}(L|_{X_0}))$$

is surjective.

There are versions of the above results where the condition on the positivity of $D|_{X_0} - aK_{X_0}$ is replaced by the condition that the stable base locus of $K_X + D$ does not contain any irreducible component of $D|_{X_0}$ [de Fernex and Hacon 2011, Theorem 4.5]. The advantage of the condition considered here is that it only requires us to know something about the special fiber X_0. This is a significant point, as after all we are trying to lift geometric properties from the special fiber to the whole space and nearby fibers of an arbitrary flat deformation.

We now come back to the original setting, and hence assume that $f : X \to T$ is a flat family of Fano varieties with $\mathbb{Q}$-factorial terminal singularities. After étale base change, we can assume that $N^1(X_t) \cong N^1(X/T)$ for every t.

Corollary 5.1 implies, by duality, that the cones of nef curves $\overline{NM}(X_t)$ are constant in the family. Combining this with Theorem 5.2, we obtain the following rigidity property of birational Mori fiber structures. Recall that we only consider Mori fiber structures that are the output of some Minimal Model Program.

Theorem 5.5. *The birational Mori fiber structures* $X_t \dashrightarrow X_t' \to Y_t'$ *are locally constant in the family* $X \to T$.

This result is implicit in [de Fernex and Hacon 2011]. As it was not explicitly stated there, we provide a proof.

Proof. Let R_0 be the extremal ray of $\overline{\mathrm{NM}}(X_0)$ corresponding to a given birational Mori fiber structure on X_0. Note that by [Birkar et al. 2010, 1.3.5] and its proof, there exists an ample $\mathbb{R}$-divisor A_0 such that the K_{X_0} Minimal Model Program with scaling of A_0 say $X_0 \dashrightarrow X_0'$ ends with Mori fiber space $X_0' \to Y_0'$ which is $(K_{X_0} + A_0)$-trivial. Notice also that if we make a general choice of A_0 in $N^1(X_0)$, then each step of this Minimal Model Program with scaling is uniquely determined since at each step there is a unique $K_{X_0} + t_i A_0$ trivial extremal ray.

We may now assume that there is an ample $\mathbb{R}$-divisor A on X such that $A_0 = A|_{X_0}$. Consider running the K_X Minimal Model Program over T with scaling of A say $X \dashrightarrow X'$. Since X is uniruled, this ends with a Mori fiber space $X' \to Y'$. By Theorem 5.2, this induces the Minimal Model Program with scaling on the fiber X_0 considered in the previous paragraph. Moreover, the Minimal Model Program on X terminates with the Mori fiber space $X' \to Y'$ at the same step when the induced Minimal Model Program on X_0 terminates with the Mori fiber space $X_0' \to Y_0'$. This implies that the birational Mori fiber structure $X_0 \dashrightarrow Y_0'$ extends to the birational Mori fiber structure $X \dashrightarrow Y'$, and thus deforms to a birational Mori fiber structure on the nearby fibers.   □

A similar application of Theorems 5.2 and 5.4 leads to the following rigidity result for the cone of moving divisors.

Theorem 5.6 [de Fernex and Hacon 2011]. *The moving cone* $\mathrm{Mov}(X_t)$ *of divisors is locally constant in the family.*

Proof. The proof is similar to the proof of Theorem 5.5 once we observe that the faces of $\mathrm{Mov}(X)$ are determined by divisorial contractions and that given an extremal contraction $X \to Z$ over T, this is divisorial if and only if the contraction on the central fiber $X_0 \to Z_0$ is divisorial.   □

Regarding the behavior of the nef cone of divisors and, more generally, of the Mori chamber decomposition of the moving cone, the question becomes however much harder. In fact, once we allow even the mildest singularities, the rigidity of the whole Mori structure only holds in small dimensions.

Theorem 5.7 [de Fernex and Hacon 2011, Theorem 6.9]. *With the notation above, assume that* X_0 *is either at most 3-dimensional, or is 4-dimensional* *and Gorenstein. Then the Mori chamber decomposition of* $\mathrm{Mov}(X_t)$ *is locally constant for t in a neighborhood of* $0 \in T$.

Totaro [2009] provides families of examples that show that this result is optimal. In particular, for every $a > b > 1$, he shows that there is a family of terminal $\mathbb{Q}$-factorial Gorenstein Fano varieties $X \to T$ such that $X_t \cong \mathbb{P}^a \times \mathbb{P}^b$ for $t \neq 0$ and $\mathrm{Nef}(X_0)$ is strictly contained in $\mathrm{Nef}(X_t)$. The reason for this is that there is a flipping contraction $X \to Z$ over T which is an isomorphism on the general fiber X_t but contracts a copy of $\mathbb{P}^a$ contained in X_0. Let $X^+ \to Z$ be the corresponding flip and fix H^+ a divisor on X^+ which is ample over T. If H is its strict transform on X, then H is negative on flipping curves and hence $H|_{X_0}$ is not ample, however $H|_{X_t} \cong H^+|_{X_t^+}$ is ample for $t \neq 0$. Therefore, the nef cone of X_0 is strictly smaller than the nef cone of X_t so that the Mori chamber decomposition of $\mathrm{Mov}(X_0)$ is finer than that of $\mathrm{Mov}(X_t)$.

The construction of this example starts from the flip from the total space of $\mathbb{O}_{\mathbb{P}^a}(-1)^{\oplus(b+1)}$ to the total space of $\mathbb{O}_{\mathbb{P}^b}(-1)^{\oplus(a+1)}$. The key idea is to interpret this local setting in terms of linear algebra, by viewing the two spaces as small resolutions of the space of linear maps of rank at most one from $\mathbb{C}^{b+1}$ to $\mathbb{C}^{a+1}$, and to use such a description to compactify the setting into a family of Fano varieties. Totaro also gives an example in dimension 4, where the generic element of the family is isomorphic to the blow-up of $\mathbb{P}^4$ along a line, and the central fiber is a Fano variety with $\mathbb{Q}$-factorial terminal singularities that is not Gorenstein.

Remark 5.8. The fact that the Mori chamber decomposition is not in general locally constant in families of Fano varieties with $\mathbb{Q}$-factorial terminal singularities is not in contradiction with the flatness of Cox rings. The point is that the flatness of such rings is to be understood only as modules, but it gives no information on the ring structure. The changes in the Mori chamber decomposition are related to jumps of the kernels of the multiplication maps.

References

[Araujo 2010] C. Araujo, "The cone of pseudo-effective divisors of log varieties after Batyrev", *Math. Z.* **264**:1 (2010), 179–193. MR 2010k:14018

[Batyrev 1992] V. V. Batyrev, "The cone of effective divisors of threefolds", pp. 337–352 in *Proceedings of the International Conference on Algebra* (Novosibirsk, 1989), vol. 3, Contemp. Math. **131**, Amer. Math. Soc., Providence, RI, 1992. MR 94f:14035

[Birkar et al. 2010] C. Birkar, P. Cascini, C. D. Hacon, and J. McKernan, "Existence of minimal models for varieties of log general type", *J. Amer. Math. Soc.* **23**:2 (2010), 405–468. MR 2011f:14023

[Boucksom et al. 2004] S. Boucksom, J.-P. Demailly, M. Paun, and T. Peternell, "The pseudo-effective cone of a compact Kähler manifold and varieties of negative Kodaira dimension", 2004. arXiv math.AG/0405285

[Campana 1991] F. Campana, "Une version géométrique généralisée du théorème du produit de Nadel", *Bull. Soc. Math. France* **119**:4 (1991), 479–493. MR 93h:14029

[Campana 1992] F. Campana, "Connexité rationnelle des variétés de Fano", *Ann. Sci. École Norm. Sup.* (4) **25**:5 (1992), 539–545. MR 93k:14050

[Clemens and Griffiths 1972] C. H. Clemens and P. A. Griffiths, "The intermediate Jacobian of the cubic threefold", *Ann. of Math.* (2) **95** (1972), 281–356. MR 46 #1796

[Cox 1995] D. A. Cox, "The homogeneous coordinate ring of a toric variety", *J. Algebraic Geom.* **4**:1 (1995), 17–50. MR 95i:14046

[de Fernex and Hacon 2011] T. de Fernex and C. D. Hacon, "Deformations of canonical pairs and Fano varieties", *J. Reine Angew. Math.* **651** (2011), 97–126. MR 2774312

[Hacon and Mckernan 2007] C. D. Hacon and J. Mckernan, "On Shokurov's rational connectedness conjecture", *Duke Math. J.* **138**:1 (2007), 119–136. MR 2008f:14030

[Hu and Keel 2000] Y. Hu and S. Keel, "Mori dream spaces and GIT", *Michigan Math. J.* **48** (2000), 331–348. MR 2001i:14059

[Iskovskih and Manin 1971] V. A. Iskovskih and J. I. Manin, "Three-dimensional quartics and counterexamples to the Lüroth problem", *Mat. Sb.* (*N.S.*) **86(128)** (1971), 140–166. MR 45 #266

[Kollár and Mori 1992] J. Kollár and S. Mori, "Classification of three-dimensional flips", *J. Amer. Math. Soc.* **5**:3 (1992), 533–703. MR 93i:14015

[Kollár et al. 1992a] J. Kollár, Y. Miyaoka, and S. Mori, "Rational connectedness and boundedness of Fano manifolds", *J. Differential Geom.* **36**:3 (1992), 765–779. MR 94g:14021

[Kollár et al. 1992b] J. Kollár, Y. Miyaoka, and S. Mori, "Rational curves on Fano varieties", pp. 100–105 in *Classification of irregular varieties* (Trento, 1990), Lecture Notes in Math. **1515**, Springer, Berlin, 1992. MR 94f:14039

[Lehmann 2008] B. Lehmann, "A cone theorem for nef curves", 2008. arXiv 0807.2294v3

[Mori 1979] S. Mori, "Projective manifolds with ample tangent bundles", *Ann. of Math.* (2) **110**:3 (1979), 593–606. MR 81j:14010

[Mori 1982] S. Mori, "Threefolds whose canonical bundles are not numerically effective", *Ann. of Math.* (2) **116**:1 (1982), 133–176. MR 84e:14032

[Nadel 1990] A. M. Nadel, "A finiteness theorem for Fano 4-folds", 1990. Unpublished.

[Nadel 1991] A. M. Nadel, "The boundedness of degree of Fano varieties with Picard number one", *J. Amer. Math. Soc.* **4**:4 (1991), 681–692. MR 93g:14048

[Ravindra and Srinivas 2006] G. V. Ravindra and V. Srinivas, "The Grothendieck–Lefschetz theorem for normal projective varieties", *J. Algebraic Geom.* **15**:3 (2006), 563–590. MR 2006m:14008

[Totaro 2009] B. Totaro, "Jumping of the nef cone for Fano varieties", 2009. To appear in *J. Algebraic Geom.* arXiv 0907.3617

[Wiśniewski 1991] J. A. Wiśniewski, "On deformation of nef values", *Duke Math. J.* **64**:2 (1991), 325–332. MR 93g:14012

[Wiśniewski 2009] J. A. Wiśniewski, "Rigidity of the Mori cone for Fano manifolds", *Bull. Lond. Math. Soc.* **41**:5 (2009), 779–781. MR 2011c:14046

[Zhang 2006] Q. Zhang, "Rational connectedness of log **Q**-Fano varieties", *J. Reine Angew. Math.* **590** (2006), 131–142. MR 2006m:14021

Department of Mathematics, University of Utah, 155 South 1400 East, Salt Lake City, UT 48112-0090, USA defernex@math.utah.edu

Department of Mathematics, University of Utah, 155 South 1400 East, Salt Lake City, UT 48112-0090, USA

hacon@math.utah.edu

Current Developments in Algebraic Geometry
MSRI Publications
Volume **59**, 2011

The Schottky problem

SAMUEL GRUSHEVSKY

In this survey we discuss some of the classical and modern methods in studying
the (Riemann–)Schottky problem, the problem of characterizing Jacobians of
curves among principally polarized abelian varieties. We present many of the
recent results in this subject, and describe some directions of current research.
This paper is based on the talk given at the "Classical algebraic geometry
today" workshop at MSRI in January 2009.

1. Introduction

The Riemann–Schottky problem is the problem of determining which complex
principally polarized abelian varieties arise as Jacobian varieties of complex
curves. The history of the problem is very long, going back to the works of Abel,
Jacobi, and Riemann. The first approach, culminating in a complete solution in
genus 4 (the first nontrivial case), was developed by Schottky and Jung [Sch88;
SJ09]. Since then a variety of different approaches to the problem have been
developed, and many geometric properties of abelian varieties in general and
Jacobian varieties in particular have been studied extensively. Numerous partial
and complete solutions to the Schottky problem have been conjectured, and some
were proven.

Research supported in part by National Science Foundation under the grant DMS-0901086/DMS-
1053313.

In this survey we will describe many of the ideas and methods that have been applied to or developed for the study of the Schottky problem. We will present some of the results, as well as various open problems and possible connections among various approaches. To keep the length of the text reasonable, the proofs for the most part will be omitted, and references will be given; when possible, we will try to indicate the general idea or philosophy behind the work done. We hope that an interested reader may consider this as an introduction to the ideas and results of the subject, and would be encouraged to explore the field in greater depth by following some of the references.

This text is in no way the first (and will certainly not be the last) survey written on the Schottky problem. Many excellent surveys, from various points of view, and emphasizing various aspects of the field, have been written, including those by Dubrovin [Dub81b], Donagi [Don88], Beauville [Bea88], Debarre [Deb95b], Taimanov [Tai97], van Geemen [vG98], Arbarello [Arb99], Buchstaber and Krichever [BK06]. A beautiful introduction is [Mum75], while many relevant results on curves, abelian varieties, and theta functions can be found in [Igu72; Mum07a; Mum07b; Mum07c; ACGH85; BL04]. The research literature on or related to the Schottky problem is vast, with many exciting results dating back to the works of Schottky, and more progress constantly happening. While we have strived to give as many references as possible, our list is by no means complete, and we apologize for any inadvertent omissions; another good source of references is [Arb99].

The structure of this work is as follows. In Section 2 we introduce most of the notation for curves, abelian varieties, and their moduli, and state the problem; in Section 3 we introduce theta constants and discuss the classical approach culminating in an explicit solution of the Schottky problem in genus 4; in Section 4 we continue the discussion of modular forms vanishing on the Jacobian locus, and describe the Schottky–Jung approach to obtaining them using the Prym construction. In Section 5 we discuss the singularities of the theta divisor for Jacobians and Pryms, and present the Andreotti–Mayer approach to the Schottky problem. In Section 6 we discuss minimal cohomology classes and the Matsusaka–Ran criterion; in Section 7 we discuss the geometry of the Kummer variety, its secants, and in particular the trisecant conjecture characterizing Jacobians, and the Prym analog. Section 8 deals with the Γ_{00} conjecture and further geometry of the $|2\Theta|$ linear system.

2. Notation: the statement of the Schottky problem

Loosely speaking, the Schottky problem is the following question: which principally polarized abelian varieties are Jacobians of curves? In this section we introduce the relevant notation and state this question more precisely. Such a brief introduction cannot do justice to the interesting and deep constructions of the moduli stacks of curves and abelian varieties; we will thus assume that the reader is in fact familiar with the ideas of moduli theory, take the existence of moduli space for granted, and will use this section primarily to fix notation.

Throughout the text we work over the field $\mathbb{C}$ of complex numbers. Unless stated otherwise, we are working in the category of smooth projective varieties.

Definition 2.1 (Complex tori and abelian varieties). A g-dimensional complex torus is a quotient of $\mathbb{C}^g$ by a full rank lattice, i.e., by a subgroup $\Lambda \subset \mathbb{C}^g$ (under addition) such that $\Lambda \sim \mathbb{Z}^{2g}$ and $\Lambda \otimes_{\mathbb{Z}} \mathbb{R} = \mathbb{R}^{2g} = \mathbb{C}^g$.

A g-dimensional complex torus is called an abelian variety if it is a projective variety. This is equivalent to the torus admitting a very ample line bundle, and it turns out that this is equivalent to the lattice Λ being conjugate under the action of $GL(g, \mathbb{C})$ to a lattice $\mathbb{Z}^g + \tau \mathbb{Z}^g$ for some symmetric complex $g \times g$ matrix τ with positive-definite imaginary part. This statement is known as Riemann's bilinear relations, see for example [BL04]. Note that for a given Λ there exist many possible τ such that Λ is isomorphic to $\mathbb{Z}^g + \tau \mathbb{Z}^g$ — this is made more precise below.

Definition 2.2 (polarization). A polarization on an abelian variety A is the first Chern class of an ample line bundle L on A, i.e., the polarization is

$$[L] := c_1(L) \in H^2(A, \mathbb{Z}) \cap H^{1,1}(A, \mathbb{C}).$$

A polarization is called principal if $h^0(A, L) = 1$ (in terms of the cohomology class, this means that the matrix is unimodular). We will denote principally polarized abelian varieties (abbreviated as ppavs) by (A, Θ), where Θ is our notation for the ample bundle, the first Chern class of which is the polarization.

We note that the polarizing line bundle L on an abelian variety can be translated by an arbitrary point of the abelian variety, preserving the class of the polarization $[L]$. However, it is customary to use the bundle rather than its class in the notations for a ppav. Given a polarization, one usually chooses a symmetric polarizing bundle, i.e., one of the 2^{2g} line bundles Θ (differing by a translation by a 2-torsion point) of the given Chern class $[\Theta]$ such that $(-1)^*\Theta = \Theta$ — where -1 is the involution of the abelian variety as a group.

Definition 2.3 (Siegel space). We denote by $\mathcal{H}_g$ the space of all symmetric $g \times g$ matrices τ with positive definite imaginary part — called the Siegel space. Given an element $\tau \in \mathcal{H}_g$, the Riemann theta function is defined by the Fourier series

$$\theta(\tau, z) := \sum_{n \in \mathbb{Z}^g} \exp(\pi i n^t \tau n + 2\pi i n^t z). \tag{1}$$

This (universal) theta function is a map $\theta : \mathcal{H}^g \times \mathbb{C}^g \to \mathbb{C}$. For fixed $\tau \in \mathcal{H}_g$ the theta function has the following automorphy property in z, easily verified from the definition:

$$\theta(\tau, z + \tau n + m) = \exp(-\pi i n^t \tau n - 2\pi i n^t z)\theta(\tau, z) \quad \forall n, m \in \mathbb{Z}^g. \quad (2)$$

It thus follows that the zero locus $\{z \in \mathbb{C}^g \mid \theta(\tau, z) = 0\}$ is invariant under the shifts by the lattice $\mathbb{Z}^g + \tau \mathbb{Z}^g$, and thus descends to a well-defined subvariety $\Theta_\tau \subset \mathbb{C}^g / (\mathbb{Z}^g + \tau \mathbb{Z}^g)$.

Definition 2.4 (moduli of ppavs). We denote by $\mathcal{A}_g$ the moduli stack of principally polarized (complex) abelian varieties (A, Θ) of dimension g. Then we have a natural map $\mathcal{H}_g \to \mathcal{A}_g$ given by

$$\tau \mapsto \left(A_\tau := \mathbb{C}^g / (\mathbb{Z}^g + \tau \mathbb{Z}^g,) \Theta_\tau := \{\theta(\tau, z) = 0\}\right).$$

It can be shown that this map exhibits $\mathcal{H}_g$ as the universal cover of $\mathcal{A}_g$. Moreover, there is a natural action of the symplectic group $\mathrm{Sp}(2g, \mathbb{Z})$ on $\mathcal{H}_g$ given by

$$\gamma \circ \tau := (\tau c + d)^{-1}(\tau a + b), \quad (3)$$

where we write $\mathrm{Sp}(2g, \mathbb{Z}) \ni \gamma = \begin{pmatrix} a & b \\ c & d \end{pmatrix}$ in $g \times g$ block form. It turns out that we have $\mathcal{A}_g = \mathcal{H}_g / \mathrm{Sp}(2g, \mathbb{Z})$, so that the stack $\mathcal{A}_g$ is in fact a global quotient of a smooth complex manifold by a discrete group. For future use we set

$$\mathcal{A}_g^{\mathrm{dec}} := \left(\bigcup_{i=1\ldots g-1} \mathcal{A}_i \times \mathcal{A}_{g-i}\right) \subset \mathcal{A}_g,$$

the locus of decomposable ppavs (products of lower-dimensional ones) — these turn out to be special in many ways, as these are the ones for which the theta divisor is reducible.

We also denote by $\mathcal{A}_g^{\mathrm{ind}} := \mathcal{A}_g \setminus \mathcal{A}_g^{\mathrm{dec}}$ its complement, the locus of indecomposable ppavs. We refer for example to [BL04] for more details on $\mathcal{A}_g$, and to [Igu72; Mum07a; Mum07b] for the theory of theta functions.

Definition 2.5 (universal family). We denote by $\mathcal{U}_g \to \mathcal{A}_g$ the universal family of ppavs, the fiber over (A, Θ) being the ppav itself. Analytically, the universal cover of $\mathcal{U}_g$ is $\mathcal{H}_g \times \mathbb{C}^g$, since the universal cover of any ppav is $\mathbb{C}^g$. Thus

$$\mathcal{U}_g = \mathcal{H}_g \times \mathbb{C}^g / (\mathrm{Sp}(2g, \mathbb{Z}) \ltimes \mathbb{Z}^{2g}),$$

where $\mathbb{Z}^{2g}$ acts by fixing $\mathcal{H}_g$ and adding lattice vectors on $\mathbb{C}^g$, and $\gamma \in \mathrm{Sp}(2g, \mathbb{Z})$ acts on $\mathbb{C}^g$ by $z \mapsto (\tau c + d)^{-1} z$.

Definition 2.6 (Jacobians). Analytically, to define the Jacobian of a Riemann surface C of genus g, one considers standard generators $A_1, \ldots, A_g, B_1, \ldots, B_g$ of the fundamental group $\pi_1(C)$ (i.e., generators such that the only relation is

$\prod_{i=1}^{g} A_i B_i A_i^{-1} B_i^{-1} = 1$). Then one chooses a basis $\omega_1, \ldots, \omega_g$ of the space of abelian differentials $H^0(C, K_C)$ dual to $\{A_i\}$, i.e., such that $\int_{A_i} \omega_j = \delta_{i,j}$. The Jacobian is then the ppav $J(C) := (A_\tau, \Theta_\tau)$ corresponding to the matrix $\{\tau_{ij}\} := \{\int_{B_i} \omega_j\} \in \mathcal{H}_g$.

Algebraically, the Jacobian of a curve C of genus g is $J(C) := \mathrm{Pic}^{g-1}(C)$, the set of linear equivalence classes of divisors of degree $g - 1$ on C. The principal polarization divisor on it is the set of effective bundles

$$\Theta := \{L \in \mathrm{Pic}^{g-1}(C) \mid h^0(C, L) \geq 1\}.$$

The Jacobian is (noncanonically) isomorphic to $\mathrm{Pic}^n(C)$; an isomorphism is defined by choosing a fixed divisor of degree $g - 1 - n$ and subtracting it. Under this identification $\mathrm{Pic}^n(C)$ has a natural principal polarization, but, unlike $\mathrm{Pic}^{g-1}(C)$, there is no natural choice of a polarization divisor. It is often convenient to view the Jacobian as $\mathrm{Pic}^0(C)$, which is naturally an abelian group (while $\mathrm{Pic}^{g-1}(C)$ is naturally a torsor over $\mathrm{Pic}^0(C)$ rather than a group in a natural way).

Definition 2.7 (moduli of curves). We denote by $\mathcal{M}_g$ the moduli stack of smooth (compact complex) curves of genus g, and by $\overline{\mathcal{M}}_g$ its Deligne–Mumford compactification [DM69] — the moduli stack of stable curves of genus g. The Torelli map is the map $J : \mathcal{M}_g \to \mathcal{A}_g$ sending C to its Jacobian $J(C)$. The Torelli theorem states that this map is injective, and we call its image the Jacobian locus $\mathcal{J}_g$ (it is sometimes called the Torelli locus in the literature).

We recall that $\dim \mathcal{M}_g = 3g - 3$ (for $g > 1$), while $\dim \mathcal{A}_g = g(g+1)/2$. Thus the dimensions coincide for $g \leq 3$, and in fact the Jacobian locus $\mathcal{J}_g$ is equal to $\mathcal{A}_g^{\mathrm{ind}}$ if and only if $g \leq 3$. For $g = 4$ we have $\dim \mathcal{M}_4 = 9 = 10 - 1 = \dim \mathcal{A}_4 - 1$, and $\mathcal{J}_4$ has been completely described by Schottky [Sch88]. (His result was proven rigorously by Igusa [Igu81a] and Freitag [Fre83], and further reformulated by Igusa [Igu81b].)

The subject of this survey is the following problem, studied by Riemann in unpublished notes, and with first published significant progress made by Schottky and Jung [Sch88; SJ09].

(Riemann)–Schottky problem: Describe the image $\mathcal{J}_g := J(\mathcal{M}_g) \subset \mathcal{A}_g$.

In many of the approaches to the Schottky problem that we will survey weak solutions to this problem will be obtained — i.e., we will get loci within $\mathcal{A}_g$, of which $\mathcal{J}_g$ would be an irreducible component.

Note that the moduli space $\mathcal{M}_g$ is not compact, and neither is the Jacobian locus $\mathcal{J}_g$. However, it was shown by Namikawa [Nam76a; Nam76b] that the Torelli map extends to a map $\bar{J} : \overline{\mathcal{M}}_g \to \overline{\mathcal{A}}_g$ from the Deligne–Mumford compactification of $\mathcal{M}_g$ to a certain (second Voronoi, but we will not need the definition here) toroidal compactification of $\mathcal{A}_g$. Thus in some sense it is more natural to consider the closed Schottky problem: that of describing the closure $\bar{J}_g \subset \bar{\mathcal{A}}_g$.

A lot of work has been done on the Schottky problem, and much progress has been made. Outwardly it may even appear that the problem has been solved completely in many ways. While we survey the many solutions or partial solutions that have indeed been developed, it will become apparent that a lot of work is still to be done, and much substantial understanding is still lacking, with many important open questions remaining. In particular, the following two questions are completely open:

Open Problem 1. Solve the Schottky problem explicitly for $g = 5$, i.e., given an explicit $\tau \in \mathcal{H}_5$, determine whether $[\tau] \in \mathcal{A}_5$ lies in $\mathcal{J}_5$.

A lot of research has been done on $\mathcal{A}_5$, and many partial solutions to the Schottky problem for $g = 5$ are available, see [Acc83; Don87a; CM08; CMF05], but a complete solution in the spirit of Schottky (similar to genus 4, see the next section) is still not known; there is no known way to answer the above question. Another question, which would be a potential application for a solution to the Schottky problem, is this:

Open Problem 2. The symmetric space $\mathcal{H}_g$ has a natural $\mathrm{Sp}(2g, \mathbb{R})$-invariant metric on it, which descends to a metric on $\mathcal{A}_g$. For g sufficiently large do there exist any complex geodesics for this metric contained in $\bar{\mathcal{J}}_g$ and intersecting $\mathcal{J}_g$?

A negative answer to this question, together with the André–Oort conjecture [And89; Oor97], would imply the Coleman conjecture [Col87], stating that for g sufficiently large there exist only finitely many $C \in \mathcal{M}_g$ such that $J(C)$ admits a complex multiplication.

If we had a good explicit solution to the Schottky problem it could allow one to check explicitly whether such geodesics may exist, but so far this has not been accomplished, though much interesting work in the area has been done [Tol87; Hai02; VZ04; MVZ06].

Questions related to the Schottky problem also seem to arise naturally in perturbative (super)string theory, and many deep mathematical questions were posed and studied by D'Hoker and Phong [DP02a; DP02b; DP02c; DP02d; DP05a; DP05b], with many more still open questions arising in recent works [CDPvG08b; Gru09; SM08; CDPvG08a; GSM10a; OPSMY10] and others.

The Schottky problem as we stated it is still a loosely phrased question — there may be different ways of describing $\mathcal{J}_g$. In the following sections we survey the many approaches to characterizing Jacobians.

3. Theta constants: the classical approach

The classical approach to the Schottky problem, initiated in the works of Riemann and Schottky, is to try to embed $\mathscr{A}_g$ in a projective space, and then try to write out the defining ideal for the image of $\mathscr{J}_g$. If successful, this would of course be a very explicit solution to the Schottky problem.

Definition 3.1 (Hodge bundle and modular forms). The Hodge rank g vector bundle $\mathbb{E}$ over $\mathscr{A}_g$ has the space of holomorphic one-forms $H^0(A, \Omega^1_A) = H^{1,0}(A)$ as the fiber over a ppav A. Its determinant line bundle $L := \det \mathbb{E}$ is called the bundle of (scalar) modular forms of weight 1, and in general we call a section of $L^{\otimes k}$ a modular form of weight k.

The sections of any line bundle on $\mathscr{A}_g$ can be pulled back to $\mathscr{H}_g$ (where the pullback of the bundle is trivial, as $\mathscr{H}_g$ is contractible). Thus the sections of any line bundle on $\mathscr{A}_g$ can be alternatively described as holomorphic functions on $\mathscr{H}_g$ satisfying certain automorphy properties under the action of the group of deck transformation $\mathrm{Sp}(2g, \mathbb{Z})$. In particular, we have

$$H^0(\mathscr{A}_g, L^{\otimes k}) = \left\{ f : \mathscr{H}_g \to \mathbb{C} \mid f(\gamma \circ \tau) = \det(\tau c + d)^k f(\tau) \right\} \tag{4}$$

for all $\gamma \in \mathrm{Sp}(2g, \mathbb{Z})$, written in $g \times g$ block form, and for all $\tau \in \mathscr{H}_g$, with the action given by (3). This is the analytic definition of modular forms, by their automorphy properties.

It can be shown (this is work of Igusa and Mumford; see [BL04] for more details) that L is an ample line bundle on $\mathscr{A}_g$, and thus its sufficiently high power gives an embedding of $\mathscr{A}_g$ into a projective space. However, the actual very ample power is quite high, and instead it is easier to work with level covers of $\mathscr{A}_g$, i.e., with certain finite covers of $\mathscr{A}_g$.

Definition 3.2 (level covers). Algebraically, for any $\ell \in \mathbb{Z}_{>0}$ we denote by $\mathscr{A}_g(\ell)$ the full level ℓ cover of $\mathscr{A}_g$, i.e., the moduli space of ppavs (A, Θ) together with a chosen symplectic (with respect to the Weil pairing) basis for the group $A[\ell]$ of points of order ℓ on the abelian variety. We denote by $\mathscr{A}_g(\ell, 2\ell)$ the theta level ℓ cover of $\mathscr{A}_g$, i.e., the moduli space of ppavs together with a chosen symplectic basis for $A[\ell]$ and also one chosen point of order 2ℓ. For ℓ divisible by two both of these are Galois covers, with the Galois groups described as follows.

The principal level ℓ subgroup of $\mathrm{Sp}(2g, \mathbb{Z})$ is defined as

$$\Gamma_g(\ell) := \left\{ \gamma = \begin{pmatrix} a & b \\ c & d \end{pmatrix} \in \mathrm{Sp}(2g, \mathbb{Z}) \mid \gamma \equiv \begin{pmatrix} 1 & 0 \\ 0 & 1 \end{pmatrix} \bmod \ell \right\}$$

while the theta level ℓ subgroup is defined as

$$\Gamma_g(\ell, 2\ell) := \left\{ \gamma \in \Gamma_g(\ell) \mid \mathrm{diag}(a^t b) \equiv \mathrm{diag}(c^t d) \equiv 0 \bmod 2\ell \right\}.$$

These are the subgroups preserving the level data as above, i.e., we have $\mathcal{A}_g(\ell) = \mathcal{H}_g / \Gamma_g(\ell)$ and $\mathcal{A}_g(\ell, 2\ell) = \mathcal{H}_g / \Gamma_g(\ell, 2\ell)$.

One then defines the level Jacobian loci $\mathcal{J}_g(\ell)$ and $\mathcal{J}_g(\ell, 2\ell)$ as the preimages of $\mathcal{J}_g$ under the corresponding level covers of $\mathcal{A}_g$.

The usefulness of these covers lies in the fact that it is much easier to construct modular forms on them (i.e., sections of the pullbacks of L to them). In particular the following construction gives a large supply of modular forms of weight $1/2$.

Definition 3.3 (theta functions with characteristics). For $\varepsilon, \delta \in \frac{1}{\ell}\mathbb{Z}^g / \mathbb{Z}^g$ (which we will also think of as a point $m = \tau\varepsilon + \delta \in A_\tau[\ell]$) the theta function with characteristic ε, δ (or m) is

$$\theta\begin{bmatrix} \varepsilon \\ \delta \end{bmatrix}(\tau, z) := \theta_m(\tau, z) := \sum_{n \in \mathbb{Z}^g} \exp\left(\pi i (n+\varepsilon)^t (\tau(n+\varepsilon) + 2(z+\delta))\right)$$

$$= \exp\left(\pi i \varepsilon^t \tau \varepsilon + 2\pi i \varepsilon^t (z+\delta)\right) \theta(\tau, z+m). \quad (5)$$

As a function of z, $\theta_m(\tau, z)$ is a section of the translate of the principal polarization Θ_τ by the point m.

Setting $z = 0$, we get theta constants of order ℓ, denoted $\theta_m(\tau) := \theta_m(\tau, 0)$, which by above formula are, up to some easy exponential factor, just the values of the Riemann theta function at points m of order ℓ.

It turns out that theta constants $\theta_m(\tau)$ are modular forms of weight $1/2$ for $\Gamma_g(2\ell, 4\ell)$, i.e., are sections in $H^0(\mathcal{A}_g(2\ell, 4\ell), L^{\otimes(1/2)})$ (the pullback to this level cover of the bundle L of modular forms has a square root, i.e., a bundle such that its tensor square is L — this is what we mean by $L^{\otimes(1/2)}$), Moreover, they define an embedding of the moduli space:

Theorem 3.4 (Igusa [Igu72] for $\ell = 4n^2$; Mumford for $\ell \geq 3$, Salvati Manni [SM94] for $\ell \geq 2$). *The map* $\Phi_\ell : \mathcal{A}_g(2\ell, 4\ell) \to \mathbb{P}^{\ell^{2g}-1}$ *given by*

$$\tau \mapsto \{\theta_m(\tau)\}_{\text{all } m \in A_\tau[\ell]}$$

is an embedding. Thus the bundle $L^{\otimes(1/2)}$ *is very ample on* $\mathcal{A}_g(2\ell, 4\ell)$. *(In fact theta constants with characteristics generate the space of its sections, and generate the ring of modular forms on* $\mathcal{A}_g(2\ell, 4\ell)$.)*

From now on we will concentrate on the simplest case, that of level $\ell = 2$. In this case the theta function $\theta_m(\tau, z)$ is an even (resp. odd) function of z if $(-1)^{4\delta^t \varepsilon} = 1$ (resp. $= -1$); we will call such $m \in A[2]$ even (resp. odd). Thus the constant $\theta_m(\tau)$ vanishes identically if (and in fact only if) $m \in A[2]^{odd}$. There are $2^{g-1}(2^g + 1)$ even theta constants, and thus we can consider the image $\Phi_2(\tau) \in \mathbb{P}^{2^{g-1}(2^g+1)-1}$; however, this is still a large space, and even for $g = 1$ the map Φ_2 is not dominant — the one defining equation for the image is in fact $\begin{bmatrix}0\\0\end{bmatrix}(\tau)^4 = \begin{bmatrix}0\\1\end{bmatrix}(\tau)^4 + \begin{bmatrix}1\\0\end{bmatrix}(\tau)^4$. For $g = 2$ the situation already gets very complicated (see [vG84; vGvdG86] for more details). Thus it is natural to try to embed a suitable cover of $\mathcal{A}_g$ in a smaller projective space.

Definition 3.5 (theta functions of the second order). For $\varepsilon \in \frac{1}{2}\mathbb{Z}^g/\mathbb{Z}^g$ the theta function of the second order with characteristics ε is

$$\Theta[\varepsilon](\tau, z) := \theta_{2\tau\varepsilon}(2\tau, 2z) \tag{6}$$

For τ fixed we have $\Theta[\varepsilon](\tau, z) \in H^0(A_\tau, 2\Theta_\tau)$, and it can be shown that in fact theta functions of the second order generate the space of sections of $2\Theta_\tau$. Noting that for any $m = \tau\varepsilon + \delta \in A_\tau[2]$ the square of the theta function with characteristic $\theta_m^2(\tau, z)$ for τ fixed is also a section of $2\Theta_\tau$, we can express it as a linear combination of $\Theta[e](\tau, z)$, The result is:

Theorem 3.6 (Riemann's bilinear addition formula).

$$\theta_m^2(\tau, z) = \sum_{\sigma \in \frac{1}{2}\mathbb{Z}^g/\mathbb{Z}^g} (-1)^{4\delta^t\sigma} \Theta[\sigma + \epsilon](\tau, 0)\Theta[\sigma](\tau, z). \tag{7}$$

Definition 3.7 (the theta map). We define the theta constants of the second order to be $\Theta[\varepsilon](\tau) := \Theta[\varepsilon](\tau, 0)$; it turns out that these are also modular forms of weight $1/2$, but for $\Gamma_g(2, 4)$, i.e., $\Theta[\varepsilon](\tau) \in H^0(\mathcal{A}_g(2, 4), L^{\otimes(1/2)})$. We can thus consider the theta map

$$Th : \tau \to \{\Theta[\varepsilon](\tau)\}_{\forall \varepsilon \in \frac{1}{2}\mathbb{Z}^g/\mathbb{Z}^g}. \tag{8}$$

Notice that Riemann's bilinear addition formula above, for $z = 0$, shows that $\Phi_2(\tau)$ can be recovered uniquely up to signs from $Th(\tau)$. Since Φ_2 is injective on $\mathcal{A}_g(4, 8)$, it follows that the map Th is finite-to-one on $\mathcal{A}_g(4, 8)$, and thus also finite-to-one on $\mathcal{A}_g(2, 4)$, its natural domain of definition. In fact it is known that $Th : \mathcal{A}_g(2, 4) \to \mathbb{P}^{2^g-1}$ is generically injective, and conjecturally it is an embedding [SM94]. Thus the classical formulation of the Schottky problem is this:

Classical Riemann–Schottky problem: Write the defining equations for

$$\overline{Th(\mathcal{J}_g(2, 4))} \subset \overline{Th(\mathcal{A}_g(2, 4))} \subset \mathbb{P}^{2^g-1}.$$

Notice that for $g = 1, 2$ the target projective space is of the same dimension as the moduli, and it is in fact known that Th is then dominant. For $g = 3$ the 6-dimensional image $\overline{Th(\mathcal{A}_g(2, 4))} = \overline{Th(\mathcal{J}_3(2, 4))}$ is a hypersurface in $\mathbb{P}^7$. The defining equation for it can be written easily in terms of polynomials of even degree in theta constants with characteristics — which can then be rewritten in terms of Th by using Riemann's bilinear addition formula. Miraculously the same kind of formula gives the one defining equation for $\overline{Th(\mathcal{J}_4(2, 4))} \subset \overline{Th(\mathcal{A}_4(2, 4))}$:

Theorem 3.8 (Schottky [Sch88]; Igusa [Igu81b], Farkas and Rauch [FR70]; see also van Geemen and van der Geer [vGvdG86]). *In genus g define the Igusa modular form to be*

$$F_g(\tau) := 2^g \sum_{m \in A[2]} \theta_m^{16}(\tau) - \left(\sum_{m \in A[2]} \theta_m^8(\tau) \right)^2. \tag{9}$$

This is a modular form with respect to the entire group $\mathrm{Sp}(2g, \mathbb{Z})$, *such that when rewritten in terms of theta constants of the second order using Riemann's bilinear addition formula* (7), *the form* F_g *is*

- *identically zero for* $g = 1, 2$;
- *the defining equation for* $\overline{Th(\mathcal{J}_3(2, 4))} = \overline{Th(\mathcal{A}_3(2, 4))} \subset \mathbb{P}^7$ *for* $g = 3$;
- *the defining equation for* $\overline{Th(\mathcal{J}_4(2, 4))} \subset \overline{Th(\mathcal{A}_4(2, 4))}$ *for* $g = 4$.

To date this remains the most explicit solution to the Schottky problem in genus 4, and no similar solution in higher genera is known or has been proposed. However, the following question, which could be a test case for tackling Open Problem 2 has not been settled:

Open Problem 3. Construct all geodesics for the invariant metric on $\mathcal{A}_4$ contained in $\overline{\mathcal{J}_4}$, and intersecting $\mathcal{J}_4$.

One of course wonders what happens for $g > 4$. It turns out that the form F_g also naturally arises in perturbative superstring theory [DP05a; GSM10a]. By interpreting F_g in terms of lattice theta functions, $F_g = \theta_{D_{16}^+} - \theta_{E_8 \times E_8}$ (see [CS99] for definitions), using some physics intuition, and with motivation from the works of Belavin, Knizhnik, and Morozov, it was conjectured by D'Hoker and Phong [DP05a] that F_g vanishes on $\mathcal{J}_g$ for any genus. If true, this would have been a very nice defining equation for the Jacobian locus (though of course more would be needed). However, the situation is more complicated.

Theorem 3.9 (Grushevsky and Salvati Manni [GSM10a]). *The modular form* F_g *does not vanish identically on* $\mathcal{J}_g$ *for any* $g \geq 5$. *In fact the zero locus of* F_5 *on* $\mathcal{J}_5$ *is the locus of trigonal curves.*

It was shown by Poor [Poo96] that for any g the form F_g vanishes on the locus of Jacobians of hyperelliptic curves, and in view of the above it is natural to ask whether it also vanishes on the trigonal locus. As a result, the following easier step on the way to solving Open Problem 1 still has not been accomplished:

Open Problem 4. Write at least one nice/invariant modular form vanishing on $\mathcal{J}_5$.

Remark 3.10. Many modular forms vanishing on $\mathcal{J}_5$ were produced by Accola [Acc83] by using the Schottky–Jung approach (see the next section of this survey). Accola showed that these modular forms give a weak solution to the Schottky problem, i.e., define a locus in $\overline{Th(\mathcal{A}_5(2,4))}$ of which the Jacobian locus is an irreducible component. However, these modular forms are not "nice" in the sense that a lot of combinatorics of theta characteristics is involved in obtaining them, and they are certainly modular forms for $\Gamma_5(2,4)$, not for all of $\mathrm{Sp}(10,\mathbb{Z})$.

Remark 3.11. The codimension of $\mathcal{J}_5$ within $\mathcal{A}_5$ is equal to 3, and it can be shown (Faber [Fab99]) that $\overline{Th(\mathcal{J}_5(2,4))} \subset \overline{Th(\mathcal{A}_5(2,4))}$ is not a complete intersection. In fact the degrees of these subvarieties of $\mathbb{P}^{31}$ can be computed [Gru04b]. One would then expect that it may be possible to check whether Accola's equations produce extra components besides the Jacobian locus, but since the defining ideal of $\overline{Th(\mathcal{A}_5(2,4))} \subset \mathbb{P}^{31}$ is not known, at the moment this seems impossible.

Instead of asking to characterize the Jacobian locus, one could ask different Schottky-type questions — to characterize the image in $\mathcal{A}_g$ of certain special subvarieties of $\mathcal{M}_g$. In sharp contrast to the original Schottky problem, the hyperelliptic Schottky problem admits a complete solution, via simple explicit equations for theta constants:

Theorem 3.12 (Mumford [Mum07b], Poor [Poo94]). *For any g there exist some (explicitly described in terms of certain combinatorics) sets of characteristics* $S_1, \ldots, S_N \subset \frac{1}{2}\mathbb{Z}^{2g}/\mathbb{Z}^{2g}$ *such that* $\tau \in \mathcal{A}_g$ *is the period matrix of a hyperelliptic Jacobian if and only if for some* $1 \le i \le N$ *we have*

$$\theta_m(\tau) = 0 \iff m \in S_i \qquad \forall m \in \tfrac{1}{2}\mathbb{Z}^{2g}/\mathbb{Z}^{2g}.$$

We remark that this solution even takes care of the locus of decomposable ppavs (for them more vanishing happens).

4. Modular forms vanishing on the Jacobian locus: the Schottky–Jung approach

Despite the fact that at the moment a solution to the classical Schottky problem in genus 5 is not known, there is a general method to produce many modular forms vanishing on the Jacobian locus. This is the original approach originated by Schottky and Jung [SJ09], and developed by Farkas and Rauch [FR70].

Definition 4.1 (Prym variety). For an étale connected double cover $\widetilde{C} \to C$ of a curve $C \in \mathcal{M}_g$ — such a cover is given by a two-torsion point $\eta \in J(C)[2] \setminus \{0\}$ — we define the Prym variety to be

$$Prym(\widetilde{C} \to C) := Prym(C, \eta) := \mathrm{Ker}_0(J(\widetilde{C}) \to J(C)) \in \mathcal{A}_{g-1},$$

where Ker_0 denotes the connected component of 0 in the kernel (which in fact has two connected components). The map $J(\widetilde{C}) \to J(C)$ here is the norm map corresponding to the cover $\widetilde{C} \to C$, so the Prym is the connected component of zero in the kernel of the norm map.

The restriction of the principal polarization $\Theta_{J(\widetilde{C})}$ to the Prym gives twice the principal polarization. However, it turns out that this polarization admits a canonical square root, which thus gives a natural principal polarization on the Prym. We denote by $\mathcal{P}_{g-1} \subset \mathcal{A}_{g-1}$ the locus of Pryms of all étale covers of curves in $\mathcal{M}_g$; the question of describing it is called the Prym–Schottky problem. See Mumford [Mum74] for a modern exposition of the Prym construction.

The idea of the Schottky–Jung approach is as follows: the curve $\widetilde{C}$, being an étale double cover, is special (i.e., for $g > 1$ the set of such double covers is a proper subvariety of $\mathcal{M}_{2g-1}$). It can be shown that this implies the vanishing of some theta constants for $J(\widetilde{C})$, see [FR70]. On the other hand, there exists a finite-to-one surjective map (such maps are called isogenies of abelian varieties) $\big(J(C) \times \mathrm{Prym}(\widetilde{C} \to C)\big) \longrightarrow J(\widetilde{C})$, and pulling back by it allows one to express the theta function on $J(\widetilde{C})$ in terms of theta functions on $J(C)$ and on the Prym. Combining this with the vanishing properties for theta constants on $J(\widetilde{C})$ gives the following:

Theorem 4.2 (Schottky–Jung proportionality [SJ09]; proven rigorously by Farkas and Rauch [Far89]). *Let τ be the period matrix of C and let π be the period matrix of the Prym for $\eta = \left[\begin{smallmatrix} 0 & 0 & \cdots & 0 \\ 1 & 0 & \cdots & 0 \end{smallmatrix}\right]$. Then for any $\varepsilon, \delta \in \frac{1}{2}\mathbb{Z}^{g-1}/\mathbb{Z}^{g-1}$ the theta constants of $J(C)$ and of the Prym are related by*

$$\theta\begin{bmatrix} \varepsilon \\ \delta \end{bmatrix}(\pi)^2 = c\theta\begin{bmatrix} 0 & \varepsilon \\ 0 & \delta \end{bmatrix}(\tau) \cdot \theta\begin{bmatrix} 0 & \varepsilon \\ 1 & \delta \end{bmatrix}(\tau), \tag{10}$$

where π is the period matrix of the Prym, τ is the period matrix of $J(C)$, and the constant c is independent of ε, δ.

To obtain the Schottky-Jung proportionality for a double cover correspond to another point η we act on the equation above by $\mathrm{Sp}(2g, \mathbb{Z})/\Gamma_g(1, 2)$. This action is transitive on the set of two-torsion points, and permutes the theta constants up to certain eighth roots of unity (see [Igu72; BL04] for precise theta transformation formulas). Analytically the situation for $\eta = \left[\begin{smallmatrix} 0 & 0 & 0 & \cdots & 0 \\ 1 & 1 & 0 & \cdots & 0 \end{smallmatrix}\right]$ was worked out by Farkas [Far89] — an extra sign depending on ε, δ is present then.

Definition 4.3 (The Schottky–Jung locus). Let I_{g-1} be the defining ideal for the image $\overline{Th(\mathcal{A}_{g-1}(2, 4))} \subset \mathbb{P}^{2^{g-1}-1}$. For any equation $F \in I_{g-1}$ let F_η be the polynomial equation on $\mathbb{P}^{2^g-1}$ obtained by using the Schottky–Jung proportionality to substitute an appropriate polynomial of degree 2 in terms of theta constants of τ for the square of any theta constant of π (as a result square roots of polynomials in theta constants may appear, and we would then take the product of the expressions for all possible choices of the square root, i.e., the norm).

Let then S_g^η be the ideal obtained from I_{g-1} in this way. The (big) Schottky–Jung locus $\mathcal{S}_g^\eta(2, 4) \subset \mathcal{A}_g(2, 4)$ is then defined to be the zero locus of S_g^η. Note that it is a priori not clear — and in fact not known — that $I_g \subset S_g^\eta$, and thus it may make a difference that we define $\mathcal{S}_g^\eta(2, 4)$ within $\mathcal{A}_g(2, 4)$, and not as a subvariety of the projective space, We then define the small Schottky–Jung locus to be

$$\mathcal{S}_g(2, 4) := \bigcap_{\eta \in \frac{1}{2}\mathbb{Z}^{2g}/\mathbb{Z}^{2g}\setminus\{0\}} \mathcal{S}_g^\eta(2, 4). \tag{11}$$

Since the action of $\mathrm{Sp}(2g, \mathbb{Z})$ permutes the different η and the ideals S_g^η, it follows that the ideal defining $\mathcal{S}_g(2, 4)$ is $\mathrm{Sp}(2g, \mathbb{Z})$-invariant, and thus the locus $\mathcal{S}_g(2, 4)$ is a preimage of some $\mathcal{S}_g \subset \mathcal{A}_g$ under the level cover.

Theorem 4.4 (van Geemen [vG84], Donagi [Don87a]). *The Jacobian locus $\mathcal{J}_g$ is an irreducible component of the small Schottky–Jung locus $\mathcal{S}_g$, and in fact $\mathcal{J}_g(2, 4)$ is an irreducible component of the big Schottky–Jung locus $\mathcal{S}_g^\eta(2, 4)$ for any η.*

In genus 4 it can in fact be shown that $\mathcal{S}_4^\eta(2, 4) = \mathcal{J}_4(2, 4)$ for one (and thus for all) η (see [Igu81a]), and thus also $\mathcal{S}_4 = \mathcal{J}_4$. However, in genus 5 it turns out that $\mathcal{J}_5(2, 4) \subsetneq \mathcal{S}_5^\eta(2, 4)$, since the latter contains the locus of intermediate Jacobians — see [Don87b]. However, it can be shown that the locus of intermediate Jacobians does not lie in $\mathcal{S}_5$, and the following bold conjecture could be made.

Conjecture 4.5 (Donagi [Don88]). *The small Schottky–Jung locus is equal to the Jacobian locus, $\mathcal{S}_g = \mathcal{J}_g$.*

Remark 4.6. The dimension of the space of étale double covers of curves of genus $g + 1$ is equal to $3g$, the same as $\dim \mathcal{M}_{g+1}$, and thus $\dim \mathcal{P}_g \le 3g$. It follows that for $g \ge 6$ the Prym locus $\mathcal{P}_g$ is a proper subvariety of $\mathcal{A}_g$ (it is in fact known that $\mathcal{P}_g$ is dense within $\mathcal{A}_g$ for $g \le 5$, see [DS81]). Thus for $g \ge 7$ one could apply the Schottky–Jung proportionality to the defining ideal for $\overline{Th(\mathcal{P}_{g-1})} \subsetneq \overline{Th(\mathcal{A}_{g-1})}$ to get further equations for $Th(\mathcal{J}_g)$. Thus the conjecture above would seem more natural for $g = 5$ and perhaps $g = 6$ than in general. On the other hand, the approaches in [vG84; Don87a] to the Schottky–Jung theory have been via degeneration, and in that case questions related to the Γ_{00} conjecture (see Section 7) appear on the boundary; it could be that the recent progress on that topic could shed more light on the Schottky–Jung locus.

To summarize, the classical Schottky–Jung approach to the Schottky results in the explicit solution for $g = 4$, a "weak" solution (i.e., up to extra components) in any genus, and conjecturally could yield a complete solution in every genus —

assuming the ideal of $\overline{Th(\mathscr{A}_g(2,4))}$ were known completely (many elements of this ideal are known, but we do not know that they generate the entire ideal, see [vG84; SM85]).

5. Singularities of the theta divisor: the Andreotti–Mayer approach

It was shown by Andreotti and Mayer [AM67] that for a generic ppav the theta divisor is smooth. The singularities of the theta divisor for a Jacobian of a curve can be described using

Theorem 5.1 (Riemann's theta singularity theorem; see [ACGH85]). *For any curve C and any $D \in J(C) = \mathrm{Pic}^{g-1}(C)$ we have* $\mathrm{mult}_D \Theta = h^0(C, D)$.

(We note that this theorem agrees with the algebraic definition of the theta divisor Θ on the Jacobian as the locus of effective divisors). This theorem was further generalized by Kempf [Kem73]. From this theorem we see that $\mathrm{Sing}\,\Theta_{J(C)} = \{D \in \mathrm{Pic}^{g-1} \mid h^0(C, D) \geq 2\}$. It follows that the theta divisor for Jacobians of curves has a large singular set:

Theorem 5.2 (Andreotti and Mayer [AM67]). *For a nonhyperelliptic curve C of genus g we have* $\dim(\mathrm{Sing}\,\Theta_{J(C)}) = g - 4$, *while for a hyperelliptic curve C this dimension is* $g - 3$.

One would thus expect that this property is special for Jacobians. However, one could not expect that it defines $\overline{\mathscr{J}}_g \subset \overline{\mathscr{A}}_g$. Indeed, note that for a decomposable ppav $(A, \Theta) = (A_1, \Theta_1) \times (A_2, \Theta_2)$ (where $(A_i, \Theta_i) \in \mathscr{A}_{g_i}$ with $g_1 + g_2 = g$) we have $\Theta = (\Theta_1 \times A_2) \cup (A_1 \times \Theta_2)$, and thus $\mathrm{Sing}\,\Theta \supset \Theta_1 \times \Theta_2$ is of dimension $g - 2$. It is thus natural to study the following notion:

Definition 5.3 (Andreotti–Mayer loci). We define the k-th Andreotti–Mayer locus to be

$$N_{k,g} := \{(A, \Theta) \in \mathscr{A}_g \mid \dim \mathrm{Sing}\,\Theta \geq k\},$$

By definition we have $N_{k,g} \subset N_{k+1,g}$. Of course we have $N_{g-1,g} = \varnothing$, and by the above we see that $\mathscr{A}_g^{\mathrm{dec}} \subset N_{g-2,g}$. It was conjectured by Arbarello and de Concini [ADC87] and proven by Ein and Lazarsfeld [EL97] that in fact $N_{g-2,g} = \mathscr{A}_g^{\mathrm{dec}}$. Thus one is led to study the next cases, which include the Jacobians:

Theorem 5.4 (Andreotti and Mayer [AM67]). $\mathscr{J}_g$ *is an irreducible component of* $N_{g-4,g}$, *while the locus of hyperelliptic Jacobians* Hyp_g *is an irreducible component of* $N_{g-3,g}$.

Thus we obtain a weak solution to the Schottky problem, and it is natural to ask to describe $(N_{g-4,g} \cap \mathscr{A}_g^{\mathrm{ind}}) \setminus \mathscr{J}_g$ and $(N_{g-3,g} \cap \mathscr{A}_g^{\mathrm{ind}}) \setminus Hyp_g$. In the simplest cases we have $N_{0,3} \cap \mathscr{A}_3^{\mathrm{ind}} = Hyp_3$, while Beauville [Bea77] showed that $N_{0,4}$ in fact has two irreducible components, one of which is $\mathcal{J}_4$, and the other one is the theta-null divisor:

$$\theta_{\text{null},g} := \left\{ \tau \mid \prod_{m \in A[2]^{even}} \theta_m(\tau) = 0 \right\} = \{ (A, \Theta) \in \mathcal{A}_g \mid A[2]^{even} \cap \Theta \neq \varnothing \}.$$

Since in the universal family of ppavs $\mathcal{U}_g \to \mathcal{A}_g$ the locus of (fiberwise) singularities of the theta divisor is given by $g+1$ equations $\theta(\tau, z) = \frac{\partial}{\partial z_1} \theta(\tau, z) = \ldots = \frac{\partial}{\partial z_g} \theta(\tau, z) = 0$, this locus has expected codimension $g+1$ in $\mathcal{U}_g$. Thus $N_{0,g}$, its image under $\mathcal{U}_g \to \mathcal{A}_g$ (with fiber dimension g), has expected codimension 1. It was shown by Beauville [Bea77] that $N_{0,g}$ is indeed always a divisor, and it was shown by Debarre [Deb92a] that for any $g \geq 4$ this divisor has two irreducible components, scheme-theoretically $N_{0,g} = \theta_{\text{null},g} + 2N'_{0,g}$.

One possible complete solution to the Schottky problem would be this:

Conjecture 5.5 (see Debarre [Deb88]). *The following set-theoretic equalities hold*:

$$N_{g-3,g} \cap \mathcal{A}_g^{ind} = Hyp_g, \qquad N_{g-4,g} \setminus \mathcal{J}_g \subset \theta_{\text{null},g},$$

In view of this conjecture it is natural to try to understand the intersection of $\overline{\mathcal{J}}_g$ with the other irreducible components of $N_{g-4,g}$ and, to start with, to study $\overline{\mathcal{J}}_g \cap \theta_{\text{null},g}$. The simplest interesting case is that of $N_{0,4} = \theta_{\text{null},4} \cup \overline{\mathcal{J}}_4$. The following result, conjectured by H. Farkas [Far06], has been proven:

Theorem 5.6 (Grushevsky and Salvati Manni [GSM08], Smith and Varley [SV10]). *The locus of Jacobians of curves of genus 4 with a vanishing theta-null is equal to the locus of 4-dimensional ppavs for which the double point singularity of the theta divisor is not ordinary (i.e., the tangent cone does not have maximal rank): this is to say*

$$\overline{\mathcal{J}}_4 \cap \theta_{\text{null},4} = \left\{ A \in \mathcal{A}_4 \mid \exists m \in A[2]^{even}; \; \theta(\tau, m) = \det_{i,j} \left. \frac{\partial^2 \theta(\tau, z)}{\partial z_i \partial z_j} \right|_{z=m} = 0 \right\}$$

$$= \{ A \in \mathcal{A}_4 \mid \exists m \in A[2]^{even} \cap \Theta; \; TC_m\Theta \text{ has rank } \leq 3 \}.$$

It is thus natural to denote the locus above by $\theta_{\text{null},4}^3$, for rank of the tangent cone being at most 3. In general:

Theorem 5.7 (Debarre [Deb92a], Grushevsky and Salvati Manni [GSM07], Smith and Varley [SV10]).

$$(\overline{\mathcal{J}}_g \cap \theta_{\text{null},g}) \subset \theta_{\text{null},g}^3 \subset \theta_{\text{null},g}^{g-1} \subset (\theta_{\text{null},g} \cap N'_0) \subset \text{Sing } N_0.$$

Here the first inclusion follows from Kempf's [Kem73] generalization of Riemann theta singularity theorem, the second inclusion is obvious by definition, the third inclusion is the content of the theorem, and the fourth inclusion is immediate since the singular locus of a reducible variety contains the intersections of its irreducible components.

At the moment this theorem seems to be the best result in trying to understand the conjecture above, and many further questions remain.

Remark 5.8. The Andreotti–Mayer loci are also of importance for the Prym–Schottky problem. There is an analog of Riemann's theta singularity theorem for Prym varieties (see [CM09]), showing that their theta divisor are also very singular, and the following result holds:

Theorem 5.9 (Debarre [Deb90]). *The Prym locus $\mathcal{P}_g$ is an irreducible component of $N_{g-6,g}$.*

It is thus also natural to try to understand the other components of $N_{g-6,g}$. Debarre ([Deb88], corollary 12.5) constructed some irreducible components of this locus, but even for the simplest case of $g = 6$ the locus $N_{0,6}$ is not completely understood.

One of the main difficulties in following this approach to the Schottky problem is the difficulty in understanding the Andreotti–Mayer loci $N_{k,g}$. In fact even their dimensions are not known, though there is this result:

Conjecture 5.10 (Ciliberto and van der Geer [CvdG00; CvdG08]). *For any $1 \le k \le g - 3$ and for any irreducible component $X \subset N_{k,g}$ such that for $(A, \Theta) \in X$ general $\mathrm{End}(A, \Theta) = \mathbb{Z}$ (in particular $X \not\subset \mathcal{A}_g^{\mathrm{dec}}$), we have*

$$\mathrm{codim}_{\mathcal{A}_g} X \ge \frac{(k+1)(k+2)}{2},$$

with equality only for components $\mathcal{J}_g \subset N_{g-4,g}$ and $\mathrm{Hyp}_g \subset N_{g-3,g}$.

Unfortunately still not much is known about this conjecture: it was shown by Mumford [Mum83] that $N_{1,g} \subsetneq N_{0,g}$, so that codim $N_{1,g} \ge 2$. Ciliberto and van der Geer's [CvdG00; CvdG08] proof of this conjecture for $k = 1$ (which shows that codim $N_{1,g} \ge 3$) is already very hard. To the best of our knowledge even the following question is open:

Open Problem 5. Can it happen that $N_{k,g} = N_{k+1,g}$ for some k, g?

Remark 5.11. We would like to mention here a related question, which does not seem to shed any new light on the Schottky problem, but is of interest in the study of $\mathcal{A}_g$. Indeed, one can study other properties of Θ; and in particular instead of looking at dim Sing Θ one could define the multiplicity loci

$$S_k := \{(A, \Theta) \in \mathcal{A}_g \mid \exists z \in A, \ \mathrm{mult}_z \Theta \ge k\}.$$

It was shown by Kollár [Kol95] that $S_{g+1} = \emptyset$, while Smith and Varley [SV96] showed that S_g is equal to the locus of products of elliptic curves

$$S_g = \mathscr{A}_1 \times \ldots \times \mathscr{A}_1.$$

However, from Riemann's theta singularity theorem it follows that the maximum multiplicity of the theta function on Jacobians is $\lfloor \frac{g+1}{2} \rfloor$, and it can be shown that this is also the case for Pryms (see [CM09]), i.e., that

$$\mathscr{J}_g \cap S_{\lfloor \frac{g+3}{2} \rfloor} = \mathscr{P}_g \cap S_{\lfloor \frac{g+3}{2} \rfloor} = \emptyset.$$

Since one expects the theta divisors of Jacobians and Prym in general to be very singular, this leads to the following natural

Conjecture 5.12.

$$\mathscr{A}_g^{\mathrm{ind}} \cap S_{\lfloor \frac{g+3}{2} \rfloor} = \emptyset.$$

For $g \leq 5$ this conjecture was shown by Casalaina and Martin [CM09] to be true, using the fact that then the Prym locus $\mathscr{P}_g$ is dense in $\mathscr{A}_g$. We refer to [CvdG08; GSM09] for more details and further conjectures on the loci of ppavs with different sorts of singularities of the theta divisor.

To summarize, the Andreotti–Mayer approach gives geometric conditions for a ppav to be a Jacobian, and the original results of Andreotti and Mayer give a geometric weak solution to the Schottky problem (though in practice it is not easy to compute $\dim \mathrm{Sing}\, \Theta_\tau$ for an explicitly given $\tau \in \mathscr{H}_g$). With recent progress, the Andreotti–Mayer approach can be improved to give a complete solution to the Schottky problem in genus 4, since it gives a geometric description of $N_{0,4} \setminus \overline{\mathscr{J}_4}$ as $\theta_{\mathrm{null},4} \setminus \theta_{\mathrm{null},4}^3$. However, already for the $g = 5$ case the situation is not completely understood, though there are compelling conjectures for arbitrary genus. The relationship between the Andreotti–Mayer approach and other properties of the theta divisor were studied in much more detail by Beauville and Debarre [BD86; Deb92c].

6. Subvarieties of a ppav: minimal cohomology classes

One striking approach to the Schottky problem stems from the observation that for a Jacobian $J(C)$ one can naturally map the symmetric product $Sym^d C$ (for $1 \leq d < g$) to $J(C) = \mathrm{Pic}^{g-1}(C)$ by fixing a divisor $D \in \mathrm{Pic}^{g-1-d}(C)$ and mapping $(p_1, \ldots, p_d) \mapsto D + \sum p_i$, The image of such a map, denoted $W^d(C) \subset J(C)$, is independent of D up to a translation, and one can compute its cohomology class

$$[W^d(C)] = \frac{[\Theta]^{g-d}}{(g-d)!} \in H^{2g-2d}(J(C)),$$

where by $[\Theta]$ we denote the cohomology class of the polarization. One can show that this cohomology class is indivisible in cohomology with $\mathbb{Z}$ coefficients, and we thus call this class minimal. Of course these subvarieties of $J(C)$ are very special — in particular $W^1(C) \simeq C$, and one can thus ask whether their existence is a special property of Jacobians.

Theorem 6.1 (Matsusaka [Mat59], Ran [Ran81]). *A ppav (A, Θ) is a Jacobian if and only if there exists a curve $C \subset A$ with $[C] = [\Theta]^{g-1}/(g-1)!$, in which case $(A, \Theta) = J(C)$.*

This gives a complete geometric solution to a weaker form of the Schottky problem: given a pair $C \subset A$, it allows us to determine whether $A = J(C)$, while given only a ppav (A, Θ), this does not provide a way to construct such a curve.

It is then natural to ask what happens for higher-dimensional subvarieties of minimal class. Debarre [Deb95a] proved that for any d the loci $W^d(C)$ and $-W^d(C)$ are the only subvarieties of $J(C)$ of the minimal class, and one can ask whether their existence also characterizes Jacobians. It turns out that for intermediate Jacobians of cubic threefolds (these are in $\mathscr{A}_5$) the Fano surface of lines also has minimal cohomology class (see [CG72; Bea82], and also [Ran82]). Motivated also by works of Ran, Debarre made the following conjecture:

Conjecture 6.2 ([Deb95a]). *If a ppav (A, Θ) has a d-dimensional subvariety of minimal class, then it is either a Jacobian of a curve or a 5-dimensional intermediate Jacobian of a cubic threefold.*

Debarre proves that this does give a weak solution to the Schottky problem, i.e., that $\mathscr{J}_g$ is an irreducible component of the locus of ppavs for which there exists a subvariety of the minimal cohomology class. While the conjecture remains open, an exciting new approach relating the minimal classes to generic vanishing was recently introduced by Pareschi and Popa [PP08b], who also made a refinement of the Conjecture 6.2 involving the theta-dual of a subvariety of a ppav.

Remark 6.3. In analogy with the situation for Jacobians, for the Prym–Schottky problem one is naturally led to look at the Abel–Prym curve

$$\tilde{C} \to J(\tilde{C}) \to Prym(\tilde{C} \to C).$$

It can be shown that this has twice the minimal cohomology class, i.e., $2\dfrac{[\Theta]^{g-1}}{(g-1)!}$, and one wonders whether an analog of Matsusaka–Ran criterion holds in this case. Welters [Wel87] showed that Pryms are an indeed an irreducible component of the locus of ppavs for which there exists a curve representing twice the minimal class, and described the other components of this locus. It is then natural to try to study maps of $Sym^d(\tilde{C})$ to the Prym, and the cohomology classes of their images, but it seems nothing is known here.

To summarize, this approach to the Schottky problem gives a complete geometric solution to the weaker version of the problem: determining whether a given ppav is the Jacobian of a given curve.

7. Projective embeddings of a ppav: the geometry of the Kummer variety

Another approach to the Schottky problem is by embedding a ppav (in fact its quotient under ± 1) into a projective space, and studying the properties of the image. We recall that by definition of a ppav (A, Θ) the line bundle Θ is ample, but since $h^0(A, \Theta) = 1$, clearly Θ is not very ample. The Lefschetz theorem (see [BL04]) states that in fact $n\Theta$ is very ample for any $n \geq 3$, while 2Θ is a very ample line bundle on $A/\pm 1$. Indeed, recall that on A_τ a basis of sections of $2\Theta_\tau$ is given by theta functions of the second order $\{\Theta[\varepsilon](\tau, z)\}$, for all $\varepsilon \in \frac{1}{2}\mathbb{Z}^g/\mathbb{Z}^g$, given by formula (6). Recall that all of these are even functions of z.

Definition 7.1 (Kummer map and Kummer variety). The Kummer map is the embedding

$$Kum : A_\tau/\pm 1 \hookrightarrow \mathbb{P}^{2^g-1}; \quad Kum(z) := \{\Theta[\varepsilon](\tau, z)\}_{\text{all } \varepsilon \in \frac{1}{2}\mathbb{Z}^g/\mathbb{Z}^g} \quad (12)$$

and we call its image $Kum(A_\tau/\pm 1) \subset \mathbb{P}^{2^g-1}$ the Kummer variety. Notice that the involution ± 1 has 2^{2g} fixed points on A, which are precisely $A[2]$, and thus the Kummer variety is singular at their images in $\mathbb{P}^{2^g-1}$.

The Kummer variety is a g-dimensional subvariety of $\mathbb{P}^{2^g-1}$ and one can ask how general it is. The following striking result shows that the Kummer image of a Jacobian is very special (to prove this one uses Riemann's theta singularity theorem to prove Weil reducibility — that the intersection of Θ with some translate is reducible — and then uses Koszul cohomology and Riemann's bilinear relation to obtain an equivalent statement in terms of the Kummer map: see [ACGH85; Tai97] for discussions):

Theorem 7.2 (Fay–Gunning trisecant identity [Fay73; Gun82a]). *For any* p, p_1, p_2, $p_3 \in C$, *the following three points on the Kummer variety are collinear:*

$$Kum(p+p_1-p_2-p_3), \quad Kum(p+p_2-p_1-p_3), \quad Kum(p+p_3-p_1-p_2) \quad (13)$$

(where we view these points as line bundles, i.e., in $\operatorname{Pic}^0(C) = J(C)$*).*

This result can be reformulated from the point of view of the geometry of the complete linear series $|2\Theta|$. Indeed, for any $z_1, z_2, z_3 \in A$ consider the vector subspace of $H^0(A, 2\Theta)$ consisting of sections vanishing at these points. Since Kum is the map given by $H^0(A, 2\Theta)$ the number of conditions imposed on a section of 2Θ by vanishing at these points is equal to the dimension of the linear span of $\{Kum(z_i)\}$. Thus the trisecant identity states that on a Jacobian if the three points are obtained from the points on C as above, the vanishing condition at these 3 points on $J(C)$ fails to impose independent conditions on $|2\Theta|$.

The trisecant identity implies that the Kummer image of a Jacobian admits a 4-dimensional family of trisecant lines (and thus is very far from being in general position in $\mathbb{P}^{2^g-1}$). It is natural to wonder whether this is a characteristic property of Kummer images of Jacobians.

Theorem 7.3 (Gunning [Gun82b]). *If for a ppav $(A, \Theta) \in \mathcal{A}_g^{\mathrm{ind}}$ there exist three fixed points p_1, p_2, $p_3 \in A$ (with some mild general position assumption) such that there exist infinitely many p such that (13) is a trisecant of the Kummer variety, then $A \in \mathcal{J}_g$ (and moreover there exists a translate of $C \subset J(C)$ such that p, p_1, p_2, p_3 are contained in the image).*

The proof of this result proceeds by reduction to the Matsusaka–Ran criterion. Similarly to that result, this theorem gives a complete solution to a weaker version of the Schottky problem: taking the Zariski closure of all p satisfying (13) already gives a curve $C \subset A$, and the criterion can be used to verify whether $(A, \Theta) = J(C)$.

It is thus natural to ask whether one could generalize the statement of the trisecant identity in such a way as to get a characterization of Jacobians not involving a curve contained in the ppav to start with, The general multisecant identity (which is unfortunately less well-known than the trisecant identity) is as follows

Theorem 7.4 (Gunning [Gun86]). *For any curve $C \in \mathcal{M}_g$, for any $1 \le k \le g$ and for any $p_1, \ldots, p_{k+2}, q_1, \ldots, q_k \in C$ the $k + 2$ points of the Kummer variety*

$$Kum\left(2p_j + \sum_{i=1}^{k} q_i - \sum_{i=1}^{k+2} p_i\right), \qquad j = 1, \ldots, k+2 \qquad (14)$$

(where we again identify $J(C) = \mathrm{Pic}^0(C)$) are linearly dependent.

Similarly to the above, this is the statement that the $k+2$ points on the Jacobian constructed as above fail to impose independent conditions on the linear system $|2\Theta|$. By fixing $p_1, \ldots, p_{k+2}$ and varying $q_1, \ldots, q_k$ we can further interpret this as the $k + 2$ fixed points

$$\left\{2p_j - \sum_{i=1}^{k+2} p_i\right\}_{j=1\ldots k+2}$$

failing to impose independent condition on $|2\Theta|$ translated by an arbitrary element $q_1 + \ldots + q_k \in W^k(C)$.

Now we recall that the map $W^g(C) \to J(C)$ is surjective, and thus the $k = g$ case of the multisecant identity above is the statement that some $g + 2$ points on $J(C)$ fail to impose independent conditions on any translate of $|2\Theta|$. This turns out to also be interpretable in terms of some addition theorems for Baker–Akhiezer functions, and the following result was conjectured by Buchstaber and Krichever [BK93; BK96]:

Theorem 7.5 (Grushevsky [Gru04a], Pareschi and Popa [PP08a]). *For any ppav* $(A, \Theta) \in \mathcal{A}_g^{\text{ind}}$ *and any* $p_1, \dots, p_{g+2} \in A$ *(in general position, which is a different condition for the two proofs!), if*

$$\forall z \in A \quad \{Kum(2p_i + z)\}_{i=1\dots g+2} \subset \mathbb{P}^{2^g-1}$$

are linearly dependent, then $A \in \mathcal{J}_g$ *(and moreover all* p_i *lie on a translate of* $C \subset J(C)$*).*

These results give a weak solution (in the sense of possible extra components that appear if we ignore the general position assumption) to the original Schottky problem — i.e., we do not start with a curve $C \subset A$ given. Still the proof proceeds by reduction to Gunning's trisecant criterion above: essentially one considers the loci of all translates of $|2\Theta|$ on which a subset of less than $g + 2$ of the given points fails to impose independent conditions, and once we get to three points not imposing independent conditions, this is a trisecant. Note that this gives a potentially new way of identifying the image of C inside $J(C)$. However, this characterization of Jacobians is still not easy to use explicitly, both because of the general position assumption and because of the necessity to choose/guess appropriate $g + 2$ points in a ppav.

Another approach to the Schottky problem based on the trisecant identity is obtained by degenerating the trisecant. Indeed, if we let all the points p and p_i vary, when all p_i come together, the trisecant will degenerate to a flex line, i.e., to a line tangent to $Kum(A)$ at a point (actually at $p - p_1$) with multiplicity 3. Varying p and $p_1 = p_2 = p_3$, we thus get a family of such flex lines, i.e., a family of partial differential equations for the Kummer image, with parameter $p - p_1$. It was shown by Welters [Wel84] that the existence of such a family of flex lines also characterizes Jacobians — this is again a solution to the weaker version of the Schottky problem, as we start with a given curve in a ppav.

Expanding these differential equations in Taylor series in $p - p_1$ near $p = p_1$ (so that the flex line goes through zero), we get an infinite sequence of partial differential equations of increasing orders for the Kummer variety at 0, essentially each equation saying that there exists a k-jet of a family of flex lines, assuming there exists a $(k-1)$-jet of a family of flexes. This sequence of equations is called the Kadomtsev–Petviashvili (KP) hierarchy. Since we know that for Jacobians there exists a family of flex lines of the Kummer variety, each term of this Taylor expansion, i.e., each equation of the KP hierarchy must be satisfied by the Kummer image of a Jacobian. We refer to [Dub81b; Tai97; BK06] for more on this circle of ideas, and for a much more detailed introduction to the integrable systems point of view on this problem.

Since we are in an algebraic setting, conversely it is natural to expect that if the Kummer variety of a ppav satisfies all the equations of the KP hierarchy, i.e., if it admits an ∞-jet of a family of flex lines, then there exists an actual family of flex lines, and thus the ppav is a Jacobian by the result of Welters. Moreover, from algebraicity one could expect that a finite number of equations from the KP hierarchy would suffice; this was proven by Arbarello and de Concini [ADC84] to indeed be the case.

Novikov conjectured that in fact the first equation of the KP hierarchy would already suffice to characterize Jacobians, i.e., that if for some (indecomposable) ppav there exists a 1-jet of a family of flex lines of the Kummer variety at 0, then the ppav is a Jacobian. Explicitly this first equation of the KP hierarchy, called the KP equation, written in terms of theta functions of the second order, has the following form [Dub81a]:

$$\left(\frac{\partial^4}{\partial U^4} + \frac{3}{4} \frac{\partial^2}{\partial V^2} - \frac{\partial^2}{\partial U \partial W} + C \right) \Theta[\varepsilon](\tau, z)|_{z=0} = 0 \quad \forall \varepsilon \in \tfrac{1}{2}\mathbb{Z}^g / \mathbb{Z}^g, \quad (15)$$

for some $U, V, W \in \mathbb{C}^g, C \in \mathbb{C}$. Essentially this equation is the condition that the Kummer image of 0 and the second and fourth derivatives of the Kummer image at 0 are linearly dependent — though the KP is more restrictive, as the coefficients of this linear dependence are not arbitrary. Novikov's conjecture was proven:

Theorem 7.6 (Shiota [Shi86]). *If the Kummer variety of a ppav $A \in \mathcal{A}_g^{\text{ind}}$ admits a 1-jet of a family of flex lines at 0, i.e., if there exist U, V, W, C such that (15) is satisfied, then $A \in \mathcal{J}_g$.*

This result is much stronger than the previous theorems for two reasons: there is no general position assumption, and to apply this theorem we do not need start with a curve in A (though we still need to choose the parameters U, V, W, C in the KP equation).

One can then wonder whether a still stronger version of this characterization of Jacobians may hold — whether the existence of just one trisecant (or one flex line) may suffice. Welters [Wel84] boldly conjectured that the existence of one trisecant of the Kummer variety already guarantees that a ppav is a Jacobian, There have been numerous results in this direction, proving the conjecture under various general position assumptions (see Debarre [Deb92b; Deb97]), and the conjecture was recently proven completely, with no general position assumption:

Theorem 7.7 (Krichever [Kri06; Kri10]). *For a ppav* $A \in \mathcal{A}_g^{\text{ind}}$, *if* $Kum(A) \subset \mathbb{P}^{2^g-1}$ *has one of the following*:

- *a trisecant line,*
- *a line tangent to it at one point, and intersecting it at another* (*this is a semidegenerate trisecant, when two points of secancy coincide*),
- *a flex line* (*this is a most degenerate trisecant when all three points of secancy coincide*),

such that none of the points of intersection of this line with the Kummer variety are in $A[2]$ (*where* $Kum(A)$ *is singular*), *then* $A \in \mathcal{J}_g$.

Here we still need to be able to choose the 3 points of trisecancy or the parameters of a line, but there is no curve involved, and for a nondegenerate trisecant we do not even need to consider the derivatives of the Kummer map. It is thus natural to expect that this result would be much harder to prove, as the jet of the curve needs to be constructed from scratch. There are many conceptual and technical difficulties in the proof, and methods of integrable systems are used extensively.

For the Prym–Schottky problem, it is then natural to wonder whether the Kummer varieties of Pryms have any special geometric properties, Indeed, the following result was obtained:

Theorem 7.8 (quadrisecant identity: Fay [Fay84], Beauville and Debarre [BD87]). *For any* $p, p_1, p_2, p_3 \in \widetilde{C} \to Prym(\widetilde{C} \to C)$ *on the Abel–Prym curve the four points of the Kummer variety*

$$\begin{aligned} Kum(p + p_1 + p_2 + p_3), \quad Kum(p + p_1 - p_2 - p_3), \\ Kum(p + p_2 - p_1 - p_3), \quad Kum(p + p_3 - p_1 - p_2) \end{aligned} \tag{16}$$

lie on a 2-plane in $\mathbb{P}^{2^g-1}$.

Remark 7.9. Note that there is something slightly puzzling here — while for the trisecant identity all the points were naturally in $J(C) = \text{Pic}^0(C)$, here the degrees of the four divisors on $\widetilde{C}$ are $2, 0, 0, 0$, so that one needs to clarify what exactly is meant by the points above. Analytically one can lift the entire discussion to the universal cover $\mathcal{H}_g \times \mathbb{C}^g$ of the universal family; we refer to [Mum74; BD87; CM09] for more on divisors on Prym varieties.

Analogously to Gunning's characterization of Jacobians by the existence of a family of trisecants, we have:

Theorem 7.10 (Debarre [Deb92c]). *If for a ppav* $(A, \Theta) \in \mathscr{A}_g^{\text{ind}}$ *there exist three fixed points* $p_1, p_2, p_3 \in A$ *(with some mild general position assumption) such that there exist infinitely many* p *such that* (16) *is a quadrisecant 2-plane of the Kummer variety, then* $A \in \overline{\mathscr{P}}_g$ *(and moreover* p, p_1, p_2, p_3 *lie on an Abel–Prym curve).*

Following the outline of our discussion for Jacobians, it is now natural to ask whether one could increase the number of points and get a suitable multisecant identity for Pryms. It appears that by generalizing the methods of [Fay84; BD87; Gun86] one could indeed construct a $(2k + 2)$-dimensional family of $(2k + 2)$-secant $2k$-planes of the Kummer images of Pryms (for comparison, Gunning's multisecant identity for Kummer varieties of Jacobians gives a $(2k+2)$-dimensional family of $(k + 2)$-secant k-planes). In particular, to try to emulate the results of [Gru04a; PP08a] one would need to consider $(2g + 2)$-secant $2g$-planes — equivalently sets of $2g + 2$ points imposing linearly dependent conditions on an arbitrary translate of $|2\Theta|$ (for comparison, for Jacobians we had $g + 2$ points). This gets very complicated, and in particular the following questions seems intriguing and wide open:

Open Problem 6. For some $A \in \mathscr{A}_g^{\text{ind}}$ suppose there exist k points

$$p_1, \ldots, p_k \in A$$

in general position imposing less than k independent conditions on any translate of $|2\Theta|$ (equivalently, such that for any $z \in A$ the Kummer images $\{Kum(p_i + z)\}$ are linearly dependent). For which maximal k does this imply that $A \in \overline{\mathscr{P}}_g$ (we know that $k = g + 2$ implies $A \in \mathscr{J}_g \subset \overline{\mathscr{P}}_g$)? It is likely that the maximum that k could be is at most $2g + 2$, so what loci within $\mathscr{A}_g^{\text{ind}}$ are characterized by this property for $g + 2 < k < 2g + 2$? Do they give examples of some natural loci $\mathscr{J}_g \subsetneq X \subsetneq \overline{\mathscr{P}}_g$, or do Pryms satisfy this condition with $k \leq g + 5$ (recall that for $g \geq 5$ we have $\dim \overline{\mathscr{P}}_g = 3g = 3 + (3g - 3) = 3 + \dim \mathcal{M}_g$)?

Of course the above characterization of Pryms by a family of quadrisecant planes makes one wonder whether one could construct an integrable hierarchy of partial differential equations satisfied by the theta function of Pryms, by allowing the quadrisecant plane to degenerate appropriately. This is indeed possible, and one can thus obtain the Novikov–Veselov, the BKP, and the Landau–Lifshitz hierarchies. One then naturally expects finitely many of the equations in these hierarchies to characterize Pryms — which is indeed the case. However, the analog of the Novikov conjecture — that the first equation of one of these hierarchies characterizes Pryms — is open, see [Tai87]. It is only known (Shiota [Shi89b; Shi89a]) that the BKP equation characterizes Pryms under a certain general position assumption. We refer to [Tai97] for a detailed survey of what is known about the Prym–Schottky problem and soliton equations.

Similarly to the trisecant conjecture for Jacobians, one can wonder whether the existence of one quadrisecant 2-plane of the Kummer variety may characterize Pryms. This turns out not to be the case: Beauville and Debarre [BD87] construct an example of $A \in \mathcal{A}_g^{\mathrm{ind}} \setminus \overline{\mathcal{P}}_g$ such that $Kum(A)$ has a quadrisecant 2-plane. Thus no analog of quadrisecant conjecture has been proposed for Pryms, though Beauville and Debarre proved in [BD87] that $\mathcal{P}_g$ is an irreducible component of the locus in $\mathcal{A}_g$ consisting of ppavs whose Kummer varieties admit a quadrisecant 2-plane.

A suitable analog of the trisecant conjecture was recently found for Pryms using ideas of integrable systems; it admits a clear geometric formulation:

Theorem 7.11 (Grushevsky and Krichever [GK10]). *If for some $A \in \mathcal{A}_g^{\mathrm{ind}}$ and some $p, p_1, p_2, p_3 \in A$ the quadrisecant condition* (16) *holds, and moreover there exists another quadrisecant given by* (16) *with p replaced by $-p$, then $A \in \overline{\mathcal{P}}_g$.*

One can think of this pair of quadrisecants in the following way: recall that $Prym(\widetilde{C} \to C)$ is the connected component of zero in the kernel of the map $J(\widetilde{C}) \to J(C)$, and that this kernel has two connected components. The theorem essentially says that both the Prym and the other component of this kernel must admit a quadrisecant.

We note that similarly to the formulation of the trisecant conjecture, this theorem is a complete (not weak) solution to the Prym–Schottky problem. There is no general position assumption here (which would result in extra components), and one does not start with the Abel–Prym curve. The proof of this result is very involved and again uses the ideas coming from integrable systems to construct first an infinitesimal jet of $\widetilde{C} \to C$ and then argue that there indeed exists such a cover.

Remark 7.12. One possible very interesting question to which one could try to apply the above theorem is to try to approach the Torelli problem for Pryms: indeed, the map from the moduli space of étale double covers of curves of genus g to $\mathcal{A}_{g-1}$ is known to be generically injective for $g \geq 7$ (see [DS81] for more on the Prym map), but it is never injective, since Donagi's tetragonal construction [Don81] can be used to construct different covers with isomorphic Pryms. It was conjectured (see Donagi [Don81], Debarre [Deb89]) that for $g \geq 11$ this was the only cause of noninjectivity of the Prym Torelli map. However, a recent preprint of Izadi and Lange [IL10] constructs examples of curves of arbitrary Clifford index (thus the tetragonal construction does not apply) such that the Prym Torelli map is not injective at the corresponding Prym. It is tempting to try to apply the above characterization of Pryms to study this problem: the data of the two quadrisecants in fact recovers the cover $\tilde{C} \to C$, not only the Prym, and thus the question is whether a given Prym may have more than one family of quadrisecants.

To summarize, the approach to the Schottky problem via the geometry of the Kummer variety yields complete (strong — no extra components) solutions to the Schottky and Prym–Schottky problems. These solutions do not require an a priori knowledge of a curve, as does the minimal cohomology class approach. However, there is still a finite amount of data involved — choosing the points of secancy. Thus given an explicit ppav it is not yet clear how to apply these solutions of the Schottky problem or to relate this approach to the Schottky–Jung approach; in particular it does not yet seem possible to apply the results in this section to solve either Open Problem 1 or Open Problem 2.

8. The Γ_{00} conjecture

In the previous section we saw that the geometry of the Kummer variety, i.e., the geometry of the $|2\Theta|$ linear system, is of great importance in studying the Schottky problem. The second order theta functions $\Theta[\varepsilon](\tau, z)$ seem to be central to the study of the Schottky problem: their values at $z = 0$, for τ varying, are theta constants giving the embedding $Th : \mathcal{A}_g(2, 4) \to \mathbb{P}^{2^g-1}$, which is the subject of the classical and the Schottky–Jung approaches (see sections 3,4); while for τ fixed and z varying theta functions of the second order define the map $Kum : A_\tau / \pm 1 \hookrightarrow \mathbb{P}^{2^g-1}$, properties of which we discussed in Section 7 (in fact the intersection of the images of Th and Kum in $\mathbb{P}^{2^g-1}$ is also of interest — see [vGvdG86]).

The difficulty in trying to relate the geometric properties of the Kummer image (e.g., the trisecants), or other properties of the theta divisor — e.g., the singularities, as in Section 5, and algebraic equation for theta constants is that it is not clear how to capture the properties of Θ and $Kum(A)$ just from $Th(A_\tau) = Kum(0)$.

Remark 8.1. An interesting problem showing that we are yet unable to translate geometric conditions for the theta divisor into modular forms is as follows. Casalaina-Martin and Friedman [CMF05; CM08] showed that intermediate Jacobians of cubic threefolds (which we already mentioned in relation to the Schottky–Jung approach) are characterized within $\mathcal{A}_5$ by the property of having a single point of multiplicity 3 on Θ (this point is thus in $A[2]^{odd}$). However, the following question remains open; solving it could also yield a solution to

Open Problem 1, as it would potentially allow one to further refine the Schottky–Jung approach in genus 5, dealing with the big Schottky–Jung locus, and in particular possibly proving Conjecture 4.5 for $g = 5$:

Open Problem 7. Determine the defining ideal for $\overline{Th(\mathcal{I}(2,4))} \subset \overline{Th(\mathcal{A}_5(2,4))}$, where $\mathcal{I} \subset \mathcal{A}_5$ denotes the locus of intermediate Jacobians of cubic threefolds.

We note that it is also very interesting to try to understand the geometry of the boundary of $\mathcal{I}$, and work on this has been done by Casalaina-Martin and Laza [CML09]; see also [GSM07].

Thus it is natural to try to study the properties of $|2\Theta|$ at $0 \in A$. Note that the statement of the trisecant conjecture explicitly excludes 0, because the Kummer variety is singular at $Kum(0)$. One thus wonders what an appropriate notion of a degenerate trisecant through $Kum(0)$ might be, and it is natural to expect that it would involve the tangent cone to the Kummer variety at 0.

Definition 8.2 (van Geemen and van der Geer [vGvdG86]). The linear system $\Gamma_{00} \subset |2\Theta|$ is defined to consist of those sections that vanish to order at least 4 at the origin:

$$\Gamma_{00} = \{f \in H^0(A, 2\Theta) \mid \text{mult}_0 \, f \geq 4\},$$

We note that since all sections of 2Θ are even functions of z, all first and third partial derivatives vanish automatically, and thus the condition for f to lie in Γ_{00} is really for the f and its second partial derivatives to vanish at zero. We recall now that theta functions of the second order generate $H^0(A, 2\Theta)$, so that

$$f(z) = \sum_{\varepsilon \in \frac{1}{2}\mathbb{Z}^g / \mathbb{Z}^g} c_\varepsilon \Theta[\varepsilon](\tau, z).$$

and thus the condition $f \in \Gamma_{00}$ is equivalent to this system of $\frac{g(g+1)}{2} + 1$ linear equations for the coefficients $\{c_\varepsilon\}$:

$$0 = \sum c_\varepsilon \Theta[\varepsilon](\tau, 0) = \sum c_\varepsilon \frac{\partial^2 \Theta[\varepsilon](\tau, z)}{\partial z_a \partial z_b}\bigg|_{z=0} \quad \text{for } 1 \leq a \leq b \leq g.$$

One can show (see [Dub81a], for example) that for any ppav in $\mathcal{A}_g^{\text{ind}}$ the rank of this linear system of equations is maximal, and thus that

$$\dim \Gamma_{00} = 2^g - \frac{g(g+1)}{2} - 1$$

on any indecomposable ppav (in particular it is empty for $g = 1, 2$, and one-dimensional for $g = 3$).

Remark 8.3. The condition $f \in \Gamma_{00}$ can actually be reformulated purely in terms of τ, not involving z. The Riemann theta function satisfies the heat equation, and so do the theta functions of the second order, for which we have

$$\frac{\partial^2}{\partial z_i \partial z_j}\Theta[\varepsilon](\tau, z) = \frac{2\pi i}{1 + \delta_{a,b}} \frac{\partial}{\partial \tau_{ab}}\Theta[\varepsilon](\tau, z)$$

(this can be easily verified from definition, by differentiating the Fourier series term by term). Thus one can study the linear system Γ_{00} purely in terms of theta constants of the second order and their derivatives (with respect to τ), which may give a hope of eventually relating this to modular forms and the Schottky–Jung approach.

A priori the base locus $Bs(\Gamma_{00})$ contains 0, and it is not clear why it would contain other points. From the description above of Γ_{00} as a system of linear equations on coefficients c_ε it follows that the condition $z \in Bs(\Gamma_{00})$ is equivalent to

$$Kum(z) \in \left\langle Kum(0), \frac{\partial}{\partial \tau_{ab}}Kum(0)\right\rangle_{\text{linear span}},$$

i.e., to the existence of $c, c_{ab} \in \mathbb{C}$ such that

$$\Theta[\varepsilon](\tau, z) = c\Theta[\varepsilon](\tau, 0) + \sum_{1 \le a \le b \le g} c_{ab}\frac{\partial\Theta[\varepsilon](\tau, 0)}{\partial \tau_{ab}} \quad \forall \varepsilon \in \tfrac{1}{2}\mathbb{Z}^g/\mathbb{Z}^g. \quad (17)$$

Indeed, $z \in Bs(\Gamma_{00})$ means that z imposes no conditions on Γ_{00}, and thus that the vanishing of $f(z)$ must be a consequence of f satisfying the linear conditions defining Γ_{00}.

We note that this condition is indeed very similar to the condition of the existence of a semidegenerate trisecant of a Kummer variety — in essence it is saying that a line through $Kum(z)$ and $Kum(0)$ lies in the tangent cone to the Kummer variety at 0. However, we are unaware of any formal relationship or implication between this statement and that of a semidegenerate trisecant through a point not of order two.

For Jacobians one can use Riemann's theta singularity theorem combined with Riemann's bilinear addition formula to show that the difference variety $C - C := \{p - q \mid p, q \in C\} \subset \text{Pic}^0(C) = J(C)$ is contained in $Bs(\Gamma_{00})$ — this is essentially due to Frobenius [Fro85], see also [vGvdG86]. Moreover, it turns out that for $p, q \in C \subset J(C)$ the rank of the matrix of coefficients c_{ab} in (17) is equal to one.

It turns out that this is the entire base locus:

Theorem 8.4 (Welters [Wel86] set-theoretically; Izadi [Iza91] scheme-theoretically). *For any $g \ge 5$ and any $C \in \mathcal{M}_g$ we have on $J(C)$ the equality $Bs(\Gamma_{00}) = C - C$.*

One then wonders whether this property characterizes Jacobians:

Conjecture 8.5 (Γ_{00} conjecture: van Geemen and van der Geer [vGvdG86]). *If* $Bs(\Gamma_{00}) \neq \{0\}$ *for some* $(A, \Theta) \in \mathcal{A}_g^{\text{ind}}$, *then* $A \in \mathcal{J}_g$.

This conjecture was proven for $g = 4$ by Izadi [Iza95], for a generic Prym for $g \geq 8$ by Izadi [Iza99], and for a generic ppav for $g = 5$ or $g \geq 14$ by Beauville, Debarre, Donagi, and van der Geer [BDDvdG88]. However, in order to obtain a solution of the Schottky problem, proving the Γ_{00} conjecture for a generic ppav is insufficient — a generic ppav is not a Jacobian anyway. At the moment there does not seem to be a promising algebro-geometric approach to the Γ_{00} conjecture in general, but the following result was recently obtained by using integrable systems methods:

Theorem 8.6 (Grushevsky [Gru10]). *If for some* $A \in \mathcal{A}_g^{\text{ind}}$ *the linear dependence* (17) *holds with* $\text{rk}(c_{ab}) = 1$, *then* $A \in \mathcal{J}_g$.

The proof of this result actually exhibits the similarity between the Γ_{00} conjecture and Krichever's characterization of Jacobians by a semidegenerate trisecant — both of these conditions turn out to imply the same differential-difference equation for the theta function (which is what eventually characterizes Jacobians analytically).

Remark 8.7. The Γ_{00} conjecture turns out to also be related to the Schottky–Jung theory. Indeed, one boundary component of the Deligne–Mumford compactification $\overline{\mathcal{M}}_g$ is the moduli space of pointed curves $\mathcal{M}_{g-1,2}$. Given a modular form vanishing identically on $\overline{\mathcal{J}}_g$, one can consider its restriction to the boundary. After a suitable blowup it means that one is considering the next term of the Taylor expansion (called Fourier–Jacobi expansion in this case) of this modular form near the boundary of $\overline{\mathcal{M}}_g$, and the vanishing of this modular form would imply — and may be implied — by the identical vanishing of this next term in the expansion along $\mathcal{M}_{g-1,2}$, which is the question of a certain modular function $F(\tau, z)$ lying in Γ_{00} for all τ (see [GSM10b; GSM10a] for a demonstration of how this works, and applications to some questions in perturbative string theory). Muñoz-Porras suggested using Γ_{00} conjecture as an approach to proving Conjecture 4.5 by such degeneration methods, but this has not yet been accomplished.

The Γ_{00} linear system also turns out to be conjecturally related to another approach to the Schottky problem. This line of investigation started probably from the work of Buser and Sarnak [BS94] who showed that for g large, the length of the shortest period of a Jacobian (i.e., the vector in the lattice $\mathbb{Z}^g + \tau\mathbb{Z}^g$ with minimal length) is much less than for a generic ppav. This immediately gives an effective combinatorial way to show that some ppavs are not Jacobians — if their periods are too long — but being an open condition of course cannot characterize Jacobians. In this direction, of finding explicit conditions guaranteeing that a ppav is *not* a Jacobian, there has also been recent progress in working with ppavs admitting extra automorphisms — note in particular Zarhin's work [Zar09b; Zar09a], where examples of ppavs with automorphisms that are not Jacobians are constructed.

The length of the shortest period was related to the value of the Seshadri constant by Lazarsfeld [Laz96], and Seshadri constants on abelian varieties were further studied by Nakamaye [Nak96] and Bauer and Szemberg [Bau98; BS97]. It turns out that for g large the Seshadri constant for any Jacobian is much smaller than for a generic ppav of that dimension. Unlike the length of the periods, which cannot possibly characterize Jacobians, it turns out that the Seshadri constant may actually give a solution at least to the hyperelliptic Schottky problem, and is related to Γ_{00}:

Theorem 8.8 (Debarre [Deb04]). *If the Γ_{00} conjecture holds, hyperelliptic Jacobians are characterized in $\mathcal{A}_g$ by the value of their Seshadri constant.*

It thus seems that the Γ_{00} conjecture approach, though not yet culminating even in a weak solution to the Schottky problem, is related to many other approaches to the Schottky problem, and may possibly serve as a tool to further study the geometric properties of the Jacobian locus $\mathcal{J}_g \subset \mathcal{A}_g$ by degeneration methods.

Acknowledgements

This paper grew out of a talk given at the "Classical algebraic geometry today" workshop during the Algebraic Geometry program at MSRI in January 2009. I am grateful to the organizers of that conference for the opportunity to present the talk and especially to Mircea Mustaţă and Mihnea Popa for the invitation and encouragement to write this survey. I would also like to thank Igor Krichever, Mihnea Popa, and Riccardo Salvati Manni for many conversations over the years, from which I learned a lot about different viewpoints and results on the Schottky problem. I am grateful to Olivier Debarre, Mihnea Popa, and the referee for carefully reading the manuscript and for numerous useful suggestions and advice. We thank Enrico Arbarello and Yuri Zarhin for drawing our attention to their works [Arb99; Zar09b; Zar09a].

References

[Acc83] R. Accola, "On defining equations for the Jacobian locus in genus five", *Proc. Amer. Math. Soc.* **89**:3 (1983), 445–448.

[Arb99] E. Arbarello, "Survey of work on the Schottky problem up to 1996", pages 287–293 and 301 in *The red book of varieties and schemes*, by D. Mumford, expanded ed., Lecture Notes in Mathematics **1358**, Springer, Berlin, 1999.

[ACGH85] E. Arbarello, M. Cornalba, P. A. Griffiths, and J. Harris, *Geometry of algebraic curves, I*, Grundlehren der Math. Wissenschaften **267**, Springer, New York, 1985.

THE SCHOTTKY PROBLEM 159

[ADC84] E. Arbarello and C. De Concini, "On a set of equations characterizing Riemann matrices", *Ann. of Math.* **120**:1 (1984), 119–140.

[ADC87] E. Arbarello and C. De Concini, "Another proof of a conjecture of S. P. Novikov on periods of abelian integrals on Riemann surfaces", *Duke Math. J.* **54**:1 (1987), 163–178.

[AM67] A. Andreotti and A. L. Mayer, "On period relations for abelian integrals on algebraic curves", *Ann. Scuola Norm. Sup. Pisa* (3) **21** (1967), 189–238.

[And89] Y. André, *G-functions and geometry*, Aspects of Mathematics E13, Vieweg, Braunschweig, 1989.

[Bau98] T. Bauer, "Seshadri constants and periods of polarized abelian varieties", *Math. Ann.* **312**:4 (1998), 607–623.

[BD86] A. Beauville and O. Debarre, "Une relation entre deux approches du problème de Schottky", *Invent. Math.* **86**:1 (1986), 195–207.

[BD87] A. Beauville and O. Debarre, "Sur le problème de Schottky pour les variétés de Prym", *Ann. Scuola Norm. Sup. Pisa Cl. Sci.* (4) **14**:4, (1988), 613–623.

[BDDvdG88] A. Beauville, O. Debarre, R. Donagi, and G. van der Geer, "Sur les functions thêta d'ordre deux et les singularités du diviseur thêta", *C. R. Acad. Sci. Paris Sér. I Math.* **307**:9 (1988), 481–484.

[Bea77] A. Beauville, "Prym varieties and the Schottky problem", *Invent. Math.* **41**:2 (1977), 149–196.

[Bea82] A. Beauville, "Sous-variétés spéciales des variétés de Prym", *Compositio Math.* **45**:3 (1982), 357–383.

[Bea88] A. Beauville, "Le problème de Schottky: une introduction", *Gaz. Math.* **37** (1988), 54–63.

[BK93] V. Buchstaber and I. Krichever, "Vector addition theorems and Baker–Akhiezer functions", *Teoret. Mat. Fiz.* **94**:2 (1993), 200–212.

[BK96] V. Buchstaber and I. Krichever, "Multidimensional vector addition theorems and the Riemann theta functions", *Internat. Math. Res. Notices* **1996**:10 (1996), 505–513.

[BK06] V. Bukhshtaber and I. Krichever, "Integrable equations, addition theorems, and the Riemann–Schottky problem", *Uspekhi Mat. Nauk* **61** (2006), 25–84.

[BL04] C. Birkenhake and H. Lange, *Complex abelian varieties*, 2nd ed., Grundlehren der Math. Wissenschaften **302**, Springer, Berlin, 2004.

[BS94] P. Buser and P. Sarnak, "On the period matrix of a Riemann surface of large genus", *Invent. Math.* **117**:1 (1994), 27–56.

[BS97] T. Bauer and T. Szemberg, "Higher order embeddings of abelian varieties", *Math. Z.* **224**:3 (1997), 449–455.

[CDPvG08a] S. Cacciatori, F. Dalla Piazza, and B. van Geemen, "Genus four superstring measures", *Lett. Math. Phys.* **85**:2–3 (2008), 185–193.

[CDPvG08b] S. Cacciatori, F. Dalla Piazza, and B. van Geemen, "Modular forms and three-loop superstring amplitudes", *Nuclear Phys. B* **800**:3 (2008), 565–590.

[CG72] H. Clemens and P. Griffiths, "The intermediate Jacobian of the cubic threefold", *Ann. of Math.*, 95 (1972), 281–356.

[CM08] S. Casalaina-Martin, "Cubic threefolds and abelian varieties of dimension five. II", *Math. Z.* **260**:1 (2008), 115–125.

[CM09] S. Casalaina-Martin, "Singularities of the Prym theta divisor", *Ann. of Math.* **170**:1 (2009), 162–204.

[CMF05] S. Casalaina-Martin and R. Friedman, "Cubic threefolds and abelian varieties of dimension five", *J. Algebraic Geom.* **14**:2 (2005), 295–326.

[CML09] S. Casalaina-Martin and R. Laza, "The moduli space of cubic threefolds via degenerations of the intermediate Jacobian", *J. Reine Angew. Math.*, 633 (2009), 29–65.

[Col87] R. Coleman, "Torsion points on curves", pages 235–247 in *Galois representations and arithmetic algebraic geometry* (Kyoto, 1985/Tokyo, 1986), Adv. Stud. Pure Math. **12**, North-Holland, Amsterdam, 1987.

[CS99] J. H. Conway and N. J. A. Sloane, *Sphere packings, lattices and groups*, 3rd ed., Grundlehren der Math. Wissenschaften **290**, Springer, New York, 1999.

[CvdG00] C. Ciliberto and G. van der Geer, "The moduli space of abelian varieties and the singularities of the theta divisor", pages 61–81 in *Surveys in differential geometry*, Surv. Differ. Geom., VII, Int. Press, Somerville, MA, 2000,

[CvdG08] C. Ciliberto and G. van der Geer, "Andreotti–Mayer loci and the Schottky problem", *Doc. Math.*, 13 (2008), 453–504.

[Deb88] O. Debarre, "Sur les variétés abéliennes dont le diviseur thêta est singulier en codimension 3", *Duke Math. J.* **57**:1 (1988), 221–273.

[Deb89] O. Debarre, "Sur le problème de Torelli pour les variétés de Prym", *Amer. J. Math.* **111**:1 (1989), 111–134.

[Deb90] O. Debarre, "Variétés de Prym et ensembles d'Andreotti et Mayer", *Duke Math. J.* **60**:3 (1990), 599–630.

[Deb92a] O. Debarre, "Le lieu des variétés abéliennes dont le diviseur thêta est singulier a deux composantes", *Ann. Sci. École Norm. Sup.* (4) **25**:6 (1992), 687–707.

[Deb92b] O. Debarre, "Trisecant lines and Jacobians", *J. Algebraic Geom.* **1**:1 (1992), 5–14.

[Deb92c] O. Debarre, *Vers une stratification de l'espace des modules des variétés abéliennes principalement polarisées*, Lecture Notes in Math. **1507**, pages 71–86, Springer, Berlin, 1992.

[Deb95a] O. Debarre, "Minimal cohomology classes and Jacobians", *J. Algebraic Geom.* **4**:2 (1995), 321–335.

[Deb95b] O. Debarre, *The Schottky problem: an update*, Math. Sci. Res. Inst. Publ. **28**, pages 57–64, Cambridge Univ. Press, Cambridge, 1995.

[Deb97] O. Debarre, "Trisecant lines and Jacobians. II", *Compositio Math.* **107**:2 (1997), 177–186.

[Deb04] O. Debarre, "Seshadri constants of abelian varieties", pages 379–394 in *The Fano Conference*. Univ. Torino, Turin, 2004.

[DM69] P. Deligne and D. Mumford, "The irreducibility of the space of curves of given genus", *Inst. Hautes Études Sci. Publ. Math.* **36** (1969), 75–109.

[Don81] R. Donagi, "The tetragonal construction", *Bull. Amer. Math. Soc.* (*N.S.*) **4**:2 (1981), 181–185.

[Don87a] R. Donagi, "Big Schottky", *Invent. Math.* **89**:3 (1987), 569–599.

[Don87b] R. Donagi, "Non-Jacobians in the Schottky loci", *Ann. of Math.* **126**:1 (1987), 193–217.

[Don88] R. Donagi, "The Schottky problem", pages 84–137 in *Theory of moduli* (Montecatini Terme, 1985), Lecture Notes in Math. **1337**, Springer, Berlin, 1988.

[DP02a] E. D'Hoker and D. H. Phong, "Two-loop superstrings, I: Main formulas", *Phys. Lett. B* **529**:3-4 (2002), 241–255.

[DP02b] E. D'Hoker and D. H. Phong, "Two-loop superstrings, II: The chiral measure on moduli space", *Nuclear Phys. B* **636**:1-2 (2002), 3–60.

[DP02c] E. D'Hoker and D. H. Phong, "Two-loop superstrings, III: Slice independence and absence of ambiguities", *Nuclear Phys. B* **636**:1-2 (2002), 61–79.

[DP02d] E. D'Hoker and D. H. Phong, "Two-loop superstrings, IV: The cosmological constant and modular forms", *Nuclear Phys. B* **639**:1-2 (2002), 129–181.

[DP05a] E. D'Hoker and D. H. Phong, "Asyzygies, modular forms, and the superstring measure, I", *Nuclear Phys. B* **710**:1-2 (2005), 58–82.

[DP05b] E. D'Hoker and D. H. Phong, "Asyzygies, modular forms, and the superstring measure, II", *Nuclear Phys. B* **710**:1-2 (2005), 83–116.

[DS81] R. Donagi and R. Smith, "The structure of the Prym map", *Acta Math.* **146**:1-2 (1981), 25–102.

[Dub81a] B. Dubrovin, "The Kadomtsev–Petviashvili equation and relations between periods of holomorphic differentials on Riemann surfaces", *Izv. Akad. Nauk SSSR Ser. Mat.* **45**:5 (1198), 1015–1028, 1981.

[Dub81b] B. A. Dubrovin, "Theta-functions and nonlinear equations", *Uspekhi Mat. Nauk* **36**:2 (1981), 11–80.

[EL97] L. Ein and R. Lazarsfeld, "Singularities of theta divisors and the birational geometry of irregular varieties", *J. Amer. Math. Soc.* **10**:1 (1997), 243–258.

[Fab99] C. Faber, "Algorithms for computing intersection numbers on moduli spaces of curves, with an application to the class of the locus of Jacobians", pages 93–109 in *New trends in algebraic geometry* (Warwick, 1996), London Math. Soc. Lecture Note Ser. **264**, Cambridge Univ. Press, Cambridge, 1999.

[Far89] H. Farkas, *Schottky–Jung theory*, Proc. Sympos. Pure Math. **49**, pages 459–483, Amer. Math. Soc., Providence, RI, 1989.

[Far06] H. Farkas, "Vanishing thetanulls and Jacobians", pages 37–53 in *The geometry of Riemann surfaces and abelian varieties*, Contemp. Math. **397**, Amer. Math. Soc,. Providence, RI, 2006.

[Fay73] J. Fay, *Theta functions on Riemann surfaces*. Lecture Notes in Mathematics, Vol. 352. Springer, Berlin, 1973.

[Fay84] J. Fay, "On the even-order vanishing of Jacobian theta functions", *Duke Math. J.* **51**:1 (1984), 109–132.

[FR70] H. Farkas and H. Rauch, "Period relations of Schottky type on Riemann surfaces", *Ann. of Math.* (2) **92** (1970), 434–461.

[Fre83] E. Freitag, "Die Irreduzibilität der Schottkyrelation (Bemerkung zu einem Satz von J. Igusa)", *Arch. Math. (Basel)* **40**:3 (1983), 255–259.

[Fro85] G. Frobenius, "Über die Constanten factoren der Thetareihen", *J. Reine Angew. Math.*, 98 (1885), 244–265.

[GK10] S. Grushevsky and I. Krichever, "Integrable discrete Schrödinger equations and a characterization of Prym varieties by a pair of quadrisecants", *Duke Math. J.*, 2010. to appear.

[Gru04a] S. Grushevsky, "Cubic equations for the hyperelliptic locus", *Asian J. Math.* **8**:1 (2004), 161–172.

[Gru04b] S. Grushevsky, "Effective algebraic Schottky problem", preprint math.AG/0403009, 2004.

[Gru09] S. Grushevsky, "Superstring scattering amplitudes in higher genus", *Comm. Math. Phys.* **287**:2 (2009), 749–767.

[Gru10] S. Grushevsky, *A special case of the Γ_{00} conjecture*, Progress in Mathematics, **280**, Birkhäuser, 2010.

[GSM07] S. Grushevsky and R. Salvati Manni, "Singularities of the theta divisor at points of order two", *Int. Math. Res. Not. IMRN*, (15):Art. ID rnm045, 15, 2007.

[GSM08] S. Grushevsky and R. Salvati Manni, "Jacobians with a vanishing theta-null in genus 4", *Israel J. Math.*, 164 (2008), 303–315.

[GSM09] S. Grushevsky and R. Salvati Manni, "The loci of abelian varieties with points of high multiplicity on the theta divisor", *Geom. Dedicata*, 139 (2009), 233–247.

[GSM10a] S. Grushevsky and R. Salvati Manni, "On the cosmological constant for the chiral superstring measure", to appear, 2010.

[GSM10b] Samuel Grushevsky and Riccardo Salvati Manni, "The vanishing of two-point functions for three-loop superstring scattering amplitudes", *Comm. Math. Phys.* **294**:2 (2010), 343–352.

[Gun82a] R. Gunning, "On generalized theta functions", *Amer. J. Math.* **104**:1 (1982), 183–208.

[Gun82b] R. Gunning, "Some curves in abelian varieties", *Invent. Math.* **66**:3 (1982), 377–389.

[Gun86] R. C. Gunning, "Some identities for abelian integrals", *Amer. J. Math.* **108**:1 (1986), 39–74.

[Hai02] R. Hain, "The rational cohomology ring of the moduli space of abelian 3-folds", *Math. Res. Lett.* **9**:4 (2002), 473–491.

[Igu72] J.-I. Igusa, *Theta functions*, Grundlehren der math. Wissenschaften **194**, Springer, New York, 1972.

[Igu81a] J.-I. Igusa, "On the irreducibility of Schottky's divisor", *J. Fac. Sci. Univ. Tokyo Sect. IA Math.* **28**:3 (1982), 531–545.

[Igu81b] J.-I. Igusa, "Schottky's invariant and quadratic forms", pages 352–362 in *E. B. Christoffel* (Aachen/Monschau, 1979), Birkhäuser, Basel, 1981.

[IL10] E. Izadi and H. Lange, "Counter-examples of high Clifford index to Prym–Torelli", arXiv 1001.3610, preprint 2010.

[Iza91] E. Izadi, "Fonctions thêta du second ordre sur la Jacobienne d'une courbe lisse", *Math. Ann.* **289**:2 (1991), 189–202.

[Iza95] E. Izadi, "The geometric structure of $\mathcal{A}_4$, the structure of the Prym map, double solids and Γ_{00}-divisors", *J. Reine Angew. Math.*, 462 (1995), 93–158.

[Iza99] E. Izadi, "Second order theta divisors on Pryms", *Bull. Soc. Math. France* **127**:1 (1999), 1–23.

[Kem73] George Kempf, "On the geometry of a theorem of Riemann", *Ann. of Math.*, 98 (1973), 178–185.

[Kol95] J. Kollár, M. B. Porter Lectures. Princeton University Press, Princeton, NJ, 1995.

[Kri06] I. Krichever, *Integrable linear equations and the Riemann–Schottky problem*, pages 497–514 *Progr. Math.* **253**, Birkhäuser, Boston, 2006.

[Kri10] I. Krichever, "Characterizing Jacobians via trisecants of the Kummer variety", *Ann. Math.*, 2010. to appear.

[Laz96] R. Lazarsfeld, "Lengths of periods and Seshadri constants of abelian varieties", *Math. Res. Lett.* **3**:4 (1996), 439–447.

[Mat59] T. Matsusaka, "On a characterization of a Jacobian variety", *Memo. Coll. Sci. Univ. Kyoto. Ser. A. Math.*, 32 (1959), 1–19.

[Mum74] D. Mumford, "Prym varieties, I", pages 325–350 in *Contributions to analysis (a collection of papers dedicated to Lipman Bers)*. Academic Press, New York, 1974.

[Mum75] D. Mumford, *Curves and their Jacobians*. The University of Michigan Press, Ann Arbor, Mich., 1975.

[Mum83] D. Mumford, "On the Kodaira dimension of the Siegel modular variety", pages 348–375 in *Algebraic geometry — open problems* (Ravello, 1982), Lecture Notes in Math. **997**, Springer, Berlin, 1983.

[Mum07a] D. Mumford, *Tata lectures on theta, I*. Birkhäuser, Boston, 2007. Reprint of the 1983 edition.

[Mum07b] D. Mumford, *Tata lectures on theta, II*. Birkhäuser, Boston, 2007. Jacobian theta functions and differential equations. Reprint of the 1984 original.

[Mum07c] D. Mumford, *Tata lectures on theta, III*. Birkhäuser, Boston, 2007. Reprint of the 1991 original.

[MVZ06] M. Möller, E. Viehweg, and K. Zuo, "Special families of curves, of abelian varieties, and of certain minimal manifolds over curves", pages 417–450 in *Global aspects of complex geometry*. Springer, Berlin, 2006.

[Nak96] M. Nakamaye, "Seshadri constants on abelian varieties", *Amer. J. Math.* **118**:3 (1996), 621–635.

[Nam76a] Y. Namikawa, "A new compactification of the Siegel space and degeneration of abelian varieties, I", *Math. Ann.* **221**:2 (1976), 97–141.

[Nam76b] Y. Namikawa, "A new compactification of the Siegel space and degeneration of abelian varieties, II", *Math. Ann.* **221**:3 (1976), 201–241.

[Oor97] F. Oort, "Canonical liftings and dense sets of CM-points", pages 228–234 in *Arithmetic geometry* (Cortona, 1994), Sympos. Math., XXXVII. Cambridge Univ. Press, Cambridge, 1997.

[OPSMY10] M. Oura, C. Poor, R. Salvati Manni, and D. Yuen, "Modular forms of weight 8", *Math. Ann.*, 2010. to appear.

[Poo94] C. Poor, "The hyperelliptic locus", *Duke Math. J.* **76**:3 (1994), 809–884.

[Poo96] C. Poor, "Schottky's form and the hyperelliptic locus", *Proc. Amer. Math. Soc.* **124**:7 (1996), 1987–1991.

[PP08a] G. Pareschi and M. Popa, "Castelnuovo theory and the geometric Schottky problem", *J. Reine Angew. Math.*, 615 (2008), 25–44.

[PP08b] G. Pareschi and M. Popa, "Generic vanishing and minimal cohomology classes on abelian varieties", *Math. Ann.* **340**:1 (2008), 209–222.

[Ran81] Z. Ran, "On subvarieties of abelian varieties", *Invent. Math.* **62**:3 (1981), 459–479.

[Ran82] Z. Ran, "A characterization of five-dimensional Jacobian varieties", *Invent. Math.* **67**:3 (1982), 395–422.

[Sch88] F. Schottky, "Zur Theorie der abelschen Functionen vor vier Variablen", *J. Reine Angew. Math.*, 102 (1888), 304–352.

[Shi86] T. Shiota, "Characterization of Jacobian varieties in terms of soliton equations", *Invent. Math.* **83**:2 (1986), 333–382.

[Shi89a] T. Shiota, "The KP equation and the Schottky problem", *Sūgaku* **41**:1 (1989), 16–33. Translated in Sugaku Expositions 3 (1990), no. 2, 183–211.

[Shi89b] T. Shiota, "Prym varieties and soliton equations", pages 407–448 in *Infinite-dimensional Lie algebras and groups*, (Luminy–Marseille, 1988), Adv. Ser. Math. Phys, **7**, World Scientific, Teaneck, NJ, 1989.

[SJ09] F. Schottky and H. Jung, "Neue Sätze uber Symmetralfunktionen und die Abel'schen
        Funktionen der Riemann'schen Theorie", *Akad. Wiss. Berlin, Phys. Math. Kl.*, pages 282–297,
        1909.

[SM85] R. Salvati Manni, "On the dimension of the vector space $\mathbb{C}[\theta_m]_4$", *Nagoya Math. J.*, 98
        (1985), 99–107.

[SM94] R. Salvati Manni, "Modular varieties with level 2 theta structure", *Amer. J. Math.* **116**:6
        (1994), 1489–1511.

[SM08] R. Salvati Manni, "Remarks on superstring amplitudes in higher genus", *Nuclear Phys. B*
        801:1-2 (2008), 163–173.

[SV96] R. Smith and R. Varley, "Multiplicity g points on theta divisors", *Duke Math. J.* **82**:2
        (1996), 319–326.

[SV10] R. Smith and R. Varley, 2010.

[Tai87] I. Taimanov, "On an analogue of the Novikov conjecture in a problem of Riemann–
        Schottky type for Prym varieties", *Dokl. Akad. Nauk SSSR* **293**:5 (1987), 1065–1068.

[Tai97] I. Taimanov, "Secants of abelian varieties, theta functions and soliton equations", *Uspekhi
        Mat. Nauk* **52**:1 (1997), 149–224.

[Tol87] D. Toledo, "Nonexistence of certain closed complex geodesics in the moduli space of
        curves", *Pacific J. Math.* **129**:1 (1987), 187–192.

[vG84] B. van Geemen, "Siegel modular forms vanishing on the moduli space of curves", *Invent.
        Math.* **78**:2 (1984), 329–349.

[vG98] B. van Geemen, *The Schottky problem and second order theta functions*, Aportaciones
        Mat. Investig. **13**, pages 41–84. Soc. Mat. Mexicana, México, 1998.

[vGvdG86] B. van Geemen and G. van der Geer, "Kummer varieties and the moduli spaces of
        abelian varieties", *Amer. J. Math.* **108**:3 (1986), 615–641.

[VZ04] E. Viehweg and K. Zuo, "A characterization of certain Shimura curves in the moduli stack
        of abelian varieties", *J. Differential Geom.* **66**:2 (2004), 233–287.

[Wel84] G. Welters, "A criterion for Jacobi varieties", *Ann. of Math.* (2) **120**:3 (1984), 497–504.

[Wel86] G. Welters, "The surface $C - C$ on Jacobi varieties and 2nd order theta functions", *Acta
        Math.* **157**:1-2 (1986), 1–22.

[Wel87] G. Welters, "Curves of twice the minimal class on principally polarized abelian varieties",
        Nederl. Akad. Wetensch. Indag. Math., **49**:1 (1987), 87–109.

[Zar09a] Yu. Zarhin, "Cubic surfaces and cubic threefolds, Jacobians and intermediate Jacobians",
        pages 687–691 in *Algebra, arithmetic, and geometry: in honor of Yu. I. Manin, Vol. II*, Progr.
        Math. **270**, Birkhäuser, Boston, 2009.

[Zar09b] Yu. Zarhin, "Absolutely simple Prymians of trigonal curves", *Tr. Mat. Inst. Steklova*, 264
        (Mnogomernaya Algebraicheskaya Geometriya) (2009), 212–223.

sam@math.sunysb.edu   *Mathematics Department, Stony Brook University,
                                 Stony Brook, NY 11790-3651, United States*

Current Developments in Algebraic Geometry
MSRI Publications
Volume 59, 2011

Interpolation

JOE HARRIS

This is an overview of interpolation problems: when, and how, do zero-dimensional schemes in projective space fail to impose independent conditions on hypersurfaces?

1. The interpolation problem

We give an overview of the exciting class of problems in algebraic geometry known as interpolation problems: basically, when points (or more generally zero-dimensional schemes) in projective space may fail to impose independent conditions on polynomials of a given degree, and by how much.

We work over an arbitrary field K. Our starting point is this elementary theorem:

Theorem 1.1. *Given any* $z_1, \ldots z_{d+1} \in K$ *and* $a_1, \ldots a_{d+1} \in K$, *there is a unique* $f \in K[z]$ *of degree at most* d *such that*

$$f(z_i) = a_i, \quad i = 1, \ldots, d+1.$$

More generally:

Theorem 1.2. *Given any* $z_1, \ldots, z_k \in K$, *natural numbers* $m_1, \ldots, m_k \in \mathbb{N}$ *with* $\sum m_i = d+1$, *and*

$$a_{i,j} \subset K, \quad 1 \leq l \leq k; \quad 0 \leq j \leq m_i - 1,$$

there is a unique $f \in K[z]$ *of degree at most* d *such that*

$$f^{(j)}(z_i) = a_{i,j} \quad \text{for all } i, j.$$

The problem we'll address here is simple: *What can we say along the same lines for polynomials in several variables?*

First, introduce some language/notation. The "starting point" statement Theorem 1.1 says that the evaluation map

$$H^0(\mathcal{O}_{\mathbb{P}^1}(d)) \to \bigoplus K_{p_i}$$

is surjective; or, equivalently,

$$h^1(\mathcal{I}_{\{p_1,\dots,p_e\}}(d)) = 0$$

for any distinct points $p_1, \dots, p_e \in \mathbb{P}^1$ whenever $e \le d + 1$. More generally, Theorem 1.2 says that

$$h^1(\mathcal{I}_{p_1}^{m_1} \cdots \mathcal{I}_{p_k}^{m_k}(d)) = 0$$

when $\sum m_i \le d+1$. To generalize this, let $\Gamma \subset \mathbb{P}^r$ be an subscheme of dimension 0 and degree n. We say that Γ *imposes independent conditions* on hypersurfaces of degree d if the evaluation map

$$\rho : H^0(\mathcal{O}_{\mathbb{P}^r}(d)) \to H^0(\mathcal{O}_\Gamma(d))$$

is surjective, that is, if

$$h^1(\mathcal{I}_\Gamma(d)) = 0;$$

we'll say it *imposes maximal conditions* if ρ has maximal rank—that is, is either injective or surjective, or equivalently if $h^0(\mathcal{I}_\Gamma(d))h^1(\mathcal{I}_\Gamma(d)) = 0$. Note that the rank of ρ is just the value of the Hilbert function of Γ at d:

$$\text{rank}(\rho) = h_\Gamma(d);$$

and we'll denote it in this way in the future.

In these terms, the starting point statement is that *any subscheme of $\mathbb{P}^1$ imposes maximal conditions on polynomials of any degree.* Accordingly, we ask in general when a zero-dimensional subscheme $\Gamma \subset \mathbb{P}^r$ may fail to impose maximal conditions, and by how much: that is, we want to

• characterize geometrically subschemes that fail to impose independent conditions; and

• say by how much they may fail: that is, how large $h^1(\mathcal{I}_\Gamma(d))$ may be (equivalently, how small $h_\Gamma(d)$ may be).

We will focus primarily on two cases: when Γ is reduced; and when Γ is a union of "fat points"—that is, the scheme

$$\Gamma = V(\mathcal{I}_{p_1}^{m_1} \cdots \mathcal{I}_{p_k}^{m_k})$$

defined by a product of powers of maximal ideals of points. Other cases have been studied, such as curvilinear schemes (zero-dimensional schemes having tangent spaces of dimension at most 1; see [Ciliberto and Miranda 1998b]), but we'll focus on these two here. (It's unreasonable to ask about arbitrary zero-dimensional subschemes $\Gamma \subset \mathbb{P}^r$, since we have no idea what they look like.)

As we'll see, these two cases give rise to very different questions and answers, but there is a common thread to both, and it is this that we hope to bring out in the course of this note.

2. Reduced schemes

In this case, the first observation is that *general points always impose maximal conditions*. So, we ask when special points may fail to impose maximal conditions, and by how much—that is, how small $h_\Gamma(d)$ can be.

In the absence of further conditions, this is trivial: $h_\Gamma(d)$ is minimal for Γ contained in a line. It's still trivial if we require Γ to be nondegenerate: the minimum then is to put $n - r + 1$ points on a line. So we typically impose a "uniformity" condition, such as linear general position—that is, we require that any $r + 1$ or fewer of the points of Γ are linearly independent. In this case, we have the fundamental

Theorem 2.1 (Castelnuovo). *If $\Gamma \subset \mathbb{P}^r$ is a collection of n points in linear general position, then*

$$h_\Gamma(d) \geq \min\{rd + 1, n\}.$$

The proof is elementary: when $n \geq rd + 1$, we exhibit hypersurfaces of degree d containing rd points of Γ and no others by the union of d hyperplanes, each spanned by r of the points of Γ. What is striking, given the apparent crudeness of the argument, is that in fact *this inequality is sharp*: configurations Γ lying on a rational normal curve $C \subset \mathbb{P}^r$ have exactly this Hilbert function.

Even more striking, though, is the converse:

Theorem 2.2 (Castelnuovo). *If $\Gamma \subset \mathbb{P}^r$ is a collection of $n \geq 2r + 3$ points in linear general position, and*

$$h_\Gamma(2) = 2r + 1,$$

then Γ is contained in a rational normal curve.

Thus we have a complete characterization of at least the extremal examples of failure to impose independent conditions. The question is, can we extend this? We believe we can. we have the

Conjecture 2.3. *For $\alpha = 1, 2, \ldots, r - 1$, if $\Gamma \subset \mathbb{P}^r$ is a collection of $n \geq 2r + 2\alpha + 1$ points in uniform position, and*

$$h_\Gamma(2) \leq 2r + \alpha,$$

then Γ is contained in a curve $C \subset \mathbb{P}^r$ of degree at most $r - 1 + \alpha$.

"Uniform position" means that, if Γ', $\Gamma'' \subset \Gamma$ are subsets of the same cardinality, then $h_{\Gamma'} = h_{\Gamma''}$. This is in some sense a strong form of linear general position: given that the points of Γ span $\mathbb{P}^r$, linear general position is tantamount to the statement that $h_{\Gamma'}(1) = h_{\Gamma''}(1)$ for subsets Γ', $\Gamma'' \subset \Gamma$ of the same cardinality. It is not very restrictive; for example, if $C \subset \mathbb{P}^{r+1}$ is any irreducible curve, the points of a general hyperplane section of C have this property [Arbarello et al. 1985].

There are a number of remarks to make about this conjecture. The first is that it is known in cases $\alpha = 2$ (Fano; see [Harris 1982]) and $\alpha = 3$ [Petrakiev 2008]. A second is that it can't be extended as stated beyond $\alpha = r - 1$: for example, configurations $\Gamma \subset \mathbb{P}^r$ contained in a rational normal surface scroll satisfy $h_{\Gamma}(2) \leq 3r$, but need not lie on a curve of small degree.

A third remark is that we know how to classify irreducible, nondegenerate subvarieties $X \subset \mathbb{P}^r$ with Hilbert function $h_X(2) = 2r + \alpha$. Thus all we have to do to prove the conjecture is to show that the intersection of the quadrics containing Γ is positive-dimensional.

Finally, and perhaps most importantly, a proof of the conjecture would yield a complete answer to the classical problem: *for which triples (n, d, g) does there exist a smooth, irreducible, nondegenerate curve $C \subset \mathbb{P}^n$ of degree d and genus g?*

We will take a moment out to describe this connection, since it's the original motivation for much of the study of Hilbert functions of points. Let $n = r + 1$, and let $C \subset \mathbb{P}^n$ be an irreducible, nondegenerate curve of degree d and genus g; let $\Gamma \subset \mathbb{P}^r$ be a general hyperplane section of C. Briefly, Castelnuovo observed that for large m,

$$g = dm - h_C(m) + 1;$$

and using the inequality

$$h_C(m) - h_C(m - 1) \leq h_{\Gamma}(m)$$

we arrive at the bound

$$g \leq \sum_{m=1}^{\infty}(d - h_{\Gamma}(m)) = \sum_{m=1}^{\infty} h^1(\mathcal{I}_{\Gamma}(m)).$$

Applying the bound in Theorem 2.1, Castelnuovo then arrives at his bound on the genus

$$g \leq \pi(d, n) = \binom{m_0}{2}(n - 1) + m_0 \epsilon,$$

where $m_0 = \left[\dfrac{d-1}{n-1}\right]$ and $\epsilon = d - 1 - m_0(n - 1)$.

Now suppose we have a curve C as above that achieves this maximal genus. Assuming $d > 2n$, then, we can apply the converse Theorem 2.2 to conclude that *C must lie on a surface $S \subset \mathbb{P}^n$ of minimal degree $n - 1$*; and indeed when we look on such surfaces we find curves of this maximal genus, showing that the bound is in fact sharp.

But this is just the beginning of the story. Assuming Conjecture 2.3, we can bound from below the Hilbert function of a configuration of points in uniform position not lying on a rational normal curve, and conclude that any curve $C \subset \mathbb{P}^n$ that does *not* lie on a surface of minimal degree must satisfy a stronger bound

$$g \leq \pi_1(d, n) \sim \frac{d^2}{2n}.$$

In other words, any curve as above with genus $g > \pi_1(d, n)$ must lie on a surface of minimal degree. Now, we know what those surfaces look like (they are either rational normal scrolls or the quadratic Veronese surface), and we know correspondingly exactly what the arithmetic genus of a curve of given degree d on such a surface may be; thus we can say exactly which g in the range $\pi_1(d, n) < g \leq \pi(d, n)$ occur as the genus of an irreducible, nondegenerate curve of degree d in $\mathbb{P}^n$.

Similarly, if we assume the conjecture in general, we can define a series of functions

$$\pi_\alpha(d, n) \sim \frac{d^2}{2(n + \alpha - 1)}, \qquad \alpha = 1, 2, \ldots, n - 1$$

such that any curve C of genus $g > \pi_\alpha(d, n)$ must lie on a surface of degree at most $n + \alpha - 2$. Again, we know what all such surfaces look like, and what may be the genera of curves on them (we're in the range $\alpha \leq n - 1$, so all such surfaces are rational or ruled), and so we can say exactly which $g > \pi_{n-1}(d, n)$ occur as the genus of an irreducible, nondegenerate curve of degree d in $\mathbb{P}^n$. Finally, I think it's the case that every genus $g \leq \pi_{n-1}(d, n)$ occurs, and in fact occurs on a K3 surface $S \subset \mathbb{P}^n$ of degree $2n - 2$.

Returning to the original question of Hilbert functions of collections of points in $\mathbb{P}^r$, we can express the bottom line as follows: Configurations $\Gamma \subset \mathbb{P}^r$ of points having small Hilbert function do so because they lie on small subvarieties $X \subset \mathbb{P}^r$—meaning, subvarieties with small Hilbert function. In this case, for small d the hypersurfaces of degree d containing Γ will just be the hypersurfaces containing X; in particular, X will be the intersection of the quadrics containing Γ.

Usually, to prove results along these lines it's enough to show the base locus $|\mathcal{I}_\Gamma(d)|$ is positive-dimensional.

3. Fat points

We now take up the second case of our general question: we let $p_1, \ldots, p_k \in \mathbb{P}^r$ be points, $m_1, \ldots, m_k \in \mathbb{N}$, and let

$$\Gamma = V(\mathcal{I}_{p_1}^{m_1} \cdots \mathcal{I}_{p_k}^{m_k}).$$

Right off the bat, we see a fundamental difference from the case of reduced points: it is *not* always the case that for $p_1, \ldots, p_k \in \mathbb{P}^r$ general, Γ imposes maximal conditions on hypersurfaces of degree d! So the first question is: assuming the points p_i are general, *for what values of the integers $r, k, m_1, \ldots, m_k$ and d does Γ fail to impose maximal conditions?*

This is a very different flavor of question, if only because the answer is numerical rather than geometric. The fact is, we don't even have a conjectured answer in general! One case where we do know the answer is the case of double points—that is, where all $m_i = 2$. Here we have:

Theorem 3.1 [Alexander and Hirschowitz 1995]. *For $p_i \in \mathbb{P}^r$ general,*

$$\Gamma = V(\mathcal{I}_{p_1}^2 \cdots \mathcal{I}_{p_k}^2)$$

imposes maximal conditions on hypersurfaces of degree d, with four exceptions:

(1) $k \geq 2$, $d = 2$;

(2) $r = 2$, $k = 5$, $d = 4$;

(3) $r = 3$, $k = 9$, $d = 4$;

(4) $r = 4$, $k = 7$, $d = 3$;

It's straightforward to see that the first three cases are counterexamples to the general statement. For example, it's three conditions for a polynomial on $\mathbb{P}^2$ to vanish to order 2, and the vector space of quadratic polynomials is six-dimensional, so we might expect that there is no conic double at each of two assigned points $p, q \in \mathbb{P}^2$, but there is: the double of the line $\overline{pq}$. Another way to say this is that if we require a quadratic polynomial to vanish to order 2 at a point $p \in \mathbb{P}^r$ and simply to vanish at another point q, it must vanish identically along the line $L = \overline{pq}$; the condition that its directional derivative at q in the direction of L also vanish is thus redundant.

Similarly, we don't expect that there should be a quartic curve in $\mathbb{P}^2$ double at five assigned points, but there is: if a quartic in $\mathbb{P}^2$ is double at four points $p_1, \ldots, p_4$ and passes through a fifth p_5, it necessarily contains the conic through all five, so one of the two additional conditions to be double at p_5 is dependent. The third example is likewise clear: since the space of quartic polynomials on $\mathbb{P}^3$ has dimension 35, and it's four conditions to vanish to order two at a point, there shouldn't be a quartic double at nine points; but the double of the quadric containing them is one such.

The last example is trickier. As in the last two, we don't expect that there will be a cubic hypersurface in $\mathbb{P}^4$ double at seven general points, but there is: the secant variety of the (unique) rational normal quartic curve passing through the seven points.

For general multiplicities m_i and general r, as we said we don't even have a conjectured answer. For $r = 2$, though, we do. To express it, we introduce some more notation:

Let $p_1, \ldots, p_k \in \mathbb{P}^2$ be general, and let

$$S = \mathrm{Bl}_{\{p_1, \ldots, p_k\}} \mathbb{P}^2$$

be the blow-up of the plane at the p_i. Let H be the divisor class of the preimage of a line in $\mathbb{P}^2$, and E_i the exceptional divisor over the point p_i. Let L be the line bundle

$$\mathcal{O}_S(dH - \sum m_i E_i)$$

on S. Then

$$h^i(L) = h^i(\mathcal{I}_{p_1}^{m_1} \cdots \mathcal{I}_{p_k}^{m_k}(d)).$$

In particular, the "expected dimension" of $h^0(L)$ is

$$\frac{(d+1)(d+2)}{2} - \sum \frac{m_i(m_i+1)}{2}$$

and this is exceeded exactly when the scheme Γ fails to impose independent conditions in degree d.

In these terms, we can interpret the basic example of conics in $\mathbb{P}^2$ double at two points p, q as saying that, since the restriction to the line $\overline{pq}$ of the line bundle $\mathcal{O}_{\mathbb{P}^2}(2)$ has degree 2, the requirement that the restriction of a section vanish four times along $\overline{pq}$ is necessarily redundant. Equivalently, if we let S be the blow-up S of $\mathbb{P}^2$ at p and q, and $D \subset S$ the proper transform of the line $\overline{pq}$, and set

$$L = \mathcal{O}_S(2H - E_p - E_q),$$

then $L|_D$ has degree -2, and from an examination of the exact sequence

$$0 \to L(-D) \to L \to L|_D \to 0$$

we see that $h^1(L|_D) \neq 0$ implies that $h^1(L) \neq 0$. A similar interpretation can be given for the example of quartics double at 5 points (the line bundle $L = \mathcal{O}_S(4H - E_1 - \cdots - E_5)$ has degree -2 on the proper transform of the conic through the five); and in general we have the

Conjecture 3.2 (Harbourne–Hirschowitz). *Let S be the blow-up of $\mathbb{P}^2$ at k general points, L any line bundle on S. Then $h^1(L) \neq 0$ if and only there is a (-1)-curve $E \subset S$ such that*

$$\deg(L|_E) \leq -2.$$

An equivalent formulation is that if $h^1(L) \neq 0$, then the base locus of the linear system $|L|$ contains a multiple (-1)-curve.

There are a number of remarks to be made here. The first is that if true, the Harbourne–Hirschowitz conjecture gives a complete answer to our question for $r = 2$: conjecturally (more about this in a moment), we know where the (-1)-curves on S are, and can check the condition $\deg(L|_E) \leq -2$.

Two cases where the conjecture is known is known are for $k \leq 9$ (in this case S has an effective anticanonical divisor), and when $\max\{m_i\} \leq 7$ (Stephanie Yang; [Yang 2007])

To explicate the conjecture, and the fact that it does answer our question, we should make a small digression to discuss our abysmal ignorance about curves of negative self-intersection on surfaces. To start, let X be any smooth, projective surface, and consider the self-intersections of curves of X; that is, set

$$\Sigma = \{(C \cdot C) : C \subset S \text{ integral}\} \subset \mathbb{Z}.$$

The first question we might ask is: *is Σ bounded below?*

The answer isn't known in characteristic 0, though János Kollár points out that there are examples in characteristic p of surfaces with integral curves of arbitrarily negative self-intersection: take B a smooth curve of genus $g \geq 2$, $S = B \times B$ and $C_n \subset S$ the graph of the nth power of Frobenius. In characteristic zero, we don't know the answer even for $X = S$ a blow-up of the plane!

We can, however, make a strong conjecture in this case. Consider an arbitrary line bundle $L = \mathcal{O}_S(dH - \sum m_i E_i)$ on a general blow-up S. The expected dimension of $h^0(L)$ is

$$\frac{(d+1)(d+2)}{2} - \sum \frac{m_i(m_i+1)}{2};$$

and the genus of a curve $C \in |L|$ is

$$\frac{(d-1)(d-2)}{2} - \sum \frac{m_i(m_i-1)}{2}.$$

If we assume the first is positive and the second nonnegative, it follows that the self-intersection of C is

$$(C \cdot C) = d^2 - \sum m_i^2 \geq -1.$$

Thus we may make the following conjecture:

Conjecture 3.3. *Let S be a general blow-up of the plane, $C \subset S$ any integral curve. Then*

$$(C \cdot C) \geq -1,$$

and if equality holds then C is a smooth rational curve.

If we believe this, the Harbourne–Hirschowitz conjecture should be equivalent to the weaker version:

Conjecture 3.4 (Harbourne–Hirschowitz; weak form). *Let S be the blow-up of $\mathbb{P}^2$ at general points, L any line bundle on S. If the linear system $|L|$ contains an integral curve, then $h^1(L) = 0$.*

If we believe the weak Harbourne–Hirschowitz, then by the calculation above Conjecture 3.3 on self-intersections of curves on S follows, and we can in turn deduce strong Harbourne–Hirschowitz. Thus, it's possible to prove that the two versions of Harbourne–Hirschowitz are equivalent. Note moreover that if we believe any version, it's possible to locate all the (-1)-curves on S, as primitive solutions of the system of equations

$$\frac{(d-1)(d-2)}{2} - \sum \frac{m_i(m_i - 1)}{2} = 0 \quad \text{and} \quad d^2 - \sum m_i^2 \geq -1.$$

Thus the condition that the line bundle L have degree -2 on a (-1)-curve $E \subset S$ is algorithmically checkable.

It's worth taking a moment to describe some approaches to Harbourne–Hirschowitz. Briefly, all approaches taken to Harbourne–Hirschowitz (in case $k > 9$) involve specialization—Ciliberto and Miranda ([Ciliberto and Miranda 2000], [Ciliberto and Miranda 1998a]) specialize a subset of the points p_i onto a line $L \subset \mathbb{P}^2$; Yang specializes the points onto a line one at a time. Either approach involves an "apparent" loss of conditions; the goal is to understand what conditions the limit of the linear series $|\mathcal{I}_\Gamma(d)|$ will satisfy beyond the obvious multiplicity ones. These questions are fascinating in their own right.

As an example: suppose $d = 4$, $k = 5$ and $(m_1, \ldots, m_5) = (1, 1, 1, 1, 3)$; suppose that $p_1, \ldots, p_4$ already lie on a line L and we specialize p_5 onto L. The limits of the curves passing through $p_1, \ldots, p_4$ and triple at p_5 will be of the form $L + C$, with C a cubic double at p_5. But there are too many of these: cubics double at p_5 form a 6-dimensional linear system, while the system of quartics passing through $p_1, \ldots, p_4$ and triple at the general p_5 is only 4-dimensional. So the question is: which cubics actually appear in the limiting curves? The answer, somewhat unexpectedly, is: cubics with a cusp at p_5, with tangent line L there.

It would be wonderful to understand better this limiting behavior. For example, does something like this occur when we specialize similarly defined linear systems on more general surfaces?

Before moving on, we should summarize one common thread running though our discussions of Castelnuovo theory and the Harbourne–Hirschowitz conjecture.

The content of the Harbourne–Hirschowitz conjectures may be thought of as this: *if general multiple points in $\mathbb{P}^2$ fail to impose maximal conditions, they do so because they lie on a "small" curve—in this case, a curve of negative self-intersection.*

It's hard to say how this might generalize to higher-dimensional space—as we said, we don't even have a conjectured answer to the question of when general multiple points impose independent conditions in general. Based on our experience in $\mathbb{P}^2$, though, we might be led to make a qualitative conjecture:

Conjecture 3.5. *Let $p_1, \ldots, p_k \in \mathbb{P}^r$ be general. If*

$$h^1(\mathscr{I}_{p_1}^{m_1} \cdots \mathscr{I}_{p_k}^{m_k}(d)) \neq 0$$

then the base locus of the linear series $|\mathscr{I}_{p_1}^{m_1} \cdots \mathscr{I}_{p_k}^{m_k}(d)|$ must have positive dimension.

4. Recasting the problem

There is a common theme to our results and conjectures so far: we believe in many cases that when a subscheme $\Gamma \subset \mathbb{P}^r$ fails to impose independent conditions on hypersurfaces of degree d—that is, has small Hilbert function $h_\Gamma(d)$—it's because it's contained in a small positive-dimensional subscheme $X \subset \mathbb{P}^r$; and moreover, in this case X will appear as the intersection of the hypersurfaces of degree d containing Γ.

So let's recast the problem: let's drop all the conditions we've put on Γ at various points above, and instead make just one assumption: that the intersection of the hypersurfaces of degree d containing Γ is zero-dimensional; in other words, Γ *is a subscheme of a complete intersection of r hypersurfaces of degree d.* We ask: what bounds can we give on $h^1(\mathscr{I}_\Gamma(d))$ (or $h_\Gamma(d)$, or $h^0(\mathscr{I}_\Gamma(d))$) under this hypothesis?

One further wrinkle: instead of specifying the degree n of Γ and asking for estimates on the size of $h^0(\mathscr{I}_\Gamma(d))$, let's turn it around: let's specify the dimension $h^0(\mathscr{I}_\Gamma(d))$, and ask for a bound on the degree of Γ. Thus, the question is:

• *Let $V \subset H^0(\mathscr{O}_{\mathbb{P}^r}(d))$ be an N-dimensional linear system of hypersurfaces of degree d, with finite intersection Γ. How large can the degree of Γ be?*

As a first example, let's try $d = 2$ and $N = r + 1$. The question is, in effect:

• *How many common zeroes can $r + 1$ quadrics in $\mathbb{P}^r$ have, if they have only finitely many common zeroes?*

• *Let $\{p_1, \ldots, p_{2^r}\} \subset \mathbb{P}^r$ be a complete intersection of quadrics in $\mathbb{P}^r$. How many of the points p_i can a quadric Q contain without containing them all?*

The first few cases can be worked out ad hoc: for example, in case $r = 2$, the answer is visibly 3. In case $r = 3$, the Cayley–Bachrach theorem [Eisenbud et al. 1996] says that any quadric containing 7 of the 8 points of a complete intersection of quadrics in $\mathbb{P}^3$ contains the eighth as well; the answer is 6. And in case $r = 4$, let $C = Q_1 \cap Q_2 \cap Q_3$. If two more quadrics had 13 common zeroes on C, they would cut out a g_3^1 on C. But C is not trigonal; thus the answer is 12.

All this leads us to the

Conjecture 4.1 (Green, Eisenbud, Harris). *If $Q_1, \ldots, Q_{r+1} \subset \mathbb{P}^r$ are linearly independent quadrics and $\Gamma = Q_1 \cap \cdots \cap Q_{r+1}$ their zero-dimensional intersection, then*

$$\deg(\Gamma) \leq 3 \cdot 2^{r-2}.$$

In fact, this is just the first case of a general conjecture about linear systems of quadrics, and of higher-degree hypersurfaces; the full statement can be found in [Eisenbud et al. 1996]. And this particular case is in fact no longer a conjecture; it's been proved by Rob Lazarsfeld, under the mild extra hypothesis that Γ is reduced.

References

[Alexander and Hirschowitz 1995] J. Alexander and A. Hirschowitz, "Polynomial interpolation in several variables", *J. Algebraic Geom.* **4**:2 (1995), 201–222. MR 96f:14065 Zbl 0829.14002

[Arbarello et al. 1985] E. Arbarello, M. Cornalba, P. A. Griffiths, and J. Harris, *Geometry of algebraic curves*, vol. I, Grundlehren der Math. Wissenschaften **267**, Springer, New York, 1985. MR 86h:14019 Zbl 0559.14017

[Ciliberto and Miranda 1998a] C. Ciliberto and R. Miranda, "Degenerations of planar linear systems", *J. Reine Angew. Math.* **501** (1998), 191–220. MR 2000m:14005 Zbl 0943.14002

[Ciliberto and Miranda 1998b] C. Ciliberto and R. Miranda, "Interpolation on curvilinear schemes", *J. Algebra* **203**:2 (1998), 677–678. MR 99b:14047 Zbl 0921.14034

[Ciliberto and Miranda 2000] C. Ciliberto and R. Miranda, "Linear systems of plane curves with base points of equal multiplicity", *Trans. Amer. Math. Soc.* **352**:9 (2000), 4037–4050. MR 2000m:14006 Zbl 0959.14015

[Eisenbud et al. 1996] D. Eisenbud, M. Green, and J. Harris, "Cayley–Bacharach theorems and conjectures", *Bull. Amer. Math. Soc.* (*N.S.*) **33**:3 (1996), 295–324. MR 97a:14059 Zbl 0871.14024

[Harris 1982] J. Harris, *Curves in projective space*, Séminaire de Mathématiques Supérieures **85**, Presses de l'Université de Montréal, Montreal, 1982. MR 84g:14024 Zbl 0511.14014

[Petrakiev 2008] I. Petrakiev, "Castelnuovo theory via Gröbner bases", *J. Reine Angew. Math.* **619** (2008), 49–73. MR 2009e:14048

[Yang 2007] S. Yang, "Linear systems in $\mathbb{P}^2$ with base points of bounded multiplicity", *J. Algebraic Geom.* **16**:1 (2007), 19–38. MR 2007i:14011 Zbl 1115.14003

harris@math.harvard.edu   *Mathematics Department, Harvard University,*
                                 1 Oxford Street, Cambridge, MA 02138, United States

Current Developments in Algebraic Geometry
MSRI Publications
Volume 59, 2011

Chow groups
and derived categories of K3 surfaces

DANIEL HUYBRECHTS

The geometry of a K3 surface (over $\mathbb{C}$ or over $\bar{\mathbb{Q}}$) is reflected by its Chow group and its bounded derived category of coherent sheaves in different ways. The Chow group can be infinite dimensional over $\mathbb{C}$ (Mumford) and is expected to inject into cohomology over $\bar{\mathbb{Q}}$ (Bloch–Beilinson). The derived category is difficult to describe explicitly, but its group of autoequivalences can be studied by means of the natural representation on cohomology. Conjecturally (Bridgeland) the kernel of this representation is generated by squares of spherical twists. The action of these spherical twists on the Chow ring can be determined explicitly by relating it to the natural subring introduced by Beauville and Voisin.

1. Introduction

In algebraic geometry a K3 surface is a smooth projective surface X over a fixed field K with trivial canonical bundle $\omega_X \simeq \Omega_X^2$ and $H^1(X, \mathcal{O}_X) = 0$. For us the field K will be either a number field, the field of algebraic numbers $\bar{\mathbb{Q}}$ or the complex number field $\mathbb{C}$. Nonprojective K3 surfaces play a central role in the theory of K3 surfaces and for some of the results that will be discussed in this text in particular, but here we will not discuss those more analytical aspects.

An explicit example of a K3 surface is provided by the Fermat quartic in $\mathbb{P}^3$ given as the zero set of the polynomial $x_0^4 + \cdots + x_3^4$. Kummer surfaces, i.e., minimal resolutions of the quotient of abelian surfaces by the sign involution, and elliptic K3 surfaces form other important classes of examples. Most of the results and questions that will be mentioned do not lose any of their interest when considered for one of theses classes of examples or any other particular K3 surface.

This text deals with three objects naturally associated with any K3 surface X:

$$\mathrm{D}^b(X), \ \mathrm{CH}^*(X) \text{ and } H^*(X, \mathbb{Z}).$$

If X is defined over $\mathbb{C}$, its *singular cohomology* $H^*(X, \mathbb{Z})$ is endowed with the intersection pairing and a natural Hodge structure. The *Chow group* $\mathrm{CH}^*(X)$

of X, defined over an arbitrary field, is a graded ring that encodes much of the algebraic geometry. The *bounded derived category* $D^b(X)$, a linear triangulated category, is a more complicated invariant and in general difficult to control.

As we will see, all three objects, $H^*(X, \mathbb{Z})$, $CH^*(X)$, and $D^b(X)$ are related to each other. On the one hand, $H^*(X, \mathbb{Z})$ as the easiest of the three can be used to capture some of the features of the other two. But on the other hand and maybe a little surprising, one can deduce from the more rigid structure of $D^b(X)$ as a linear triangulated category interesting information about cycles on X, i.e., about some aspects of $CH^*(X)$.

This text is based on my talk at the conference *Classical Algebraic Geometry Today* at MSRI in January 2009 and is meant as a nontechnical introduction to the standard techniques in the area. At the same time it surveys recent developments and presents some new results on a question on symplectomorphisms that was raised in this talk (see Section 6). I wish to thank the organizers for the invitation to a very stimulating conference.

2. Cohomology of K3 surfaces

The second singular cohomology of a complex K3 surface is endowed with the additional structure of a weight two Hodge structure and the intersection pairing. The global Torelli theorem shows that it determines the K3 surface uniquely. We briefly recall the main features of this Hodge structure and of its extension to the Mukai lattice which governs the derived category of the K3 surface. For the general theory of complex K3 surfaces see [Barth et al. 2004] or [Beauville et al. 1985], for example. In this section all K3 surfaces are defined over $\mathbb{C}$.

2.1. To any complex K3 surface X we can associate the singular cohomology $H^*(X, \mathbb{Z})$ (of the underlying complex or topological manifold). Clearly, $H^0(X, \mathbb{Z}) \simeq H^4(X, \mathbb{Z}) \simeq \mathbb{Z}$. Hodge decomposition yields

$$H^1(X, \mathbb{C}) \simeq H^{1,0}(X) \oplus H^{0,1}(X) = 0,$$

since by assumption $H^{0,1}(X) \simeq H^1(X, \mathcal{O}_X) = 0$, and hence $H^1(X, \mathbb{Z}) = 0$. One can also show $H^3(X, \mathbb{Z}) = 0$. Thus, the only interesting cohomology group is $H^2(X, \mathbb{Z})$ which together with the intersection pairing is abstractly isomorphic to the unique even unimodular lattice of signature $(3, 19)$ given by $U^{\oplus 3} \oplus E_8(-1)^{\oplus 2}$. Here, U is the hyperbolic plane and $E_8(-1)$ is the standard root lattice E_8 changed by a sign. Thus, the full cohomology $H^*(X, \mathbb{Z})$ endowed with the intersection pairing is isomorphic to $U^{\oplus 4} \oplus E_8(-1)^{\oplus 2}$.

For later use we introduce $\widetilde{H}(X, \mathbb{Z})$, which denotes $H^*(X, \mathbb{Z})$ with the Mukai paring, i.e., with a sign change in the pairing between H^0 and H^4. Note that as abstract lattices $H^*(X, \mathbb{Z})$ and $\widetilde{H}(X, \mathbb{Z})$ are isomorphic.

2.2. The complex structure of the K3 surface X induces a weight two Hodge structure on $H^2(X, \mathbb{Z})$ given explicitly by the decomposition

$$H^2(X, \mathbb{C}) = H^{2,0}(X) \oplus H^{1,1}(X) \oplus H^{0,2}(X).$$

It is determined by the complex line $H^{2,0}(X) \subset H^2(X, \mathbb{C})$ which is spanned by a trivializing section of ω_X and by requiring the decomposition to be orthogonal with respect to the intersection pairing. This natural Hodge structure induces at the same time a weight two Hodge structure on the Mukai lattice $\widetilde{H}(X, \mathbb{Z})$ by setting $\widetilde{H}^{2,0}(X) = H^{2,0}(X)$ and requiring $(H^0 \oplus H^4)(X, \mathbb{C}) \subset \widetilde{H}^{1,1}(X)$.

The global Torelli theorem and its derived version, due to Piatetski-Shapiro and Shafarevich [1971] on the one hand and Mukai and Orlov on the other, can be stated as follows. For complex projective K3 surfaces X and X' one has:

i) There exists an isomorphism $X \simeq X'$ (over $\mathbb{C}$) if and only if there exists an isometry of Hodge structures $H^2(X, \mathbb{Z}) \simeq H^2(X', \mathbb{Z})$.

ii) There exists a $\mathbb{C}$-linear exact equivalence $\mathrm{D}^b(X) \simeq \mathrm{D}^b(X')$ if and only if there exists an isometry of Hodge structures $\widetilde{H}(X, \mathbb{Z}) \simeq \widetilde{H}(X', \mathbb{Z})$.

Note that for purely lattice theoretical reasons the weight two Hodge structures $\widetilde{H}(X, \mathbb{Z})$ and $\widetilde{H}(X', \mathbb{Z})$ are isometric if and only if their transcendental parts (see 2.3) are.

2.3. The Hodge index theorem shows that the intersection pairing on $H^{1,1}(X, \mathbb{R})$ has signature $(1, 19)$. Thus the cone of classes α with $\alpha^2 > 0$ decomposes into two connected components. The connected component $\mathscr{C}_X$ containing the Kähler cone $\mathscr{K}_X$ (the cone of all Kähler classes) is called the positive cone. Note that for the Mukai lattice $\widetilde{H}(X, \mathbb{Z})$ the set of real $(1, 1)$-classes of positive square is connected.

The Néron–Severi group $\mathrm{NS}(X)$ is identified with $H^{1,1}(X) \cap H^2(X, \mathbb{Z})$ and its rank is the Picard number $\rho(X)$. Since X is projective, the intersection form on $\mathrm{NS}(X)_{\mathbb{R}}$ has signature $(1, \rho(X) - 1)$. The transcendental lattice $T(X)$ is by definition the orthogonal complement of $\mathrm{NS}(X) \subset H^2(X, \mathbb{Z})$. Hence, $H^2(X, \mathbb{Q}) = \mathrm{NS}(X)_{\mathbb{Q}} \oplus T(X)_{\mathbb{Q}}$ which can be read as an orthogonal decomposition of weight two rational Hodge structures (but in general not over $\mathbb{Z}$). Note that $T(X)_{\mathbb{Q}}$ cannot be decomposed further, it is an irreducible Hodge structure. The ample cone is the intersection of the Kähler cone $\mathscr{K}_X$ with $\mathrm{NS}(X)_{\mathbb{R}}$ and is spanned by ample line bundles.

Analogously, one has the extended Néron–Severi group

$$\widetilde{\mathrm{NS}}(X) := \widetilde{H}^{1,1}(X) \cap \widetilde{H}(X, \mathbb{Z}) = \mathrm{NS}(X) \oplus (H^0 \oplus H^4)(X, \mathbb{Z}).$$

Note that $\widetilde{\mathrm{NS}}(X)$ is simply the lattice of all algebraic classes. More precisely, $\widetilde{\mathrm{NS}}(X)$ can be seen as the image of the cycle map $\mathrm{CH}^*(X) \longrightarrow H^*(X, \mathbb{Z})$ or the set of all Mukai vectors $v(E) = \mathrm{ch}(E).\sqrt{\mathrm{td}(X)} = \mathrm{ch}(E).(1, 0, 1)$ with $E \in D^b(X)$. Note that the transcendental lattice in $\widetilde{H}(X, \mathbb{Z})$ coincides with $T(X)$.

2.4. The so-called (-2)-classes, i.e., integral $(1, 1)$-classes δ with $\delta^2 = -2$, play a central role in the classical theory as well as in the modern part related to derived categories and Chow groups.

Classically, one considers the set Δ_X of (-2)-classes in $\mathrm{NS}(X)$. For instance, every smooth rational curve $\mathbb{P}^1 \simeq C \subset X$ defines by adjunction a (-2)-class, hence C is called a (-2)-curve. Examples of (-2)-classes in the extended Néron–Severi lattice $\widetilde{\mathrm{NS}}(X)$ are provided by the Mukai vector $v(E)$ of spherical objects $E \in D^b(X)$ (see 4.2 and 5.1). Note that $v(\mathcal{O}_C) \neq [C]$, but $v(\mathcal{O}_C(-1)) = [C]$. For later use we introduce $\widetilde{\Delta}_X$ as the set of (-2)-classes in $\widetilde{\mathrm{NS}}(X)$.

Clearly, an ample or, more generally, a Kähler class has positive intersection with all effective curves and with (-2)-curves in particular. Conversely, one knows that every class $\alpha \in \mathcal{C}_X$ with $(\alpha.C) > 0$ for all (-2)-curves is a Kähler class (cf. [Barth et al. 2004, VIII, Corollary 3.9]).

To any (-2)-class δ one associates the reflection $s_\delta : \alpha \longmapsto \alpha + (\alpha.\delta)\delta$ which is an orthogonal transformation of the lattice also preserving the Hodge structure. The Weyl group is by definition the subgroup of the orthogonal group generated by the reflections s_δ. So one has two groups

$$W_X \subset O(H^2(X, \mathbb{Z})) \quad \text{and} \quad \widetilde{W}_X \subset O(\widetilde{H}(X, \mathbb{Z})).$$

The union of hyperplanes $\bigcup_{\delta \in \Delta_X} \delta^\perp$ is locally finite in the interior of $\mathcal{C}_X$ and endows $\mathcal{C}_X$ with a chamber structure. The Weyl group W_X acts simply transitively on the set of chambers and the Kähler cone is one of the chambers. The action of W_X on $\mathrm{NS}(X)_\mathbb{R} \cap \mathcal{C}_X$ can be studied analogously. It can also be shown that the reflections $s_{[C]}$ with $C \subset X$ a smooth rational curve generate W_X.

Another part of the global Torelli theorem complementing i) in 2.2 says that a nontrivial automorphism $f \in \mathrm{Aut}(X)$ acts always nontrivially on $H^2(X, \mathbb{Z})$. Moreover, any Hodge isometry of $H^2(X, \mathbb{Z})$ preserving the positive cone is induced by an automorphism up to the action of W_X. In fact, Piatetski-Shapiro and Shafarevich also showed that the action on $\mathrm{NS}(X)$ is essentially enough to determine f. More precisely, one knows that the natural homomorphism

$$\mathrm{Aut}(X) \longrightarrow O(\mathrm{NS}(X))/W_X$$

has finite kernel and cokernel. Roughly, the kernel is finite because an automorphism that leaves invariant a polarization is an isometry of the underlying hyperkähler structure and these isometries form a compact group. For the finiteness of the cokernel note that some high power of any automorphism f always acts trivially on $T(X)$.

The extended Néron–Severi group plays also the role of a period domain for the space of stability conditions on $D^b(X)$ (see 4.5). For this consider the open set $\mathscr{P}(X) \subset \widetilde{NS}(X)_{\mathbb{C}}$ of vectors whose real and imaginary parts span a positively oriented positive plane. Then let $\mathscr{P}_0(X) \subset \mathscr{P}(X)$ be the complement of the union of all codimension two sets $\delta^{\perp}$ with $\delta \in \widetilde{NS}(X)$ and $\delta^2 = -2$ (or, equivalently, $\delta = v(E)$ for some spherical object $E \in D^b(X)$ as we will explain later):

$$\mathscr{P}_0(X) := \mathscr{P}(X) \setminus \bigcup_{\delta \in \widetilde{\Delta}_X} \delta^{\perp}.$$

Since the signature of the intersection form on $\widetilde{NS}(X)$ is $(2, \rho(X))$, the set $\mathscr{P}_0(X)$ is connected. Its fundamental group $\pi_1(\mathscr{P}_0(X))$ is generated by loops around each $\delta^{\perp}$ and the one induced by the natural $\mathbb{C}^*$-action.

3. Chow ring

We now turn to the second object that can naturally be associated with any K3 surface X defined over an arbitrary field K, the Chow group $\mathrm{CH}^*(X)$. For a separably closed field like $\bar{\mathbb{Q}}$ or $\mathbb{C}$ it is torsion free due to a theorem of Roitman [1980] and for number fields we will simply ignore everything that is related to the possible occurrence of torsion. The standard reference for Chow groups is [Fulton 1998]. For the interplay between Hodge theory and Chow groups see [Voisin 2002], for example.

3.1. The Chow group $\mathrm{CH}^*(X)$ of a K3 surface (over K) is the group of cycles modulo rational equivalence. Thus, $\mathrm{CH}^0(X) \simeq \mathbb{Z}$ (generated by $[X]$) and $\mathrm{CH}^1(X) = \mathrm{Pic}(X)$. The interesting part is $\mathrm{CH}^2(X)$ which behaves differently for $K = \bar{\mathbb{Q}}$ and $K = \mathbb{C}$. Let us begin with the following celebrated result.

Theorem 3.2 [Mumford 1968]. *If $K = \mathbb{C}$, then $\mathrm{CH}^2(X)$ is infinite dimensional.*

(A priori $\mathrm{CH}^2(X)$ is simply a group, so one needs to explain what it means that $\mathrm{CH}^2(X)$ is infinite dimensional. A first and very weak version says that $\dim_{\mathbb{Q}} \mathrm{CH}^2(X)_{\mathbb{Q}} = \infty$. For a more geometrical and more precise definition of infinite dimensionality see e.g. [Voisin 2002, Chapter 22].)

For $K = \bar{\mathbb{Q}}$ the situation is expected to be different. The Bloch–Beilinson conjectures lead one to the following conjecture for K3 surfaces.

Conjecture 3.3. *If K is a number field or $K = \mathbb{Q}$, then $\mathrm{CH}^2(X)_{\mathbb{Q}} = \mathbb{Q}$.*

So, if X is a K3 surface defined over $\bar{\mathbb{Q}}$, then one expects $\dim_{\mathbb{Q}} \mathrm{CH}^2(X)_{\mathbb{Q}} = 1$, whereas for the complex K3 surface $X_{\mathbb{C}}$ obtained by base change from X one knows $\dim_{\mathbb{Q}} \mathrm{CH}^2(X_{\mathbb{C}})_{\mathbb{Q}} = \infty$. To the best of my knowledge not a single example of a K3 surface X defined over $\bar{\mathbb{Q}}$ is known where finite dimensionality of $\mathrm{CH}^2(X)_{\mathbb{Q}}$ could be verified.

Also note that the Picard group does not change under base change from $\bar{\mathbb{Q}}$ to $\mathbb{C}$, i.e., for X defined over $\bar{\mathbb{Q}}$ one has $\text{Pic}(X) \simeq \text{Pic}(X_{\mathbb{C}})$ (see 5.4). But over the actual field of definition of X, which is a number field in this case, the Picard group can be strictly smaller.

The central argument in Mumford's proof is that an irreducible component of the closed subset of effective cycles in X^n rationally equivalent to a given cycle must be proper, due to the existence of a nontrivial regular two-form on X, and that a countable union of those cannot cover X^n if the base field is not countable. This idea was later formalized and has led to many more results proving nontriviality of cycles under nonvanishing hypotheses on the nonalgebraic part of the cohomology (see e.g. [Voisin 2002, Chapter 22]). There is also a more arithmetic approach to produce arbitrarily many nontrivial classes in $\text{CH}^2(X)$ for a complex K3 surface X which proceeds via curves over finitely generated field extensions of $\bar{\mathbb{Q}}$ and embeddings of their function fields into $\mathbb{C}$. See [Green et al. 2004], for example.

The degree of a cycle induces a homomorphism $\text{CH}^2(X) \longrightarrow \mathbb{Z}$ and its kernel $\text{CH}^2(X)_0$ is the group of homologically (or algebraically) trivial classes. Thus, the Bloch–Beilinson conjecture for a K3 surface X over $\bar{\mathbb{Q}}$ says that $\text{CH}^2(X)_0 = 0$ or, equivalently, that

$$\text{CH}^*(X) \simeq \widetilde{\text{NS}}(X_{\mathbb{C}}) \hookrightarrow \tilde{H}(X_{\mathbb{C}}, \mathbb{Z}).$$

3.4. The main results presented in my talk were triggered by the paper [Beauville and Voisin 2004] on a certain natural subring of $\text{CH}^*(X)$. They show in particular that for a complex K3 surface X there is a natural class $c_X \in \text{CH}^2(X)$ of degree one with the following properties:

i) $c_X = [x]$ for any point $x \in X$ contained in a (possibly singular) rational curve $C \subset X$.

ii) $c_1(L)^2 \in \mathbb{Z}c_X$ for any $L \in \text{Pic}(X)$.

iii) $c_2(X) = 24c_X$.

Let us introduce

$$R(X) := \text{CH}^0(X) \oplus \text{CH}^1(X) \oplus \mathbb{Z}c_X.$$

Then ii) shows that $R(X)$ is a subring of $\text{CH}^*(X)$. A different way of expressing ii) and iii) together is to say that for any $L \in \text{Pic}(X)$ the Mukai vector

$$v^{\text{CH}}(L) = \text{ch}(L)\sqrt{\text{td}(X)}$$

is contained in $R(X)$ (see 4.1). It will be in this form that the results of Beauville and Voisin can be generalized in a very natural form to the derived context (Theorem 5.3).

Note that the cycle map induces an isomorphism $R(X) \simeq \widetilde{NS}(X)$ and that for a K3 surface X over $\bar{\mathbb{Q}}$ the Bloch–Beilinson conjecture can be expressed by saying that base change yields an isomorphism $CH^*(X) \simeq R(X_{\mathbb{C}})$.

So, the natural filtration $CH^*(X)_0 \subset CH^*(X)$ (see also below) with quotient $\widetilde{NS}(X)$ admits a split given by $R(X)$. This can be written as

$$CH^*(X) = R(X) \oplus CH^*(X)_0$$

and seems to be a special feature of K3 surfaces and higher-dimensional symplectic varieties. For instance, in [Beauville 2007] it was conjectured that any relation between $c_1(L_i)$ of line bundles L_i on an irreducible symplectic variety X in $H^*(X)$ also holds in $CH^*(X)$. The conjecture was completed to also incorporate Chern classes of X and proved for low-dimensional Hilbert schemes of K3 surfaces in [Voisin 2008]. See also the more recent thesis [Ferretti 2009] which deals with double EPW sextics, which are special deformations of four-dimensional Hilbert schemes.

3.5. The Bloch–Beilinson conjectures also predict for smooth projective varieties X the existence of a functorial filtration

$$0 = F^{p+1}CH^p(X) \subset F^p CH^p(X) \subset \cdots \subset F^1 CH^p(X) \subset F^0 CH^p(X)$$

whose first step F^1 is simply the kernel of the cycle map. Natural candidates for such a filtration were studied e.g. by Green, Griffiths, Jannsen, Lewis, Murre, and S. Saito (see [Green and Griffiths 2003] and the references therein).

For a surface X the interesting part of this filtration is $0 \subset \ker(alb_X) \subset CH^2(X)_0 \subset CH^2(X)$. Here $alb_X : CH^2(X)_0 \longrightarrow Alb(X)$ denotes the Albanese map.

A cycle $\Gamma \in CH^2(X \times X)$ naturally acts on cohomology and on the Chow group. We write $[\Gamma]_*^{i,0}$ for the induced endomorphism of $H^0(X, \Omega_X^i)$ and $[\Gamma]_*$ for the action on $CH^2(X)$. The latter respects the natural filtration $\ker(alb_X) \subset CH^2(X)_0 \subset CH^2(X)$ and thus induces an endomorphism $gr[\Gamma]_*$ of the graded object $\ker(alb_X) \oplus Alb(X) \oplus \mathbb{Z}$.

The following is also a consequence of Bloch's conjecture; see [Bloch 1980] or [Voisin 2002, Chapter 11], not completely unrelated to Conjecture 3.3.

Conjecture 3.6. $[\Gamma]_*^{2,0} = 0$ *if and only if* $gr[\Gamma]_* = 0$ *on* $\ker(alb_X)$.

It is known that this conjecture is implied by the Bloch–Beilinson conjecture for $X \times X$ when X and Γ are defined over $\bar{\mathbb{Q}}$. But otherwise, very little is known about it. Note that the analogous statement $[\Gamma]_*^{1,0} = 0$ if and only if $gr[\Gamma]_* = 0$ on $Alb(X)$ holds true by definition of the Albanese.

For K3 surfaces the Albanese map is trivial and so the Bloch–Beilinson filtration for K3 surfaces is simply $0 \subset \ker(alb_X) = CH^2(X)_0 \subset CH^2(X)$. In particular Conjecture 3.6 for a K3 surface becomes: $[\Gamma]_*^{2,0} = 0$ if and only if $\mathrm{gr}[\Gamma]_* = 0$ on $\mathrm{CH}^2(X)_0$. In this form the conjecture seems out of reach for the time being, but the following special case seems more accessible and we will explain in Section 6 to what extend derived techniques can be useful to answer it.

Conjecture 3.7. *Let $f \in \mathrm{Aut}(X)$ be a symplectomorphism of a complex projective K3 surface X, i.e., $f^* = \mathrm{id}$ on $H^{2,0}(X)$. Then $f^* = \mathrm{id}$ on $\mathrm{CH}^2(X)$.*

Remark 3.8. Note that the converse is true: If $f \in \mathrm{Aut}(X)$ acts as id on $\mathrm{CH}^2(X)$, then f is a symplectomorphism. This is reminiscent of a consequence of the global Torelli theorem which for a complex projective K3 surface X states:

$$f = \mathrm{id} \iff f^* = \mathrm{id} \text{ on the Chow ring}(!) \mathrm{CH}^*(X).$$

4. Derived category

The Chow group $\mathrm{CH}^*(X)$ is the space of cycles divided by rational equivalence. Equivalently, one could take the abelian or derived category of coherent sheaves on X and pass to the Grothendieck K-groups. It turns out that considering the more rigid structure of a category that lies behind the Chow group can lead to new insight. See [Huybrechts 2006] for a general introduction to derived categories and for more references to the original literature.

4.1. For a K3 surface X over a field K the category $\mathrm{Coh}(X)$ of coherent sheaves on X is a K-linear abelian category and its *bounded derived category*, denoted $\mathrm{D}^b(X)$, is a K-linear triangulated category.

If $E^\bullet$ is an object of $\mathrm{D}^b(X)$, its *Mukai vector* $v(E^\bullet) = \sum(-1)^i v(E^i) = \sum(-1)^i v(\mathcal{H}^i(E^\bullet)) \in \widetilde{\mathrm{NS}}(X) \subset \widetilde{H}(X, \mathbb{Z})$ is well defined. By abuse of notation, we will write the Mukai vector as a map

$$v : \mathrm{D}^b(X) \longrightarrow \widetilde{\mathrm{NS}}(X).$$

Since the Chern character of a coherent sheaf and the Todd genus of X exist as classes in $\mathrm{CH}^*(X)$, the Mukai vector with values in $\mathrm{CH}^*(X)$ can also be defined. This will be written as

$$v^{\mathrm{CH}} : \mathrm{D}^b(X) \longrightarrow \mathrm{CH}^*(X).$$

(It is a special feature of K3 surfaces that the Chern character really is integral.)

Note that $\mathrm{CH}^*(X)$ can also be understood as the Grothendieck K-group of the abelian category $\mathrm{Coh}(X)$ or of the triangulated category $\mathrm{D}^b(X)$, i.e., $K(X) \simeq K(\mathrm{Coh}(X)) \simeq K(\mathrm{D}^b(X)) \simeq \mathrm{CH}^*(X)$. (In order to exclude any torsion phenomena we assume here that K is algebraically closed, i.e., $K = \mathbb{C}$ or $K = \bar{\mathbb{Q}}$, or, alternatively, pass to the associated $\mathbb{Q}$-vector spaces.)

Clearly, the lift of a class in $\mathrm{CH}^*(X)$ to an object in $\mathrm{D}^b(X)$ is never unique. Of course, for certain classes there are natural choices; for instance, $v^{\mathrm{CH}}(L)$ naturally lifts to L which is a spherical object (see below).

4.2. Due to a result of Orlov, every K-linear exact equivalence

$$\Phi : \mathrm{D}^b(X) \xrightarrow{\sim} \mathrm{D}^b(X')$$

between the derived categories of two smooth projective varieties is a Fourier–Mukai transform, i.e., there exists a unique object $\mathscr{E} \in \mathrm{D}^b(X \times X')$, the kernel, such that Φ is isomorphic to the functor $\Phi_{\mathscr{E}} = p_*(q^*(\ )\otimes\mathscr{E})$. Here p_*, q^*, and $\otimes$ are derived functors. It is known that if X is a K3 surface also X' is one.

It would be very interesting to use Orlov's result to deduce the existence of objects in $\mathrm{D}^b(X \times X')$ that are otherwise difficult to describe. However, we are not aware of any nontrivial example of a functor that can be shown to be an equivalence, or even just fully faithful, without actually describing it as a Fourier–Mukai transform.

Here is a list of essentially all known (auto)equivalences for K3 surfaces:

i) Any isomorphism $f : X \xrightarrow{\sim} X'$ induces an exact equivalence

$$f_* : \mathrm{D}^b(X) \xrightarrow{\sim} \mathrm{D}^b(X')$$

with Fourier–Mukai kernel the structure sheaf $\mathcal{O}_{\Gamma_f}$ of the graph $\Gamma_f \subset X \times X'$ of f.

ii) The tensor product $L \otimes (\ )$ for a line bundle $L \in \mathrm{Pic}(X)$ defines an autoequivalence of $\mathrm{D}^b(X)$ with Fourier–Mukai kernel $\Delta_* L$.

iii) An object $E \in \mathrm{D}^b(X)$ is called *spherical* if $\mathrm{Ext}^*(E, E) \simeq H^*(S^2, K)$ as graded vector spaces. The *spherical twist*

$$T_E : \mathrm{D}^b(X) \xrightarrow{\sim} \mathrm{D}^b(X)$$

associated with it is the Fourier–Mukai equivalence whose kernel is given as the cone of the trace map

$$E^* \boxtimes E \longrightarrow (E^* \boxtimes E)|_\Delta \xrightarrow{\sim} \Delta_*(E^* \otimes E) \longrightarrow \mathcal{O}_\Delta.$$

(For examples of spherical objects see 5.1.)

iv) If X' is a fine projective moduli space of stable sheaves and $\dim(X') = 2$, then the universal family $\mathscr{E}$ on $X \times X'$ (unique up to a twist with a line bundle on X') can be taken as the kernel of an equivalence $\mathrm{D}^b(X) \xrightarrow{\sim} \mathrm{D}^b(X')$.

4.3. Writing an equivalence as a Fourier–Mukai transform allows one to associate directly to any autoequivalence $\Phi : \mathrm{D}^b(X) \xrightarrow{\sim} \mathrm{D}^b(X)$ of a complex K3 surface X an isomorphism

$$\Phi^H : \widetilde{H}(X, \mathbb{Z}) \xrightarrow{\sim} \widetilde{H}(X, \mathbb{Z})$$

which in terms of the Fourier–Mukai kernel $\mathcal{E}$ is given by $\alpha \longmapsto p_*(q^*\alpha.v(\mathcal{E}))$. As was observed by Mukai, this isomorphism is defined over $\mathbb{Z}$ and not only over $\mathbb{Q}$. Moreover, it preserves the Mukai pairing and the natural weight two Hodge structure, i.e., it is an integral Hodge isometry of $\widetilde{H}(X, \mathbb{Z})$. As above, $v(\mathcal{E})$ denotes the Mukai vector $v(\mathcal{E}) = \mathrm{ch}(\mathcal{E})\sqrt{\mathrm{td}(X \times X)}$.

Clearly, the latter makes also sense in $\mathrm{CH}^*(X \times X)$ and so one can as well associate to the equivalence Φ a group automorphism

$$\Phi^{\mathrm{CH}} : \mathrm{CH}^*(X) \xrightarrow{\sim} \mathrm{CH}^*(X).$$

The reason why the usual Chern character is replaced by the Mukai vector is the Grothendieck–Riemann–Roch formula. With this definition of Φ^H and Φ^{CH} one finds that $\Phi^H(v(E)) = v(\Phi(E))$ and $\Phi^{\mathrm{CH}}(v^{\mathrm{CH}}(E)) = v^{\mathrm{CH}}(\Phi(E))$ for all $E \in \mathrm{D}^{\mathrm{b}}(X)$.

Note that Φ^H and Φ^{CH} do not preserve, in general, neither the multiplicative structure nor the grading of $\widetilde{H}(X, \mathbb{Z})$ or $\mathrm{CH}^*(X)$.

The derived category $\mathrm{D}^{\mathrm{b}}(X)$ is difficult to describe in concrete terms. Its group of autoequivalences, however, seems more accessible. So let $\mathrm{Aut}(\mathrm{D}^{\mathrm{b}}(X))$ denote the group of all K-linear exact equivalences $\Phi : \mathrm{D}^{\mathrm{b}}(X) \xrightarrow{\sim} \mathrm{D}^{\mathrm{b}}(X)$ up to isomorphism. Then $\Phi \longmapsto \Phi^H$ and $\Phi \longmapsto \Phi^{\mathrm{CH}}$ define the two representations

$$\rho^H : \mathrm{Aut}(\mathrm{D}^{\mathrm{b}}(X)) \longrightarrow \mathrm{O}(\widetilde{H}(X, \mathbb{Z})) \quad \text{and} \quad \rho^{\mathrm{CH}} : \mathrm{Aut}(\mathrm{D}^{\mathrm{b}}(X)) \longrightarrow \mathrm{Aut}(\mathrm{CH}^*(X)).$$

Here, $\mathrm{O}(\widetilde{H}(X, \mathbb{Z}))$ is the group of all integral Hodge isometries of the weight two Hodge structure defined on the Mukai lattice $\widetilde{H}(X, \mathbb{Z})$ and $\mathrm{Aut}(\mathrm{CH}^*(X))$ denotes simply the group of all automorphisms of the additive group $\mathrm{CH}^*(X)$.

Although $\mathrm{CH}^*(X)$ is a much bigger group than $\widetilde{H}(X, \mathbb{Z})$, at least over $K = \mathbb{C}$, both representations carry essentially the same information. More precisely one can prove (see [Huybrechts 2010]):

Theorem 4.4. $\ker(\rho^H) = \ker(\rho^{\mathrm{CH}})$.

In the following we will explain what is known about this kernel and the images of the representations ρ^H and ρ^{CH}.

4.5. Due to the existence of the many spherical objects in $\mathrm{D}^{\mathrm{b}}(X)$ and their associated spherical twists, the kernel $\ker(\rho^H) = \ker(\rho^{\mathrm{CH}})$ has a rather intriguing structure. Let us be a bit more precise: If $E \in \mathrm{D}^{\mathrm{b}}(X)$ is spherical, then T_E^H is the reflection s_δ in the hyperplane orthogonal to $\delta := v(E)$. Hence, the square T_E^2 is an element in $\ker(\rho^H)$ which is easily shown to be nontrivial.

Due to the existence of the many spherical objects on any K3 surface (all line bundles are spherical) and the complicated relations between them, the group generated by all T_E^2 is a very interesting object. In fact, conjecturally $\ker(\rho^H)$

is generated by the T_E^2's and the double shift. This and the expected relations between the spherical twists are expressed by the following conjecture:

Conjecture 4.6 [Bridgeland 2008]. $\ker(\rho^H) = \ker(\rho^{CH}) \simeq \pi_1(\mathcal{P}_0(X))$.

For the definition of $\mathcal{P}_0(X)$ see 2.4. The fundamental group of $\mathcal{P}_0(X)$ is generated by loops around each $\delta^\perp$ and the generator of $\pi_1(\mathcal{P}(X)) \simeq \mathbb{Z}$. The latter is naturally lifted to the autoequivalence given by the double shift $E \longmapsto E[2]$.

Since each (-2)-vector δ can be written as $\delta = v(E)$ for some spherical object, one can lift the loop around $\delta^\perp$ to T_E^2. However, the spherical object E is by no means unique. Just choose any other spherical object F and consider $T_F^2(E)$ which has the same Mukai vector as E. Even for a Mukai vector $v = (r, \ell, s)$ with $r > 0$ there is in general more than one spherical bundle(!) E with $v(E) = v$ (see 5.1).

Nevertheless, Bridgeland does construct a group homomorphism

$$\pi_1(\mathcal{P}_0(X)) \longrightarrow \ker(\rho^H) \subset \text{Aut}(D^b(X)).$$

The injectivity of this map is equivalent to the simply connectedness of the distinguished component $\Sigma(X) \subset \text{Stab}(X)$ of stability conditions considered by Bridgeland. If $\Sigma(X)$ is the only connected component, then the surjectivity would follow.

Note that, although $\ker(\rho^H)$ is by definition not visible on $\widetilde{H}(X, \mathbb{Z})$ and by Theorem 4.4 also not on $\text{CH}^*(X)$, it still seems to be governed by the Hodge structure of $\widetilde{H}(X, \mathbb{Z})$. Is this in any way reminiscent of the Bloch conjecture (see 3.5)?

4.7. On the other hand, the image of ρ^H is well understood which is (see [Huybrechts et al. 2009]):

Theorem 4.8. *The image of* $\rho^H : \text{Aut}(D^b(X)) \longrightarrow O(\widetilde{H}(X, \mathbb{Z}))$ *is the group* $O_+(\widetilde{H}(X, \mathbb{Z}))$ *of all Hodge isometries leaving invariant the natural orientation of the space of positive directions.*

Recall that the Mukai pairing has signature $(4, 20)$. The classes $\text{Re}(\sigma)$, $\text{Im}(\sigma)$, $1 - \omega^2/2$, ω, where $0 \neq \sigma \in H^{2,0}(X)$ and $\omega \in \mathcal{K}_X$ an ample class, span a real subspace V of dimension four which is positive definite with respect to the Mukai pairing. Using orthogonal projection, the orientations of V and $\Phi^H(V)$ can be compared.

To show that $\text{Im}(\rho^H)$ has at most index two in $O(\widetilde{H}(X, \mathbb{Z}))$ uses techniques of Mukai and Orlov and was observed by Hosono, Lian, Oguiso, Yau [Hosono et al. 2004] and Ploog. As it turned out, the difficult part is to prove that the index is exactly two. This was predicted by Szendrői, based on considerations in mirror symmetry, and recently proved in a joint work with Macrì and Stellari [Huybrechts et al. 2009].

Let us now turn to the image of ρ^{CH}. The only additional structure the Chow group $\mathrm{CH}^*(X)$ seems to have is the subring $R(X) \subset \mathrm{CH}^*(X)$ (see 3.4). And indeed, this subring is preserved under derived equivalences (see [Huybrechts 2010]):

Theorem 4.9. *If* $\rho(X) \geq 2$ *and* $\Phi \in \mathrm{Aut}(\mathrm{D}^b(X))$, *then* Φ^H *preserves the subring* $R(X) \subset \mathrm{CH}^*(X)$.

In other words, autoequivalences (and in fact equivalences) respect the direct sum decomposition $\mathrm{CH}^*(X) = R(X) \oplus \mathrm{CH}^*(X)_0$ (see 3.4).

The assumption on the Picard rank should eventually be removed, but as for questions concerning potential density of rational points the Picard rank one case is indeed more complicated.

Clearly, the action of Φ^{CH} on $R(X)$ can be completely recovered from the action of Φ^H on $\widetilde{\mathrm{NS}}(X)$. On the other hand, according to the Bloch conjecture (see 3.5) the action of Φ^{CH} on $\mathrm{CH}^*(X)_0$ should be governed by the action of Φ^H on the transcendental part $T(X)$. Note that for $K = \bar{\mathbb{Q}}$ one expects $\mathrm{CH}^*(X)_0 = 0$, so nothing interesting can be expected in this case. However, for $K = \mathbb{C}$ well-known arguments show that $\Phi^H \neq \mathrm{id}$ on $T(X)$ implies $\Phi^{\mathrm{CH}} \neq \mathrm{id}$ on $\mathrm{CH}^*(X)_0$ (see [Voisin 2002]). As usual, it is the converse that is much harder to come by. Let us nevertheless rephrase the Bloch conjecture once more for this case.

Conjecture 4.10. *Suppose* $\Phi^H = \mathrm{id}$ *on* $T(X)$. *Then* $\Phi^{\mathrm{CH}} = \mathrm{id}$ *on* $\mathrm{CH}^*(X)_0$.

By Theorem 4.4 one has $\Phi^{\mathrm{CH}} = \mathrm{id}$ under the stronger assumption $\Phi^H = \mathrm{id}$ not only on $T(X)$ but on all of $\widetilde{H}(X, \mathbb{Z})$. The special case of $\Phi = f_*$ will be discussed in more detail in Section 6

Note that even if the conjecture can be proved we would still not know how to describe the image of ρ^{CH}. It seems, $\mathrm{CH}^*(X)$ has just not enough structure that could be used to determine explicitly which automorphisms are induced by derived equivalences.

5. Chern classes of spherical objects

It has become clear that spherical objects and the associated spherical twists play a central role in the description of $\mathrm{Aut}(\mathrm{D}^b(X))$. Together with automorphisms of X itself and orthogonal transformations of $\widetilde{H}$ coming from universal families of stable bundles, they determine the action of $\mathrm{Aut}(\mathrm{D}^b(X))$ on $\widetilde{H}(X, \mathbb{Z})$. The description of the kernel of ρ^{CH} should only involve squares of spherical twists by Conjecture 4.6.

5.1. It is time to give more examples of spherical objects.

i) Every line bundle $L \in \text{Pic}(X)$ is a spherical object in $D^b(X)$ with Mukai vector $v = (1, \ell, \ell^2/2 + 1)$ where $\ell = c_1(L)$. Note that the spherical twist T_L has nothing to do with the equivalence given by the tensor product with L. Also the relation between T_L and, say, T_{L^2}, is not obvious.

ii) If $C \subset X$ is a smooth irreducible rational curve, then all $\mathcal{O}_C(i)$ are spherical objects with Mukai vector $v = (0, [C], i + 1)$. The spherical twist $T_{\mathcal{O}_C(-1)}$ induces the reflection $s_{[C]}$ on $\tilde{H}(X, \mathbb{Z})$, an element of the Weyl group W_X.

iii) Any simple vector bundle E which is also rigid, i.e., $\text{Ext}^1(E, E) = 0$, is spherical. This generalizes i). Note that rigid torsion free sheaves are automatically locally free (see [Mukai 1987, Proposition 2.14]). Let $v = (r, \ell, s) \in \tilde{N}S(X)$ be a (-2)-class with $r > 0$ and H be a fixed polarization. Then due to a result of Mukai there exists a unique rigid bundle E with $v(E) = v$ which is slope stable with respect to H (see [Huybrechts and Lehn 2010, Theorem 6.16] for the uniqueness). However, varying H usually leads to (finitely many) different spherical bundles realizing v. They should be considered as nonseparated points in the moduli space of simple bundles (on deformations of X). This can be made precise by saying that for two different spherical bundles E_1 and E_2 with $v(E_1) = v(E_2)$ there always exists a nontrivial homomorphism $E_1 \twoheadrightarrow E_2$.

5.2. The Mukai vector $v(E)$ of a spherical object $E \in D^b(X)$ is an integral $(1, 1)$-class of square -2 and every such class can be lifted to a spherical object. For the Mukai vectors in $\text{CH}^*(X)$ we have:

Theorem 5.3 [Huybrechts 2010]. *If $\rho(X) \geq 2$ and $E \in D^b(X)$ is spherical, then $v^{\text{CH}}(E) \in R(X)$.*

In particular, two nonisomorphic spherical bundles realizing the same Mukai vector in $\tilde{H}(X, \mathbb{Z})$ are also not distinguished by their Mukai vectors in $\text{CH}^*(X)$. Again, the result should hold without the assumption on the Picard group.

This theorem is first proved for spherical bundles by using Lazarsfeld's technique to show that primitive ample curves on K3 surfaces are Brill–Noether general [Lazarsfeld 1986] and the Bogomolov–Mumford theorem on the existence of rational curves in ample linear systems [Mori and Mukai 1983] (which is also at the core of [Beauville and Voisin 2004]). Then one uses Theorem 4.4 to generalize this to spherical objects realizing the Mukai vector of a spherical bundle. For this step one observes that knowing the Mukai vector of the Fourier–Mukai kernel of T_E in $\text{CH}^*(X \times X)$ allows one to determine $v^{\text{CH}}(E)$.

Actually Theorem 5.3 is proved first and Theorem 4.9 is a consequence of it. Indeed, if $\Phi : D^b(X) \xrightarrow{\sim} D^b(X)$ is an equivalence, then for a spherical object $E \in D^b(X)$ the image $\Phi(E)$ is again spherical. Since $v^{\text{CH}}(\Phi(E)) = \Phi^{\text{CH}}(v^{\text{CH}}(E))$,

Theorem 5.3 shows that Φ^{CH} sends Mukai vectors of spherical objects, in particular of line bundles, to classes in $R(X)$. Clearly, $R(X)$ is generated as a group
by the $v^{\mathrm{CH}}(L)$ with $L \in \mathrm{Pic}(X)$ which then proves Theorem 4.9.

5.4. The true reason behind Theorem 5.3 and in fact behind most of the results in
[Beauville and Voisin 2004] is the general philosophy that every rigid geometric
object on a variety X is already defined over the smallest algebraically closed field
of definition of X. This is then combined with the Bloch–Beilinson conjecture
which for X defined over $\bar{\mathbb{Q}}$ predicts that $R(X_{\mathbb{C}}) = \mathrm{CH}^*(X)$.

To make this more precise consider a K3 surface X over $\bar{\mathbb{Q}}$ and the associated
complex K3 surface $X_{\mathbb{C}}$. An object $E \in \mathrm{D}^b(X_{\mathbb{C}})$ is defined over $\bar{\mathbb{Q}}$ if there exists
an object $F \in \mathrm{D}^b(X)$ such that its base-change to $X_{\mathbb{C}}$ is isomorphic to E. We
write this as $E \simeq F_{\mathbb{C}}$.

The pull-back yields an injection of rings $\mathrm{CH}^*(X) \hookrightarrow \mathrm{CH}^*(X_{\mathbb{C}})$ and if $E \in$
$\mathrm{D}^b(X_{\mathbb{C}})$ is defined over $\bar{\mathbb{Q}}$ its Mukai vector $v^{\mathrm{CH}}(E)$ is contained in the image of
this map. Now, if we can show that $\mathrm{CH}^*(X) = R(X_{\mathbb{C}})$, then the Mukai vector of
every $E \in \mathrm{D}^b(X_{\mathbb{C}})$ defined over $\bar{\mathbb{Q}}$ is contained in $R(X_{\mathbb{C}})$.

Eventually one observes that spherical objects on $X_{\mathbb{C}}$ are defined over $\bar{\mathbb{Q}}$. For
line bundles $L \in \mathrm{Pic}(X_{\mathbb{C}})$ this is well-known, i.e., $\mathrm{Pic}(X) \simeq \mathrm{Pic}(X_{\mathbb{C}})$. Indeed, the
Picard functor is defined over $\bar{\mathbb{Q}}$ (or in fact over the field of definition of X) and
therefore the set of connected components of the Picard scheme does not change
under base change. The Picard scheme of a K3 surface is zero-dimensional, a
connected component consists of one closed point and, therefore, base change
identifies the set of closed points. For the algebraically closed field $\bar{\mathbb{Q}}$ the set
of closed points of the Picard scheme of X is the Picard group of X which thus
does not get bigger under base change e.g. to $\mathbb{C}$.

For general spherical objects in $\mathrm{D}^b(X_{\mathbb{C}})$ the proof uses results of Inaba and
Lieblich (see [Inaba 2002], for instance) on the representability of the functor
of complexes (with vanishing negative Ext's) by an algebraic space. This is
technically more involved, but the underlying idea is just the same as for the
case of line bundles.

6. Automorphisms acting on the Chow ring

We come back to the question raised as Conjecture 3.7. So suppose $f \in \mathrm{Aut}(X)$
is an automorphism of a complex projective K3 surface X with $f^*\sigma = \sigma$ where
σ is a trivializing section of the canonical bundle ω_X. In other words, the Hodge
isometry f^* of $H^2(X, \mathbb{Z})$ (or of $\tilde{H}(X, \mathbb{Z})$) is the identity on $H^{0,2}(X) = \tilde{H}^{0,2}(X)$
or, equivalently, on the transcendental lattice $T(X)$. What can we say about the
action induced by f on $\mathrm{CH}^2(X)$? Obviously, the question makes sense for K3
surfaces defined over other fields, say $\bar{\mathbb{Q}}$, but $\mathbb{C}$ is the most interesting case (at

least in characteristic zero) and for $\bar{\mathbb{Q}}$ the answer should be without any interest due to the Bloch–Beilinson conjecture.

In this section we will explain that the techniques of the earlier sections and of [Huybrechts 2010] can be combined with results of Kneser on the orthogonal group of lattices to prove Conjecture 3.7 under some additional assumptions on the Picard group of X.

6.1. Suppose $f \in \mathrm{Aut}(X)$ is a nontrivial symplectomorphism, i.e., $f^*\sigma = \sigma$. If f has finite order n, then $n = 2, \ldots, 7$, or 8. This is a result from [Nikulin 1980] and follows from the holomorphic fixed point formula (see [Mukai 1988]). Moreover, in this case f has only finitely many fixed points, all isolated, and depending on n the number of fixed points is 8, 6, 4, 4, 2, 3, 2, respectively. The minimal resolution of the quotient $Y \longrightarrow \bar{X} := X/\langle f \rangle$ yields again a K3 surface Y. Thus, for symplectomorphisms of finite order Conjecture 3.7 is equivalent to the bijectivity of the natural map $\mathrm{CH}^2(Y)_{\mathbb{Q}} \longrightarrow \mathrm{CH}^2(X)_{\mathbb{Q}}$. Due to a result of Nikulin the action of a symplectomorphism f of finite order on $H^2(X, \mathbb{Z})$ is as an abstract lattice automorphism independent of f and depends only on the order. For prime order 2, 3, 5, and 7 it was explicitly described and studied in [van Geemen and Sarti 2007; Garbagnati and Sarti 2007]. For example, for a symplectic involution the fixed part in $H^2(X, \mathbb{Z})$ has rank 14. The moduli space of K3 surfaces X endowed with a symplectic involution is of dimension 11 and the Picard group of the generic member contains $E_8(-2)$ as a primitive sublattice of corank one.

Explicit examples of symplectomorphisms are easy to construct. For example, $(x_0 : x_1 : x_2 : x_3) \mapsto (-x_0 : -x_1 : x_2 : x_3)$ defines a symplectic involution on the Fermat quartic $X_0 \subset \mathbb{P}^3$. On an elliptic K3 surface with two sections one can use fiberwise addition to produce symplectomorphisms.

6.2. The orthogonal group of a unimodular lattice Λ has been investigated in detail in [Wall 1963]. Subsequently, there have been many attempts to generalize some of his results to nonunimodular lattices. Of course, often new techniques are required in the more general setting and some of the results do not hold any longer.

The article [Kneser 1981] turned out to be particularly relevant for our purpose. Before we can state Kneser's result we need to recall a few notions. First, the Witt index of a lattice Λ is the maximal dimension of an isotropic subspace in $\Lambda_{\mathbb{R}}$. So, if Λ is nondegenerate of signature (p, q), then the Witt index is $\min\{p, q\}$. The p-rank $\mathrm{rk}_p(\Lambda)$ of Λ is the maximal rank of a sublattice $\Lambda' \subset \Lambda$ whose discriminant is not divisible by p.

Recall that every orthogonal transformation of the real vector space $\Lambda_{\mathbb{R}}$ can be written as a composition of reflections. The spinor norm of a reflection with respect to a vector $v \in \Lambda_{\mathbb{R}}$ is defined as $-(v, v)/2$ in $\mathbb{R}^*/\mathbb{R}^{*2}$. In particular, a reflection s_δ for a (-2)-class $\delta \in \Lambda$ has trivial spinor norm. The spinor norm for reflections is extended multiplicatively to a homomorphism $O(\Lambda) \longrightarrow \{\pm 1\}$.

The following is a classical result due to Kneser, motivated by work of Ebeling, which does not seem widely known.

Theorem 6.3. *Let Λ be an even nondegenerate lattice of Witt index at least two such that Λ represents -2. Suppose $\mathrm{rk}_2(\Lambda) \geq 6$ and $\mathrm{rk}_3(\Lambda) \geq 5$. Then every $g \in SO(\Lambda)$ with $g = \mathrm{id}$ on Λ^*/Λ and trivial spinor norm can be written as a composition of an even number of reflections $\prod s_{\delta_i}$ with (-2)-classes $\delta_i \in \Lambda$.*

By using that a (-2)-reflection has determinant -1 and trivial spinor norm and discriminant, Kneser's result can be rephrased as follows: Under the above conditions on Λ the Weyl group W_Λ of Λ is given by

$$W_\Lambda = \ker \left(O(\Lambda) \longrightarrow \{\pm 1\} \times O(\Lambda^*/\Lambda) \right). \tag{6-1}$$

The assumption on rk_2 and rk_3 can be replaced by assuming that the reduction mod 2 resp. 3 are not of a very particular type. For instance, for $p = 2$ one has to exclude the case $\bar{x}_1 \bar{x}_2$, $\bar{x}_1 \bar{x}_2 + \bar{x}_3^2$, and $\bar{x}_1 \bar{x}_2 + \bar{x}_3 \bar{x}_4 + \bar{x}_5^2$. See [Kneser 1981] or details.

6.4. Kneser's result can never be applied to the Néron–Severi lattice $NS(X)$ of a K3 surface X, because its Witt index is one. But the extended Néron–Severi lattice $\widetilde{NS}(X) \simeq NS(X) \oplus U$ has Witt index two. The conditions on rk_2 and rk_3 for $\widetilde{NS}(X)$ become $\mathrm{rk}_2(NS(X)) \geq 4$ and $\mathrm{rk}_3(NS(X)) \geq 3$. This leads to the main result of this section.

Theorem 6.5. *Suppose $\mathrm{rk}_2(NS(X)) \geq 4$ and $\mathrm{rk}_3(NS(X)) \geq 3$. Then any symplectomorphism $f \in \mathrm{Aut}(X)$ acts trivially on $CH^2(X)$.*

Proof. First note that the discriminant of an orthogonal transformation of a unimodular lattice is always trivial and that the discriminant groups of $NS(X)$ and $T(X)$ are naturally identified. Since a symplectomorphism acts as id on $T(X)$, its discriminant on $NS(X)$ is also trivial. Note that a (-2)-reflection s_δ has also trivial discriminant and spinor norm 1. Its determinant is -1.

Let now $\delta_0 := (1, 0, -1)$, which is a class of square $\delta_0^2 = 2$ (and not -2). So the induced reflection s_δ has spinor norm and determinant both equal to -1. Its discriminant is trivial. To a symplectomorphism f we associate the orthogonal transformation g_f as follows. It is f_* if the spinor norm of f_* is 1 and $s_{\delta_0} \circ f_*$ otherwise. Then g_f has trivial spinor norm and trivial discriminant, By Equation (6-1) this shows $g_f \in \widetilde{W}_X$, i.e., f_* or $s_{\delta_0} \circ f_*$ is of the form $\prod s_{\delta_i}$ with (-2)-classes δ_i. Writing $\delta_i = v(E_i)$ with spherical E_i allows one to interpret the right hand side as $\prod T_{E_i}^H$.

Clearly, the $T_{E_i}^H$ preserve the orientation of the four positive directions and so does f_*. But s_{δ_0} does not, which proves a posteriori that the spinor norm of f_* must always be trivial: $g_f = f_*$.

Thus, $f_* = \prod T_{E_i}^H$ and hence we proved that under the assumptions on NS(X) the action of the symplectomorphism f on $\tilde{H}(X, \mathbb{Z})$ coincides with the action of the autoequivalence

$$\Phi := \prod T_{E_i}.$$

But by Theorem 4.4 their actions then coincide also on CH$^*(X)$. To conclude, use Theorem 5.3 which shows that the action of Φ on CH$^2(X)_0$ is trivial. $\square$

Remark 6.6. The proof actually shows that the image of the subgroup of those $\Phi \in \text{Aut}(\text{D}^b(X))$ acting trivially on $T(X)$ (the "symplectic equivalences") in O($\widetilde{\text{NS}}(X)$) is $\tilde{W}_X$, i.e., coincides with the image of the subgroup spanned by all spherical twists T_E.

Unfortunately, Theorem 6.5 does not cover the generic case of symplecto-morphisms of finite order. For example, the Néron–Severi group of a generic K3 surface endowed with a symplectic involution is up to index two isomorphic to $\mathbb{Z}\ell \oplus E_8(-2)$ (see [van Geemen and Sarti 2007]). Whatever the square of ℓ is, the extended Néron–Severi lattice $\widetilde{\text{NS}}(X)$ will have rk$_2 = 2$ and indeed its reduction mod 2 is of the type $\bar{x}_1\bar{x}_2$ explicitly excluded in Kneser's result and its refinement alluded to above.

Example 6.7. By a result from [Morrison 1984] one knows that for Picard rank 19 or 20 the Néron–Severi group NS(X) contains $E_8(-1)^{\oplus 2}$ and hence the assumptions of Theorem 6.5 are satisfied (by far). In particular, our result applies to the members X_t of the Dwork family $\sum x_i^4 + t \prod x_i$ in $\mathbb{P}^3$, so in particular to the Fermat quartic itself. We can conclude that all symplectic automorphisms of X_t act trivially on CH$^2(X_t)$. For the symplectic automorphisms given by multiplication with roots of unities this was proved by different methods already in [Chatzistamatiou 2009]. To come back to the explicit example mentioned before: The involution of the Fermat quartic X_0 given by

$$(x_0 : x_1 : x_2 : x_3) \longmapsto (-x_0 : -x_1 : x_2 : x_3)$$

acts trivially on CH$^2(X)$.

Although K3 surfaces X with a symplectomorphism f and a Néron–Severi group satisfying the assumptions of Theorem 6.5 are dense in the moduli space of all (X, f) without any condition on the Néron–Severi group, this is not enough to prove Bloch's conjecture for all (X, f).

References

[Barth et al. 2004] W. P. Barth, K. Hulek, C. A. M. Peters, and A. Van de Ven, *Compact complex surfaces*, Second ed., Ergebnisse der Math. (3) **4**, Springer, Berlin, 2004. MR 2004m:14070 Zbl 1036.14016

[Beauville 2007] A. Beauville, "On the splitting of the Bloch–Beilinson filtration", pp. 38–53 in *Algebraic cycles and motives*, vol. 2, London Math. Soc. Lecture Note Ser. **344**, Cambridge Univ. Press, Cambridge, 2007. MR 2009c:14007 Zbl 1130.14006

[Beauville and Voisin 2004] A. Beauville and C. Voisin, "On the Chow ring of a K3 surface", *J. Algebraic Geom.* **13**:3 (2004), 417–426. MR 2005b:14011 Zbl 1069.14006

[Beauville et al. 1985] A. Beauville, J.-P. Bourguignon, and M. Demazure (editors), *Géométrie des surfaces K3: modules et périodes* (Sém. Palaiseau 1981/1982), Astérisque **126**, Soc. Math. de France, Paris, 1985.

[Bloch 1980] S. Bloch, *Lectures on algebraic cycles*, Duke University Mathematics Series, IV, Duke University Mathematics Department, Durham, N.C., 1980. MR 82e:14012 Zbl 0436.14003

[Bridgeland 2008] T. Bridgeland, "Stability conditions on K3 surfaces", *Duke Math. J.* **141**:2 (2008), 241–291. MR 2009b:14030 Zbl 1138.14022

[Chatzistamatiou 2009] A. Chatzistamatiou, "First coniveau notch of the Dwork family and its mirror", *Math. Res. Lett.* **16**:4 (2009), 563–575. MR 2010f:11107 Zbl 1184.14071

[Ferretti 2009] A. Ferretti, "The Chow ring of double EPW sextics", 2009. arXiv 0907.5381v1

[Fulton 1998] W. Fulton, *Intersection theory*, Second ed., Ergebnisse der Math. (3) **2**, Springer, Berlin, 1998. MR 99d:14003 Zbl 0885.14002

[Garbagnati and Sarti 2007] A. Garbagnati and A. Sarti, "Symplectic automorphisms of prime order on K3 surfaces", *J. Algebra* **318**:1 (2007), 323–350. MR 2008j:14070 Zbl 1129.14049

[van Geemen and Sarti 2007] B. van Geemen and A. Sarti, "Nikulin involutions on K3 surfaces", *Math. Z.* **255**:4 (2007), 731–753. MR 2007j:14057 Zbl 1141.14022

[Green and Griffiths 2003] M. Green and P. Griffiths, "Hodge-theoretic invariants for algebraic cycles", *Int. Math. Res. Not.* **2003**:9 (2003), 477–510. MR 2004g:19003 Zbl 1049.14002

[Green et al. 2004] M. Green, P. A. Griffiths, and K. H. Paranjape, "Cycles over fields of transcendence degree 1", *Michigan Math. J.* **52**:1 (2004), 181–187. MR 2005f:14019 Zbl 1058.14009

[Hosono et al. 2004] S. Hosono, B. H. Lian, K. Oguiso, and S.-T. Yau, "Autoequivalences of derived category of a $K3$ surface and monodromy transformations", *J. Algebraic Geom.* **13**:3 (2004), 513–545. MR 2005f:14076 Zbl 1070.14042

[Huybrechts 2006] D. Huybrechts, *Fourier-Mukai transforms in algebraic geometry*, Oxford University Press, Oxford, 2006. MR 2007f:14013 Zbl 1095.14002

[Huybrechts 2010] D. Huybrechts, "Chow groups of K3 surfaces and spherical objects", *J. Eur. Math. Soc. (JEMS)* **12**:6 (2010), 1533–1551. MR 2734351 Zbl 1206.14032

[Huybrechts and Lehn 2010] D. Huybrechts and M. Lehn, *The geometry of moduli spaces of sheaves*, Second ed., Cambridge Mathematical Library, Cambridge University Press, Cambridge, 2010. MR 2011e:14017 Zbl 1206.14027

[Huybrechts et al. 2009] D. Huybrechts, E. Macrì, and P. Stellari, "Derived equivalences of $K3$ surfaces and orientation", *Duke Math. J.* **149**:3 (2009), 461–507. MR 2010j:14075 Zbl 05611496

[Inaba 2002] M.-a. Inaba, "Toward a definition of moduli of complexes of coherent sheaves on a projective scheme", *J. Math. Kyoto Univ.* **42**:2 (2002), 317–329. MR 2004e:14022 Zbl 1063.14013

[Kneser 1981] M. Kneser, "Erzeugung ganzzahliger orthogonaler Gruppen durch Spiegelungen", *Math. Ann.* **255**:4 (1981), 453–462. MR 82i:10026 Zbl 0439.10016

[Lazarsfeld 1986] R. Lazarsfeld, "Brill-Noether-Petri without degenerations", *J. Differential Geom.* **23**:3 (1986), 299–307. MR 88b:14019 Zbl 0608.14026

[Mori and Mukai 1983] S. Mori and S. Mukai, "The uniruledness of the moduli space of curves of genus 11", pp. 334–353 in *Algebraic geometry (Tokyo/Kyoto, 1982)*, Lecture Notes in Math. **1016**, Springer, Berlin, 1983. MR 85b:14033 Zbl 0557.14015

[Morrison 1984] D. R. Morrison, "On $K3$ surfaces with large Picard number", *Invent. Math.* **75**:1 (1984), 105–121. MR 85j:14071 Zbl 0509.14034

[Mukai 1987] S. Mukai, "On the moduli space of bundles on K3 surfaces, I", pp. 341–413 in *Vector bundles on algebraic varieties* (Bombay, 1984), Tata Inst. Fund. Res. Stud. Math. **11**, Tata Inst. Fund. Res., Bombay, 1987. MR 88i:14036 Zbl 0674.14023

[Mukai 1988] S. Mukai, "Finite groups of automorphisms of K3 surfaces and the Mathieu group", *Invent. Math.* **94**:1 (1988), 183–221. MR 90b:32053 Zbl 0705.14045

[Mumford 1968] D. Mumford, "Rational equivalence of 0-cycles on surfaces", *J. Math. Kyoto Univ.* **9** (1968), 195–204. MR 40 #2673 Zbl 0184.46603

[Nikulin 1980] V. Nikulin, "Finite automorphism groups of Kähler K3 surfaces", *Trans. Mosc. Math. Soc.* **2** (1980), 71–135.

[Pyatetskij-Shapiro and Shafarevich 1971] I. Pyatetskij-Shapiro and I. Shafarevich, "Torelli's theorem for algebraic surfaces of type K3", *Izv. Akad. Nauk SSSR, Ser. Mat.* **35** (1971), 530–572. In Russian.

[Rojtman 1980] A. A. Rojtman, "The torsion of the group of 0-cycles modulo rational equivalence", *Ann. of Math.* (2) **111**:3 (1980), 553–569. MR 81g:14003 Zbl 0504.14006

[Voisin 2002] C. Voisin, *Théorie de Hodge et géométrie algébrique complexe*, Cours Spécialisés **10**, Société Mathématique de France, Paris, 2002. MR 2005c:32024a Zbl 1032.14001

[Voisin 2008] C. Voisin, "On the Chow ring of certain algebraic hyper-Kähler manifolds", *Pure Appl. Math. Q.* **4**:3, part 2 (2008), 613–649. MR 2009m:14004

[Wall 1963] C. T. C. Wall, "On the orthogonal groups of unimodular quadratic forms, II", *J. Reine Angew. Math.* **213** (1963), 122–136. MR 27 #5732 Zbl 0135.08802

huybrech@math.uni-bonn.de   *Mathematisches Institut, Universität Bonn,*
                             Endenicher Allee 60, 53115 Bonn, Germany

Current Developments in Algebraic Geometry
MSRI Publications
Volume 59, 2011

Geometry of varieties of minimal rational tangents

JUN-MUK HWANG

We present the theory of varieties of minimal rational tangents (VMRT), with
an emphasis on its own structural aspect, rather than applications to concrete
problems in algebraic geometry. Our point of view is based on differential
geometry, in particular, Cartan's method of equivalence. We explain various
aspects of the theory, starting with the relevant basic concepts in differential
geometry and then relating them to VMRT. Several open problems are proposed, which are natural from the view point of understanding the geometry
of VMRT itself.

1. Introduction

The concept of varieties of minimal rational tangents (VMRT) on uniruled
projective manifolds first appeared as a tool to study the deformation of Hermitian
symmetric spaces [Hwang and Mok 1998]. For many classical examples of
uniruled manifolds, VMRT is a very natural geometric object associated to low
degree rational curves, and as such, it had been studied and used long before
its formal definition appeared in that reference. At a more conceptual level,
namely, as a tool to investigate unknown varieties, it had been already used in
[Mok 1988] for manifolds with nonnegative curvature. However, in the context
of that work, its very special relation with the curvature property of the Kähler
metric somewhat overshadowed its role as an algebro-geometric object, so it had
not been considered for general uniruled manifolds. Thus it is fair to say that
the concept as an independent geometric object defined on uniruled projective
manifolds really originated from [Hwang and Mok 1998]. Shortly after this
formal debut, numerous examples of its applications to classical problems of
algebraic geometry were discovered. In the early MSRI survey [Hwang and Mok
1999], written only a couple of years after the first discovery, one can already
find a substantial list of problems in a wide range of topics, which can be solved
by the help of VMRT.

Supported by National Researcher Program 2010-0020413 of NRF and MEST..

Keywords: varieties of minimal rational tangents, equivalence problem, cone structure,
  G-structure, distribution, foliation.

Since the beginning VMRT has been studied exclusively in relation with some classical problems, namely, problems which do not involve VMRT itself explicitly. In particular, this is the case for most of my collaboration with N. Mok. In other words, VMRT has mostly served as a *tool* to study uniruled manifolds. However, after more than a decade's service, I believe it is time to give due recognition and it is not unreasonable to start to regard VMRT itself as a central object of research. The purpose of this exposition is to introduce and advertise this new viewpoint. In fact, the title of the current article, as opposed to that of my old survey [Hwang 2001], is deliberately chosen to emphasize this shift of perspective.

As a result, in this article, I intentionally avoided talking about applications. Also, only a minimal number of examples are given. This omission is happily justifiable by the appearance of the excellent survey paper [Mok 2008a], which covers many recent applications. The old surveys [Hwang and Mok 1999] and [Hwang 2001] as well as [Mok 2008a] all emphasize the applications to concrete geometric problems. The reader is encouraged to look at these surveys for explicit examples and applications.

There is another reason that I believe it is justifiable to give such an emphasis on the theoretical aspect of the theory. After seeing many applications of the techniques developed so far, it seems to me that we need a considerable advancement of the structural theory of VMRT itself, to enhance the applicability of the theory to a wider class of geometric problems. With this motivation in mind, I will propose several open problems of this sort throughout the article, which I believe are not only natural, but will be useful in applications.

Most of these open problems are likely to be of less interest unless one believes that VMRT itself is an interesting object. In this regard, part of my aim is to advertise VMRT, trying to convince the reader that the subject is exciting and amusing. In other words, by presenting these open problems, I hope to transfer to the reader the kind of perspective I have about this subject. The reader is encouraged to try to think about the meaning of the open problems and why they are interesting, to understand the underlying philosophy.

The basic framework of my presentation is essentially differential geometric, belonging to Cartanian geometry. Since this is an article for algebraic geometers, very little knowledge of differential geometry will be assumed. Essentially all differential geometric concepts are explained from the very definition. This differential geometric framework has been in the background of most of my joint work with N. Mok, but has not been explicitly explained in publications so far. The basic idea is that VMRT is a special kind of cone structure and one of the key issue is to understand what is special about it. In this article, we will mostly concentrate on the existence of a characteristic connection among the special properties. Schematically, we may put it as

{ cone strucures } ⊃ { characteristic connections } ⊃ { VMRT }.

Some of the discussion below works for cone structures, some for characteristic connections and some for VMRT.

For me, the most amusing aspect of the study of VMRT is the interaction, or rather the fusion, of algebraic geometry and differential geometry. I hope that this expository article helps algebraic geometers to become more familiar with the concepts and methods originating from differential geometry. Most of the sections start with an introduction of certain differential geometric concepts and then mix them with the algebraic geometry of rational curves.

In another direction, although it is written for algebraic geometers, I hope this article will attract differential geometers, especially those working on Cartanian geometry, to problems arising from the algebraic geometry of rational curves. Many of the problems I propose have differential geometric components. Moreover, I think the theory of VMRT provides a lot of new examples of geometric structures which are highly interesting from a differential geometric point of view.

2. Preliminaries on distributions

Throughout the paper, we will work over the complex numbers. All differential geometric objects are holomorphic. In this section, we collect some terms and facts about distributions. These will be used throughout the paper.

A *distribution* D on a complex manifold U means a vector subbundle $D \subset T(U)$ of the tangent bundle. In particular, $T(U)/D$ is locally free.

The *Frobenius tensor* of the distribution D is the homomorphism of vector bundles $\beta : \bigwedge^2 D \to T(U)/D$ defined by

$$\beta(v, w) = [\tilde{v}, \tilde{w}] \mod D,$$

where for $x \in U$ and two vectors $v, w \in D_x$, $\tilde{v}$ and $\tilde{w}$ are local sections of D extending v and w in a neighborhood of x, and $[\tilde{v}, \tilde{w}]$ denotes the bracket of $\tilde{v}$ and $\tilde{w}$ as holomorphic vector fields. It is easy to see that $\beta(v, w)$ does not depend on the choice of the extensions $\tilde{v}, \tilde{w}$. By the Frobenius theorem, if the Frobenius tensor is identically zero then the distribution comes from a foliation, i.e., a partition of U into complex submanifolds whose tangent spaces correspond to D. In this case, we say that the distribution is *integrable*.

For each $x \in U$ define

$$Ch(D)_x := \{v \in D_x, \ \beta(v, w) = 0 \text{ for all } w \in D_x.\}.$$

In a Zariski open subset $U' \subset U$,

$$\{Ch(D)_x, x \in U'\}$$

defines a distribution, called the *Cauchy characteristic* of D and denoted by $Ch(D)$. This distribution is always integrable.

Given a distribution D on U, its *first derived system*, denoted by ∂D, is the distribution defined on a Zariski open subset of U whose associated sheaf corresponds to $\mathcal{O}(D) + [\mathcal{O}(D), \mathcal{O}(D)]$. Define successively

$$\partial^1 D := \partial D, \quad \partial^k D := \partial(\partial^{k-1}D).$$

There exists some ℓ such that $\partial^\ell D = \partial^{\ell+1}D$ so that the Frobenius tensor of $\partial^\ell D$ is zero. The foliation on a Zariski open subset of U determined by $\partial^\ell D$ is called the *foliation generated by* D. We say that the distribution is *bracket-generating* if $\partial^\ell D = T(U')$ on some Zariski open subset U' and $\ell > 0$.

Let $f : M \to B$ be a holomorphic submersion between two complex manifolds, i.e., $df : T(M) \to f^*T(B)$ is a surjective bundle homomorphism. The distribution $\operatorname{Ker} df$ on M is integrable and the corresponding foliation of M has the fibers of f as leaves. Given a distribution D on B, we have a distribution $f^{-1}D$ on M, called the *inverse-image of the distribution* D, given by the subbundle $df^{-1}(f^*D)$ of $T(M)$ where $f^*D \subset f^*T(B)$ is the pull-back of $D \subset T(B)$. It is clear that $\operatorname{Ker} df \subset Ch(f^{-1}D)$.

3. Equivalence of cone structures

A well-known philosophy, going back to Klein's Erlangen program, is that the fundamental problem in any area of geometry is the study of invariant properties under equivalence relations. Algebraic geometry is no exception. In classical projective geometry, the most fundamental equivalence relation is the equivalence of two subvarieties of projective space under a projective transformation, or more generally, the equivalence of two families of subvarieties under a family of projective transformations. One possible formulation of this equivalence relation is the following.

Definition 3.1. Let U and U' be a (connected) complex manifold. Let $\mathcal{V}$ and $\mathcal{V}'$ be vector bundles on U and U', respectively, and let $\mathbb{P}\mathcal{V}$ and $\mathbb{P}\mathcal{V}'$ be their projectivizations, as sets of 1-dimensional subspaces in the fibers. Given (not necessarily irreducible) subvarieties $\mathscr{C} \subset \mathbb{P}\mathcal{V}$ and $\mathscr{C}' \subset \mathbb{P}\mathcal{V}'$ of pure dimension, surjective over U and U' respectively, we say that $\mathscr{C}$ and $\mathscr{C}'$ are *equivalent as families of projective subvarieties* if there exist a biholomorphic map $\varphi : U \to U'$

and a projective bundle isomorphism $\psi : \mathbb{P}\mathcal{V} \to \mathbb{P}\mathcal{V}'$ with a commuting diagram

$$
\begin{array}{ccc}
\mathbb{P}\mathcal{V} & \xrightarrow{\psi} & \mathbb{P}\mathcal{V}' \\
\downarrow & & \downarrow \\
U & \xrightarrow{\phi} & U'
\end{array}
$$

such that $\psi(\mathcal{C}) = \mathcal{C}'$. For a point $x \in U$ and a point $x' \in U'$, we say that the family $\mathcal{C}$ at x is *locally equivalent* to the family $\mathcal{C}'$ at x', if there exist a neighborhood $W \subset U$ of x and a neighborhood $W' \subset U'$ of x' such that the restriction $\mathcal{C}|_W \subset \mathbb{P}\mathcal{V}|_W$ is equivalent to the restriction $\mathcal{C}'|_{W'} \subset \mathbb{P}\mathcal{V}'|_{W'}$.

Suppose that $\mathcal{V}$ and $\mathcal{V}'$ in Definition 3.1 are the tangent bundles $T(U)$ and $T(U')$. Then we have the following finer equivalence relation.

Definition 3.2. For a complex manifold U, a subvariety of pure dimension $\mathcal{C} \subset \mathbb{P}T(U)$ which is surjective over the base U will be called a *cone structure* on U. Here, we do not assume that $\mathcal{C}$ is irreducible. The fiber dimension of the projection $\mathcal{C} \to U$, i.e., $\dim \mathcal{C} - \dim U$, will be called the *projective rank* of the cone structure. The *rank* of the cone structure is the projective rank of the cone structure plus one. A cone structure $\mathcal{C} \subset \mathbb{P}T(U)$ on U and a cone structure $\mathcal{C}' \subset \mathbb{P}T(U')$ on U' are *equivalent as cone structures* if there exists a biholomorphic map $\varphi : U \to U'$ such that the projective bundle isomorphism $\psi : \mathbb{P}T(U) \to \mathbb{P}T(U')$ induced by the differential $d\varphi : T(U) \to T(U')$ of φ

$$
\begin{array}{ccc}
\mathbb{P}T(U) & \xrightarrow{\psi = d\varphi} & \mathbb{P}T(U') \\
\downarrow & & \downarrow \\
U & \xrightarrow{\phi} & U'
\end{array}
$$

satisfies $\psi(\mathcal{C}) = \mathcal{C}'$. For a point $x \in U$ and a point $x' \in U'$, we say that the cone structure $\mathcal{C}$ at x is *locally equivalent* to the cone structure $\mathcal{C}'$ at x', if there exist a neighborhood $W \subset U$ of x and a neighborhood $W' \subset U'$ such that the restriction $\mathcal{C}|_W \subset \mathbb{P}T(W)$ is equivalent as cone structures to the restriction $\mathcal{C}'|_{W'} \subset \mathbb{P}T(W')$.

Notice the essential difference between Definitions 3.1 and 3.2: the projective bundle isomorphism ψ is arbitrary in the former as long as it is compatible with the map φ while ψ is completely determined by φ in the latter. Since ψ comes from the derivative of φ in Definition 3.2, the equivalence of cone structures has features of differential geometry as well as algebraic geometry. Let us look at two classical examples.

Example 3.3. A cone structure $\mathcal{C} \subset \mathbb{P}T(U)$ where each fiber $\mathcal{C}_x$, $x \in U$, is a linear subspace of $\mathbb{P}T_x(U)$ is equivalent to a distribution on U. The rank of the distribution (as a subbundle of $T(U)$) is equal to the rank of the cone structure.

Two such cone structures $\mathscr{C} \subset \mathbb{P}T(U)$ and $\mathscr{C}' \subset \mathbb{P}T(U')$ of the same projective rank on complex manifolds U, U' of the same dimension are always locally equivalent as families of projective subvarieties. Their local equivalence as cone structures is much more subtle. For example, an integrable distribution cannot be locally equivalent to a non-integrable one.

Example 3.4. A cone structure $\mathscr{C} \subset \mathbb{P}T(U)$ where each fiber $\mathscr{C}_x$, $x \in U$, is a nonsingular quadric hypersurface of $\mathbb{P}T_x(U)$ is called a *conformal structure* on U. Locally, a conformal structure is determined by a nondegenerate holomorphic symmetric bilinear form $g : S^2 T(U) \to \mathcal{O}_U$, i.e., a holomorphic Riemannian metric, up to multiplication by nowhere-zero holomorphic functions. Two conformal structures $\mathscr{C} \subset \mathbb{P}T(U)$ and $\mathscr{C}' \subset \mathbb{P}T(U')$ on complex manifolds U and U' of the same dimension are always locally equivalent as families of projective subvarieties. They are locally equivalent as cone structures if the associated holomorphic Riemannian metrics are conformally isometric. The study of this equivalence relation is the subject of conformal geometry, an active area of research in differential geometry.

A more "modern" version of Definition 3.1 is the equivalence of families of polarized projective varieties. Let us assume that the family is smooth for simplicity. One possible formulation is as follows.

Definition 3.5. A *polarized family* $f : M \to U$ is just a smooth projective morphism between two complex manifolds M and U with a line bundle L on M which is f-ample. Here we assume that U is connected, M is of pure dimension, but not necessarily connected. The line bundle L is called a *polarization*. Two polarized families $f : M \to U$ with a polarization L and $f' : M' \to U'$ with a polarization L' are *equivalent as polarized families* if there exist a biholomorphism $\varphi : U \to U'$ and a biholomorphism $\psi : M \to M'$ satisfying $\varphi \circ f = f' \circ \psi$ and $L \cong \psi^* L'$.

In Definition 3.5, taking a sufficiently high power $L^{\otimes m}$ to make it f-very-ample, we get an embedding $M \to \mathbb{P}(f_* L^{\otimes m})^*$ whose image $\mathscr{C}$ is a family of projective subvarieties. The equivalence in Definition 3.5 implies the equivalence in the sense of Definition 3.1 for this family $\mathscr{C}$ of projective subvarieties. The more intrinsic formulation of Definition 3.5 is often more convenient than the classical version in Definition 3.1. Analogously, sometimes it is convenient to have a more intrinsic formulation of Definition 3.2 as follows.

Definition 3.6. Given a polarized family $f : M \to U$ with a polarization L, a distribution $\mathscr{J} \subset T(M)$ on M is called a *precone* structure if $\operatorname{Ker} df \subset \mathscr{J}$ and the quotient bundle $\mathscr{J}/\operatorname{Ker} df$ is a line bundle isomorphic to the dual line bundle L^* of L. Two precone structures $(f : M \to U, L, \mathscr{J})$ and $(f' : M' \to U', L', \mathscr{J}')$ are equivalent if they are equivalent as polarized families in the sense of Definition 3.5 such that the differential $d\psi : T(M) \to T(M')$ sends the distribution $\mathcal{J}$ to $\mathcal{J}'$.

The relation between Definition 3.2 and Definition 3.6 is given by the following proposition, which is essentially [Yamaguchi 1982, Lemma 1.5], attributed to N. Tanaka.

Proposition 3.7. *Given a precone structure, $(f : M \to U, L, \mathcal{J})$, define a morphism $\tau : M \to \mathbb{P}T(U)$ by*

$$\text{for each } \alpha \in M, \quad \tau(\alpha) := df(\mathcal{J}_\alpha),$$

called the tangent morphism. The image of τ determines a cone structure $\mathscr{C} := \tau(M) \subset \mathbb{P}T(U)$. Moreover, when $\mathcal{O}(1)$ is the relative hyperplane bundle on $\mathbb{P}T(U)$, the polarization L is isomorphic to $\tau^* \mathcal{O}(1)$, implying that τ is a finite morphism over its image. The rank of this cone structure is equal to the rank of the distribution $\mathcal{J}$.

Note that although $M \to U$ in Definition 3.6 is assumed to be a smooth morphism, the induced cone structure $\mathscr{C} \to U$ by the tangent morphism is not necessarily smooth. This is one advantage of Definition 3.6, in the sense that a cone structure coming from a precone structure has a hidden regularity. The VMRT structure in the next section is such an example. It is easy to see that when a cone structure $\mathscr{C} \to U$ is smooth, it comes from a precone structure:

Proposition 3.8. *Given a cone structure $\mathscr{C} \subset \mathbb{P}T(U)$ such that the projection $f : \mathscr{C} \to U$ is a smooth morphism, the distribution $\mathcal{J}$ on $\mathscr{C}$ defined by*

$$\text{for each } \alpha \in \mathscr{C}, \quad \mathcal{J}_\alpha := df_\alpha^{-1}(\hat{\alpha})$$

where $\hat{\alpha} \subset T_x(U)$, $x = f(\alpha)$, is the 1-dimensional subspace corresponding to α, is a precone structure on $\mathscr{C}$. The cone structure induced by this precone structure via Proposition 3.7 agrees with the original cone structure $\mathscr{C} \subset \mathbb{P}T(U)$.

4. Varieties of minimal rational tangents

Now we define the cone structure which is our main interest.

Definition 4.1. A rational curve $C \subset X$ on a projective manifold X is *free* if under the normalization $\nu : \mathbb{P}_1 \to C$, the vector bundle $\nu^* T(X)$ is nef. The *normalized space of free rational curves* on X, to be denoted by $\mathrm{FRC}(X)$, is a smooth scheme with countably many components, by [Kollár 1996, II.3]. We have the universal family $\mathrm{Univ}(X)$ with a $\mathbb{P}_1$-bundle structure $\mathrm{Univ}(X) \to \mathrm{FRC}(X)$ and the evaluation morphism $\mathrm{Univ}(X) \to X$.

Note that $\mathrm{FRC}(X) \neq \varnothing$ if and only if X is a uniruled projective manifold.

Definition 4.2. Let X be a uniruled projective manifold. An irreducible component $\mathcal{K}$ of $\mathrm{FRC}(X)$ is called a *minimal component* if for the universal family $\rho : \mathcal{U} \to \mathcal{K}$ and $\mu : \mathcal{U} \to X$ obtained by restricting $\mathrm{Univ}(X)$ to $\mathcal{K}$, the morphism μ is generically projective, i.e., the fiber $\mathcal{K}_x := \mu^{-1}(x)$ over a general point $x \in X$ is projective. A member of $\mathcal{K}$ is called a *minimal free rational curve*.

Proposition 4.3. *For a minimal component $\mathcal{K}$, there exists a Zariski open subset $X_o \subset X$ such that*

(i) *each fiber $\mu^{-1}(x)$, $x \in X_o$, is smooth;*

(ii) $\mathrm{Ker}\, d\mu \cap \mathrm{Ker}\, d\rho = 0$ *at every point of $\mu^{-1}(X_o)$; and*

(iii) *the dual bundle L of the line bundle $\mathrm{Ker}\, d\rho \subset T(\mathcal{U})$ is μ-ample on $\mu^{-1}(X_o)$.*

In particular, the distribution $\mathcal{J} := \mathrm{Ker}\, d\mu + \mathrm{Ker}\, d\rho$ defines a precone structure on the family $\mu|_{\mu^{-1}(X_o)} : \mu^{-1}(X_o) \to X_o$ of smooth projective varieties with the polarization L.

Proof. Part (i) is [Kollár 1996, Corollary II.3.11.5]. Parts (ii) and (iii) follow from [Kebekus 2002, Theorem 3.4]. $\qquad\qquad\qquad\qquad\qquad\qquad\qquad\qquad\quad$ $\square$

Definition 4.4. The cone structure $\mathcal{C} \subset \mathbb{P}T(X_o)$ associated to the precone structure of Proposition 4.3 via Proposition 3.7 is called the *family of varieties of minimal rational tangents* (in short, VMRT) of the minimal component $\mathcal{K}$. Its fiber $\mathcal{C}_x$ at $x \in X_o$ is called the *variety of minimal rational tangents at x*. For each $x \in X_o$, the restriction $\tau_x : \mu^{-1}(x) \to \mathcal{C}_x$ of the tangent morphism $\tau : \mu^{-1}(X_o) \to \mathcal{C}$ defined in Proposition 3.7 is called the *tangent morphism at x*.

This cone structure, VMRT, is our main interest. Before going into the study of VMRT in detail, let us give at least one reason why it is interesting to consider the equivalence problem for such a cone structure. The following is the main result of [Hwang and Mok 2001].

Theorem 4.5. *Let X and X' be two Fano manifolds of Picard number 1. Let $\mathcal{K}$ and $\mathcal{K}'$ be minimal components on X and X', respectively, with associated VMRT $\mathcal{C}$ and $\mathcal{C}'$). Assume that the VMRT $\mathcal{C}_x \subset \mathbb{P}T_x(X)$ at a general point $x \in X$ is not a finite union of linear subspaces. Suppose $\mathcal{C}$ at some point $x \in X_o$ is locally equivalent as cone structures to $\mathcal{C}'$ at some point $x' \in X_o'$. Then X and X' are biregular.*

To be precise, in [Hwang and Mok 2001], Theorem 4.5 is proved under the stronger assumption that $\mathcal{C}_x \subset \mathbb{P}T_x(X)$ has generically finite Gauss map. However, one can extend the argument to the above form by using results from [Hwang and Mok 2004]. See [Mok 2008a, Theorem 9] for a discussion of this extension.

Theorem 4.5 is not just of theoretical interest. It is used to identify certain Fano manifolds of Picard number 1 in a number of classical problems. See [Hwang and Mok 1999], [Hwang 2001] and [Mok 2008a] for concrete examples.

The VMRT is an algebraic object defined as a quasi-projective variety in $\mathbb{P}T(X)$. It is convenient to introduce a local version of this definition:

Definition 4.6. A cone structure $\mathscr{C}' \subset \mathbb{P}T(U)$ on a complex manifold U is a *VMRT structure* if there exists a VMRT $\mathscr{C} \subset \mathbb{P}T(X_o)$ as in Definition 4.4 such that $\mathscr{C}' \to U$ is locally equivalent as cone structures to $\mathscr{C} \to X_o$ at every point of U.

This is not a truly local definition. It is introduced merely for linguistic convenience. A truly local definition of VMRT structure as a cone structure with certain distinguished differential geometric properties is still lacking. One special property is obvious: from the very definition, it is provided with a connection in the following sense.

Definition 4.7. Let $\mathscr{J} \subset T(M)$ be a precone structure on a polarized family $(f : M \to U, L)$. A *connection* of the precone structure is a line subbundle $F \subset \mathscr{J}$ with an isomorphism $F \cong L^*$ splitting the exact sequence

$$0 \longrightarrow \operatorname{Ker} df \longrightarrow \mathscr{J} \longrightarrow L^* \longrightarrow 0.$$

By abuse of terminology, we will also say that F is a connection for the cone structure $\mathscr{C} \subset \mathbb{P}T(U)$ induced by the precone structure. When $\mathscr{C} = \mathbb{P}T(U)$, a connection is called a *projective connection* on U.

From the fact that the set of splittings of the exact sequence in Definition 4.7 is $H^0(M, \operatorname{Ker} df \otimes L)$, we have

Proposition 4.8. *Given a precone structure on $M \to U$, if the fiber M_x at some $x \in U$ satisfies $H^0(M_x, T(M_x) \otimes L) = 0$, then a connection is unique if it exists.*

A VMRT structure is naturally equipped with a connection given by $\operatorname{Ker} d\rho$ of Proposition 4.3. This connection has a distinguished property.

Definition 4.9. A connection $F \subset \mathscr{J}$ in Definition 4.7 is a *characteristic connection* if $F \subset Ch(\partial \mathscr{J})$ on an open subset of M.

The following result is in [Hwang and Mok 2004, Proposition 8].

Proposition 4.10. *In Proposition 4.3, the distribution $\partial \mathscr{J}$ on $\mu^{-1}(X_o)$ is of the form $\rho^{-1} D$ for some distribution D on $\mathscr{K}$. In particular, the line subbundle $\operatorname{Ker} d\rho$ is a characteristic connection of the precone structure.*

The existence of the characteristic connection is a key property of VMRT structure. Most of the algebraic geometric applications of the local differential geometry of VMRT come from this property. There are other local differential geometric properties of VMRT. For example, the admissibility condition of [Bernstein and Gindikin 2003] holds for minimal free rational curves, which can be interpreted as a property of the cone structure. However, I feel that the investigation of these additional properties is not yet mature enough to be discussed here.

An important property of a characteristic connection is the relation with the projective differential geometry of the fibers of the cone structure. Let us start the discussion by recalling some definitions from projective geometry.

Definition 4.11. For each point $v \in \mathbb{P}V$, let $\hat{v} \subset V$ be the 1-dimensional subspace corresponding to v. Let $Z \subset \mathbb{P}V$ be a projective subvariety and let $\hat{Z} \subset V$ be the affine cone of Z. Denote by $Sm(Z)$ the smooth locus of Z. For each $z \in Sm(Z)$, let $\hat{T}_z(Z) \subset V$ be the *affine tangent space* to Z at z, i.e., the affine cone of the projective tangent space to Z at z:

$$\hat{T}_z(Z) = T_{z'}(\hat{Z}) \text{ for any } z' \in \hat{z} \setminus \{0\}.$$

Let $\mathbf{Gr}(p, V)$ be the Grassmannian of p-dimensional subspaces of V. The *Gauss map* of Z is the morphism $\gamma : Sm(Z) \to \mathbf{Gr}(\dim \hat{Z}, V)$, defined by $\gamma(z) := \hat{T}_z(Z)$. The *second fundamental form* at $z \in Sm(Z)$ is the derivative of the Gauss map at z defined as the homomorphism

$$II_z(Z) : S^2 T_z(Z) \to T_z(\mathbb{P}V)/T_z(Z).$$

The following is an immediate property of having a connection. It follows essentially from [Hwang and Mok 2004, Proposition 1].

Proposition 4.12. *Let* $(f : M \to U, \mathcal{F})$ *be a precone structure with a connection. For a point* $\alpha \in M$ *where the tangent morphism* $\tau : M \to \mathscr{C} \subset \mathbb{P}T(U)$ *in Proposition 3.7 is immersive,*

$$df_\alpha(\partial \mathcal{F}) = \hat{T}_{\tau(\alpha)}(\mathscr{C}_x)$$

with $x = f(\alpha)$.

The following proposition is proved in [Hwang and Mok 2004, Proposition 2], where a precise meaning of "describes the second fundamental form" is given.

Proposition 4.13. *Let* $\mathcal{F}$ *be a precone structure on* $f : M \to U$ *with characteristic connection* F. *Then the Frobenius tensor of* $\partial \mathcal{F}$ *at a point* $\alpha \in M$ *describes the second fundamental form of the projective variety* $\mathscr{C}_x \subset \mathbb{P}T_x(U)$, $x = f(\alpha)$ *at the point* $\tau(\alpha)$, *via Proposition 4.12. In particular, the second fundamental form remains unchanged along a leaf of* $F \subset Ch(\partial \mathcal{F})$.

Proposition 4.13 gives a necessary condition for a polarized family to admit a precone structure with characteristic connection.

One reason the characteristic connection is important is its uniqueness under a mild assumption. The following is in [Hwang and Mok 2004, Proposition 3].

Proposition 4.14. *Let* $\mathscr{C} \subset \mathbb{P}T(U)$ *be a cone structure associated to a precone structure* $(M, \mathscr{F})$ *such that a general fiber* $\mathscr{C}_x \subset \mathbb{P}T_x(U)$ *has generically finite Gauss map. If* $(M, \mathscr{F})$ *has a characteristic connection* F, *then* $F = Ch(\partial \mathscr{F})$ *on an open subset in* M. *In particular, a characteristic connection is unique if it exists.*

The condition that $\mathscr{C}_x \subset \mathbb{P}T_x(U)$ has generically finite Gauss map holds as long as $\mathscr{C}_x$ is smooth and its components are not linear subspaces. The smoothness of $\mathscr{C}_x$ holds in most natural examples, as discussed below. The non-linearity condition will be discussed in Section 7. One can say that the uniqueness of characteristic connection holds in all essential cases.

Regarding the smoothness of VMRT, the following has been one of the most tantalizing questions.

Problem 4.15. In Definition 4.4, is the tangent morphism τ_x at a general point x an immersion? Is it an embedding?

The immersiveness of τ_x at a point $\alpha \in \mathscr{U}$ can be interpreted as a geometric property of the rational curve $\rho(\alpha)$.

Definition 4.16. A free rational curve $C \subset X$ is *standard* if under the normalization $\nu : \mathbb{P}_1 \to C$,

$$\nu^* T(X) \cong \mathcal{O}(2) \oplus \mathcal{O}(1)^p \oplus \mathcal{O}^{\dim X - p - 1}$$

where p is the nonnegative integer satisfying $(-K_X) \cdot C = p + 2$.

The following is a consequence of Mori's bend-and-break argument and basic deformation theory of rational curves (cf. [Hwang 2001, Proposition 1.4]).

Proposition 4.17. *Given a minimal component* $\mathscr{K}$, *a general member of* $\mathscr{K}$ *is a standard rational curve. For such a standard rational curve, the integer* p *in Definition 4.16 is the dimension of* $\mathscr{K}_x$ *and is equal to the projective rank of the VMRT. For a point* $x \in X_o$, *the tangent morphism* τ_x *is immersive at* $\alpha \in \mathscr{K}_x$ *if and only if* $\rho(\alpha) \in \mathscr{K}$ *corresponds to a standard rational curve.*

Thus checking the immersiveness of τ_x is equivalent to showing that all members of $\mathscr{K}_x$ are standard. Although many people believe that this is true for a general point x, no plausible approach has been suggested up to now.

Toward the injectivity of τ_x, the best result so far is the following result of [Hwang and Mok 2004].

Theorem 4.18. *The tangent morphism* $\tau : \mu^{-1}(X_o) \to \mathscr{C}$ *is birational. Consequently,* τ_x *is birational for a general point* x.

Another result on the injectivity is Proposition 7.7 discussed below, in the particular case when the components of $\mathscr{C}_x$ are linear subspaces. There is also a study of the injectivity of τ_x in [Kebekus and Kovács 2004], relating the problem to the existence of certain singular rational curves.

As these discussions show, Problem 4.15 is quite difficult and its solution will be very important in this subject. On the other hand, since it holds in all concrete examples, sometimes it is OK to work under the assumption that it is true. More precisely, it is meaningful to work with projective manifolds and minimal components whose VMRT is smooth.

The uniqueness of the characteristic connection in Proposition 4.14 suggests the following stronger uniqueness question.

Problem 4.19. Can a polarized family $(f : \mathscr{C} \to U, L)$ have two distinct precone structures inducing non-equivalent VMRT structures on U?

We will see below an example where this is not unique, i.e., two VMRT's with the same underlying polarized family (Example 5.9). However, this example is very special. It is likely that there are many examples of polarized families for which uniqueness holds. Some cases will be discussed in Theorem 5.11 and Theorem 5.12. One can also ask the following weaker question.

Problem 4.20. For a polarized family $(f : \mathscr{C} \to U, L)$, can there exist a positive dimensional family of precone structures inducing locally non-equivalent VMRT structures on U?

Problem 4.20 is closely related to the deformation of Fano manifolds of Picard number 1 via Theorem 4.5.

5. Isotrivial VMRT

In this section, we will discuss a special class of cone structures, for which there exists a good differential geometric tool to study the equivalence problem. Let us start by recalling the relevant notion in differential geometry. Chapter VII of [Sternberg 1983] is a good reference.

Definition 5.1. Fix a vector space V. For a complex manifold U of dimension equal to dim V, its *frame bundle* $\mathbf{Fr}(U)$ is a $\mathbf{GL}(V)$-principal fiber bundle with the fiber at $x \in U$ defined by

$$\mathbf{Fr}_x(U) := \mathrm{Isom}(V, T_x(U))$$

the set of isomorphisms from V to $T_x(U)$. Given an algebraic subgroup $G \subset \mathbf{GL}(V)$, a *G-structure* on U means a G-principal subbundle $\mathscr{G} \subset \mathbf{Fr}(U)$. Two G-structures $\mathscr{G} \subset \mathbf{Fr}(U)$ and $\mathscr{G}' \subset \mathbf{Fr}(U')$ are *equivalent* if there is a biholomorphic map $\varphi : U \to U'$ whose differential $\varphi_* : \mathbf{Fr}(U) \to \mathbf{Fr}(U')$ sends $\mathscr{G}$ to $\mathscr{G}'$. The local equivalence of G-structures is defined similarly.

As a trivial example:

Example 5.2. The tangent bundle $T(V)$ of a vector space V is naturally isomorphic to $V \times V$. The frame bundle $\mathbf{Fr}(V)$ is naturally isomorphic to $\mathbf{GL}(V) \times V$. For any $G \subset \mathbf{GL}(V)$, we have a natural G-structure

$$G \times V \subset \mathbf{GL}(V) \times V = \mathbf{Fr}(V)$$

on the manifold V. This is called the *flat* G-structure on V. A G-structure $\mathscr{G}$ on a manifold is said to be *locally flat* if it is locally equivalent to the flat G-structure.

Many classical geometric structures in differential geometry are G-structures for various choices of G. For this reason, the equivalence problem for G-structures has been studied extensively. For the following special class of cone structures, the equivalence problem can be reduced to that of certain G-structures.

Definition 5.3. Let $Z \subset \mathbb{P}V$ be a projective subvariety. A cone structure $\mathscr{C} \subset \mathbb{P}T(U)$ is Z-*isotrivial* if the fiber $\mathscr{C}_x \subset \mathbb{P}T_x(U)$ at each $x \in U$ is isomorphic to $Z \subset \mathbb{P}V$.

The simplest example is the following analog of Example 5.2.

Example 5.4. The projectivized tangent bundle $\mathbb{P}T(V)$ of a vector space V is naturally isomorphic to $\mathbb{P}V \times V$. For a given subvariety $Z \subset \mathbb{P}V$, we have a natural cone structure

$$Z \times V \subset \mathbb{P}V \times V = \mathbb{P}T(V)$$

on the manifold V. This will be called the Z-*isotrivial flat cone structure on V.

The equivalence problem for isotrivial cone structures can be reduced to that of G-structures.

Definition 5.5. When $\mathscr{C} \subset \mathbb{P}T(U)$ is a Z-isotrivial cone structure, the subbundle $\mathscr{G} \subset \mathbf{Fr}(U)$ with the fiber at x defined by

$$\mathscr{G}_x := \{h \in \mathrm{Isom}(V, T_x(U)), \; h(\hat{Z}) = \hat{\mathscr{C}}_x\}$$

is a G-structure with $G = \mathrm{Aut}(\hat{Z}) \subset \mathbf{GL}(V)$, the group of linear automorphisms of $\hat{Z} \subset V$. This is called the G-*structure induced by the isotrivial cone structure*.

For example, Example 5.2 is the G-structure induced by Example 5.4. It is easy to see that two Z-isotrivial cone structures are locally equivalent as cone structures if and only if the G-structures induced by them are locally equivalent. Thus we can use the the theory of G-structures to study isotrivial cone structures. However, this does not mean that the theory of isotrivial cone structures can be completely reduced to the theory of G-structures: it is a highly non-trivial problem to translate conditions on an isotrivial cone structure into the language of G-structures. The problem gets more serious when we consider isotrivial VMRT structures.

Let us say that a VMRT for a uniruled manifold is Z-isotrivial if it is a Z-isotrivial cone structure at a general point. The first question one can ask is whether there is any restriction on the subvariety $Z \subset \mathbb{P}V$ for the existence of a Z-isotrivial VMRT on a uniruled projective manifold. Problem 4.15 suggests that $Z \subset \mathbb{P}V$ should be nonsingular. The next example shows that this is the only necessary condition, if Z is irreducible.

Example 5.6. Let $Z \subset \mathbb{P}_{n-1} \subset \mathbb{P}_n$ be a nonsingular irreducible projective variety contained in a hyperplane. Let $X_Z \to \mathbb{P}_n$ be the blow-up of $\mathbb{P}_n$ with center Z. Let $\mathcal{K}_Z$ be the family of curves on X_Z which are proper transforms of lines in $\mathbb{P}^n$ intersecting Z. Then $\mathcal{K}_Z$ determines a minimal component of X_Z with Z-isotrivial VMRT. In fact, the VMRT at a general point is locally equivalent to Example 5.4.

The situation is quite different when Z is reducible. The construction of Example 5.6 does not work, because $\mathcal{K}_Z$ there would be reducible. In fact, the following problem has not been studied.

Problem 5.7. Given a nonsingular variety $Z \subset \mathbb{P}V$ with more than one irreducible component, does there exist a uniruled projective manifold X with Z-isotrivial VMRT?

Going back to the irreducible case, the most basic question one can ask about isotrivial VMRT is the following.

Problem 5.8. Let $Z \subset \mathbb{P}V$ be an irreducible nonsingular variety. Let X be an n-dimensional uniruled projective manifold with Z-isotrivial VMRT. Is the VMRT at a general point of X locally equivalent to that of Example 5.6?

Recall that $Z \subset \mathbb{P}V$ is degenerate if it is contained in a hyperplane of $\mathbb{P}V$ and nondegenerate otherwise. When $Z \subset \mathbb{P}V$ is degenerate, there are many examples where the answer is negative, as will be seen in Section 6. Even for a nondegenerate $Z \subset \mathbb{P}V$, the answer to Problem 5.8 is not always affirmative.

Example 5.9. Let W be a 2ℓ-dimensional complex vector space with a symplectic form. Fix an integer k, $1 < k < \ell$ and let S be the variety of all k-dimensional isotropic subspaces of W. S is a uniruled homogeneous projective manifold. There is a unique minimal component consisting of all lines on S under the Plücker embedding. The VMRT is Z-isotrivial where Z is the projectivization of the vector bundle $\mathcal{O}(-1)^{2\ell-2k} \oplus \mathcal{O}(-2)$ on $\mathbb{P}_{k-1}$ embedded by the dual tautological bundle of the projective bundle (cf. Proposition 3.2.1 of [Hwang and Mok 2005]).

Let us denote it by $Z \subset \mathbb{P}V$. There is a distinguished hypersurface $R \subset Z$ corresponding to $\mathbb{P}\mathcal{O}(-1)^{2\ell-2k}$. Let D be the linear span of R in V. This D defines a distribution on S which is not integrable (cf. Section 4 of [Hwang and Mok 2005]). However, the corresponding distribution on X_Z of Example 5.6 is integrable. Thus VMRT of S cannot be locally equivalent to that of Example 5.6.

For Z in Example 5.9 or degenerate $Z \subset \mathbb{P}(V)$, the group $\mathrm{Aut}(\hat{Z}) \subset \mathfrak{gl}(V)$ is not reductive. Thus it is reasonable to refine Problem 5.8 to

Problem 5.10. Let $Z \subset \mathbb{P}V$ be an irreducible nonsingular subvariety such that $\mathrm{Aut}(\hat{Z}) \subset \mathbf{GL}(V)$ is reductive. Let X be an n-dimensional projective manifold with Z-isotrivial VMRT. Is the VMRT locally equivalent to that of Example 5.6?

What is nice about Problem 5.10 is that we have a classical differential geometric tool to check local flatness. In fact, given a Z-isotrivial cone structure $\mathscr{C} \subset \mathbb{P}T(U)$, we get an induced G-structure where $G = \mathrm{Aut}(\hat{Z})$. The flatness of a G-structure for a reductive group G can be checked by the vanishing of certain curvature tensors (cf. [Hwang and Mok 1997]). Thus Problem 5.10 is reduced to checking the vanishing of the curvature tensors using properties of VMRT structures.

There are two classes of examples for which Problem 5.10 has been answered in the affirmative. The first one are those covered by the next theorem of Mok [2008b].

Theorem 5.11. *Let S be an n-dimensional irreducible Hermitian symmetric space of compact type with a base point $o \in S$. There exists a unique minimal component on S. Let $\mathscr{C}_o \subset \mathbb{P}T_o(S)$ be the VMRT at o. If the projective variety $Z \subset \mathbb{P}V$ is isomorphic to $\mathscr{C}_o \subset \mathbb{P}T_o(S)$, then Problem 5.10 has an affirmative answer.*

For example when S is the n-dimensional quadric hypersurface, $Z \subset \mathbb{P}V$ is just an $(n-2)$-dimensional non-singular quadric hypersurface. Then $\mathscr{C}_x \subset \mathbb{P}T_x(X)$ in Problem 5.10 defines a conformal structure at the general point of X. In this case, Theorem 5.11 says that this conformal structure is locally flat.

Let us recall Mok's strategy for the proof of Theorem 5.11. The main point is to show that the G-structure which is defined at the general point of X can be extended to a G-structure in a *neighborhood* of a standard rational curve, by exploiting Proposition 4.13 and the fact that $Z \subset \mathbb{P}V$ in Theorem 5.11 is determined by the second fundamental form. Once this extension is obtained, one can deduce the flatness by applying [Hwang and Mok 1997], which shows the vanishing of the curvature tensor from *global* information of the tangent bundle of X on the standard rational curve.

It is very difficult to apply Mok's approach to other $Z \subset \mathbb{P}V$. Since the projective variety $Z \subset \mathbb{P}V$ treated in Theorem 5.11 is the highest weight variety associated to an irreducible representation, one would hope that a similar approach holds for the highest weight variety $Z \subset \mathbb{P}V$ associated to other irreducible representation. However, this is not possible. In fact, [Hwang and Mok 1997] shows that the only irreducible G-structure which can be extended to a neighborhood of a standard rational curve is the one covered by Theorem 5.11.

The other class of examples for which an affirmative answer to Problem 5.10 is known belong to the opposite case when the projective automorphism group of $Z \subset \mathbb{P}V$ is 0-dimensional, i.e. when $\mathrm{Aut}_o(\hat{Z}) = \mathbb{C}^*$. In this case, we cannot use Mok's approach, i.e., the G-structure with $G = \mathbb{C}^*$ *cannot* be extended to a neighborhood of a standard rational curve. One can see this as follows. Suppose it is possible to extend the G-structure to a neighborhood U of a standard rational curve. For simplicity, let us assume that the automorphism group of $Z \subset \mathbb{P}V$ is trivial. In $\mathbb{P}T(U)$ we have a submanifold $\mathscr{C} \subset \mathbb{P}T(U)$ with each fiber $\mathscr{C}_x \subset \mathbb{P}T_x(U)$ isomorphic to $Z \subset \mathbb{P}_{n-1}$ and since the automorphism group is trivial, we get a unique trivialization of the projective bundle $\mathbb{P}T(U)$. But on a standard rational curve, $T(U)$ splits into $\mathbb{O}(2) \oplus \mathbb{O}(1)^p \oplus \mathbb{O}^{n-1-p}$ for some $p \geq 0$, a contradiction.

In this case, the flatness of the G-structure, or the vanishing of the corresponding curvature tensors, must be proved only at general points. In other words, it must come from the flatness of cone structures satisfying certain differential geometric condition. In this direction, we have the following result from [Hwang 2010, Theorem 1.11].

Theorem 5.12. *Assume that $Z \subset \mathbb{P}V$ satisfies the following conditions.*

(1) *Z is nonsingular and linearly normal, i.e., $H^0(Z, \mathbb{O}(1)) = V^*$.*

(2) *The variety of tangent lines to Z, defined as a subvariety of $\mathbf{Gr}(2, V) \subset \mathbb{P}(\wedge^2 V)$, is nondegenerate in $\mathbb{P}(\wedge^2 V)$.*

(3) *$H^0(Z, T(Z) \otimes \mathbb{O}(1)) = H^0(Z, \mathrm{ad}(T(Z)) \otimes \mathbb{O}(1)) = 0$ where $\mathrm{ad}(T(Z))$ denotes the bundle of traceless endomorphisms of the tangent bundle of Z.*

Then a Z-isotrivial cone structure with characteristic connection is locally equivalent to that of Example 5.6.

Note that nondegeneracy and $H^0(Z, T(Z) \otimes \mathbb{O}(1)) = 0$ imply that the projective automorphism group of $Z \subset \mathbb{P}V$ is 0-dimensional.

Let us recall the strategy of the proof of Theorem 5.12 in [Hwang 2010]. Given a Z-isotrivial cone structure $\mathscr{C} \subset \mathbb{P}T(U)$ with $\mathrm{Aut}_o(\hat{Z}) = \mathbb{C}^*$, we have a uniquely determined trivialization $\bar{\theta} : \mathbb{P}T(U) \cong \mathbb{P}V \times U$ with $\bar{\theta}(\mathscr{C}) = Z \times U$. Up to multiplication by a scalar function, this means a trivialization $\theta : T(U) \cong V \times U$, $\theta(\hat{\mathscr{C}}) = \hat{Z} \times U$. Such a trivialization θ determines an affine connection on U

and consequently a projective connection on $\mathbb{P}T(U)$. This projective connection is tangent to $\mathscr{C} \subset \mathbb{P}T(U)$ from the way θ is chosen. In particular, it induces a natural connection on the cone structure $\mathscr{C}$. One difficulty is that there is no reason why this natural connection on $\mathscr{C}$ agrees with the characteristic connection on $\mathscr{C}$ whose existence is assumed in Theorem 5.12. This difficulty is avoided by the first condition in Theorem 5.12 (3) via Proposition 4.8. In particular, the connection on $\mathscr{C}$ induced by θ is a characteristic connection. This enables us to relate the projective geometry of $\mathscr{C}_x$ to the curvature tensors of the G-structure. The rest of the conditions in Theorem 5.12 are used to derive the vanishing of the curvature tensors via this relation.

The conditions of Theorem 5.12 are rather restrictive, although some examples of $Z \subset \mathbb{P}V$ satisfying them are given in [Hwang 2010]. The main remaining problem is how to weaken these conditions, by using more properties of VMRT-structures, other than the existence of a characteristic connection.

Another interesting problem, in view of Theorem 5.12, is to prove a local version of Theorem 5.11, namely, the local flatness of a Z-isotrivial cone structure, for the same Z as in Theorem 5.11, with some additional differential geometric conditions on the cone structure, which always holds for VMRT structures.

6. Distribution spanned by VMRT

When the fibers of a cone structure are degenerate, the cone structure defines a non-trivial distribution as follows.

Definition 6.1. Let $\mathscr{C} \subset \mathbb{P}T(U)$ be a cone structure. For each $x \in U$, let $D_x \subset T_x(U)$ be the vector space spanned by the cone $\hat{\mathscr{C}}_x$. Let $U_o \subset U$ be the open subset where the dimension of D_x is constant. Then $\{D_x, x \in U_o\}$ determine a distribution on U_o, denoted by $\mathrm{Dist}(\mathscr{C})$ and called the *distribution spanned by* $\mathscr{C}$.

When $\mathscr{C}$ admits a characteristic connection or a VMRT structure, $\mathrm{Dist}(\mathscr{C})$ has a special feature. There is an intricate relation between the projective geometry of the fibers $\mathscr{C}_x \subset \mathbb{P}T_x(U)$ and the Frobenius tensor of $\mathrm{Dist}(\mathscr{C})$. The following was first proved in [Hwang and Mok 1998, Proposition 10], when $\mathscr{C}$ is a VMRT structure, by a more geometric argument using a family of standard rational curves.

Theorem 6.2. *Let $M \to U$ be a precone structure with a characteristic connection and $\tau : M \to \mathbb{P}T(U)$ be the tangent morphism in Proposition 3.7. Let $D := \mathrm{Dist}(\mathscr{C})$, $\mathscr{C} = \tau(M)$, and let $\beta : \bigwedge^2 D \to T(U)/D$ be the Frobenius tensor of D. For a general point $x \in U$, any point $\alpha \in M_x$ and a tangent vector $v \in T_\alpha(M_x)$, let $a \in D_x$ be a vector belonging to $\tau(\alpha) \subset \mathbb{P}T_x(U)$ and $b \in D_x$ be a vector proportional to the image of $d\tau(v) \in T(\mathscr{C}_x)$ in $T_x(U)$. Then the Frobenius tensor satisfies $\beta(a \wedge b) = 0$.*

Proof. Let $\pi : M \to U$ be the natural projection and $\mathcal{J}$ be the distribution of the precone structure. The distribution $\pi^{-1}D$ on M contains the distribution $\partial\mathcal{J}$ by Proposition 4.12. The vertical distribution $\text{Ker}\, d\pi$ is in $Ch(\pi^{-1}(D))$. Let

$$\delta : \textstyle\bigwedge^2 (\pi^{-1}D) \to T(M)/\pi^{-1}D$$

be the Frobenius tensor of $\pi^{-1}D$. Then for any $w_1, w_2 \in (\pi^{-1}D)_\alpha \subset T_\alpha(M)$,

$$\beta(d\pi(w_1), d\pi(w_2)) = d\pi(\delta(w_1, w_2))$$

where $d\pi$ on the right-hand side refers to the natural map

$$T(M)/\pi^{-1}D \to T(U)/D$$

induced by $d\pi : T(M) \to T(U)$. For any $a \in \hat{\alpha} \in T_x(U)$ and $b \in \hat{T}_\alpha(\mathcal{C}_x) \subset T_x(U)$, we have their lifts $w_1, w_2 \in (\pi^{-1}D)_\alpha$ with $d\pi(w_1) = a, d\pi(w_2) = b$ such that $w_1 \in F_\alpha$ where F is the characteristic connection and $w_2 \in (\partial\mathcal{J})_\alpha$ by Proposition 4.12. By the definition of a characteristic connection, if $\lambda : \bigwedge^2(\partial\mathcal{J}) \to T(\mathcal{C})/(\partial\mathcal{J})$ is the Frobenius tensor of $\partial\mathcal{J}$, then $\lambda(w_1, w_2) = 0$. Since $\partial\mathcal{J}$ is a sub-distribution of $\pi^{-1}D$, we have $\delta|_{\wedge^2(\partial\mathcal{J})} = \lambda \mod \pi^{-1}D$. It follows that $\delta(w_1, w_2) = 0$ and consequently, $\beta(a, b) = 0$. $\square$

In other words, at a general point $x \in X$, if $H \subset \bigwedge^2 D_x$ denotes the linear span of the points in $\mathbb{P}\bigwedge^2 D_x$ corresponding to the bivectors given by the tangent lines of $\mathcal{C}_x \subset \mathbb{P}T_x(U)$, then H is in the kernel of the Frobenius tensor of $D = \text{Dist}(\mathcal{C})$. This gives non-trivial information about the Frobenius tensor of the distribution spanned by a VMRT structure. Are there more restrictions on the Frobenius tensor enforced by a VMRT structure? This is a very interesting question to study. More specifically, one can ask the following.

Problem 6.3. Given a nondegenerate nonsingular projective variety $Z \subset \mathbb{P}V$, let $H \subset \bigwedge^2 V$ be the linear span of the variety of tangent lines to Z. Then does there exist a uniruled manifold X with VMRT $\mathcal{C} \subset \mathbb{P}T(X_o)$ such that for a general $x \in X_o$, $\mathcal{C}_x \subset \mathbb{P}\text{Dist}(\mathcal{C})_x$ is isomorphic to $Z \subset \mathbb{P}V$ and the kernel of the Frobenius tensor of $\text{Dist}(\mathcal{C})$ is precisely H?

Problem 6.3 is trivial if $H = \bigwedge^2 V$. As was noticed in [Hwang and Mok 1999, Proposition 1.3.2], this is the case if $\dim Z > \frac{1}{2}\dim V - 1$. Thus we may assume that $\dim Z \le \frac{1}{2}\dim V - 1$ in Problem 6.3. A special case of Problem 6.3 is when $Z \subset \mathbb{P}V$ is the highest weight variety associated to an irreducible representation. Even in this case, the answer is unknown in general. A known example is when Z comes from the VMRT of the rational homogenous space G/P associated with a long simple root, as explained in [Hwang and Mok 2002, Proposition 5]. In this case, the condition that the Frobenius tensor is determined by H has to do with the finiteness condition in Serre's presentation of simple Lie algebras.

Now let us turn our attention from Dist($\mathscr{C}$) to the foliation it generates, in the sense explained in Section 2. When the cone structure is a VMRT on a uniruled projective manifold, the leaves of this foliation have a strong algebraic property. To explain this, we will give a general construction of a foliation of a uniruled projective manifold by members of a component of FRC(X), the space of free rational curves on X. Firstly, recall the following basic fact.

Proposition 6.4. *Let X^o be an irreducible quasi-projective variety. Suppose that for each irreducible subvariety $W \subset X^o$, we have associated a subvariety $C_W \subset X^o$ with finitely many components such that each irreducible component of C_W contains W and if $W \subset W'$, then $C_W \subset C_{W'}$. We say that an irreducible subvariety W is saturated if $C_W = W$. For each $x \in X^o$, define*

$$Z_x := \text{the intersection of all saturated subvarieties through } x.$$

Then the followings hold.

(i) *Z_x is irreducible and saturated.*

(ii) *Let $Z^0 = \{x\}$ and let Z^{i+1} be a component of C_{Z^i} for $i \geq 0$. Then $Z^n = Z_x$ where $n = \dim X$.*

(iii) *There exists a Zariski open subset $X^* \subset X^o$ and a foliation on X^* whose leaves are algebraic such that the leaf through a very general point $x \in X^*$ is $Z_x \cap X^*$.*

Proof. For (i), it suffices to show that each component Y of Z_x containing x is saturated. Suppose that W is a saturated subvariety through x. Then $Y \subset W$, hence $C_Y \subset C_W = W$. It follows that C_Y is contained in any saturated variety through x, implying that $Y \subset C_Y \subset Z_x$. But each component of C_Y contains Y. Consequently, $C_Y = Y$ and Y is saturated, implying $Z_x = Y$. For (ii), note that if W is not saturated, then every component of C_W has dimension strictly bigger than $\dim W$. Thus Z^n must be a saturated variety containing x, implying $Z_x \subset Z^n$. On the other hand, if W is any saturated variety through x, then $Z^1 \subset C_x \subset C_W = W$. Thus we get inductively $Z^i \subset C_{Z^i} \subset C_W = W$. It follows that $Z^n \subset W$ for any saturated subvariety W through x, showing $Z^n \subset Z_x$. For (iii), since the collection of subvarieties Z_x cover X^o, there must be a flat family of subvarieties whose very general member is of the form Z_x for some $x \in X^o$ and the members of the family cover a Zariski dense open subset in X^o. To show (iii), it suffices to see that for two Z_x and Z_y in this family, if $y \in Z_x$ then $Z_y = Z_x$. From the definition of Z_y, we get $Z_y \subset Z_x$. Then we get equality by flatness of the family. $\qquad\square$

Definition 6.5. Let $\mathscr{H}$ be a component of FRC(X). Let $X^o \subset X$ be a Zariski open subset in the union of members of $\mathscr{H}$. Given an irreducible subvariety

$W \subset X$ with $W \cap X^o \neq \varnothing$, define

$$C_W := \text{ closure of } \bigcup_{C \in \mathcal{K}, C \cap W \neq \varnothing} C.$$

Since the evaluation morphism for $\mathcal{K}$ is smooth, each component of C_W contains W. If W' is another irreducible subvariety of X with $W \subset W'$, then $C_W \subset C_{W'}$. Thus we can apply Proposition 6.4 to get a foliation $\mathcal{F}$ on a Zariski open subset $X^* \subset X$ such that for a very general $x \in X^*$, the leaf through x is $Z_x \cap X^*$. In fact, we can choose X^* such that $\text{codim}(X \setminus X^*) \geq 2$. This foliation F is called the *foliation generated by* $\mathcal{K}$.

The construction of the foliation generated by a minimal component $\mathcal{K}$ is purely algebro-geometric. We want to relate it to a differential geometric construction.

Proposition 6.6. *Let X be a uniruled projective manifold and $\mathscr{C} \subset \mathbb{P}T(X)$ the VMRT associated to a minimal component $\mathcal{K}$. Then the foliation generated by* $\text{Dist}(\mathscr{C})$ *in the sense of Section 2 coincides, on a dense open subset of X, with the foliation generated by $\mathcal{K}$ in the sense of Definition 6.5. In particular, for the normalization of a member $\nu : \mathbb{P}_1 \to X$ of $\mathcal{K}$ through a very general point $x \in X$, if $f^*T(X) = P \oplus N$ is the decomposition into an ample vector bundle P and a trivial vector bundle N on $\mathbb{P}_1$, then the fiber P_x lies in the tangent space $T_x(Z_x)$.*

Proof. Let $\mathcal{F}$ be the foliation generated by $\mathcal{K}$ and let $\mathcal{F}'$ be the foliation generated by $\text{Dist}(\mathscr{C})$. For a very general point $x \in X$, the leaf of $\mathcal{F}$ corresponds to an open subset in Z_x. Let $\mathcal{L}$ be the leaf of $\mathcal{F}'$ through x. From the construction of Z_x in Proposition 6.4 (iii), the germ of Z_x must be contained in $\mathcal{L}$. Thus $\mathcal{F}$ is a foliation contained in $\mathcal{F}'$. On the other hand, the tangent space to Z_x at a general point must contain all vectors tangent to members of $\mathcal{K}_x$. This implies that $\mathcal{F}'$ is a foliation contained in $\mathcal{F}$. We conclude that the two foliations coincide. The second statement follows because P_x must lie in $\text{Dist}(\mathscr{C})_x$. $\qquad \square$

To my knowledge, the next theorem has not appeared in print.

Theorem 6.7. *Let $\mathcal{F}$ be the foliation defined in Proposition 6.6, extended to a foliation on a maximal open subset of X and let $\mathcal{L}$ be a general leaf of $\mathcal{F}$. Then for a general point $x \in \mathcal{L}$, all members of $\mathcal{K}_x$ lie in $\mathcal{L}$. In particular, if $\tilde{\mathcal{L}}$ is a desingularization of $\bar{\mathcal{L}}$, then $\tilde{\mathcal{L}}$ is a uniruled projective manifold and there exists a minimal component $\tilde{\mathcal{K}}$ with a natural identification $\tilde{\mathcal{K}}_x = \mathcal{K}_x$ for a general point $x \in \mathcal{L}$. Consequently, the VMRT-structure $\mathscr{C}$ of X restricted $\mathcal{L}$, i.e., $\mathscr{C} \cap \mathbb{P}T(\mathcal{L})$ is equivalent to a VMRT-structure of the manifold $\tilde{\mathcal{L}}$.*

Proof. From the construction of the foliation generated by $\mathcal{K}$ in Definition 6.5, there exists a rational map $\eta : X \to B$ surjective over a projective manifold B such that the fiber of η through a very general point $x \in X$ corresponds to Z_x.

Let $\tilde{X} \subset X \times B$ be the graph of η with the birational morphism $p_1 : \tilde{X} \to X$ and the morphism $p_2 : \tilde{X} \to B$ which is an elimination of the indeterminacy locus of η.

We claim that the proper transforms of members of $\mathcal{H}$ to $\tilde{X}$, which intersect the exceptional divisors of p_1 do not cover $\tilde{X}$. Suppose not. An exceptional divisor E of p_1 is covered by curves which are contracted by p_1 but not contracted by p_2. Thus we get a 1-dimensional family $\{C_t \subset X, t \in \Delta\}$ of members of $\mathcal{H}$ passing through general points $x_t \in C_t$ with a common point $y \in \cap C_t$ such that $C_t \subset Z_{x_t}$ with $Z_{x_t} \neq Z_{x_s}$ for $t \neq s \in \Delta$. Let $f_t : \mathbb{P}_1 \to C_t$ be the normalization with $f_t(o) = y$ and $f_t(\infty) = x_t$ for two fixed points $o, \infty \in \mathbb{P}_1$. If $\sigma_t \in H^0(\mathbb{P}_1, f_t^* T(X))$ is the infinitesimal deformation of f_t, then $\sigma_t(o) = 0$. By Proposition 6.6, we see that $\sigma(\infty) \in T_{x_t}(Z_{x_t})$ for all $t \in \Delta$. But this is contradiction, because the deformation C_t moves out of Z_{x_t}. This verifies the claim.

Since a general fiber of p_2 is smooth, the complement of the exceptional divisors of p_1 in the general fiber of p_2 is sent into a general leaf $\mathcal{L}$ in X. By the claim, all members of $\mathcal{H}_x$ for a general point $x \in X$ lie on the leaf $\mathcal{L}$ through x, completing the proof. $\qquad \square$

The following is essentially [Hwang and Mok 1998, Proposition 13].

Proposition 6.8. *Let X be a uniruled projective manifold of Picard number 1. Then the distribution spanned by the VMRT of a minimal component is bracket-generating.*

Proof. Suppose it is not bracket-generating. Then in the proof of Theorem 6.7, $\dim B > 0$. Choose a hypersurface $H \subset B$ to get the hypersurface $p_1(p_2^{-1}(H)) \subset X$. Let C be a general member of $\mathcal{H}$ whose proper transform in $\tilde{X}$ is disjoint from $p_2^{-1}(H)$. By the proof of Theorem 6.7, the proper transform of C is disjoint from the exceptional divisors of p_1. It follows that C is disjoint from the divisor $p_1(p_2^{-1}(H))$ in X, a contradiction to the fact that all effective divisors on X are ample. $\qquad \square$

Even when the Picard number of X is bigger than 1, Theorem 6.7 implies that to study the VMRT-structure, it is necessary to study the VMRT-structure of the desingularized leaf closure $\tilde{\mathcal{L}}$. This justifies that when studying VMRT, it makes sense to assume that the distribution spanned by the cone is bracket-generating.

7. Linear VMRT

In this section, we will look at the case when the fiber $\mathcal{C}_x$ of a cone structure $\mathcal{C} \subset \mathbb{P}T(U)$ is a union of linear subspaces. To start with, we need some differential geometric concepts.

Definition 7.1. Recall that a connection on the full cone structure $\mathscr{C} = \mathbb{P}T(U)$ is called a *projective connection* on U. It is automatically a characteristic connection. We say that a projective connection is *locally flat* if it is locally equivalent to the one on $\mathbb{P}_n$ induced by the family $\mathscr{K}$ of lines in $\mathbb{P}_n$ where $n = \dim U$.

Definition 7.2. A *web* (of rank m) on a complex manifold U is a finite collection $\{F_1, \ldots, F_\ell\}$ of foliations (of rank m) on U. A cone structure $\mathscr{C} \subset \mathbb{P}T(U)$ is a *linear cone structure* if each fiber $\mathscr{C}_x$ is a union of linear subspaces. A web defines a linear cone structure $\mathscr{C} := \mathbb{P}F_1 \cup \cdots \cup \mathbb{P}F_\ell$.

Proposition 7.3. *Let $\mathscr{C} \subset \mathbb{P}T(U)$ be a linear cone structure. Shrinking U, assume that $\mathscr{C} = \mathscr{C}_1 \cup \cdots \cup \mathscr{C}_\ell$ such that each $\mathscr{C}_i$ is just a distribution. If $\mathscr{C}$ has a characteristic connection, then each $\mathscr{C}_i$ is integrable and $\mathscr{C}$ defines a web structure. Moreover, each leaf of the web has a projective connection. Conversely, a web with a projective connection on each leaf gives rise to a cone structure with characteristic connection whose fibers are union of linear subspaces.*

The only non-trivial part in Proposition 7.3 is the integrability of each distribution $\mathscr{C}_i$ when there is a characteristic connection. This is a direct consequence of Theorem 6.2.

Now when the cone structure is a VMRT structure, something special happens, namely, the projective connection becomes locally flat. In fact, the following general structure theorem is proved in [Araujo 2006, Theorem 3.1] and [Hwang 2007, Proposition 1].

Proposition 7.4. *Let X be a uniruled manifold and $\mathscr{C} \subset \mathbb{P}T(X_o)$ be the VMRT of a minimal component $\mathscr{K}$. Assume that $\mathscr{C}$ is a linear cone structure. Then there exists a normal variety $\tilde{X}$ with a finite holomorphic map $\eta : \tilde{X} \to X$ and a dense open subset $\tilde{X}_o$ of $\tilde{X}$ equipped with a proper holomorphic map $\varphi : \tilde{X}_o \to T$ such that each fiber of φ is biregular to $\mathbb{P}_k$ where k is the rank of the VMRT and each member of $\mathscr{K}_x$ for a general $x \in X$ is the image of a line in some fibers of φ. Moreover, for each $t \in T$, let $P_t := \eta(\varphi^{-1}(t))$ be a subvariety in X. Then P_t is an immersed submanifold with trivial normal bundle in X, $\rho|_{\varphi^{-1}(t)}$ is its normalization, and for two distinct points $t_1 \neq t_2 \in T$, the two subvarieties P_{t_1} and P_{t_2} are distinct.*

One may say that a linear VMRT defines a web structure whose "leaves" are immersed projective spaces with trivial normal bundles. Proposition 6.8 has the following consequence.

Proposition 7.5. *In the situation of Proposition 7.4, assume that X has Picard number 1. Then the morphism η has degree > 1.*

The most interesting case of a linear VMRT is one with projective rank 0, i.e., when the VMRT at a general point is finite. In all known examples of linear

VMRT on uniruled manifolds of Picard number 1, the VMRT turns out to be of projective rank 0. This leads to the following question.

Problem 7.6. In the situation of Proposition 7.4, suppose the degree of μ is bigger than 1. Is $\dim \mathcal{C}_x = 0$ at a general point x?

Regarding Problem 7.6, there is at least a restriction on $\dim \mathcal{C}_x$. The following was in [Hwang 2007, Proposition 2].

Proposition 7.7. *In the case of Proposition 7.4, $\mathcal{C}_x$ is smooth. In other words, components of $\mathcal{C}_x$ are disjoint from each other. In particular, if the degree of μ is bigger than 1, then $2 \dim \mathcal{C}_x \leq \dim X - 2$.*

The geometric idea behind this result is as follows. From Proposition 7.4, in an unramified cover of a neighborhood of $\mathbb{P}_k$, we have a foliation with leaves isomorphic to $\mathbb{P}_k$. If two different components of $\mathcal{C}_x$ intersect, then one of the foliations defines in the leaf $\mathbb{P}_k$ of the other foliation a positive-dimensional subvariety with trivial normal bundle, a contradiction to the ampleness of the tangent bundle of $\mathbb{P}_k$.

When X is embedded in projective space such that members of $\mathcal{K}$ are lines, one can go one step further from Proposition 7.7: Theorem 1.1 of [Novelli and Occhetta 2011] excludes the case when $2 \dim \mathcal{C}_x = \dim X - 2$ under this assumption. Their argument seems difficult to generalize to arbitrary uniruled projective manifolds.

In Theorem 4.5, the case of linear VMRT was excluded. When VMRT is linear, a counterexample can be constructed.

Example 7.8. Note that when $f : X \to Y$ is a finite morphism between two Fano manifolds and Y has a minimal component $\mathcal{K}_Y$ with VMRT of projective rank 0, i.e., the VMRT at a general point is finite, then the inverse images of members of $\mathcal{K}_Y$ under f form a minimal component $\mathcal{K}_X$ on X with VMRT of projective rank 0. See, for example, [Hwang and Mok 2003, Proposition 6] for a proof. Thus to get a counterexample to an analogue of Theorem 4.5 for VMRT of projective rank 0, it suffices to provide a finite morphism $f : X \to Y$ between two non-isomorphic Fano manifolds such that Y has VMRT of projective rank 0. Such an example is given by [Schuhmann 1999, Example 1.1]. More precisely, let $Y \subset \mathbb{P}_4$ be a cubic threefold. Let $X_1 \subset \mathbb{P}_5$ be the cone over Y and $X_2 \subset \mathbb{P}_5$ be a quadric hypersurface such that the intersection $X = X_1 \cap X_2$ is a smooth threefold in $\mathbb{P}_5$. Then X is a Fano threefold of index 1 and projection from the vertex of the cone X_1 onto Y induces a finite morphism from X to Y.

It is natural to ask what partial result toward Theorem 4.5 holds when VMRT is linear. We hope that the following has an affirmative answer.

Problem 7.9. Let X be a uniruled projective manifold of Picard number 1 with linear VMRT. Let $\varphi : U \to U'$ be a biholomorphic map between two connected open subsets in X such that $d\varphi$ sends $\mathscr{C}|_U$ to $\mathscr{C}|_{U'}$. Does φ extend to a biregular automorphism $\tilde{\varphi} : X \to X$?

We point out that U and U' in Problem 7.9 are open subsets in the classical topology. In fact, if U and U' are Zariski open and φ is birational, it is easy to show that φ extends to a biregular morphism $\tilde{\varphi}$, as explained in [Hwang and Mok 2001, Proposition 4.4].

Problem 7.9 can be viewed as a generalization of the Liouville theorem in conformal geometry (e.g., [Dubrovin et al. 1984, 15.2]) which says that for the flat conformal model of dimension ≥ 3 a local conformal transformation comes from a global conformal transformation. There are only a few examples where the answer to Problem 7.9 is known.

When $X \subset \mathbb{P}_{n+1}$ is a smooth hypersurface of degree n and $\mathscr{K}$ is a family of lines covering X, the VMRT has projective rank 0 (cf. [Hwang 2001, 1.4.2]). In fact, the ideal defining these finite points in $\mathbb{P}T_x(X)$ is given by the complete intersection of homogeneous polynomials of degree $2, 3, \ldots, n$. The quadric polynomial corresponds to the second fundamental form of the hypersurface X at x and the cubic polynomial corresponds to the Fubini cubic form in the language of [Jensen and Musso 1994]. In particular, the second fundamental form and the Fubini cubic form are determined by VMRT. Thus by the result of [Jensen and Musso 1994] or [Sasaki 1988], we have an affirmative answer for Problem 7.9 when $X \subset \mathbb{P}_{n+1}$ is a hypersurface of degree n.

One can also ask the same question for the hypersurface $X \subset \mathbb{P}_{n+1}$ of degree $n + 1$. For $\mathscr{K}$ consisting of conics covering X, then the VMRT has projective rank 0. However, for this example, it is still unknown whether Problem 7.9 has an affirmative answer.

Another example of Problem 7.9 is Mukai–Umemura threefolds in [Mukai and Umemura 1983]. Recall that these are Fano threefolds of Picard number 1, which are quasi-homogeneous under the three-dimensional Lie group $SL(3, \mathbb{C})$. In fact, they are equivariant compactifications of $SL(3, \mathbb{C})/\mathbf{O}$ and $SL(3, \mathbb{C})/\mathbf{I}$ where $\mathbf{O}$ and $\mathbf{I}$ denote the octahedral and icosahedral groups, respectively. The choice of a Cartan subgroup of $SL(3, \mathbb{C})$ determines a rational curve on X, whose orbits under $SL(3, \mathbb{C})$ give a minimal component $\mathscr{K}$. The VMRT at a base point $x \in X$ in the open orbit is given by the orbit of the Cartan subalgebra by the action of $\mathbf{O}$ or $\mathbf{I}$. Using this one can explicitly describe the web structure in a neighborhood of x, from which one can check that Problem 7.9 has an affirmative answer.

There are many examples with VMRT of projective rank 0; see [Hwang and Mok 2003]. For example, all Fano threefolds of Picard number 1, other than $\mathbb{P}_3$ and the quadric threefold in $\mathbb{P}_4$, provide such examples. For most of these examples, Problem 7.9 is still open.

On the other hand, one may wonder whether a counterexample to Problem 7.9 can be constructed in a way analogous to Example 7.8. This is not the case. It is related to the following well-known problem.

Problem 7.10. Let X be a Fano manifold of Picard number 1 different from projective space. If $f : X \to X$ is a finite self-morphism, should f be bijective?

An affirmative answer is known for Problem 7.10 when X has linear VMRT. This was proved when the projective rank is 0, by [Hwang and Mok 2003, Corollary 3]. The proof for any projective rank, which also gives a simpler and different proof for projective rank 0 case, is given in [Hwang and Nakayama 2011, Theorem 1.3].

8. Symmetries of cone structures

An important component of any equivalence problem is its symmetries, i.e., the self-equivalence, or the automorphisms, of the geometric structure. In the study of continuous symmetries, the investigation of the Lie algebra of the symmetry group is an efficient method. In this section, we present the theory of the local symmetries of the cone structure. More precisely, for a given cone structure $\mathscr{C} \subset \mathbb{P}T(U)$, we want to understand the Lie algebra of germs of holomorphic vector fields at a point $x \in U$ which preserve the cone structure in the following sense.

Definition 8.1. Given a G-structure $\mathscr{G} \subset \mathbf{Fr}(U)$ (resp. a cone structure $\mathscr{C} \subset \mathbb{P}T(U)$), a holomorphic vector field σ on U *preserves* the G-structure (resp. cone structure) if the induced vector field $\tilde{\sigma}$ on $\mathbf{Fr}(U)$ (resp. on $\mathbb{P}T(U)$) is tangent to the subvariety $\mathscr{G}$ (resp. $\mathscr{C}$).

A convenient notion in studying symmetries of G-structures is the following.

Definition 8.2. Let V be a vector space. Let $\mathfrak{gl}(V)$ be the Lie algebra of endomorphisms of V. Given a Lie subalgebra $\mathfrak{g} \subset \mathfrak{gl}(V)$, its m-th *prolongation* is the subspace $\mathfrak{g}^{(m)} \subset \mathrm{Hom}(S^{m+1}V, V)$ consisting of multi-linear homomorphisms $\sigma : S^{m+1}V \to V$ such that for any fixed $v_1, \ldots, v_m \in V$, the endomorphism

$$v \in V \mapsto \sigma(v, v_1, \ldots, v_m) \in V$$

belongs to $\mathfrak{g}$.

Lemma 8.3. *The following properties are immediate.*

(i) $\mathfrak{g}^{(0)} = \mathfrak{g}$.

(ii) *If $\mathfrak{g}^{(m)} = 0$ for some $m \geq 0$, then $\mathfrak{g}^{(m+1)} = 0$.*

(iii) *If $\mathfrak{h} \subset \mathfrak{g} \subset \mathfrak{gl}(V)$ is a Lie subalgebra, then $\mathfrak{h}^{(m)} \subset \mathfrak{g}^{(m)}$ for each $m \geq 0$.*

This is related to the symmetries of G-structures as follows, which is explained well in [Yamaguchi 1993, Section 2.1].

Proposition 8.4. *Let $G \subset \mathbf{GL}(V)$ be a connected algebraic subgroup and $\mathfrak{g} \subset \mathfrak{gl}(V)$ be its Lie algebra. Let $\mathscr{G} \subset \mathbf{Fr}(U)$ be a G-structure. For a point $x \in U$, let $\mathfrak{f}$ be the Lie algebra of all germs of holomorphic vector fields at x which preserve $\mathscr{G}$. Let $\mathfrak{f}^k$ be the Lie subalgebra of $\mathfrak{f}$ consisting of vector fields which vanish at x to order $\geq k+1$ for some integer $k \geq -1$. For each $k \geq 0$, regard the quotient space $\mathfrak{f}^k/\mathfrak{f}^{k+1}$ as a subspace of $\mathrm{Hom}(S^{k+1}V, V)$ by taking the leading coefficients of the Taylor expansion of the vector fields at x. Then*

$$\mathfrak{f}^k/\mathfrak{f}^{k+1} \subseteq \mathfrak{g}^{(k)}$$

and equality holds for a locally flat G-structure.

In other words, the prolongations of $\mathfrak{g}$ are the graded pieces of the Lie algebra of infinitesimal symmetries of the G-structure. The prolongations of subalgebras of $\mathfrak{gl}(V)$ have been much studied in differential geometry. When $\mathfrak{g}$ is reductive, the theory of Lie algebras and their representations is particularly powerful and one can use it to get a good understanding of the prolongations. A fundamental result is the following result stated by E. Cartan [1909], with modern proofs in [Kobayashi and Nagano 1965] and [Singer and Sternberg 1965].

Theorem 8.5. *Let $\mathfrak{g} \subset \mathfrak{gl}(V)$ be an irreducible representation of a Lie algebra $\mathfrak{g}$. Then $\mathfrak{g}^{(2)} = 0$ unless $\mathfrak{g} = \mathfrak{gl}(V)$, $\mathfrak{sl}(V)$, $\mathfrak{csp}(V)$, or $\mathfrak{sp}(V)$, where in the last two cases V is an even-dimensional vector space provided with a symplectic form.*

Now we want to modify these notions on G-structures to cone structures.

Definition 8.6. *Let $Z \subset \mathbb{P}V$ be a projective subvariety and let $\mathrm{aut}(\hat{Z}) \subset \mathfrak{gl}(V)$ be the Lie algebra of $\mathrm{Aut}(\hat{Z})$. In other words,*

$$\mathrm{aut}(\hat{Z}) = \{A \in \mathfrak{gl}(V), \text{ for each smooth point } z \in Z, A(\hat{z}) \subset \hat{T}_z(Z).\}.$$

Let $\mathscr{C}$ be a Z-isotrivial cone structure. It is easy to see that a germ of a vector field at a point of U preserves the cone structure if and only if it preserves the induced G-structure. Thus to understand the infinitesimal symmetries of a Z-isotrivial cone structure it suffices to understand the prolongations of $\mathrm{aut}(\hat{Z})$. This is true even for non-isotrivial cone structures:

Proposition 8.7. *Let $\mathscr{C} \subset \mathbb{P}T(U)$ be a smooth cone structure. For a point $x \in U$, let $\mathfrak{f}$ be the Lie algebra of all germs of holomorphic vector fields at x which preserve $\mathscr{C}$. Let $\mathfrak{f}^k$ be the Lie subalgebra of $\mathfrak{f}$ consisting of vector fields which vanish at x to order $\geq k+1$ for some integer $k \geq -1$. Let $V = T_x(U)$ and $Z \subset \mathbb{P}V$ be the fiber of $\mathscr{C}$ at x. For each $k \geq 0$, regard the quotient space $\mathfrak{f}^k/\mathfrak{f}^{k+1}$*

as a subspace of $\mathrm{Hom}(S^{k+1}V, V)$ *by taking the leading coefficients of the Taylor expansion of the vector fields at x. Then*

$$\mathfrak{f}^k/\mathfrak{f}^{k+1} \subset \mathrm{aut}(\hat{Z})^{(k)}.$$

This is checked in [Hwang and Mok 2005, Proposition 1.2.1]. The argument is analogous to the case of G-structures, i.e., Proposition 8.4.

This gives the hope that the theory of G-structures provides an effective tool to study the symmetries of cone structures. However, the classical theory of G-structures is not very powerful in dealing with non-reductive groups and it is more efficient to work with the cone structure directly in many cases. Technically, one has to replace the use of Lie theory by projective geometry of the fibers of the cone structure to investigate the symmetry. One example of this approach is in [HM05] Section 1. In particular, the following generalization of Theorem 8.5 is proved there.

Theorem 8.8. *Let* $Z \subset \mathbb{P}V$ *be an irreducible nonsingular nondegenerate subvariety. Then* $\mathrm{aut}(\hat{Z})^{(2)} = 0$ *unless* $Z = \mathbb{P}V$.

The assumption of irreducibility and nondegeneracy in Theorem 8.8 is necessary: just consider a linear subspace $Z \subset \mathbb{P}V$. It is easy to check that $\mathrm{aut}(\hat{Z})^{(m)} \neq 0$ for all $m \geq 0$. Since nonsingularity is also a reasonable condition in view of Problem 4.15, Theorem 8.8 is a fairly satisfactory result.

Theorem 8.8 implies Theorem 8.5. In fact, let $Z \subset \mathbb{P}V$ be the highest weight variety of $\mathfrak{g} \subset \mathfrak{gl}(V)$ in Theorem 8.5 so that $\mathfrak{g} \subset \mathrm{aut}(\hat{Z})$. If $\mathfrak{g}^{(2)} \neq 0$, then $\mathrm{aut}(\hat{Z})^{(2)} \neq 0$ by Lemma 8.3. Since Z is nonsingular and nondegenerate, $Z = \mathbb{P}V$. But it is well-known that an irreducible Lie subalgebra of $\mathfrak{gl}(V)$ whose highest weight variety is $\mathbb{P}V$ is one of the four listed in Theorem 8.5.

The proof of Theorem 8.8 is quite different from the old proofs of Theorem 8.5. Since $\mathrm{aut}(\hat{Z})$ is a priori not reductive, Lie theory is not so helpful in the proof. One has to replace Lie theory by projective geometry of $Z \subset \mathbb{P}V$. The proof in [Hwang and Mok 2005] involves a complicated induction, using the theory of VMRT.

Theorem 8.8 implies, by Proposition 8.4, that the symmetry group of a cone structure is finite dimensional if a fiber of $\mathscr{C} \to U$ is irreducible, nonsingular and nondegenerate with rank $<$ dim U. In fact, the dimension of the group must be bounded by

$$\dim V + \dim \mathfrak{gl}(V) + \dim \mathfrak{gl}(V)^{(1)}.$$

It is natural to ask for the following extension of Theorem 8.8.

Problem 8.9. Classify all nonsingular linearly normal subvarieties $Z \subset \mathbb{P}V$ with $\mathrm{aut}(\hat{Z})^{(1)} \neq 0$.

The additional assumption of linear normality, i.e., $H^0(Z, \mathcal{O}(1)) = V^*$, is added to simplify the problem. Under this condition, Theorem 1.1.3 of [Hwang and Mok 2005] says that

$$\dim \mathrm{aut}(Z)^{(1)} \le \dim V$$

and Z must be a quasi-homogeneous variety. When Z is a homogeneous variety, we have the following classification result of [Kobayashi and Nagano 1964].

Theorem 8.10. *Let $Z \subset \mathbb{P}V$ be the highest weight variety of an irreducible representation. Then $\mathrm{aut}(\hat{Z})^{(1)} \ne 0$ if and only if Z is the highest weight variety of the isotropy representation of an irreducible Hermitian symmetric space of compact type, i.e., $Z \subset \mathbb{P}V$ in Theorem 5.11.*

There are non-homogeneous examples of $Z \subset \mathbb{P}V$ with $\mathrm{aut}(\hat{Z})^{(1)} \ne 0$. See, e.g., [Hwang and Mok 2005, Propositions 4.2.3 and 7.2.3]. Their automorphism groups are not reductive, making it hard to approach Problem 8.9 by Lie theory. It seems to require a good amount of classical projective algebraic geometry.

By Proposition 8.7, we can reformulate Theorem 8.8 in terms of symmetries of cone structures:

Theorem 8.11. *Let $\mathscr{C} \subset \mathbb{P}T(U)$ be a cone structure with irreducible nonsingular and nondegenerate fibers. Suppose there exists a nonzero element of $\mathfrak{f}^2$ in the notation of Proposition 8.7 which preserves the cone structure. Then $\mathscr{C} = \mathbb{P}T(U)$.*

When stated this way, further questions arise. It is very natural to replace the nondegeneracy assumption of Theorem 8.11 by a bracket-generating condition. For example one can raise the following question, as a generalization of Theorem 8.11.

Problem 8.12. Let $\mathscr{C} \subset \mathbb{P}T(U)$ be a cone structure with irreducible nonsingular fibers such that the distribution $\mathrm{Dist}(\mathscr{C})$ spanned by $\mathscr{C}$ is bracket-generating. Suppose there exists a nonzero element of $\mathfrak{f}^2$ in the notation of Proposition 8.7. What are the possible fibers of $\mathscr{C}$?

There are serious difficulties in generalizing the method of the proof of Theorem 8.11 in the direction of Problem 8.12. The differential geometric problem of Theorem 8.11 has been reduced to purely projective geometric problem of Theorem 8.8 by Proposition 8.7. So far, the differential geometric theory needed to make such a reduction for Problem 8.12 has not been fully developed.

One may wonder why in this section we consider a general cone structure. Maybe by restricting to the more special case of cone structures with characteristic connection or VMRT-structures we can get better results? It is likely that such a restriction gives a non-trivial improvement. This direction has not been pursued so far. I hope that a more refined theory can be developed by such an approach, leading to a better formulation of Problems 8.9 and 8.12.

References

[Araujo 2006] C. Araujo, "Rational curves of minimal degree and characterizations of projective spaces", *Math. Ann.* **335**:4 (2006), 937–951.

[Bernstein and Gindikin 2003] J. Bernstein and S. Gindikin, "Notes on integral geometry for manifolds of curves", pp. 57–80 in *Lie groups and symmetric spaces*, Amer. Math. Soc. Transl. Ser. 2 **210**, Amer. Math. Soc., Providence, RI, 2003.

[Cartan 1909] E. Cartan, "Les groupes de transformations continus, infinis, simples", *Ann. Sci. École Norm. Sup.* (3) **26** (1909), 93–161.

[Dubrovin et al. 1984] B. A. Dubrovin, A. T. Fomenko, and S. P. Novikov, *Modern Geometry I*, Grad. Texts in Math. **93**, Springer, Berlin, 1984.

[Hwang 2001] J.-M. Hwang, "Geometry of minimal rational curves on Fano manifolds", pp. 335–393 in *School on Vanishing Theorems and Effective Results in Algebraic Geometry* (Trieste, 2000), ICTP Lect. Notes **6**, Abdus Salam Int. Cent. Theoret. Phys., Trieste, 2001.

[Hwang 2007] J.-M. Hwang, "Deformation of holomorphic maps onto Fano manifolds of second and fourth Betti numbers 1", *Ann. Inst. Fourier (Grenoble)* **57**:3 (2007), 815–823.

[Hwang 2010] J.-M. Hwang, "Equivalence problem for minimal rational curves with isotrivial varieties of minimal rational tangents", *Ann. Sci. Éc. Norm. Supér.* (4) **43**:4 (2010), 607–620.

[Hwang and Mok 1997] J.-M. Hwang and N. Mok, "Uniruled projective manifolds with irreducible reductive *G*-structures", *J. Reine Angew. Math.* **490** (1997), 55–64.

[Hwang and Mok 1998] J.-M. Hwang and N. Mok, "Rigidity of irreducible Hermitian symmetric spaces of the compact type under Kähler deformation", *Invent. Math.* **131**:2 (1998), 393–418.

[Hwang and Mok 1999] J.-M. Hwang and N. Mok, "Varieties of minimal rational tangents on uniruled projective manifolds", pp. 351–389 in *Several complex variables* (Berkeley, 1995–1996), Math. Sci. Res. Inst. Publ. **37**, Cambridge Univ. Press, Cambridge, 1999.

[Hwang and Mok 2001] J.-M. Hwang and N. Mok, "Cartan–Fubini type extension of holomorphic maps for Fano manifolds of Picard number 1", *J. Math. Pures Appl.* (9) **80**:6 (2001), 563–575.

[Hwang and Mok 2002] J.-M. Hwang and N. Mok, "Deformation rigidity of the rational homogeneous space associated to a long simple root", *Ann. Sci. École Norm. Sup.* (4) **35**:2 (2002), 173–184.

[Hwang and Mok 2003] J.-M. Hwang and N. Mok, "Finite morphisms onto Fano manifolds of Picard number 1 which have rational curves with trivial normal bundles", *J. Algebraic Geom.* **12**:4 (2003), 627–651.

[Hwang and Mok 2004] J.-M. Hwang and N. Mok, "Birationality of the tangent map for minimal rational curves", *Asian J. Math.* **8**:1 (2004), 51–63.

[Hwang and Mok 2005] J.-M. Hwang and N. Mok, "Prolongations of infinitesimal linear automorphisms of projective varieties and rigidity of rational homogeneous spaces of Picard number 1 under Kähler deformation", *Invent. Math.* **160**:3 (2005), 591–645.

[Hwang and Nakayama 2011] J.-M. Hwang and N. Nakayama, "On endomorphisms of Fano manifolds of Picard number 1", *Pure Appl. Math. Q.* **7**:4 (2011), 1407–1426.

[Jensen and Musso 1994] G. R. Jensen and E. Musso, "Rigidity of hypersurfaces in complex projective space", *Ann. Sci. École Norm. Sup.* (4) **27**:2 (1994), 227–248.

[Kebekus 2002] S. Kebekus, "Families of singular rational curves", *J. Algebraic Geom.* **11**:2 (2002), 245–256.

[Kebekus and Kovács 2004] S. Kebekus and S. J. Kovács, "Are rational curves determined by tangent vectors?", *Ann. Inst. Fourier (Grenoble)* **54**:1 (2004), 53–79.

[Kobayashi and Nagano 1964] S. Kobayashi and T. Nagano, "On filtered Lie algebras and geometric structures, I", *J. Math. Mech.* **13** (1964), 875–907.

[Kobayashi and Nagano 1965] S. Kobayashi and T. Nagano, "On filtered Lie algebras and geometric structures, III", *J. Math. Mech.* **14** (1965), 679–706.

[Kollár 1996] J. Kollár, *Rational curves on algebraic varieties*, Ergebnisse der Mathematik und ihrer Grenzgebiete (3) **32**, Springer, Berlin, 1996.

[Mok 1988] N. Mok, "The uniformization theorem for compact Kähler manifolds of nonnegative holomorphic bisectional curvature", *J. Differential Geom.* **27**:2 (1988), 179–214.

[Mok 2008a] N. Mok, "Geometric structures on uniruled projective manifolds defined by their varieties of minimal rational tangents", pp. 151–205 in *Géométrie différentielle, physique mathématique, mathématiques et société*, vol. 2, Astérisque **322**, Soc. Math. de France, Paris, 2008.

[Mok 2008b] N. Mok, "Recognizing certain rational homogeneous manifolds of Picard number 1 from their varieties of minimal rational tangents", pp. 41–61 in *Third International Congress of Chinese Mathematicians*, vol. 2, AMS/IP Stud. Adv. Math. **42**, pt. 1, Amer. Math. Soc., Providence, RI, 2008.

[Mukai and Umemura 1983] S. Mukai and H. Umemura, "Minimal rational threefolds", pp. 490–518 in *Algebraic geometry* (Tokyo/Kyoto, 1982), Lecture Notes in Math. **1016**, Springer, Berlin, 1983.

[Novelli and Occhetta 2011] C. Novelli and G. Occhetta, "Projective manifolds containing a large linear subspace with nef normal bundle", *Mich. Math. J.* **60**:2 (2011), 441–462.

[Sasaki 1988] T. Sasaki, "On the projective geometry of hypersurfaces", pp. 115–161 in *Équations différentielles dans le champ complexe* (Strasbourg, 1985), vol. III, Univ. Louis Pasteur, Strasbourg, 1988.

[Schuhmann 1999] C. Schuhmann, "Morphisms between Fano threefolds", *J. Algebraic Geom.* **8**:2 (1999), 221–244.

[Singer and Sternberg 1965] I. M. Singer and S. Sternberg, "The infinite groups of Lie and Cartan, I: The transitive groups", *J. Analyse Math.* **15** (1965), 1–114.

[Sternberg 1983] S. Sternberg, *Lectures on differential geometry*, 2nd ed., Chelsea, New York, 1983.

[Yamaguchi 1982] K. Yamaguchi, "Contact geometry of higher order", *Japan. J. Math. (N.S.)* **8**:1 (1982), 109–176.

[Yamaguchi 1993] K. Yamaguchi, "Differential systems associated with simple graded Lie algebras", pp. 413–494 in *Progress in differential geometry*, Adv. Stud. Pure Math. **22**, Math. Soc. Japan, Tokyo, 1993.

jmhwang@kias.re.kr   *Korea Institute for Advanced Study, Hoegiro 87,*
Seoul, 130-722, Korea

Current Developments in Algebraic Geometry
MSRI Publications
Volume **59**, 2011

Quotients by finite equivalence relations

JÁNOS KOLLÁR
APPENDIX BY CLAUDIU RAICU

We study the existence of geometric quotients by finite set-theoretic equivalence relations. We show that such geometric quotients always exist in positive
characteristic but not in characteristic 0. The appendix gives some examples
of unexpected behavior for scheme-theoretic equivalence relations.

Let $f : X \to Y$ be a finite morphism of schemes. Given Y, one can easily
describe X by the coherent sheaf of algebras $f_*\mathcal{O}_X$. Here our main interest is
the converse. Given X, what kind of data do we need to construct Y? For this
question, the surjectivity of f is indispensable.

The fiber product $X \times_Y X \subset X \times X$ defines an equivalence relation on X,
and one might hope to reconstruct Y as the quotient of X by this equivalence
relation. Our main interest is in the cases when f is not flat. A typical example
we have in mind is when Y is not normal and X is its normalization. In these
cases, the fiber product $X \times_Y X$ can be rather complicated. Even if Y and X are
pure-dimensional and CM, $X \times_Y X$ can have irreducible components of different
dimension and its connected components need not be pure-dimensional. None
of these difficulties appear if f is flat [Raynaud 1967; SGA 3 1970] or if Y is
normal (Lemma 21).

Finite equivalence relations appear in moduli problems in two ways. First,
it is frequently easier to construct or to understand the normalization $\bar{M}$ of
a moduli space M. Then one needs to construct M as a quotient of $\bar{M}$ by a
finite equivalence relation. This method was used in [Kollár 1997] and finite
equivalence relations led to some unsolved problems in [Viehweg 1995, Section
9,5]; see also [Kollár 2011].

Second, in order to compactify moduli spaces of varieties, one usually needs
nonnormal objects. The methods of the minimal model program seem to apply
naturally to their normalizations. It is quite subtle to descend information from
the normalization to the nonnormal variety, see [Kollár 2012, Chapter 5].

In Sections 1, 2, 3 and 6 of this article we give many examples, review (and
correct) known results and pose some questions. New results concerning finite
equivalence relations are in Sections 4 and 5 and in the Appendix.

1. Definition of equivalence relations

Definition 1 (equivalence relations). Let X be an S-scheme and $\sigma : R \to X \times_S X$ a morphism (or $\sigma_1, \sigma_2 : R \rightrightarrows X$ a pair of morphisms). We say that R is an *equivalence relation* on X if, for every scheme $T \to S$, we get a (set-theoretic) equivalence relation

$$\sigma(T) : \mathrm{Mor}_S(T, R) \hookrightarrow \mathrm{Mor}_S(T, X) \times \mathrm{Mor}_S(T, X).$$

Equivalently, the following conditions hold:

(1) σ is a monomorphism (Definition 31).

(2) (reflexive) R contains the diagonal Δ_X.

(3) (symmetric) There is an involution τ_R on R such that $\tau_{X \times X} \circ \sigma \circ \tau_R = \sigma$, where $\tau_{X \times X}$ denotes the involution which interchanges the two factors of $X \times_S X$.

(4) (transitive) For $1 \le i < j \le 3$ set $X_i := X$ and let $R_{ij} := R$ when it maps to $X_i \times_S X_j$. Then the coordinate projection of $R_{12} \times_{X_2} R_{23}$ to $X_1 \times_S X_3$ factors through R_{13}:

$$R_{12} \times_{X_2} R_{23} \to R_{13} \xrightarrow{\pi_{13}} X_1 \times_S X_3.$$

We say that $\sigma_1, \sigma_2 : R \rightrightarrows X$ is a *finite* equivalence relation if the maps σ_1, σ_2 are finite. In this case, $\sigma : R \to X \times_S X$ is also finite, hence a closed embedding (Definition 31).

Definition 2 (set-theoretic equivalence relations). Let X and R be reduced S-schemes. We say that a morphism $\sigma : R \to X \times_S X$ is a *set-theoretic equivalence relation* on X if, for every geometric point $\mathrm{Spec}\, K \to S$, we get an equivalence relation on K-points

$$\sigma(K) : \mathrm{Mor}_S(\mathrm{Spec}\, K, R) \hookrightarrow \mathrm{Mor}_S(\mathrm{Spec}\, K, X) \times \mathrm{Mor}_S(\mathrm{Spec}\, K, X).$$

Equivalently:

(1) σ is geometrically injective.

(2) (reflexive) R contains the diagonal Δ_X.

(3) (symmetric) There is an involution τ_R on R such that $\tau_{X \times X} \circ \sigma \circ \tau_R = \sigma$, where $\tau_{X \times X}$ denotes the involution which interchanges the two factors of $X \times_S X$.

(4) (transitive) For $1 \le i < j \le 3$ set $X_i := X$ and let $R_{ij} := R$ when it maps to $X_i \times_S X_j$. Then the coordinate projection of $\mathrm{red}(R_{12} \times_{X_2} R_{23})$ to $X_1 \times_S X_3$ factors through R_{13}:

$$\mathrm{red}(R_{12} \times_{X_2} R_{23}) \to R_{13} \xrightarrow{\pi_{13}} X_1 \times_S X_3.$$

Note that the fiber product need not be reduced, and taking the reduced structure above is essential, as shown by Example 3.

It is sometimes convenient to consider finite morphisms $p : R \to X \times_S X$ such that the injection $i : p(R) \hookrightarrow X \times_S X$ is a set-theoretic equivalence relation. Such a $p : R \to X \times_S X$ is called a *set-theoretic pre-equivalence relation*.

Example 3. On $X := \mathbb{C}^2$ consider the $\mathbb{Z}/2$-action $(x, y) \mapsto (-x, -y)$. This can be given by a set-theoretic equivalence relation $R \subset X_{x_1,y_1} \times X_{x_2,y_2}$ defined by the ideal

$$(x_1-x_2, y_1-y_2) \cap (x_1+x_2, y_1+y_2) = (x_1^2-x_2^2, y_1^2-y_2^2, x_1y_1-x_2y_2, x_1y_2-x_2y_1)$$

in $\mathbb{C}[x_1, y_1, x_2, y_2]$. We claim that this is *not* an equivalence relation. The problem is transitivity. The defining ideal of $R_{12} \times_{X_2} R_{23}$ in $\mathbb{C}[x_1, y_1, x_2, y_2, x_3, y_3]$ is

$$\left(x_1^2 - x_2^2, y_1^2 - y_2^2, x_1y_1 - x_2y_2, x_1y_2 - x_2y_1,\right.$$
$$\left.x_2^2 - x_3^2, y_2^2 - y_3^2, x_2y_2 - x_3y_3, x_2y_3 - x_3y_2\right).$$

This contains $(x_1^2 - x_3^2, y_1^2 - y_3^2, x_1y_1 - x_3y_3)$ but it does not contain $x_1y_3 - x_3y_1$. Thus there is no map $R_{12} \times_{X_2} R_{23} \to R_{13}$. Note, however, that the problem is easy to remedy. Let $R^* \subset X \times X$ be defined by the ideal

$$(x_1^2 - x_2^2, y_1^2 - y_2^2, x_1y_1 - x_2y_2) \subset \mathbb{C}[x_1, y_1, x_2, y_2].$$

We see that R^* defines an equivalence relation. The difference between R and R^* is one embedded point at the origin.

Definition 4 (categorical and geometric quotients). Given two morphisms

$$\sigma_1, \sigma_2 : R \rightrightarrows X,$$

there is at most one scheme $q : X \to (X/R)^{cat}$ such that $q \circ \sigma_1 = q \circ \sigma_2$ and q is universal with this property. We call $(X/R)^{cat}$ the *categorical quotient* (or *coequalizer*) of $\sigma_1, \sigma_2 : R \rightrightarrows X$.

The categorical quotient is easy to construct in the affine case. Given $\sigma_1, \sigma_2 : R \rightrightarrows X$, the categorical quotient $(X/R)^{cat}$ is the spectrum of the S-algebra

$$\ker\left[\mathcal{O}_X \xrightarrow{\sigma_1^* - \sigma_2^*} \mathcal{O}_R\right].$$

Let $\sigma_1, \sigma_2 . R \rightrightarrows X$ be a finite equivalence relation. We say that $q : X \to Y$ is a *geometric quotient* of X by R if

(1) $q : X \to Y$ is the categorical quotient $q : X \to (X/R)^{cat}$,

(2) $q : X \to Y$ is finite, and

(3) for every geometric point $\operatorname{Spec} K \to S$, the fibers of $q_K : X_K(K) \to Y_K(K)$ are the $\sigma\big(R_K(K)\big)$-equivalence classes of $X_K(K)$.

The geometric quotient is denoted by X/R.

The main example to keep in mind is the following, which easily follows from Lemma 17 and the construction of $(X/R)^{cat}$ for affine schemes.

Example 5. Let $f : X \to Y$ be a finite and surjective morphism. Set $R :=$ red$(X \times_Y X) \subset X \times X$ and let $\sigma_i : R \to X$ denote the coordinate projections. Then the geometric quotient X/R exists and $X/R \to Y$ is a finite and universal homeomorphism (Definition 32). Therefore, if X is the normalization of Y, then X/R is the weak normalization of Y. (See [Kollár 1996, Section 7.2] for basic results on seminormal and weakly normal schemes.)

By taking the reduced structure of $X \times_Y X$ above, we chose to focus on the set-theoretic properties of Y. However, as Example 16 shows, even if X, Y and $X \times_Y X$ are all reduced, $X/R \to Y$ need not be an isomorphism. Thus X and $X \times_Y X$ do not determine Y uniquely.

In Section 2 we give examples of finite, set-theoretic equivalence relations $R \rightrightarrows X$ such that the categorical quotient $(X/R)^{cat}$ is non-Noetherian and there is no geometric quotient. This can happen even when X is very nice, for instance a smooth variety over $\mathbb{C}$. Some elementary results about the existence of geometric quotients are discussed in Section 3.

An inductive plan to construct geometric quotients is outlined in Section 4. As an application, we prove in Section 5 the following:

Theorem 6. *Let S be a Noetherian $\mathbb{F}_p$-scheme and X an algebraic space which is essentially of finite type over S. Let $R \rightrightarrows X$ be a finite, set-theoretic equivalence relation. Then the geometric quotient X/R exists.*

Remark 7. There are many algebraic spaces which are not of finite type and such that the Frobenius map $F^q : X \to X^{(q)}$ is finite. By a result of Kunz (see [Matsumura 1980, p. 302]) such algebraic spaces are excellent. As the proof shows, Theorem 6 remains valid for algebraic spaces satisfying this property.

In the Appendix, C. Raicu constructs finite scheme-theoretic equivalence relations R on $X = \mathbb{A}^2$ (in any characteristic) such that the geometric quotient X/R exists yet R is strictly smaller than the fiber product $X \times_{X/R} X$. Closely related examples are in [Venken 1971; Philippe 1973].

In characteristic zero, this leaves open the following:

Question 8. Let $R \subset X \times X$ be a scheme-theoretic equivalence relation such that the coordinate projections $R \rightrightarrows X$ are finite.

Is there a geometric quotient X/R?

A special case of the quotient problem, called gluing or pinching, is discussed in Section 6. This follows [Artin 1970], [Ferrand 2003] (which is based on an unpublished manuscript from 1970) and [Raoult 1974].

2. First examples

The next examples show that in many cases, the categorical quotient of a very nice scheme X can be non-Noetherian. We start with a nonreduced example and then we build it up to smooth ones.

Example 9. Let k be a field and consider $k[x, \epsilon]$, where $\epsilon^2 = 0$. Set

$$g_1\big(a(x)+\epsilon b(x)\big)=a(x)+\epsilon b(x) \quad \text{and} \quad g_2\big(a(x)+\epsilon b(x)\big)=a(x)+\epsilon\big(b(x)+a'(x)\big).$$

If char $k = 0$, the coequalizer is the spectrum of

$$\ker\left[k[x, \epsilon] \xrightarrow{g_1-g_2} k[x, \epsilon]\right] = k+\epsilon k[x].$$

Note that $k + \epsilon k[x]$ is not Noetherian and its only prime ideal is $\epsilon k[x]$.

If char $k = p$ then the coequalizer is the spectrum of the finitely generated k-algebra

$$\ker\left[k[x, \epsilon] \xrightarrow{g_1-g_2} k[x, \epsilon]\right] = k[x^p]+\epsilon k[x].$$

It is not surprising that set-theoretic equivalence relations behave badly on nonreduced schemes. However, the above example is easy to realize on reduced and even on smooth schemes.

Example 10. (Compare [Holmann 1963, p. 342].) Let $p_i : Z \to Y_i$ be finite morphisms for $i = 1, 2$. We can construct out of them an equivalence relation on $Y_1 \sqcup Y_2$, where R is the union of the diagonal with two copies of Z, one of which maps as

$$(p_1, p_2) : Z \to Y_1 \times Y_2 \subset (Y_1 \sqcup Y_2) \times (Y_1 \sqcup Y_2),$$

the other its symmetric pair. The categorical quotient $((Y_1 \sqcup Y_2)/R)^{cat}$ is also the universal push-out of $Y_1 \xleftarrow{p_1} Z \xrightarrow{p_2} Y_2$. If Z and the Y_i are affine over S, then it is the spectrum of the S-algebra

$$\ker\left[\mathcal{O}_{Y_1} + \mathcal{O}_{Y_2} \xrightarrow{p_1^*-p_2^*} \mathcal{O}_Z\right].$$

For the first example let $Y_1 \cong Y_2 := \operatorname{Spec} k[x, y^2, y^3]$ and $Z := \operatorname{Spec} k[u, v]$ with p_i given by

$$p_1^* : (x, y^2, y^3) \mapsto (u, v^2, v^3) \quad \text{and} \quad p_2^* : (x, y^2, y^3) \mapsto (u+v, v^2, v^3).$$

Since the p_i^* are injective, the categorical quotient is the spectrum of the k-algebra $k[u, v^2, v^3] \cap k[u+v, v^2, v^3]$. Note that

$$k[u, v^2, v^3] = \big\{f_0(u)+\textstyle\sum_{i\geq 2} v^i f_i(u) : f_i \in k[u]\big\},$$

$$k[u+v, v^2, v^3] = \big\{f_0(u)+vf_0'(u)+\textstyle\sum_{i\geq 2} v^i f_i(u) : f_i \in k[u]\big\}.$$

As in Example 9, if char $k = 0$ then the categorical quotient is the spectrum of the non-Noetherian algebra $k + \sum_{n \geq 2} v^n k[u]$. If char $k = p$ then the geometric quotient is given by the finitely generated k-algebra

$$k[u^p] + \sum_{n \geq 2} v^n k[u].$$

This example can be embedded into a set-theoretic equivalence relation on a smooth variety.

Example 11. Let $Y_1 \cong Y_2 := \mathbb{A}^3_{xyz}$, $Z := \mathbb{A}^2_{uv}$ and

$$p_1^* : (x_1, y_1, z_1) \mapsto (u, v^2, v^3) \quad \text{and} \quad p_2^* : (x_2, y_2, z_2) \mapsto (u + v, v^2, v^3).$$

By the previous computations, in characteristic zero the categorical quotient is given by

$$k + (y_1, z_1) + (y_2, z_2) \subset k[x_1, y_1, z_1] + k[x_2, y_2, z_2],$$

where (y_i, z_i) denotes the ideal $(y_i, z_i) \subset k[x_i, y_i, z_i]$. A minimal generating set is given by

$$y_1 x_1^m, z_1 x_1^m, y_2 x_2^m, z_2 x_2^m : m = 0, 1, 2, \ldots$$

In positive characteristic the categorical quotient is given by

$$k[x_1^p, x_2^p] + (y_1, z_1) + (y_2, z_2) \subset k[x_1, y_1, z_1] + k[x_2, y_2, z_2].$$

A minimal generating set is given by

$$x_1^p, x_2^p, y_1 x_1^m, z_1 x_1^m, y_2 x_2^m, z_2 x_2^m : m = 0, 1, \ldots, p - 1.$$

Example 12. The following example, based on [Nagata 1969], shows that even for rings of invariants of finite group actions some finiteness assumption on X is necessary in order to obtain geometric quotients.

Let k be a field of characteristic $p > 0$ and $K := k(x_1, x_2, \ldots)$, where the x_i are algebraically independent over k. Let

$$D := \sum_i x_{i+1} \frac{\partial}{\partial x_i} \quad \text{be a derivation of } K.$$

Let $F := \{ f \in K \mid D(f) = 0 \}$ be the subfield of "constants". Set

$$R = K + \epsilon K \quad \text{where } \epsilon^2 = 0 \quad \text{and} \quad \sigma : f + \epsilon g \mapsto f + \epsilon(g + D(f)).$$

R is a local Artin ring. It is easy to check that σ is an automorphism of R of order p. The fixed ring is $R^\sigma = F + \epsilon K$. Its maximal ideal is $m := (\epsilon K)$ and generating sets of m correspond to F-vector space bases of K. Next we show the x_i are linearly independent over F, which implies that R^σ is not Noetherian.

Assume that we have a relation

$$\sum_{i \leq n} f_i x_i = 0.$$

We may assume that $f_n = 1$ and $f_i \in F \cap k(x_1, \ldots, x_r)$ for some r. Apply D to get that

$$0 = \sum_{i \leq n} f_i D(x_i) = \sum_{i \leq n} f_i x_{i+1}.$$

Repeating s times gives that $\sum_{i \leq n} f_i x_{i+s} = 0$ or, equivalently,

$$x_{n+s} = - \sum_{i \leq n-1} f_i x_{i+s}.$$

This is impossible if $n + s > r$; a contradiction.

It is easy to see that R is not a submodule of any finitely generated R^σ-module.

Example 13. This example of [Nagarajan 1968] gives a 2-dimensional regular local ring R and an automorphism of order 2 such that the ring of invariants is not Noetherian.

Let k be a field of characteristic 2 and $K := k(x_1, y_1, x_2, y_2, \ldots)$, where the x_i, y_i are algebraically independent over k. Let $R := K[[u, v]]$ be the power series ring in 2 variables. Note that R is a 2-dimensional regular local ring, but it is not essentially of finite type over k. Define a derivation of K to R by

$$D_K := \sum_i v(x_{i+1}u + y_{i+1}v) \frac{\partial}{\partial x_i} + u(x_{i+1}u + y_{i+1}v) \frac{\partial}{\partial y_i}.$$

This extends to a derivation of R to R by setting

$$D_R|_K = D_K \quad \text{and} \quad D_R(u) = D_R(v) = 0.$$

Note that $D_R \circ D_R = 0$, thus $\sigma : r \mapsto r + D_R(r)$ is an order 2 automorphism of R. We claim that the ring of invariants R^σ is not Noetherian.

To see this, note first that $x_i u + y_i v \in R^\sigma$ for every i.

Claim. For every n, $x_{n+1}u + y_{n+1}v \notin (x_1 u + y_1 v, \ldots, x_n u + y_n v) R^\sigma$.

Proof. Assume the contrary and write

$$x_{n+1}u + y_{n+1}v = \sum_{i \leq n} r_i (x_i u + y_i v), \quad \text{where } r_i \in R^\sigma.$$

Working modulo $(u, v)^2$ and gathering the terms involving u, we get an equality

$$x_{n+1} \equiv \sum_{i \leq n} r_i x_i \quad \text{mod } R^\sigma \cap (u, v)R.$$

Applying D_R and again gathering the terms involving u we obtain

$$x_{n+2} \equiv \sum_{i \le n} r_i x_{i+1} \quad \text{modulo } R^\sigma \cap (u, v)R.$$

Repeating this s times gives

$$x_{n+s+1} = \sum_{i \le n} \bar{r}_i x_{i+s}, \quad \text{where } \bar{r}_i \in K.$$

Since the $\bar{r}_i$ involve only finitely many variables, we get a contradiction for large s. Thus

$$(x_1 u + y_1 v) \subset (x_1 u + y_1 v, x_2 u + y_2 v)$$

$$\subset (x_1 u + y_1 v, x_2 u + y_2 v, x_3 u + y_3 v) \subset \cdots$$

is an infinite increasing sequence of ideals in R^σ. □

The next examples show that, if S is a smooth projective surface, then a geometric quotient S/R can be nonprojective (but proper) and if X is a smooth proper 3-fold, X/R can be an algebraic space which is not a scheme.

Example 14. 1. Let C, D be smooth projective curves and S the blow up of $C \times D$ at a point (c, d). Let $C_1 \subset S$ be the birational transform of $C \times \{d\}$, $C_2 := C \times \{d'\}$ for some $d' \ne d$ and $\mathbb{P}^1 \cong E \subset S$ the exceptional curve.

Fix an isomorphism $\sigma : C_1 \cong C_2$. This generates an equivalence relation R which is the identity on $S \setminus (C_1 \cup C_2)$. As we will see in Proposition 33, S/R is a surface of finite type. Note however that the image of E in S/R is numerically equivalent to 0, thus S/R is not quasiprojective. Indeed, let M be any line bundle on S/R. Then $\pi^* M$ is a line bundle on S such that $(C_1 \cdot \pi^* M) = (C_2 \cdot \pi^* M)$. Since C_2 is numerically equivalent to $C_1 + E$, this implies that $(E \cdot \pi^* M) = 0$.

2. Take $S \cong \mathbb{P}^2$ and $Z := (x(y^2 - xz) = 0)$. Fix an isomorphism of the line $(x = 0)$ and the conic $(y^2 - xz = 0)$ which is the identity on their intersection. As before, this generates an equivalence relation R which is the identity on their complement. By Proposition 33, $\mathbb{P}^2/R$ exists as a scheme but it is not projective.

Indeed, if M is a line bundle on $\mathbb{P}^2/R$ then $\pi^* M$ is a line bundle on $\mathbb{P}^2$ whose degree on a line is the same as its degree on a conic. Thus $\pi^* M \cong \mathcal{O}_{\mathbb{P}^2}$ and so M is not ample.

3. Let $S = S_1 \amalg S_2 \cong \mathbb{P}^2 \times \{1, 2\}$ be 2 copies of $\mathbb{P}^2$. Let $E \subset \mathbb{P}^2$ be a smooth cubic. For a point $p \in E$, let $\sigma_p : E \times \{1\} \to E \times \{2\}$ be the identity composed with translation by $p \in E$. As before, this generates an equivalence relation R which is the identity on their complement.

Let M be a line bundle on S/R. Then $\pi^* M|_{S_i} \cong \mathcal{O}_{\mathbb{P}^2}(m_i)$ for some $m_i > 0$, and we conclude that

$$\mathcal{O}_{\mathbb{P}^2}(m_1)|E \cong \tau_p^*(\mathcal{O}_{\mathbb{P}^2}(m_2)|E).$$

This holds if and only if $m_1 = m_2$ and $p \in E$ is a $3m_1$-torsion point. Thus the projectivity of S/R depends very subtly on the gluing map σ_p.

Example 15. Hironaka's example in [Hartshorne 1977, B.3.4.1] gives a smooth, proper threefold X and two curves $\mathbb{P}^1 \cong C_1 \cong C_2 \subset X$ such that $C_1 + C_2$ is homologous to 0. Let $g : C_1 \cong C_2$ be an isomorphism and R the corresponding equivalence relation.

We claim that there is no quasiprojective open subset $U \subset X$ which intersects both C_1 and C_2. Assume to the contrary that U is such. Then there is an ample divisor $H_U \subset U$ which intersects both curves but does not contain either. Its closure $H \subset X$ is a Cartier divisor which intersects both curves but does not contain either. Thus $H \cdot (C_1 + C_2) > 0$, a contradiction.

This shows that if $p \in X/R$ is on the image of C_i then p does not have any affine open neighborhood since the preimage of an affine set by a finite morphism is again affine. Thus X/R is not a scheme.

Example 16. [Lipman 1975] Fix a field k and let $a_1, \ldots, a_n \in k$ be different elements. Set

$$A := k[x, y]/\prod_i (x - a_i y).$$

Then $Y := \operatorname{Spec} A$ is n lines through the origin. Let $f : X \to Y$ its normalization. Thus $X = \amalg_i \operatorname{Spec} k[x, y]/(x - a_i y)$. Note that

$$k[x, y]/(x - a_i y) \otimes_A k[x, y]/(x - a_j y) = \begin{cases} k[x, y]/(x - a_i y) & \text{if } a_i = a_j, \\ k & \text{if } a_i \neq a_j. \end{cases}$$

Thus $X \times_Y X$ is reduced. It is the union of the diagonal Δ_X and of $f^{-1}(0, 0) \times f^{-1}(0, 0)$. Thus $X/(X \times_Y X)$ is a seminormal scheme which is isomorphic to the n coordinate axes in $\mathbb{A}^n$. For $n \geq 3$, it is not isomorphic to Y.

One can also get similar examples where Y is integral. Indeed, let $Y \subset \mathbb{A}^2$ be any plane curve whose only singularities are ordinary multiple points and let $f : X \to Y$ be its normalization. By the above computations, $X \times_Y X$ is reduced and $X/(X \times_Y X)$ is the seminormalization of Y.

If Y is a reduced scheme with normalization $\bar{Y} \to Y$, then, as we see in Lemma 17, the geometric quotient $\bar{Y}/(\bar{Y} \times_Y \bar{Y})$ exists. It coincides with the strict closure considered in [Lipman 1971]. The curve case was introduced in [Arf 1948].

The related Lipschitz closure is studied in [Pham 1971] and [Lipman 1975].

3. Basic results

In this section we prove some basic existence results for geometric quotients.

Lemma 17. *Let S be a Noetherian scheme. Assume that X is finite over S and let $p_1, p_2 : R \rightrightarrows X$ be a finite, set-theoretic equivalence relation over S. Then the geometric quotient X/R exists.*

Proof. Since $X \to S$ is affine, the categorical quotient is the spectrum of the $\mathcal{O}_S$-algebra

$$\ker\left[\mathcal{O}_X \xrightarrow{p_1^* - p_2^*} \mathcal{O}_R\right].$$

This kernel is a submodule of the finite $\mathcal{O}_S$-algebra $\mathcal{O}_X$, hence itself a finite $\mathcal{O}_S$-algebra. The only question is about the geometric fibers of $X \to (X/R)^{cat}$. Pick any $s \in S$. Taking the kernel commutes with flat base scheme extensions. Thus we may assume that S is complete, local with closed point s and algebraically closed residue field $k(s)$. We need to show that the reduced fiber of $(X/R)^{cat} \to S$ over s is naturally isomorphic to red $X_s / $ red R_s.

If $U \to S$ is any finite map then $\mathcal{O}_{\text{red}\, U_s}$ is a sum of $m(U)$ copies of $k(s)$ for some $m(U) < \infty$. U has $m(U)$ connected components $\{U_i : i = 1, \ldots, m(U)\}$ and each $U_i \to S$ is finite. Thus $U \to S$ uniquely factors as

$$U \xrightarrow{g} \amalg_{m(U)} S \to S \quad \text{such that} \quad g_s : \text{red}\, U_s \xrightarrow{\cong} \amalg_{m(U)} \operatorname{Spec} k(s)$$

is an isomorphism, where $\amalg_m S$ denotes the disjoint union of m copies of S.

Applying this to $X \to S$ and $R \to S$, we obtain a commutative diagram

$$
\begin{array}{ccc}
R & \overset{p_1, p_2}{\rightrightarrows} & X \\
\downarrow & & \downarrow \\
\amalg_{m(R)} S & \overset{p_1(s), p_2(s)}{\rightrightarrows} & \amalg_{m(X)} S.
\end{array}
$$

Passing to global sections we get

$$
\begin{array}{ccc}
\mathcal{O}_X & \overset{p_1^* - p_2^*}{\longrightarrow} & \mathcal{O}_R \\
\uparrow & & \uparrow \\
\mathcal{O}_S^{m(X)} & \overset{p_1^*(s) - p_2^*(s)}{\longrightarrow} & \mathcal{O}_S^{m(R)}.
\end{array}
$$

The kernel of $p_1^*(s) - p_2^*(s)$ is $m := |X_s / R_s|$ copies of $\mathcal{O}_S$, hence we obtain a factorization

$$(X/R)^{cat} \to \amalg_m S \to S \quad \text{such that} \quad \text{red}(X/R)^{cat}_s \to \amalg_m \operatorname{Spec} k(s)$$

is an isomorphism. $\square$

For later reference, we record the following straightforward consequence.

Corollary 18. *Let $R \rightrightarrows X$ be a finite, set-theoretic equivalence relation such that X/R exists. Let $Z \subset X$ be a closed R-invariant subscheme. Then $Z/R|_Z$ exists and $Z/R|_Z \rightarrow X/R$ is a finite and universal homeomorphism (Definition 32) onto its image.* □

Example 19. Even in nice situations, $Z/R|_Z \rightarrow X/R$ need not be a closed embedding, as shown by the following examples.

(19.1) Set $X := \mathbb{A}^2_{xy} \amalg \mathbb{A}^2_{uv}$ and let R be the equivalence relation that identifies the x-axis with the u-axis.

Let $Z = (y = x^2) \amalg (v = u^2)$. In $Z/R|_Z$ the two components intersect at a node, but the image of Z in X/R has a tacnode.

In this example the problem is clearly caused by ignoring the scheme structure of $R|_Z$. As the next example shows, similar phenomena happen even if $R|_Z$ is reduced.

(19.2) Set $Y := (xyz = 0) \subset \mathbb{A}^3$. Let X be the normalization of Y and $R := X \times_Y X$. Set $W := (x + y + z = 0) \subset Y$ and let $Z \subset X$ be the preimage of W. As computed in Example 16, R and $R|_Z$ are both reduced, $Z/R|_Z$ is the seminormalization of W and $Z/R|_Z \rightarrow W$ is not an isomorphism.

Remark 20. The following putative counterexample to Lemma 17 is proposed in [Białynicki-Birula 2004, 6.2]. Consider the diagram

$$
\begin{array}{ccc}
\operatorname{Spec} k[x, y] & \xrightarrow{p_1} & \operatorname{Spec} k[x, y^2, y^3] \\
{\scriptstyle p_2} \downarrow & & \downarrow {\scriptstyle q_2} \\
\operatorname{Spec} k[x + y, x + x^2, y^2, y^3] & \xrightarrow{q_1} & \operatorname{Spec} k[x + x^2, xy^2, xy^3, y^2, y^3]
\end{array}
\qquad (20.1)
$$

It is easy to see that the p_i are homeomorphisms but $q_2 p_1 = q_1 p_2$ maps $(0, 0)$ and $(-1, 0)$ to the same point. If (20.1) were a universal push-out, one would get a counterexample to Lemma 17. However, it is not a universal push-out. Indeed,

$$
\begin{aligned}
\tfrac{1}{3}(x + y)^3 + \tfrac{1}{2}(x + y)^2 &= \left(\tfrac{1}{3}x^3 + \tfrac{1}{2}x^2\right) + (x^2 + x)y & + xy^2 + \tfrac{1}{2}y^2 + \tfrac{1}{3}y^3 \\
&= -\left(\tfrac{2}{3}x^3 + \tfrac{1}{2}x^2\right) + (x^2 + x)(x + y) + xy^2 + \tfrac{1}{2}y^2 + \tfrac{1}{3}y^3
\end{aligned}
$$

shows that $\tfrac{2}{3}x^3 + \tfrac{1}{2}x^2$ is also in the intersection

$$
k[x, y^2, y^3] \cap k[x + y, x + x^2, y^2, y^3].
$$

Another case where X/R is easy to obtain is the following.

Lemma 21. *Let $p_1, p_2 : R \rightrightarrows X$ be a finite, set-theoretic equivalence relation where X is normal, Noetherian and X, R are both pure-dimensional. Assume*

(1) *X is defined over a field of characteristic 0, or*

(2) *X is essentially of finite type over S, or*

(3) X *is defined over a field of characteristic* $p > 0$ *and the Frobenius map* $F^p : X \to X^{(p)}$ *of §34 is finite.*

Then the geometric quotient X/R *exists as an algebraic space.* X/R *is normal, Noetherian and essentially of finite type over* S *in case* (2).

Proof. Thus U_x/G_x exists and it is easy to see that the U_x/G_x give étale charts for X/G.

In the general case, it is enough to construct the quotient when X is irreducible. Let m be the separable degree of the projections $\sigma_i : R \to X$.

Consider the m-fold product $X \times \cdots \times X$ with coordinate projections π_i. Let R_{ij} (resp. Δ_{ij}) denote the preimage of R (resp. of the diagonal) under (π_i, π_j). A geometric point of $\bigcap_{ij} R_{ij}$ is a sequence of geometric points $(x_1, \ldots, x_m)$ such that any 2 are R-equivalent and a geometric point of $\bigcap_{ij} R_{ij} \setminus \cup_{ij} \Delta_{ij}$ is a sequence $(x_1, \ldots, x_m)$ that constitutes a whole R-equivalence class. Let X' be the normalization of the closure of $\bigcap_{ij} R_{ij} \setminus \cup_{ij} \Delta_{ij}$. Note that every $\pi_\ell : \bigcap_{ij} R_{ij} \to X$ is finite, hence the projections $\pi'_\ell : X' \to X$ are finite.

The symmetric group S_m acts on $X \times \cdots \times X$ by permuting the factors and this lifts to an S_m-action on X'. Over a dense open subset of X, the S_m-orbits on the geometric points of X' are exactly the R-equivalence classes.

Let $X^* \subset X'/S_m \times X$ be the image of X' under the diagonal map.

By construction, $X^* \to X$ is finite and one-to-one on geometric points over an open set. Since X is normal, $X^* \cong X$ in characteristic 0 and $X^* \to X$ is purely inseparable in positive characteristic.

In characteristic 0, we thus have a morphism $X \to X'/S_m$ whose geometric fibers are exactly the R-equivalence classes. Thus $X'/S_m = X/R$.

Essentially the same works in positive characteristic, see Section 5 for details. $\qquad \square$

Lemma 22. *Let* $p_1, p_2 : R \rightrightarrows X$ *be a finite, set-theoretic equivalence relation such that* $(X/R)^{cat}$ *exists.*

(1) *If* X *is normal and* X, R *are pure-dimensional then* $(X/R)^{cat}$ *is also normal.*

(2) *If* X *is seminormal then* $(X/R)^{cat}$ *is also seminormal.*

Proof. In the first case, let $Z \to (X/R)^{cat}$ be a finite morphism which is an isomorphism at all generic points of $(X/R)^{cat}$. Since X is normal, $\pi : X \to (X/R)^{cat}$ lifts to $\pi_Z : X \to Z$. By assumption, $\pi_Z \circ p_1$ equals $\pi_Z \circ p_2$ at all generic points of R and R is reduced. Thus $\pi_Z \circ p_1 = \pi_Z \circ p_2$. The universal property of categorical quotients gives $(X/R)^{cat} \to Z$, hence $Z = (X/R)^{cat}$ and $(X/R)^{cat}$ is normal.

In the second case, let $Z \to (X/R)^{cat}$ be a finite morphism which is a universal homeomorphism; see 32. As before, we get liftings $\pi_Z \circ p_1, \pi_Z \circ p_2 : R \rightrightarrows$

$X \to Z$ which agree on closed points. Since R is reduced, we conclude that $\pi_Z \circ p_1 = \pi_Z \circ p_2$, thus $(X/R)^{cat}$ is seminormal. ☐

The following result goes back at least to E. Noether.

Proposition 23. *Let A be a Noetherian ring, R a Noetherian A-algebra and G a finite group of A-automorphisms of R. Let $R^G \subset R$ denote the subalgebra of G-invariant elements. Assume that*

(1) *$|G|$ is invertible in A, or*

(2) *R is essentially of finite type over A, or*

(3) *R is finite over $A[R^p]$ for every prime p that divides $|G|$.*

Then R^G is Noetherian and R is finite over R^G.

Proof. Assume first that R is a localization of a finitely generated A algebra $A[r_1, \ldots, r_m] \subset R$. We may assume that G permutes the r_j. Let σ_{ij} denote the jth elementary symmetric polynomial of the $\{g(r_i) : g \in G\}$. Then

$$A[\sigma_{ij}] \subset A[r_1, \ldots, r_m]^G \subset R^G$$

and, with $n := |G|$, each r_i satisfies the equation

$$r_i^n - \sigma_{i1} r_i^{n-1} + \sigma_{i2} r_i^{n-2} - + \cdots = 0.$$

Thus $A[r_1, \ldots, r_m]$ is integral over $A[\sigma_{ij}]$, and therefore also over the larger ring $A[r_1, \ldots, r_m]^G$.

By assumption $R = U^{-1} A[r_1, \ldots, r_m]$, where U is a subgroup of units in $A[r_1, \ldots, r_m]$. We may assume that U is G-invariant. If $r/u \in R$, where $r \in A[r_1, \ldots, r_m]$ and u a unit in $A[r_1, \ldots, r_m]$, then

$$\frac{r}{u} = \frac{r \prod_{g \neq 1} g(u)}{u \prod_{g \neq 1} g(u)},$$

where the product is over the nonidentity elements of G. Thus $r/u = r'/u'$, where $r' \in A[r_1, \ldots, r_m]$ and u' is a G-invariant unit in $A[r_1, \ldots, r_m]$. Therefore,

$$R = (U^G)^{-1} A[r_1, \ldots, r_m] \quad \text{is finite over} \quad (U^G)^{-1} A[\sigma_{ij}].$$

Since R^G is an $(U^G)^{-1} A[\sigma_{ij}]$-submodule of R, it is also finite over $(U^G)^{-1} A[\sigma_{ij}]$, hence the localization of a finitely generated algebra.

Assume next that $|G|$ is invertible in A. We claim that $JR \cap R^G = J$ for any ideal $J \subset R^G$. Indeed, if $a_i \in R^G$, $r_i \in R$ and $\sum r_i a_i \in R^G$ then

$$|G| \cdot \sum_i r_i a_i = \sum_{g \in G} \sum_i g(r_i) g(a_i) = \sum_i a_i \sum_{g \in G} g(r_i) \in \sum_i a_i R^G.$$

If $|G|$ is invertible, this gives

$$R^G \cap \sum a_i R = \sum a_i R^G.$$

Thus the map $J \mapsto JR$ from the ideals of R^G to the ideals of R is an injection which preserves inclusions. Therefore R^G is Noetherian if R is.

If R is an integral domain, then R is finite over R^G by Lemma 24. The general case, which we do not use, is left to the reader.

The arguments in case (3) are quite involved; see [Fogarty 1980].   □

Lemma 24. *Let R be an integral domain and G a finite group of automorphisms of R. Then R is contained in a finite R^G-module. Thus, if R^G is Noetherian, then R is finite over R^G.*

Proof. Let $K \supset R$ and $K^G \supset R^G$ denote the quotient fields. K/K^G is a Galois extension with group G. Pick $r_1, \dots, r_n \in R$ that form a K^G-basis of K. Then any $r \in R$ can be written as

$$r = \sum_i a_i r_i, \quad \text{where } a_i \in K^G.$$

Applying any $g \in G$ to it, we get a system of equations

$$\sum_i g(r_i)a_i = g(r) \quad \text{for } g \in G.$$

We can view these as linear equations with unknowns a_i. The system determinant is $D := \det_{i,g}\big(g(r_i)\big)$, which is nonzero since its square is the discriminant of K/K^G. The value of D is G-invariant up to sign; thus D^2 is G-invariant hence in R^G. By Kramer's rule, $a_i \in D^{-2}R^G$, hence $R \subset D^{-2}\sum_i r_i R^G$.   □

In the opposite case, when the equivalence relation is nontrivial only on a proper subscheme, we have the following general result.

Proposition 25. *Let X be a reduced scheme, $Z \subset X$ a closed, reduced subscheme and $R \rightrightarrows X$ a finite, set-theoretic equivalence relation. Assume that R is the identity on $R \setminus Z$ and that the geometric quotient $Z/R|_Z$ exists. Then X/R exists and is given by the universal push-out diagram*

$$
\begin{array}{ccc}
Z & \hookrightarrow & X \\
\downarrow & & \downarrow \\
Z/R|_Z & \hookrightarrow & X/R.
\end{array}
$$

Proof. Let Y denote the universal push-out (Theorem 38). Then $X \to Y$ is finite and so X/R exists and we have a natural map $X/R \to Y$ by Lemma 17. On the other hand, there is a natural map $Z/R|_Z \to X/R$ by Corollary 18, hence the universal property of the push-out gives the inverse $Y \to X/R$.   □

4. Inductive plan for constructing quotients

Definition 26. Let $R \rightrightarrows X$ be a finite, set-theoretic equivalence relation and $g : Y \to X$ a finite morphism. Then

$$g^* R := R \times_{(X \times X)} (Y \times Y) \rightrightarrows Y$$

defines a finite, set-theoretic equivalence relation on Y. It is called the *pull-back* of $R \rightrightarrows X$. (Strictly speaking, it should be denoted by $(g \times g)^* R$.)

Note that the $g^* R$-equivalence classes on the geometric points of Y map injectively to the R-equivalence classes on the geometric points of X.

If X/R exists then, by Lemma 17, $Y/g^* R$ also exists and the natural morphism $Y/g^* R \to X/R$ is injective on geometric points. If, in addition, g is surjective then $Y/g^* R \to X/R$ is a finite and universal homeomorphism; see Definition 32. Thus, if X is seminormal and the characteristic is 0, then $Y/g^* R \cong X/R$.

Let $h : X \to Z$ be a finite morphism. If the geometric fibers of h are subsets of R-equivalence classes, then the composite $R \rightrightarrows X \to Z$ defines a finite, set-theoretic pre-equivalence relation

$$h_* R := (h \times h)(R) \subset Z \times Z,$$

called the *push forward* of $R \rightrightarrows X$. If Z/R exists, then, by Lemma 17, X/R also exists and the natural morphism $X/R \to Z/R$ is a finite and universal homeomorphism.

Lemma 27. *Let X be weakly normal, excellent and $R \rightrightarrows X$ a finite, set-theoretic equivalence relation. Let $\pi : X^n \to X$ be the normalization and $R^n \rightrightarrows X^n$ the pull back of R to X^n. If X^n/R^n exists then X/R also exists and $X/R = X^n/R^n$.*

Proof. Let $X^* \subset (X^n/R^n) \times_S X$ be the image of X^n under the diagonal morphism. Since $X^n \to X$ is a finite surjection, X^* is a closed subscheme of $(X^n/R^n) \times_S X$ and $X^* \to X$ is a finite surjection. Moreover, for any geometric point $\bar{x} \to X$, its preimages $\bar{x}_i \to X^n$ are R^n-equivalent, hence they map to the same point in $(X^n/R^n) \times_S X$. Thus $X^* \to X$ is finite and one-to-one on geometric points, so it is a finite and universal homeomorphism; see Definition 32. $X^n \to X$ is a local isomorphism at the generic point of every irreducible component of X, hence $X^* \to X$ is also a local isomorphism at the generic point of every irreducible component of X. Since X is weakly normal, $X^* \cong X$ and we have a morphism $X \to X^n/R^n$ and thus $X/R = X^n/R^n$. $\square$

Lemma 28. *Let X be normal and of pure dimension d. Let $\sigma : R \rightrightarrows X$ be a finite, set-theoretic equivalence relation and $R^d \subset R$ its d-dimensional part. Then $\sigma^d : R^d \rightrightarrows X$ is also an equivalence relation.*

Proof. The only question is transitivity. Since X is normal, the maps σ^d : $R^d \rightrightarrows X$ are both universally open by Chevalley's criterion; see [EGA IV-3 1966, IV.14.4.4]. Thus the fiber product $R^d \times_X R^d \to X$ is also universally open and hence its irreducible components have pure dimension d. $\square$

Example 29. Let C be a curve with an involution τ. Pick $p, q \in C$ with q different from p and $\tau(p)$. Let C' be the nodal curve obtained from C by identifying p and q. The equivalence relation generated by τ on C' consists of the diagonal, the graph of τ plus the pairs $\big(\tau(p), \tau(q)\big)$ and $\big(\tau(q), \tau(p)\big)$. The 1-dimensional parts of the equivalence relation do not form an equivalence relation.

§30 (Inductive plan). Let X be an excellent scheme that satisfies one of the conditions of Lemma 21. and $R \rightrightarrows X$ a finite, set-theoretic equivalence relation. We aim to construct the geometric quotient X/R in two steps. First we construct a space that, roughly speaking, should be the normalization of X/R and then we try to go from the normalization to the geometric quotient itself.

Step 1. Let $X^n \to X$ be the normalization of X and $R^n \rightrightarrows X^n$ the pull back of R to X^n. Set $d = \dim X$ and let $X^{nd} \subset X^n$ (resp. $R^{nd} \subset R^n$) denote the union of the d-dimensional irreducible components. By Lemma 28, $R^{nd} \rightrightarrows X^{nd}$ is a pure-dimensional, finite, set-theoretic equivalence relation and the geometric quotient X^{nd}/R^{nd} exists by Lemma 21.

There is a closed, reduced subscheme $Z \subset X^n$ of dimension $< d$ such that Z is closed under R^n and the two equivalence relations

$$R^n|_{X^n \setminus Z} \quad \text{and} \quad R^{nd}|_{X^n \setminus Z} \quad \text{coincide.}$$

Let $Z_1 \subset X^{nd}/R^{nd}$ denote the image of Z. $R^n|_Z \rightrightarrows Z$ gives a finite set-theoretic equivalence relation on Z. Since the geometric fibers of $Z \to Z_1$ are subsets of R^n-equivalence classes, by Definition 26, the composite maps $R^n|_Z \rightrightarrows Z \to Z_1$ define a finite set-theoretic pre-equivalence relation on Z_1.

Step 2. In order to go from X^{nd}/R^{nd} to X/R, we make the following

Inductive assumption (30.2.1). The geometric quotient $Z_1/\big(R^n|_Z\big)$ exists.

Then, by Proposition 25, X^n/R^n exists and is given as the universal push-out of the following diagram:

$$\begin{array}{ccc} Z_1 & \hookrightarrow & X^n/R^{nd} \\ \downarrow & & \downarrow \\ Z_1/\big(R^n|_Z\big) & \hookrightarrow & X^n/R^n. \end{array}$$

As in Lemma 27, let $X^* \subset \left(X^n/R^n\right) \times_S X$ be the image of X^n under the diagonal morphism. We have established that $X^* \to X$ is a finite and universal homeomorphism (Definition 32) sitting in the following diagram:

$$
\begin{array}{ccc}
Z_1 & \hookrightarrow & X^n/R^{nd} \leftarrow X^n \\
\downarrow & & \downarrow \nearrow \downarrow \searrow \\
Z_1/\left(R^n|_Z\right) & \to & X^n/R^n \leftarrow X^* \to X
\end{array}
\qquad (30.2.2)
$$

There are now two ways to proceed.

Positive characteristic (30.2.3). Most finite and universal homeomorphisms can be inverted, up to a power of the Frobenius (Proposition 35), and so we obtain a morphism

$$
X \to \left(X^*\right)^{(q)} \to \left(X^n/R^n\right)^{(q)}
$$

for some $q = p^m$. X/R is then obtained using Lemma 17. This is discussed in Section 5.

In this case the inductive assumption (30.2.1) poses no extra problems.

Zero characteristic (30.2.4). As the examples of Section 2 show, finite and universal homeomorphisms cause a substantial problem. The easiest way to overcome these difficulties is to assume to start with that X is seminormal. In this case, by Lemma 27, we obtain $X/R = X^n/R^n$.

Unfortunately, the inductive assumption (30.2.1) becomes quite restrictive. By construction Z_1 is reduced, but it need not be seminormal in general. Thus we get the induction going only if we can guarantee that Z_1 is seminormal. Note that, because of the inductive set-up, seminormality needs to hold not only for X and Z_1, but on further schemes that one obtains in applying the inductive proof to $R^n|_Z \rightrightarrows Z_1$, and so on.

It turns out, however, that the above inductive plan works when gluing semi-log-canonical schemes. See [Kollár 2012, Chapters 5 and 8].

Definition 31. A morphism of schemes $f : X \to Y$ is a *monomorphism* if for every scheme Z the induced map of sets $\mathrm{Mor}(Z, X) \to \mathrm{Mor}(Z, Y)$ is an injection.

By [EGA IV-4 1967, IV.17.2.6] this is equivalent to assuming that f is universally injective and unramified.

A proper monomorphism $f : Y \to X$ is a closed embedding. Indeed, a proper monomorphism is injective on geometric points, hence finite. Thus it is a closed embedding if and only if $\mathcal{O}_X \to f_*\mathcal{O}_Y$ is onto. By the Nakayama lemma this is equivalent to $f_x : f^{-1}(x) \to x$ being an isomorphism for every $x \in f(Y)$. By passing to geometric points, we are down to the case when $X = \mathrm{Spec}\, k$, k is algebraically closed and $Y = \mathrm{Spec}\, A$, where A is an Artin k-algebra.

If $A \neq k$, there are at least 2 different k maps $A \to k[\epsilon]$; thus Spec $A \to$ Spec k is not a monomorphism.

Definition 32. We say that a morphism of schemes $g : U \to V$ is a *universal homeomorphism* if it is a homeomorphism and for every $W \to V$ the induced morphism $U \times_V W \to W$ is again a homeomorphism. The definition extends to morphisms of algebraic spaces the usual way [Knutson 1971, II.3].

A simple example of a homeomorphism which is not a universal homeomorphism is Spec $K \to$ Spec L, where L/K is a finite field extension and $L \neq K$. A more interesting example is given by the normalization of the nodal curve $(y^2 = x^2(x + 1))$ with one of the preimages of the node removed:

$$\mathbb{A}^1 \setminus \{-1\} \to (y^2 = x^2(x + 1)) \quad \text{given by} \quad t \mapsto (t^2 - 1, t(t^2 - 1)).$$

When g is finite, the notion is pretty much set-theoretic since a continuous proper map of topological spaces which is injective and surjective is a homeomorphism. Thus we see that for a finite and surjective morphism of algebraic spaces $g : U \to V$ the following are equivalent (see [Grothendieck 1971, I.3.7–8]):

(1) g is a universal homeomorphism.

(2) g is surjective and universally injective.

(3) For every $v \in V$ the fiber $g^{-1}(v)$ has a single point v' and $k(v')$ is a purely inseparable field extension of $k(v)$.

(4) g is surjective and injective on geometric points.

One of the most important properties of these morphisms is that taking the fiber product induces an equivalence between the categories

$$(\text{étale morphisms: } * \to V) \xrightarrow{* \mapsto * \times_V U} (\text{étale morphisms: } * \to U).$$

See [SGA 1 1971, IX.4.10] for a proof. We do not use this in the sequel.

In low dimensions one can start the method of §30 and it gives the following. These results are sufficient to deal with the moduli problem for surfaces.

Proposition 33. *Let S be a Noetherian scheme over a field of characteristic 0 and X an algebraic space of finite type over S. Let $R \rightrightarrows X$ be a finite, set-theoretic equivalence relation. Assume that*

(1) *X is 1-dimensional and reduced, or*

(2) *X is 2-dimensional and seminormal, or*

(3) *X is 3-dimensional, normal and there is a closed, seminormal $Z \subset X$ such that R is the identity on $X \setminus Z$.*

Then the geometric quotient X/R exists.

Proof. Consider first the case when $\dim X = 1$. Let $\pi : X^n \to X$ be the normalization. We construct X^n / R^{nd} as in §30. Note that since Z is zero-dimensional, it is finite over S. Let $V \subset S$ be its image. Next we make a different choice for Z_1. Instead, we take a subscheme $Z_2 \subset X^n / R^{nd}$ whose support is Z_1 such that the pull back of its ideal sheaf $I(Z_2)$ to X^n is a subsheaf of the inverse image sheaf $\pi^{-1} \mathcal{O}_X \subset \mathcal{O}_{X^n}$.

Then we consider the push-out diagram

$$V \leftarrow Z_2 \hookrightarrow X^n / R^{nd}$$

with universal push-out Y. Then $X \to Y$ is a finite morphism and X/R exists by Lemma 17.

The case when $\dim X = 2$ and X is seminormal is a direct consequence of (30.2.4) since the inductive assumption (30.2.1) is guaranteed by item (1) of Proposition 33.

If $\dim X = 3$, then X is already normal and Z is seminormal by assumption. Thus $Z/(R|_Z)$ exists by Proposition 33(2). Therefore X/R is given by the push-out of $Z/(R|_Z) \leftarrow Z \hookrightarrow X$. $\qquad\square$

5. Quotients in positive characteristic

The main result of this section is the proof of Theorem 6.

§34 (Geometric Frobenius morphism [SGA 5 1977, XIV]). Let S be an $\mathbb{F}_p$-scheme. Fix $q = p^r$ for some natural number r. Then $a \mapsto a^q$ defines an $\mathbb{F}_p$-morphism $F^q : S \to S$. This can be extended to polynomials by the formula

$$f = \sum a_I x^I \mapsto f^{(q)} := \sum a_I^q x^I.$$

Let $U = \operatorname{Spec} R$ be an affine scheme over S. Write

$$R = \mathcal{O}_S[x_1, \ldots, x_m]/(f_1, \ldots, f_n)$$

and set

$$R^{(q)} := \mathcal{O}_S[x_1^{(q)}, \ldots, x_m^{(q)}]/(f_1^{(q)}, \ldots, f_n^{(q)}) \quad \text{and} \quad U^{(q)} := \operatorname{Spec} R^{(q)},$$

where the $x_i^{(q)}$ are new variables. Thus we have a surjection $R^{(q)} \twoheadrightarrow R^q \subset R$, where R^q denotes the S-algebra generated by the q-th powers of all elements. $R^{(q)} \twoheadrightarrow R^q$ is an isomorphism if and only if R has no nilpotents.

There are natural morphisms

$$F^q : U \to U^{(q)} \quad \text{and} \quad (F^q)^* : R^{(q)} \to R \quad \text{given by} \quad (F^q)^*(x_i^{(q)}) = x_i^q.$$

It is easy to see that these are independent of the choices made. Thus F^q gives a natural transformation from algebraic spaces over S to algebraic spaces over S. One can define $X^{(q)}$ intrinsically as

$$X^{(q)} = X \times_{S, F^q} S.$$

If X is an algebraic space which is essentially of finite type over $\mathbb{F}_p$ then $F^q : X \to X^{(q)}$ is a finite and universal homeomorphism.

For us the most important feature of the Frobenius morphism is the following universal property:

Proposition 35. *Let S be a scheme essentially of finite type over $\mathbb{F}_p$ and X, Y algebraic spaces which are essentially of finite type over S. Let $g : X \to Y$ be a finite and universal homeomorphism. Then for $q = p^r \gg 1$ the map F^q can be factored as*

$$F^q : X \xrightarrow{g} Y \xrightarrow{\bar{g}} X^{(q)}.$$

Moreover, for large enough q (depending on $g : X \to Y$), there is a functorial choice of the factorization in the sense that if

$$
\begin{array}{ccc}
X_1 & \xrightarrow{g_1} & Y_1 \\
\downarrow & & \downarrow \\
X_2 & \xrightarrow{g_2} & Y_2
\end{array}
$$

is a commutative diagram where the g_i are finite and universal homeomorphisms, then, for $q \gg 1$ (depending on the $g_i : X_i \to Y_i$) the factorization gives a commutative diagram

$$
\begin{array}{ccccc}
X_1 & \xrightarrow{g_1} & Y_1 & \xrightarrow{\bar{g}_1} & X_1^{(q)} \\
\downarrow & & \downarrow & & \downarrow \\
X_2 & \xrightarrow{g_2} & Y_2 & \xrightarrow{\bar{g}_2} & X_2^{(q)}.
\end{array}
$$

Proof. It is sufficient to construct the functorial choice of the factorization in case X and Y are affine schemes over an affine scheme $\operatorname{Spec} C$. Thus we have a ring homomorphism $g^* : A \to B$, where A and B are finitely generated C-algebras. We can decompose g^* into $A \twoheadrightarrow B_1$ and $B_1 \hookrightarrow B$. We deal with them separately.

First consider $B_1 \subset B$. In this case there is no choice involved and we need to show that there is a q such that $B^q \subset B_1$, where B^q denotes the C-algebra generated by the q-th powers of all elements. The proof is by Noetherian induction.

First consider the case when B is Artinian. The residue field of B is finite and purely inseparable over the residue field of B_1. For large enough q, taking qth powers kills all the nilpotents, thus B^q is contained in a field of representatives of B_1.

In the general, we can use the Artinian case over the generic points to obtain that $B_1 \subset B_1 B^q$ is an isomorphism at all generic points for $q \gg 1$. Let $I \subset B_1$

denote the conductor of this extension. That is, $I B_1 B^q = I$. By induction we know that there is a q' such that $(B_1 B^q / I)^{q'} \subset B_1 / I$. Thus we get that

$$B^{(qq')} \to B^{qq'} \subset (B_1 B^q)^{q'} \subset B_1 + I B_1 B^q = B_1.$$

Next consider $A \twoheadrightarrow B_1$. Here we have to make a good choice. The kernel is a nilpotent ideal $I \subset A$, say $I^m = 0$. Choose q' such that $q' \geq m$. For $b_1 \in B_1$ let $b_1' \in A$ be any preimage. Then $(b_1')^{q'}$ depends only on b_1. The map

$$b_1 \mapsto (b_1')^{q'} \quad \text{defines a factorization} \quad B_1^{(q')} \to A \to B_1.$$

Combining the map $B^{(q)} \to B_1$ with $B_1^{(q')} \to A$ we obtain $B^{(qq')} \to A$. $\qquad\square$

§36 (Proof of Theorem 6). The question is local on S, hence we may assume that S is affine. X and R are defined over a finitely generated subring of $\mathcal{O}_S$, hence we may assume that S is of finite type over $\mathbb{F}_p$.

The proof is by induction on $\dim X$. We follow the inductive plan in §30 and use its notation.

If $\dim X = 0$ then X is finite over S and the assertion follows from Lemma 17.

In going from dimension $d-1$ to d, the assumption (30.2.1) holds by induction. Thus (30.2.3) shows that X^n / R^n exists.

Let $X^* \subset (X^n / R^n) \times_S X$ be the image of X^n under the diagonal morphism. As we noted in §30, $X^* \to X$ is a finite and universal homeomorphism. Thus, by Proposition 35, there is a factorization

$$X^* \to X \to X^{*(q)} \to (X^n / R^n)^{(q)}.$$

Here $X \to (X^n / R^n)^{(q)}$ is finite and R is an equivalence relation on X over the base scheme $(X^n / R^n)^{(q)}$. Hence, by Lemma 17, the geometric quotient X / R exists. $\qquad\square$

Remark 37. Some of the scheme-theoretic aspects of the purely inseparable case are treated in [Ekedahl 1987] and [SGA 3 1970, Exposé V].

6. Gluing or pinching

The aim of this section is to give an elementary proof of the following.

Theorem 38 [Artin 1970, Theorem 3.1]. *Let X be a Noetherian algebraic space over a Noetherian base scheme A. Let $Z \subset X$ be a closed subspace. Let $g : Z \to V$ be a finite surjection. Then there is a universal push-out diagram of algebraic spaces*

$$\begin{array}{ccc} Z & \hookrightarrow & X \\ g\downarrow & & \downarrow \pi \\ V & \hookrightarrow & Y := X/(Z \to V) \end{array}$$

Furthermore:

(1) Y is a Noetherian algebraic space over A.

(2) $V \to Y$ is a closed embedding and $Z = \pi^{-1}(V)$.

(3) The natural map $\ker[\mathcal{O}_Y \to \mathcal{O}_V] \to \pi_* \ker[\mathcal{O}_X \to \mathcal{O}_Z]$ is an isomorphism.

(4) if X is of finite type over A then so is Y.

Remark 39. If X is of finite type over A and A itself is of finite type over a field or an excellent Dedekind ring, then this is an easy consequence of the contraction results [Artin 1970, Theorem 3.1]. The more general case above follows using the later approximation results [Popescu 1986]. The main point of [Artin 1970] is to understand the case when $Z \to V$ is proper but not finite. This is much harder than the finite case we are dealing with. An elementary approach following [Ferrand 2003] and [Raoult 1974] is discussed below.

§40. The affine case of Theorem 38 is simple algebra. Indeed, let $q : \mathcal{O}_X \to \mathcal{O}_Z$ be the restriction. By Theorem 41, $q^{-1}(\mathcal{O}_V)$ is Noetherian; set $Y := \operatorname{Spec} q^{-1}(\mathcal{O}_V)$.

If $\bar{r}_i \in \mathcal{O}_X / I(Z)$ generate $\mathcal{O}_X / I(Z)$ as an $\mathcal{O}_V$-module then $r_i \in \mathcal{O}_X$ and $I(Z)$ generate $\mathcal{O}_X$ as a $q^{-1}(\mathcal{O}_V)$-module. Since $I(Z) \subset q^{-1}(\mathcal{O}_V)$, we obtain that $r_i \in \mathcal{O}_X$ and $1 \in \mathcal{O}_X$ generate $\mathcal{O}_X$ as a $q^{-1}(\mathcal{O}_V)$-module. Applying Theorem 41 to $R_1 = \mathcal{O}_X$ and $R_2 = q^{-1}(\mathcal{O}_V)$ gives the rest. $\qquad\qquad\square$

For the proof of the following result, see [Matsumura 1986, Theorem 3.7] and the proof of Proposition 23.

Theorem 41 (Eakin and Nagata). *Let $R_1 \supset R_2$ be A-algebras with A Noetherian. Assume that R_1 is finite over R_2.*

(1) If R_1 is Noetherian then so is R_2.

(2) If R_1 is a finitely generated A-algebra then so is R_2. $\qquad\qquad\square$

Gluing for algebraic spaces, following [Raoult 1974], is easier than the quasi-projective case.

§42 (Proof of Theorem 38). For every $p \in V$ we construct a commutative diagram

$$
\begin{array}{ccccc}
V_p & \xleftarrow{g_p} & Z_p & \to & X_p \\
{\scriptstyle \tau_V}\downarrow & & \downarrow{\scriptstyle \tau_Z} & & \downarrow{\scriptstyle \tau_X} \\
V & \xleftarrow{g} & Z & \to & X
\end{array}
$$

where

(1) V_p, Z_p, X_p are affine,

(2) g_p is finite and $Z_p \to X_p$ is a closed embedding,

(3) V_p (resp. Z_p, X_p) is an étale neighborhood of p (resp. $g^{-1}(p)$) and

(4) both squares are fiber products.

Affine gluing (§40) then gives $Y_p := X_p/(Z_p \to V_p)$ and Lemma 44 shows that the Y_p are étale charts on $Y = X/(Z \to V)$.

Start with affine, étale neighborhoods $V_1 \to V$ of p and $X_1 \to X$ of $g^{-1}(p)$. Set $Z_1 := Z \times_X X_1 \subset X_1$. By §43 we may assume that there is a connected component $(Z \times_V V_1)^\circ \subset Z \times_V V_1$ and a (necessarily étale) morphism $(Z \times_V V_1)^\circ \to Z_1$. In general there is no étale neighborhood $X' \to X_1$ extending $(Z \times_V V_1)^\circ \to Z_1$, but there is an affine, étale neighborhood $X_2 \to X_1$ extending $(Z \times_V V_1)^\circ \to Z_1$ over a Zariski neighborhood of $g^{-1}(p)$ (§43).

Thus we have affine, étale neighborhoods $V_2 \to V$ of p, $X_2 \to X$ of $g^{-1}(p)$ and an open embedding $Z \times_X X_2 \hookrightarrow Z_2 := Z \times_V V_2$. Our only remaining problem is that $Z_2 \neq Z \times_X X_2$, hence Z_2 is not a subscheme of X_2. We achieve this by further shrinking V_2 and X_2.

The complement $B_2 := Z_2 \setminus Z \times_X X_2$ is closed, thus $g(B_2) \subset V_2$ is a closed subset not containing p. Pick $\phi \in \Gamma(\mathcal{O}_{V_2})$ that vanishes on $g(B_2)$ such that $\phi(p) \neq 0$. Then $\phi \circ g$ is a function on Z_2 that vanishes on B_2 but is nowhere zero on $g^{-1}(p)$. We can thus extend $\phi \circ g$ to a function Φ on X_2. Thus $V_P := V_2 \setminus (\phi = 0)$, $Z_P := Z_2 \setminus (\phi \circ g = 0)$ and $X_P := X_2 \setminus (\Phi = 0)$ have the required properties. $\qquad\square$

§43. During the proof we have used two basic properties of étale neighborhoods.

First, if $\pi : X \to Y$ is finite then for every étale neighborhood $(u \in U) \to (x \in X)$ there is an étale neighborhood $(v \in V) \to (\pi(x) \in Y)$ and a connected component $(v' \in V') \subset X \times_Y V$ such that there is a lifting $(v' \in V') \to (u \in U)$.

Second, if $\pi : X \to Y$ is a closed embedding, $U \to X$ is étale and $P \subset U$ is a finite set of points then we can find an étale $V \to Y$ such that $P \subset V$ and there is an open embedding $(P \subset X \times_Y V) \to (P \subset U)$.

For proofs see [Milne 1980, 3.14 and 4.2–3].

The next result shows that gluing commutes with flat morphisms.

Lemma 44. *For $i = 1, 2$, let X_i be Noetherian affine A-schemes, $Z_i \subset X_i$ closed subschemes and $g_i : Z_i \to V_i$ finite surjections with universal push-outs Y_i. Assume that in the diagram below both squares are fiber products.*

$$
\begin{array}{ccccc}
V_1 & \overset{g_1}{\leftarrow} & Z_1 & \to & X_1 \\
\downarrow & & \downarrow & & \downarrow \\
V_2 & \overset{g_2}{\leftarrow} & Z_2 & \to & X_2
\end{array}
$$

(1) *If the vertical maps are flat then $Y_1 \to Y_2$ is also flat.*

(2) *If the vertical maps are smooth then $Y_1 \to Y_2$ is also smooth.*

Proof. We may assume that all occurring schemes are affine. Set $R_i := \mathcal{O}_{X_i}$, $I_i := I(Z_i)$ and $S_i := \mathcal{O}_{V_i}$. Thus we have $I_i \subset R_i$ and $S_i \subset R_i/I_i$. Furthermore, R_1 is flat over R_2, $I_1 = I_2 R_1$ and S_1 is flat over S_2. We may also assume that R_2 is local. The key point is the isomorphism

$$(R_1/I_1) \cong (R_2/I_2) \otimes_{R_2} R_1 \cong (R_2/I_2) \otimes_{S_2} S_1. \tag{44.3}$$

This isomorphism is not naturally given; see Remark 45.

We check the local criterion of flatness in [Matsumura 1986, Theorem 22.3]. The first condition we need is that $q_1^{-1}(S_1)/I_1 \cong S_1$ be flat over $q_2^{-1}(S_2)/I_2 \cong S_2$. This holds by assumption. Second, we need that the maps

$$\left(I_2^n/I_2^{n+1}\right) \otimes_{S_2} S_1 \to I_2^n R_1/I_2^{n+1} R_1$$

be isomorphisms. Since R_1 is flat over R_2, the right hand side is isomorphic to

$$\left(I_2^n/I_2^{n+1}\right) \otimes_{R_2/I_2} (R_1/I_1).$$

Using (44.3), we get that

$$\left(I_2^n/I_2^{n+1}\right) \otimes_{R_2/I_2} (R_1/I_1) \cong \left(I_2^n/I_2^{n+1}\right) \otimes_{R_2/I_2} (R_2/I_2) \otimes_{S_2} S_1 \cong \left(I_2^n/I_2^{n+1}\right) \otimes_{S_2} S_1.$$

This settles flatness. In order to prove the smooth case, we just need to check that the fibers of $Y_1 \to Y_2$ are smooth. Outside $V_1 \to V_2$ we have the same fibers as before and $V_1 \to V_2$ is smooth by assumption. $\square$

Remark 45. There is some subtlety in Lemma 44. Consider the simple case when X_2 is a smooth curve over a field k, $Z_2 = \{p, q\}$ two k-points and $V_2 = \operatorname{Spec} k$. Then Y_2 is a nodal curve where p and q are identified.

Let now $X_1 = X_2 \times \{0, 1\}$ as 2 disjoint copies. Then Z_1 consists of 4 points p_0, q_0, p_1, q_1 and V_1 is 2 copies of $\operatorname{Spec} k$. There are two distinct way to arrange g_1. Namely,

- either $g_1'(p_0) = g_1'(q_0)$ and $g_1'(p_1) = g_1'(q_1)$ and then Y_1' consists of 2 disjoint nodal curves,

- or $g_1''(p_0) = g_1''(q_1)$ and $g_1''(p_1) = g_1''(q_0)$ and then Y_1'' consists of a connected curve with 2 nodes and 2 irreducible components.

Both of these are étale double covers of Y_2.

As in §42, the next lemma will be used to reduce quasiprojective gluing to the affine case.

Lemma 46. *Let X be an A-scheme, $Z \subset X$ a closed subscheme and $g : Z \to V$ a finite surjection.*

Let $P \subset V$ be a finite subset and assume that there are open affine subsets $P \subset V_1 \subset V$ and $g^{-1}(P) \subset X_1 \subset X$.

Then there are open affine subsets $P \subset V_P \subset V_1$ and $g^{-1}(P) \subset X_P \subset X_1$ such that g restricts to a finite morphism $g : Z \cap X_P \to V_P$.

Proof. There is an affine subset $g^{-1}(P) \subset X_2 \subset X_1$ such that $g^{-1}(V \setminus V_1)$ is disjoint from X_2. Thus g maps $Z \cap X_2$ to V_1. The problem is that $(Z \cap X_2) \to V_1$ is only quasi finite in general. The set $Z \setminus X_2$ is closed in X and so $g(Z \setminus X_2)$ is closed in V. Since V_1 is affine, there is a function f_P on V_1 which vanishes on $g(Z \setminus X_2) \cap V_1$ but does not vanish on P. Then $f_P \circ g$ is a function on $g^{-1}(V_1)$ which vanishes on $(Z \setminus X_2) \cap g^{-1}(V_1)$ but does not vanish at any point of $g^{-1}(P)$. Since $Z \cap X_1$ is affine, $f_P \circ g$ can be extended to a regular function F_P on X_2.

Set $V_P := V_1 \setminus (f_P = 0)$ and $X_P := X_2 \setminus (F_P = 0)$. The restriction $(Z \cap X_P) \to V_P$ is finite since, by construction, $X_P \cap Z$ is the preimage of V_P. $\qquad\square$

Definition 47. We say that an algebraic space X has the *Chevalley–Kleiman property* if X is separated and every finite subscheme is contained in an open affine subscheme. In particular, X is necessarily a scheme.

These methods give the following interesting corollary.

Corollary 48. *Let $\pi : X \to Y$ be a finite and surjective morphism of separated, excellent algebraic spaces. Then X has the Chevalley–Kleiman property if and only if Y has.*

Proof. Assume that Y has the Chevalley–Kleiman property and let $P \subset X$ be a finite subset. Since $\pi(P) \subset Y$ is finite, there is an open affine subset $Y_P \subset Y$ containing $\pi(P)$. Then $g^{-1}(Y_P) \subset X$ is an open affine subset containing P.

Conversely, assume that X has the Chevalley–Kleiman property. By the already established direction, we may assume that X is normal. Next let Y^n be the normalization of Y. Then $X \to Y^n$ is finite and dominant. Fix irreducible components $X_1 \subset X$ and $Y_1 \subset Y^n$ such that the induced map $X_1 \to Y_1$ is finite and dominant. Let $\pi'_1 : X'_1 \to X_1 \to Y_1$ be the Galois closure of X_1/Y_1 with Galois group G. We already know that X'_1 has the Chevalley–Kleiman property, hence there is an open affine subset $X'_P \subset X'_1$ containing $(\pi'_1)^{-1}(P)$. Then $U'_P := \bigcap_{g \in G} g(X'_P) \subset X'_1$ is affine, Galois invariant and $(\pi'_1)^{-1}\big(\pi'_1(U'_P)\big) = U'_P$.

Thus $U'_P \to \pi'_1(U'_P)$ is finite and, by Chevalley's theorem [Hartshorne 1977, Exercise III.4.2], $\pi'_1(U'_P) \subset Y_1$ is an open affine subset containing P. Thus Y^n has the Chevalley–Kleiman property.

Next consider the normalization map $g : Y^n \to \operatorname{red} Y$. There are lower-dimensional closed subschemes $P \subset V \subset \operatorname{red} Y$ and $Z := g^{-1}(V) \subset Y^n$ such that $g : Y^n \setminus Z \cong \operatorname{red} Y \setminus V$ is an isomorphism. By induction on the dimension, V has the Chevalley–Kleiman property.

By Lemma 46 there are open affine subsets $P \subset V_P \subset V$ and $g^{-1}(P) \subset Y^n_P \subset Y^n$ such that g restricts to a finite morphism $g : Z \cap Y^n_P \to V_P$. Thus, by §40,

$g(Y_P^n) \subset \text{red } Y$ is open, affine and it contains P. Thus red Y has the Chevalley–Kleiman property.

Finally, red $Y \to Y$ is a homeomorphism, thus if $U \subset \text{red } Y$ is an affine open subset and $U' \subset Y$ the "same" open subset of Y then U' is also affine by Chevalley's theorem and so Y has the Chevalley–Kleiman property. □

Example 49. Let E be an elliptic curve and set $S := E \times \mathbb{P}^1$. Pick a general $p \in E$ and $g : E \times \{0, 1\} \to E$ be the identity on $E_0 := E \times \{0\}$ and translation by $-p$ on $E_1 := E \times \{1\}$. Where are the affine charts on the quotient Y?

If $P_i \subset E_i$ are 0-cycles then there is an ample divisor H on S such that $(H \cdot E_i) = P_i$ if and only if $\mathcal{O}_{E_0}(P_0) = \mathcal{O}_{E_1}(P_1)$ under the identity map $E_0 \cong E_1$.

Pick any $a, b \in E_0$ and let $a + p, b + p \in E_1$ be obtained by translation by p. Assume next that $2a + b = a + p + 2(b + p)$, or, equivalently, that $3p = a - b$. Let $H(a, b)$ be an ample divisor on S such that $H(a, b) \cap E_0 = \{a, b\}$ and $H(a, b) \cap E_1 = \{a + p, b + p\}$. Then $U(a, b) := S \setminus H(a, b)$ is affine and g maps $E_i \cap U(a, b)$ isomorphically onto $E \setminus \{a, b\}$ for $i = 0, 1$. As we vary a, b (subject to $3p = a - b$) we get an affine covering of Y.

Note however that the curves $H(a, b)$ do not give Cartier divisors on Y. In fact, for nontorsion $p \in E$, every line bundle on Y pulls back from the nodal curve obtained from the $\mathbb{P}^1$ factor by gluing the points 0 and 1 together.

Appendix by Claudiu Raicu

§50. Let A be a noetherian commutative ring and $X = \mathbb{A}_S^n$ the n-dimensional affine space over $S = \text{Spec } A$. Then $\mathcal{O}_X \simeq A[\boldsymbol{x}]$, where $\boldsymbol{x} = (x_1, \ldots, x_n)$. To give a finite equivalence relation $R \subset X \times_S X$ is equivalent to giving an ideal $I(\boldsymbol{x}, \boldsymbol{y}) \subset A[\boldsymbol{x}, \boldsymbol{y}]$ which satisfies the following properties:

(1) (reflexivity) $I(\boldsymbol{x}, \boldsymbol{y}) \subset (x_1 - y_1, \ldots, x_n - y_n)$.

(2) (symmetry) $I(\boldsymbol{x}, \boldsymbol{y}) = I(\boldsymbol{y}, \boldsymbol{x})$.

(3) (transitivity) $I(\boldsymbol{x}, \boldsymbol{z}) \subset I(\boldsymbol{x}, \boldsymbol{y}) + I(\boldsymbol{y}, \boldsymbol{z})$ in $A[\boldsymbol{x}, \boldsymbol{y}, \boldsymbol{z}]$.

(4) (finiteness) $A[\boldsymbol{x}, \boldsymbol{y}]/I(\boldsymbol{x}, \boldsymbol{y})$ is finite over $A[\boldsymbol{x}]$.

Suppose now that we have an ideal $I(\boldsymbol{x}, \boldsymbol{y})$ satisfying these four conditions and let R be the equivalence relation it defines. If the geometric quotient exists, then by Definition 4 it is of the form $\text{Spec } A[f_1, \ldots, f_m]$ for some polynomials $f_1, \ldots, f_m \in A[\boldsymbol{x}]$. It follows that

$$I(\boldsymbol{x}, \boldsymbol{y}) \supset (f_i(\boldsymbol{x}) - f_i(\boldsymbol{y}) : i = 1, 2, \ldots, m)$$

and R is said to be *effective* if and only if equality holds.

We are mainly interested in the case when A is $\mathbb{Z}$ or some field k and I is homogeneous. Consider an ideal $I(\boldsymbol{x}, \boldsymbol{y}) = (J(\boldsymbol{x}, \boldsymbol{y}), f(\boldsymbol{x}, \boldsymbol{y}))$, where J is an ideal of the form

$$J(x, y) = (f_i(x) - f_i(y) : i = 1, 2, \ldots, m),$$

with homogeneous $f_i \in A[x]$ such that $A[x]$ is a finite module over $A[f_1, \ldots, f_m]$ and $f \in A[x, y]$ a homogeneous polynomial that satisfies the cocycle condition

$$f(x, y) + f(y, z) - f(x, z) \in J(x, y) + J(y, z) \subset A[x, y, z]. \tag{50.1}$$

The reason we call (50.1) a cocycle condition is the following. If we let $B = A[f_1, \ldots, f_m]$, $C = A[x]$ and consider the complex (starting in degree zero)

$$C \to C \otimes_B C \to \cdots \to C^{\otimes_B m} \to \cdots \tag{50.2}$$

with differentials given by the formula

$$d_{m-1}(c_1 \otimes c_2 \otimes \cdots \otimes c_m) = \sum_{i=1}^{m+1} (-1)^i c_1 \otimes \cdots \otimes c_{i-1} \otimes 1 \otimes c_i \otimes \cdots \otimes c_m,$$

then $C \otimes_B C \simeq A[x, y]/J(x, y)$, $C \otimes_B C \otimes_B C \simeq A[x, y, z]/(J(x, y) + J(y, z))$, and if the polynomial $f(x, y)$ satisfies (50.1), then its class in $C \otimes_B C$ is a 1-cocycle in the complex (50.2).

Any ideal $I(x, y)$ defined as above is the ideal of a finite equivalence relation (though the geometric quotient can be different from B). To show that the equivalence relation it defines is noneffective it suffices to check that $f(x, y)$ is not congruent to a difference modulo $J(x, y)$. This can be done using a computer algebra system by computing the finite A-module U of homogeneous forms of the same degree as f which are congruent to differences modulo J, and checking that f is not contained in U. We used Macaulay 2 to check that the following example gives a noneffective equivalence relation (we took $A = \mathbb{Z}$ and $n = 2$):

$$f_1(x) = x_1^2, \quad f_2(x) = x_1 x_2 - x_2^2, \quad f_3(x) = x_2^3,$$
$$f(x, y) = (x_1 y_2 - x_2 y_1) y_2^3,$$
$$I(x, y) = (x_1^2 - y_1^2, x_1 x_2 - x_2^2 - y_1 y_2 + y_2^2, x_2^3 - y_2^3, (x_1 y_2 - x_2 y_1) y_2^3).$$

We also claim that this example remains noneffective after any base change. Indeed, the Λ-module V generated by the forms of degree $5(= \deg(f))$ in I and the differences $g(x) - g(y)$ with g homogeneous of degree 5, is a direct summand in U. Elements of V correspond to 0-coboundaries in (50.2). The module W consisting of elements of $k[x, y]_5$ whose classes in $k[x, y]/J$ are 1-cocycles is also a direct summand in U. The quotient W/V is a free $\mathbb{Z}$-module H generated by the class of $f(x, y)$. This shows that $W = V \oplus H$, hence for any field k we have $W_k = V_k \oplus H_k$, where for an abelian group G we let $G_k = G \otimes_{\mathbb{Z}} k$. If we denote by d_i^k the differentials in the complex obtained from (50.2) by base changing from $\mathbb{Z}$ to k, then we get that $\operatorname{im} d_0^k = V_k$ and $\ker d_1^k \supset W_k$. It follows that the nonzero elements of H_k will represent nonzero cohomology classes in (50.2) for any field k, hence our example is indeed universal.

By [Raicu 2010, Lemma 4.3], all homogeneous noneffective equivalence relations are contained in a homogeneous noneffective equivalence relation constructed as above.

In the positive direction, we have the following result in the toric case, where a *toric equivalence relation* (over a field k) is a scheme-theoretic equivalence relation R on a (not necessarily normal) toric variety X/k that is invariant under the diagonal action of the torus.

Theorem 51 [Raicu 2010, Theorem 4.1]. *Let k be a field, X/k an affine toric variety, and R a toric equivalence relation on X. Then there exists an affine toric variety Y/k together with a toric map $X \to Y$ such that $R \simeq X \times_Y X$.*

Notice that we do not require the equivalence relation to be finite.

Acknowledgements

We thank D. Eisenbud, M. Hashimoto, C. Huneke, K. Kurano, M. Lieblich, J. McKernan, M. Mustaţă, P. Roberts, Ch. Rotthaus, D. Rydh and R. Skjelnes for useful comments, corrections and references. Partial financial support was provided by the NSF under grant number DMS-0758275.

References

[Arf 1948] C. Arf, "Une interprétation algébrique de la suite des ordres de multiplicité d'une branche algébrique", *Proc. London Math. Soc.* (2) **50** (1948), 256–287.

[Artin 1970] M. Artin, "Algebraization of formal moduli, II: Existence of modifications", *Ann. of Math.* (2) **91** (1970), 88–135.

[Białynicki-Birula 2004] A. Białynicki-Birula, "Finite equivalence relations on algebraic varieties and hidden symmetries", *Transform. Groups* **9**:4 (2004), 311–326.

[EGA IV-3 1966] A. Grothendieck, "Éléments de géométrie algébrique, IV: étude locale des schémas et des morphismes de schémas (troisième partie)", *Inst. Hautes Études Sci. Publ. Math.* **28** (1966), 5–255.

[EGA IV-4 1967] A. Grothendieck, "Éléments de géométrie algébrique, IV: étude locale des schémas et des morphismes de schémas (quatrième partie)", *Inst. Hautes Études Sci. Publ. Math.* **32** (1967), 1–361. Zbl 0153.22301

[Ekedahl 1987] T. Ekedahl, "Foliations and inseparable morphisms", pp. 139–149 in *Algebraic geometry* (Brunswick, ME, 1985), Proc. Sympos. Pure Math. **46**, Amer. Math. Soc., Providence, RI, 1987.

[Ferrand 2003] D. Ferrand, "Conducteur, descente et pincement", *Bull. Soc. Math. France* **131**:4 (2003), 553–585.

[Fogarty 1980] J. Fogarty, "Kähler differentials and Hilbert's fourteenth problem for finite groups", *Amer. J. Math.* **102**:6 (1980), 1159–1175.

[Grothendieck 1971] A. Grothendieck, *Eléments de géométrie algébrique, I: Le langage des schémas*, 2nd ed., Grundlehren der Mathematischen Wissenschaften **166**, Springer, Berlin, 1971.

[Hartshorne 1977] R. Hartshorne, *Algebraic geometry*, Graduate Texts in Math. **52**, Springer, New York, 1977.

[Holmann 1963] H. Holmann, "Komplexe Räume mit komplexen Transformations-gruppen", *Math. Ann.* **150** (1963), 327–360.

[Knutson 1971] D. Knutson, *Algebraic spaces*, Lecture Notes in Mathematics **203**, Springer, Berlin, 1971.

[Kollár 1996] J. Kollár, *Rational curves on algebraic varieties*, Ergebnisse der Mathematik und ihrer Grenzgebiete (3) **32**, Springer, Berlin, 1996.

[Kollár 1997] J. Kollár, "Quotient spaces modulo algebraic groups", *Ann. of Math.* (2) **145**:1 (1997), 33–79.

[Kollár 2011] J. Kollár, "Two examples of surfaces with normal crossing singularities", *Sci. China Math.* **54**:8 (2011), 1707–1712.

[Kollár 2012] J. Kollár, *Singularities of the minimal model program*, Cambridge Univ. Press, 2012. To appear.

[Lipman 1971] J. Lipman, "Stable ideals and Arf rings", *Amer. J. Math.* **93** (1971), 649–685.

[Lipman 1975] J. Lipman, "Relative Lipschitz-saturation", *Amer. J. Math.* **97**:3 (1975), 791–813.

[Matsumura 1980] H. Matsumura, *Commutative algebra*, 2nd ed., Mathematics Lecture Note Series **56**, Benjamin/Cummings, Reading, MA, 1980.

[Matsumura 1986] H. Matsumura, *Commutative ring theory*, Cambridge Studies in Advanced Mathematics **8**, Cambridge University Press, Cambridge, 1986.

[Milne 1980] J. S. Milne, *Étale cohomology*, Princeton Mathematical Series **33**, Princeton University Press, Princeton, N.J., 1980.

[Nagarajan 1968] K. R. Nagarajan, "Groups acting on Noetherian rings", *Nieuw Arch. Wisk.* (3) **16** (1968), 25–29.

[Nagata 1969] M. Nagata, "Some questions on rational actions of groups", pp. 323–334 in *Algebraic geometry: papers presented at the International Colloquium* (Bombay, 1968), Tata Institute, Bombay, 1969.

[Pham 1971] F. Pham, "Fractions lipschitziennes et saturation de Zariski des algèbres analytiques complexes", in *Actes du Congrès International des Mathématiciens* (Nice, 1970), vol. 2, Gauthier-Villars, Paris, 1971. Exposé d'un travail fait avec Bernard Teisser: "Fractions lipschitziennes d'un algèbre analytique complexe et saturation de Zariski", Centre Math. École Polytech., Paris, 1969, pp. 649–654.

[Philippe 1973] A. Philippe, "Morphisme net d'anneaux, et descente", *Bull. Sci. Math.* (2) **97** (1973), 57–64.

[Popescu 1986] D. Popescu, "General Néron desingularization and approximation", *Nagoya Math. J.* **104** (1986), 85–115.

[Raicu 2010] C. Raicu, "Affine toric equivalence relations are effective", *Proc. Amer. Math. Soc.* **138**:11 (2010), 3835–3847.

[Raoult 1974] J.-C. Raoult, "Compactification des espaces algébriques", *C. R. Acad. Sci. Paris Sér. A* **278** (1974), 867–869.

[Raynaud 1967] M. Raynaud, "Passage au quotient par une relation d'équivalence plate", pp. 78–85 in *Proc. Conf. Local Fields* (Driebergen, 1966), Springer, Berlin, 1967.

[SGA 1 1971] A. Grothendieck and M. Raynaud, *Séminaire de Géométrie Algébrique du Bois Marie* 1960/61: *Revêtements étales et groupe fondamental* (SGA 1), Lecture Notes in Mathematics **224**, Springer, Berlin, 1971. Updated and annotated reprint, Soc. Math. de France, Paris, 2003. Zbl 0234.14002

[SGA 3 1970] M. Demazure and A. Grothendieck (editors), *Schémas en groupes, I–III* (Séminaire de Géométrie Algébrique du Bois Marie 1962/64 = SGA 3), Lecture Notes in Math. **151–153**, Springer, Berlin, 1970.

[SGA 5 1977] P. Deligne, *Cohomologie l-adique et fonctions L* (Séminaire de Géométrie Algébrique du Bois-Marie = SGA 5), Lecture Notes in Math. **589**, Springer, Berlin, 1977.

[Venken 1971] J. Venken, "Non effectivité de la descente de modules plats par un morphisme fini d'anneaux locaux artiniens", *C. R. Acad. Sci. Paris Sér. A-B* **272** (1971), A1553–A1554.

[Viehweg 1995] E. Viehweg, *Quasi-projective moduli for polarized manifolds*, Ergebnisse der Mathematik und ihrer Grenzgebiete (3) **30**, Springer, Berlin, 1995.

kollar@math.princeton.edu   *Department of Mathematics, Princeton University, Princeton, NJ 08544, United States*

claudiu@math.berkeley.edu   *Department of Mathematics, University of California, Berkeley, CA 94720-3840, United States*

                               Institute of Mathematics "Simion Stoilow" of the Romanian Academy

Current Developments in Algebraic Geometry
MSRI Publications
Volume 59, 2011

Higher-dimensional analogues of K3 surfaces

KIERAN G. O'GRADY

A Kähler manifold X is *hyperkähler* if it is simply connected and carries a holomorphic symplectic form whose cohomology class spans $H^{2,0}(X)$. A hyperkähler manifold of dimension 2 is a K3 surface. In many respects higher-dimensional hyperkähler manifolds behave like K3 surfaces: they are the higher dimensional analogues of K3 surfaces of the title. In each dimension greater than 2 there is more than one deformation class of hyperkähler manifolds. One deformation class of dimension $2n$ is that of the Hilbert scheme $S^{[n]}$ where S is a K3 surface. We will present a program which aims to prove that a numerical K3$^{[2]}$ is a deformation of K3$^{[2]}$ — a numerical K3$^{[2]}$ is a hyperkähler 4-fold 4 such that there is an isomorphism of abelian groups $H^2(X; \mathbb{Z}) \xrightarrow{\sim} H^2(\text{K3}^{[2]}; \mathbb{Z})$ compatible with the polynomials given by 4-tuple cup-product.

0. Introduction

K3 surfaces were known classically as complex smooth projective surfaces whose generic hyperplane section is a canonically embedded curve; an example is provided by a smooth quartic surface in $\mathbb{P}^3$. One naturally encounters K3's in the Enriques–Kodaira classification of compact complex surfaces: they are defined to be compact Kähler surfaces with trivial canonical bundle and vanishing first Betti number. Here are a few among the wonderful properties of K3's:

(1) [Kodaira 1964] Any two K3 surfaces are deformation equivalent — thus they are all deformations of a quartic surface.

Supported by Cofinanziamento M.U.R. 2008-2009.

(2) The Kähler cone of a K3 surface X is described as follows. Let $\omega \in H_{\mathbb{R}}^{1,1}(X)$ be one Kähler class and $\mathcal{N}_X$ be the set of nodal classes

$$\mathcal{N}_X := \left\{\alpha \in H_{\mathbb{Z}}^{1,1}(X) \mid \alpha \cdot \alpha = -2,\ \alpha \cdot \omega > 0\right\}. \qquad (0.0.1)$$

The Kähler cone $\mathcal{K}_X$ is given by

$$\mathcal{K}_X := \left\{\alpha \in H_{\mathbb{R}}^{1,1}(X) \mid \alpha \cdot \alpha > 0 \text{ and } \alpha \cdot \beta > 0 \text{ for all } \beta \in \mathcal{N}_X\right\}. \qquad (0.0.2)$$

(3) [Piatetski-Shapiro and Shafarevich 1971; Burns and Rapoport 1975; Looijenga and Peters 1980/81] Weak and strong global Torelli hold. The weak version states that two K3 surfaces X, Y are isomorphic if and only if there exists an integral isomorphism of Hodge structures $f : H^2(X) \xrightarrow{\sim} H^2(Y)$ which is an isometry (with respect to the intersection forms), the strong version states that f is induced by an isomorphism $\phi : Y \xrightarrow{\sim} X$ if and only if it maps effective divisors to effective divisors.[1]

The higher-dimensional complex manifolds closest to K3 surfaces are *hyperkähler manifolds* (HK); they are defined to be simply connected Kähler manifolds with $H^{2,0}$ spanned by the class of a holomorphic *symplectic* form. The terminology originates from riemannian geometry: Yau's solution of Calabi's conjecture gives that every Kähler class ω on a HK manifold contains a Kähler metric g with holonomy the compact symplectic group. There is a sphere S^2 (the pure quaternions of norm 1) parametrizing complex structures for which g is a Kähler metric — the *twistor family* associated to g; it plays a key role in the general theory of HK manifolds.[2] Notice that a HK manifold has trivial canonical bundle and is of even dimension. An example of Beauville [1983] is the Douady space $S^{[n]}$ parametrizing length-n analytic subsets of a K3 surface S — it has dimension $2n$. (Of course $S^{[n]}$ is a Hilbert scheme if S is projective.) We mention right away two results which suggest that HK manifolds might behave like K3's. Let X be HK:

(a) By a theorem of Bogomolov [1978] deformations of X are unobstructed;[3] that is, the deformation space $\mathrm{Def}(X)$ is smooth of the expected dimension $H^1(T_X)$.

(b) Since the sheaf map $T_X \to \Omega_X^1$ given by contraction with a holomorphic symplectic form is an isomorphism it follows that the differential of the

---

[1]Effective divisors have a purely Hodge-theoretic description once we have located one Kähler class.

[2]Hyperkähler manifolds are also known as *irreducible symplectic*.

[3]The obstruction space $H^2(T_X)$ might be nonzero — for example, if X is a generalized Kummer. See Section 1.1.

weight-2 period map

$$H^1(T_X) \longrightarrow \text{Hom}(H^{2,0}(X), H^{1,1}(X)) \tag{0.0.3}$$

is injective, i.e., infinitesimal Torelli holds.

Assuming (a) we may prove that the generic deformation of X has $h_{\mathbb{Z}}^{1,1} = 0$ arguing as follows. A given $\alpha \in H^1(\Omega_X^1)$ remains of type $(1, 1)$ to first order in the direction determined by $\kappa \in H^1(T_X)$ if and only if $\text{Tr}(\kappa \cup \alpha) = 0$ (Griffiths). On the other hand if $\alpha \neq 0$ the map

$$H^1(T_X) \to H^2(\mathcal{O}_X), \quad \kappa \mapsto \text{Tr}(\kappa \cup \alpha) \tag{0.0.4}$$

is surjective by Serre duality; it follows that α does not remain of type $(1, 1)$ on a generic deformation X_t of X (of course what we denote by α is actually the class $\alpha_t \in H^2(X_t)$ obtained from α by Gauss–Manin parallel transport). Item (b) above suggests that the weight-2 Hodge structure of X might capture much of the geometry of X. One is naturally led to ask whether analogues of properties (1)–(3) above hold for higher-dimensional HK manifolds. Let us first discuss (1).

In each (even) dimension greater than 2 we know of two distinct deformation classes of HK manifolds, with one extra deformation class in dimensions 6 and 10. The known examples are distinguished up to deformation by the isomorphism class of their integral weight-2 cohomology group equipped with the top cup-product form — we might name these, together with the dimension, the *basic discrete data* of a HK manifold. Huybrechts [2003b] has shown that the set of deformation classes of HK's with assigned discrete data is finite. In this paper we will present a program which aims to prove that a HK whose discrete data are isomorphic to those of K3$^{[2]}$ is in fact a deformation of K3$^{[2]}$. For the reader's convenience we spell out the meaning of the previous sentence. A *numerical* (K3)$^{[2]}$ is a HK 4-fold X such that there exists an isomorphism of abelian groups $\psi : H^2(X; \mathbb{Z}) \xrightarrow{\sim} H^2(S^{[2]}; \mathbb{Z})$ (here S is a K3) for which

$$\int_X \alpha^4 = \int_{S^{[2]}} \psi(\alpha)^4 \quad \text{for all } \alpha \in H^2(X; \mathbb{Z}). \tag{0.0.5}$$

Our program aims to prove that a numerical K3$^{[2]}$ is a deformation of K3$^{[2]}$. What about analogues of properties (2) and (3) above? We start with (3), global Torelli. On the weight-2 cohomology of a HK there is a natural quadratic form (named after Beauville and Bogomolov) and hence one may formulate a statement — call it *naive Torelli* — analogous to the weak global Torelli statement. The key claim in such a naive Torelli is that if two HK's have Hodge-isometric H^2's then they are bimeromorphic (one cannot require that they be isomorphic; see [Debarre 1984]). However it has been known for some time [Namikawa 2002; Markman 2010] that naive Torelli is false for HK's belonging to certain deformation classes. Recently Verbitsky [2009] proposed a proof of a suitable version of global Torelli valid for arbitrary HK's (see also [Huybrechts 2011]); that result together with Markman's monodromy computations [2010] implies that naive Torelli holds for deformations of $K3^{[p^k+1]}$ where p is a prime. To sum up: an appropriate version of global Torelli holds for any deformation class of HK's. Regarding item (2): Huybrechts [2003c] and Boucksom [2001] have given a description of the Kähler cone in terms of intersections with rational curves (meaning curves with vanishing geometric genus), but that is not a purely Hodge-theoretic description. Hassett and Tschinkel [2001] have formulated a conjectural Hodge-theoretic description of the ample cone of a deformation of $K3^{[2]}$ and they have proved that the divisors satisfying their criterion are indeed ample.

The paper is organized as follows. Following a brief section devoted to the known examples of HK's we introduce basic results on topology and the Kähler cone of a HK in Section 2. After that we will present examples of *explicit* locally complete families of *projective* higher-dimensional HK's. These are analogues of the explicit families of projective K3's such as double covers of $\mathbb{P}^2$ branched over a sextic curve, quartic surfaces in $\mathbb{P}^3$ etc. (The list goes on for quite a few values of the degree, thanks to Mukai, but there are theoretical reasons [Gritsenko et al. 2007] why it should stop before degree 80, more or less.) In particular we will introduce double EPW-sextics, these are double covers of special sextic hypersurfaces in $\mathbb{P}^5$; they play a key rôle in our program for proving that a numerical $K3^{[2]}$ is a deformation of $K3^{[2]}$. The last section is devoted to that program: we discuss what has been proved and what is left to be proved.

1. Examples

The surprising topological properties of HK manifolds (see Section 2.1) led Bogomolov [1978] to state erroneously that no higher-dimensional (dim > 2) HK exists. Some time later Fujiki [1983] realized that $K3^{[2]}$ is a higher-dimensional HK manifold.[4] Beauville [1983] then showed that $K3^{[n]}$ is a HK manifold; moreover by constructing generalized Kummers he exhibited another deformation class of HK manifolds in each even dimension greater that 2. In [O'Grady 1999; 2003] we exhibited two "sporadic" deformation classes, one in dimension 6 the other in dimension 10. No other deformation classes are known other than those mentioned above.

1.1. *Beauville*. Beauville discovered another class of $2n$-dimensional HK manifolds besides $(K3)^{[n]}$: generalized Kummers associated to a 2-dimensional

---

[4]Fujiki described $K3^{[2]}$ not as a Douady space but as the blow-up of the diagonal in the symmetric square of a K3 surface.

compact complex torus. Before defining generalized Kummers we recall that the Douady space $W^{[n]}$ comes with a cycle (Hilbert–Chow) map

$$W^{[n]} \xrightarrow{\kappa_n} W^{(n)}, \quad [Z] \mapsto \sum_{p \in W} \ell(\mathcal{O}_{Z,p})p, \tag{1.1.1}$$

where $W^{(n)}$ is the symmetric product of W. Now suppose that T is a 2-dimensional compact complex torus. We have the summation map $\sigma_n : W^{(n)} \to W$. Composing the two maps above (with $n+1$ replacing n) we get a locally (in the classical topology) trivial fibration $\sigma_{n+1} \circ \kappa_{n+1} : W^{[n+1]} \to W$. The $2n$-dimensional *generalized Kummer* associated to T is

$$K^{[n]}T := (\sigma_{n+1} \circ \kappa_{n+1})^{-1}(0). \tag{1.1.2}$$

The name is justified by the observation that if $n = 1$ then $K^{[1]}T$ is the Kummer surface associated to T (and hence a K3). Beauville [1983] proved that $K^{[n]}(T)$ is a HK manifold. Moreover if $n \geq 2$ then

$$b_2((\mathrm{K3})^{[n]}) = 23, \quad b_2(K^{[n]}T) = 7. \tag{1.1.3}$$

In particular $(\mathrm{K3})^{[n]}$ and $K^{[n]}T$ are not deformation equivalent as soon as $n \geq 2$. The second cohomology of these manifolds is described as follows. Let W be a compact complex surface. There is a "symmetrization map"

$$\mu_n : H^2(W; \mathbb{Z}) \longrightarrow H^2(W^{(n)}; \mathbb{Z}) \tag{1.1.4}$$

characterized by the following property. Let $\rho_n : W^n \to W^{(n)}$ be the quotient map and $\pi_i : W^n \to W$ be the i-th projection: then

$$\rho_n^* \circ \mu_n(\alpha) = \sum_{i=1}^{n} \pi_i^* \alpha, \quad \alpha \in H^2(W; \mathbb{Z}). \tag{1.1.5}$$

Composing with κ_n^* and extending scalars one gets an injection of integral Hodge structures

$$\tilde{\mu}_n := \kappa_n^* \circ \mu_n : H^2(W; \mathbb{C}) \longrightarrow H^2(W^{[n]}; \mathbb{C}). \tag{1.1.6}$$

This map is not surjective unless $n = 1$; we are missing the Poincaré dual of the exceptional set of κ_n, that is,

$$\Delta_n := \{[Z] \in W^{[n]} \mid Z \text{ is nonreduced}\}. \tag{1.1.7}$$

It is known that Δ_n is a prime divisor and that it is divisible[5] by 2 in $\mathrm{Pic}(W^{[n]})$:

$$\mathcal{O}_{W^{[n]}}(\Delta_n) \cong L_n^{\otimes 2}, \quad L_n \in \mathrm{Pic}(W^{[n]}). \tag{1.1.8}$$

---

[5]If $n = 2$ Equation (1.1.8) follows from existence of the double cover $Bl_{\mathrm{diag}}(S^2) \to S^{[2]}$ ramified over Δ_2.

Let $\xi_n := c_1(L_n)$; one has

$$H^2(W^{[n]}; \mathbb{Z}) = \tilde{\mu}_n H^2(W; \mathbb{Z}) \oplus \mathbb{Z}\xi_n \quad \text{if } H_1(W) = 0. \tag{1.1.9}$$

That describes $H^2((K3)^{[n]})$. Beauville proved that an analogous result holds for generalized Kummers, namely we have an isomorphism

$$H^2(T; \mathbb{Z}) \oplus \mathbb{Z} \xrightarrow{\sim} H^2(K^{[n]}T; \mathbb{Z}), \quad (\alpha, k) \mapsto (\tilde{\mu}_{n+1}(\alpha) + k\xi_{n+1})|_{K^{[n]}T}. \tag{1.1.10}$$

This description of the H^2 gives the following interesting result: if $n \geq 2$ the generic deformation of $S^{[n]}$ where S is a K3 is not isomorphic to $T^{[n]}$ for some other K3 surface T. In fact every deformation of $S^{[n]}$ obtained by deforming S keeps ξ_n of type $(1, 1)$, while, as noticed previously, the generic deformation of a HK manifold has no nontrivial integral $(1, 1)$-classes. (Notice that if S is a surface of general type then every deformation of $S^{[n]}$ is indeed obtained by deforming S, see [Fantechi 1995].)

1.2. *Mukai and beyond.* Mukai [1984; 1987b; 1987a] and Tyurin [1987] analyzed moduli spaces of semistable sheaves on projective K3's and abelian surfaces and obtained other examples of HK manifolds. Let S be a projective K3 and $\mathcal{M}$ the moduli space of $\mathcal{O}_S(1)$-semistable pure sheaves on S with assigned Chern character — by results of Gieseker, Maruyama, and Simpson, $\mathcal{M}$ has a natural structure of projective scheme. A nonzero canonical form on S induces a holomorphic symplectic 2-form on the open $\mathcal{M}^s \subset \mathcal{M}$ parametrizing stable sheaves (notice that $\mathcal{M}^s$ is smooth; see [Mukai 1984]). If $\mathcal{M}^s = \mathcal{M}$ then $\mathcal{M}$ is a HK variety;[6] in general it is not isomorphic (nor birational) to $(K3)^{[n]}$, but it can be deformed to $(K3)^{[n]}$ (here $2n = \dim \mathcal{M}$). See [Göttsche and Huybrechts 1996; O'Grady 1997; Yoshioka 1999]. Notice that $S^{[n]}$ may be viewed as a particular case of Mukai's construction by identifying it with the moduli space of rank-1 semistable sheaves on S with $c_1 = 0$ and $c_2 = n$. Notice also that these moduli spaces give explicit deformations of $(K3)^{[n]}$ which are not $(K3)^{[n]}$. Similarly one may consider moduli spaces of semistable sheaves on an abelian surface A: in the case when $\mathcal{M} = \mathcal{M}^s$ one gets deformations of the generalized Kummer. To be precise, it is not $\mathcal{M}$ which is a deformation of a generalized Kummer but rather one of its Beauville–Bogomolov factors. Explicitly we consider the map

$$\mathcal{M}(A) \xrightarrow{\alpha} A \times \widehat{A}, \quad [F] \mapsto \big(\mathrm{alb}(c_2(F) - c_2(F_0)), [\det F \otimes (\det F_0)^{-1}]\big), \tag{1.2.1}$$

where $[F_0] \in \mathcal{M}$ is a "reference" point and $\mathrm{alb} : CH_0^{\mathrm{hom}}(A) \to A$ is the Albanese map. Then α is a locally (classical topology) trivial fibration; Yoshioka [2001] proved that the fibers of α are deformations of a generalized Kummer. What can we say about moduli spaces such that $\mathcal{M} \neq \mathcal{M}^s$? The locus $(\mathcal{M} \setminus \mathcal{M}^s)$ parametrizing

---

[6]A HK variety is a projective HK manifold.

S-equivalence classes of semistable nonstable sheaves is the singular locus of $\mathcal{M}$ except for pathological choices of Chern character which do not give anything particularly interesting; thus we assume that $(\mathcal{M} \setminus \mathcal{M}^s)$ is the singular locus of $\mathcal{M}$. A natural question is the following: does there exist a crepant desingularization $\widetilde{\mathcal{M}} \to \mathcal{M}$? We constructed such a desingularization in [O'Grady 1999; 2003] (see also [Lehn and Sorger 2006]) for the moduli space $\mathcal{M}_4(S)$ of semistable rank-2 sheaves on a K3 surface S with $c_1 = 0$ and $c_2 = 4$ and for the moduli space $\mathcal{M}_2(A)$ of semistable sheaves on an abelian surface A with $c_1 = 0$ and $c_2 = 2$; the singularities of the moduli spaces are the same in both cases and both moduli spaces have dimension 10. Let M_{10} be our desingularization of $\mathcal{M}_4(S)$ where S is a K3. Since the resolution is crepant Mukai's holomorphic symplectic form on $(\mathcal{M}(S) \setminus \mathcal{M}(S)^s)$ extends to a holomorphic symplectic form on M_{10}. We proved in [O'Grady 1999] that M_{10} is HK; that is, it is simply connected and $h^{2,0}(M_{10}) = 1$. Moreover M_{10} is not a deformation of one of Beauville's examples because $b_2(M_{10}) = 24$. (We proved that $b_2(M_{10}) \geq 24$; later Rapagnetta [2008] proved that equality holds.) Next let A be an abelian surface and $\widetilde{\mathcal{M}}_2(A) \to \mathcal{M}_2(A)$ be our desingularization. Composing the map (1.2.1) for $\mathcal{M}(A) = \mathcal{M}_2(A)$ with the desingularization map we get a locally (in the classical topology) trivial fibration $\widetilde{\mathfrak{a}} : \widetilde{\mathcal{M}}_2(A) \to A \times \widehat{A}$; let M_6 be any fiber of $\widetilde{\mathfrak{a}}$. We proved in [O'Grady 1999] that M_6 is HK and that $b_2(M_6) = 8$; thus M_6 is not a deformation of one of Beauville's examples. We point out that while all Betti and Hodge numbers of Beauville's examples are known [Göttsche 1994] the same is not true of our examples (Rapagnetta [2007] computed the Euler characteristic of M_6). Of course there are examples of moduli spaces $\mathcal{M}$ with $\mathcal{M} \neq \mathcal{M}^s$ in any even dimension; one would like to desingularize them and produce many more deformation classes of HK manifolds. Kaledin, Lehn, and Sorger [Kaledin et al. 2006] have determined exactly when the moduli space has a crepant desingularization. Combining their results with those of [Perego and Rapagnetta 2010] one gets that if there is a crepant desingularization then it is a deformation of M_{10} if the surface is a K3, while in the case of an abelian surface the fibers of map (1.2.1) composed with the desingularization map are deformations of M_6.[7] In fact all known examples of HK manifolds are deformations either of Beauville's examples or of ours.

1.3. Mukai flops. Let X be a HK manifold of dimension $2n$ containing a submanifold Z isomorphic to $\mathbb{P}^n$. The *Mukai flop of Z* (introduced in [Mukai 1984]) is a bimeromorphic map $X \dashrightarrow X^\vee$ which is an isomorphism away from Z and

---

[7]To be precise, their result holds if the polarization of the surface is "generic" relative to the chosen Chern character; with this hypothesis the singular locus of $\mathcal{M}$ is, so to speak, as small as possible

replaces Z by the dual plane $Z^\vee := (\mathbb{P}^n)^\vee$. Explicitly let $\tau : \widetilde{X} \to X$ be the blow-up of Z and $E \subset \widetilde{X}$ be the exceptional divisor. Since Z is Lagrangian the symplectic form on X defines an isomorphism $N_{Z/X} \cong \Omega_Z = \Omega_{\mathbb{P}^n}$. Thus

$$E \cong \mathbb{P}(N_{Z/X}) = \mathbb{P}(\Omega_{\mathbb{P}^n}) \subset \mathbb{P}^n \times (\mathbb{P}^n)^\vee. \tag{1.3.1}$$

Hence E is a $\mathbb{P}^{n-1}$-fibration in two different ways: we have $\pi : E \to \mathbb{P}^n$, i.e., the restriction of τ to E and $\rho : E \to (\mathbb{P}^n)^\vee$. A straightforward computation shows that the restriction of $N_{E/\widetilde{X}}$ to a fiber of ρ is $\mathcal{O}_{\mathbb{P}^{n-1}}(-1)$. By the Fujiki–Nakano contractibility criterion there exists a proper map $\tau^\vee : \widetilde{X} \to X^\vee$ to a complex manifold $X^\vee$ which is an isomorphism outside E and which restricts to ρ on E. Clearly $\tau^\vee(E)$ is naturally identified with $Z^\vee$ and we have a bimeromorphic map $X \dashrightarrow X^\vee$ which defines an isomorphism $(X \setminus Z) \xrightarrow{\sim} (X^\vee \setminus Z^\vee)$. Summarizing, we have the commutative diagram

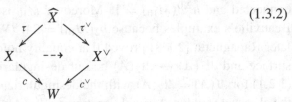

$$\tag{1.3.2}$$

where $c : X \to W$ and $c^\vee : X^\vee \to W$ are the contractions of Z and $Z^\vee$ respectively — see the Introduction of [Wierzba and Wiśniewski 2003]. It follows that $X^\vee$ is simply connected and a holomorphic symplectic form on X gives a holomorphic symplectic form on $X^\vee$ spanning $H^0(\Omega_{X^\vee}^2)$; thus $X^\vee$ is HK if it is Kähler. We give an example with X and $X^\vee$ projective. Let $f : S \to \mathbb{P}^2$ be a double cover branched over a smooth sextic and $\mathcal{O}_S(1) := f^*\mathcal{O}_{\mathbb{P}^2}(1)$: thus S is a K3 of degree 2. Let $X := S^{[2]}$ and $\mathcal{M}$ be the moduli space of pure 1-dimensional $\mathcal{O}_S(1)$-semistable sheaves on S with typical member $\iota_*\mathcal{L}$ where $\iota : C \hookrightarrow S$ is the inclusion of $C \in |\mathcal{O}_S(1)|$ and $\mathcal{L}$ is a line bundle on C of degree 2. We have a natural rational map

$$\phi : S^{[2]} \dashrightarrow \mathcal{M} \tag{1.3.3}$$

which associates to $[W] \in S^{[2]}$ the sheaf $\iota_*\mathcal{L}$ where C is the unique curve containing W (uniqueness requires W to be generic!) and $\mathcal{L} := \mathcal{O}_C(W)$. If every divisor in $|\mathcal{O}_S(1)|$ is prime (i.e., the branch curve of f has no tritangents) then $\mathcal{M}$ is smooth (projective) and the rational map ϕ is identified with the flop of

$$Z := \{f^{-1}(p) \mid p \in \mathbb{P}^2\}. \tag{1.3.4}$$

Wierzba and Wiśniewsky [2003] have proved that any birational map between HK four-folds is a composition of Mukai flops. In higher dimensions Mukai [1984] defined more general flops in which the indeterminacy locus is a fibration in projective spaces. Markman [2001] constructed *stratified Mukai flops*.

2. General theory

It is fair to state that there are three main ingredients in the general theory of HK manifolds developed by Bogomolov, Beauville, Fujiki, Huybrechts and others:

(1) Deformations are unobstructed (Bogomolov's Theorem).

(2) The canonical Bogomolov–Beauville quadratic form on H^2 of a HK manifold (see the next subsection).

(3) Existence of the twistor family on a HK manifold equipped with a Kähler class: this is a consequence of Yau's solution of Calabi's conjecture.

2.1. *Topology.* Let X be a HK-manifold of dimension $2n$. Beauville [1983] and Fujiki [1987] proved that there exist an integral indivisible quadratic form

$$q_X : H^2(X) \to \mathbb{C} \qquad (2.1.1)$$

(cohomology is with complex coefficients) and $c_X \in \mathbb{Q}_+$ such that

$$\int_X \alpha^{2n} = c_X \frac{(2n)!}{n! 2^n} q_X(\alpha)^n, \qquad \alpha \in H^2(X). \qquad (2.1.2)$$

This equation determines c_X and q_X with no ambiguity unless n is even. If n is even then q_X is determined up to ± 1: one singles out one of the two choices by imposing the inequality $q_X(\sigma + \overline{\sigma}) > 0$ for σ a holomorphic symplectic form. The *Beauville–Bogomolov* form and the *Fujiki constant* of X are q_X and c_X respectively. We note that the equation in (2.1.2) is equivalent (by polarization) to

$$\int_X \alpha_1 \wedge \cdots \wedge \alpha_{2n}$$
$$= c_X \sum_{\sigma \in \mathfrak{R}_{2n}} (\alpha_{\sigma(1)}, \alpha_{\sigma(2)})_X \cdot (\alpha_{\sigma(3)}, \alpha_{\sigma(4)})_X \cdots (\alpha_{\sigma(2n-1)}, \alpha_{\sigma(2n)})_X, \qquad (2.1.3)$$

where $(\cdot, \cdot)_X$ is the symmetric bilinear form associated to q_X and $\mathfrak{R}_{2n}$ is a set of representatives for the left cosets of the subgroup $\mathscr{G}_{2n} < \mathscr{S}_{2n}$ of permutations of $\{1, \ldots, 2n\}$ generated by transpositions $(2i - 1, 2i)$ and by products of transpositions $(2i - 1, 2j - 1)(2i, 2j)$ — in other words in the right-hand side of (2.1.3) we avoid repeating addends which are equal.[8] The existence of q_X, c_X is by no means trivial; we sketch a proof. Let $f : \mathscr{X} \to T$ be a deformation of X representing $\mathrm{Def}(X)$; more precisely letting $X_t := f^{-1}\{t\}$ for $t \in T$, we are given $0 \in T$, an isomorphism $X_0 \xrightarrow{\sim} X$ and the induced map of germs $(T, 0) \to \mathrm{Def}(X)$ is an isomorphism. In particular T is smooth in 0 and hence we may assume that it is a polydisk. The Gauss–Manin connection defines an

---

[8]In defining c_X we have introduced a normalization which is not standard in order to avoid a combinatorial factor in (2.1.3).

integral isomorphism $\phi_t : H^2(X) \xrightarrow{\sim} H^2(X_t)$. The *local period map* of X is given by

$$T \xrightarrow{\pi} \mathbb{P}(H^2(X)), \quad t \mapsto \phi_t^{-1} H^{2,0}(X_t). \tag{2.1.4}$$

By infinitesimal Torelli — see (0.0.3) — Im π is an analytic hypersurface in an open (classical topology) neighborhood of $\pi(0)$ and hence its Zariski closure $V = \overline{\mathrm{Im}\,\pi}$ is either all of $\mathbb{P}(H^2(X))$ or a hypersurface. One shows that the latter holds by considering the (nonzero) degree-$2n$ homogeneous polynomial

$$H^2(X) \xrightarrow{G} \mathbb{C}, \quad \alpha \mapsto \int_X \alpha^{2n}. \tag{2.1.5}$$

In fact if $\sigma_t \in H^{2,0}(X_t)$ then

$$\int_{X_t} \sigma_t^{2n} = 0 \tag{2.1.6}$$

by type consideration and it follows by Gauss–Manin parallel transport that G vanishes on V. Thus $I(V) = (F)$ where F is an irreducible homogeneous polynomial. By considering the derivative of the period map (0.0.3) one checks easily that V is not a hyperplane and hence deg $F \geq 2$. On the other hand type consideration gives something stronger than (2.1.6), namely

$$\int_{X_t} \sigma_t^{n+1} \wedge \alpha_1 \cdots \wedge \alpha_{n-1} = 0, \quad \alpha_1, \ldots, \alpha_{n-1} \in H^2(X_t). \tag{2.1.7}$$

It follows that all the derivatives of G up to order $(n-1)$ included vanish on V. Since deg $G = 2n$ and deg $F \geq 2$ it follows that $G = c \cdot F^n$ and deg $F = 2$. By integrality of G there exists $\lambda \in \mathbb{C}^*$ such that $c_X := \lambda c$ is rational positive, $q_X := \lambda \cdot F$ is integral indivisible and (2.1.2) is satisfied.

Of course if X is a K3 then q_X is the intersection form of X (and $c_X = 1$). In general q_X gives $H^2(X; \mathbb{Z})$ a structure of lattice just as in the well-known case of K3 surfaces. Suppose that X and Y are deformation equivalent HK-manifolds: it follows from (2.1.2) that $c_X = c_Y$ and the lattices $H^2(X; \mathbb{Z})$, $H^2(Y; \mathbb{Z})$ are isometric (see the comment following (2.1.2) if n is even). Consider the case when $X = (K3)^{[n]}$; then $\tilde{\mu}_n$ is an isometry, $\xi_n \perp \mathrm{Im}\,\tilde{\mu}_n$ and $q_X(\xi_n) = -2(n-1)$, i.e.,

$$H^2(S^{[n]}; \mathbb{Z}) \cong U^3 \widehat{\oplus} E_8\langle -1 \rangle^2 \widehat{\oplus} \langle -2(n-1) \rangle \tag{2.1.8}$$

where $\widehat{\oplus}$ denotes orthogonal direct sum, U is the hyperbolic plane and $E_8\langle -1 \rangle$ is the unique rank-8 negative definite unimodular even lattice. Moreover the Fujiki constant is

$$c_{S^{[n]}} = 1. \tag{2.1.9}$$

In [Rapagnetta 2008] the reader will find the B-B quadratic form and Fujiki constant of the other known deformation classes of HK manifolds.

Remark 2.1. Let X be a HK manifold of dimension $2n$ and $\omega \in H_{\mathbb{R}}^{1,1}(X)$ be a Kähler class.

(1) Equation (2.1.2) gives that, with respect to $(\cdot, \cdot)_X$

$$H^{p,q}(X) \perp H^{p',q'}(X) \quad \text{unless } (p', q') = (2 - p, 2 - q). \qquad (2.1.10)$$

(2) $q_X(\omega) > 0$. In fact let σ be generator of $H^{2,0}(X)$; by (2.1.3) and item (1) above we have

$$0 < \int_X \sigma^{n-1} \wedge \overline{\sigma}^{n-1} \wedge \omega^2 = c_X(n-1)! \, (\sigma, \overline{\sigma})_X q_X(\omega). \qquad (2.1.11)$$

Since $c_X > 0$ and $(\sigma, \overline{\sigma})_X > 0$ we get $q_X(\omega) > 0$ as claimed.

(3) The index of q_X is $(3, b_2(X) - 3)$ (i.e., that is the index of its restriction to $H^2(X; \mathbb{R})$). In fact applying (2.1.3) to $\alpha_1 = \cdots = \alpha_{2n-1} = \omega$ and arbitrary α_{2n} we get that $\omega^{\perp}$ is equal to the primitive cohomology $H_{pr}^2(X)$ (primitive with respect to ω). On the other hand (2.1.3) with $\alpha_1 = \cdots = \alpha_{2n-2} = \omega$ and $\alpha_{2n-1}, \alpha_{2n} \in \omega^{\perp}$ gives that a positive multiple of $q_X|_{\omega^{\perp}}$ is equal to the standard quadratic form on $H_{pr}^2(X)$. By the Hodge index Theorem it follows that the restriction of q_X to $\omega^{\perp} \cap H^2(X; \mathbb{R})$ has index $(2, b_2(X) - 3)$. Since $q_X(\omega) > 0$ it follows that q_X has index $(3, b_2(X) - 3)$.

(4) Let D be an effective divisor on X; then $(\omega, D)_X > 0$. (Of course $(\omega, D)_X$ denotes $(\omega, c_1(\mathcal{O}_X(D)))_X$.) In fact the inequality follows from the inequality $\int_D \omega^{2n-1} > 0$ together with (2.1.3) and item (2) above.

(5) Let $f : X \dashrightarrow Y$ be a birational map where Y is a HK manifold. Since X and Y have trivial canonical bundle f defines an isomorphism $U \xrightarrow{\sim} V$ where $U \subset X$ and $V \subset Y$ are open sets with complements of codimension at least 2. It follows that f induces an isomorphism $f^* : H^2(Y; \mathbb{Z}) \xrightarrow{\sim} H^2(X; \mathbb{Z})$; f^* is an isometry of lattices, see Lemma 2.6 of [Huybrechts 1999].

The proof of existence of q_X and c_X may be adapted to prove the following useful generalization of (2.1.2).

Proposition 2.2. *Let X be a HK manifold of dimension $2n$. Let $\mathscr{X} \to T$ be a representative of the deformation space of X. Suppose that*

$$\gamma \in H_{\mathbb{R}}^{p,p}(X)$$

is a nonzero class which remains of type (p, p) under Gauss–Manin parallel transport (such as the Chern class $c_p(X)$). Then p is even and moreover there exists $c_\gamma \in \mathbb{R}$ such that

$$\int_X \gamma \wedge \alpha^{2n-p} = c_\gamma q_X(\alpha)^{n-p/2}. \qquad (2.1.12)$$

Our next topic is Verbitsky's theorem. Let X be a HK-manifold of dimension $2n$. Our sketch proof of (2.1.2) shows that

$$\alpha \in H^2(X) \text{ and } q_X(\alpha) = 0 \implies \alpha^{n+1} = 0 \text{ in } H^{2n+2}(X). \qquad (2.1.13)$$

In fact, using the notation in the proof of (2.1.2), we have $0 = \sigma_t^{n+1} \in H^{2n+2}(X_t)$; hence, by Gauss–Manin transport we get $0 = (\psi_t^{-1}\sigma_t)^{n+1} \in H^{2n+2}(X)$. Since the set $\{\psi_t^{-1}\sigma_t \mid t \in T\}$ is Zariski dense in the zero-set $V(q_X) \subset H^2(X)$ we get (2.1.13). Let $I \subset \text{Sym}^\bullet H^2(X)$ be the ideal generated by α^{n+1} where $\alpha \in H^2(X)$ and $q_X(\alpha) = 0$:

$$I := \langle \{\alpha^{n+1} \mid \alpha \in H^2(X), \quad q_X(\alpha) = 0\} \rangle. \qquad (2.1.14)$$

By (2.1.13) we have a natural map of $\mathbb{C}$-algebras

$$\text{Sym}^\bullet H^2(X)/I \longrightarrow H^\bullet(X). \qquad (2.1.15)$$

Theorem 2.3 [Verbitsky 1996] (see also [Bogomolov 1996]). *The map (2.1.15) is injective.*

In particular we get that cup-product defines an injection

$$\bigoplus_{q=0}^{n} \text{Sym}^q H^2(X) \hookrightarrow H^\bullet(X). \qquad (2.1.16)$$

S. M. Salamon proved that there is a nontrivial linear constraint on the Betti numbers of a compact Kähler manifold carrying a holomorphic symplectic form (for example a HK manifold); the proof consists in a clever application of the Hirzebruch–Riemann–Roch formula to the sheaves Ω_X^p and the observation that the symplectic form induces an isomorphism $\Omega_X^p \cong \Omega_X^{2n-p}$ where $2n = \dim X$.[9]

Theorem 2.4 [Salamon 1996]. *Let X be a compact Kähler manifold of dimension $2n$ carrying a holomorphic symplectic form. Then*

$$nb_{2n}(X) = 2 \sum_{i=1}^{2n} (-1)^i (3i^2 - n) b_{2n-i}(X). \qquad (2.1.17)$$

The following corollary of Verbitsky's and Salamon's results was obtained by Beauville (unpublished) and Guan [2001].

Corollary 2.5 (Beauville and Guan). *Let X be a HK 4-fold. Then $b_2(X) \leq 23$. If equality holds then $b_3(X) = 0$ and moreover the map*

$$\text{Sym}^2 H^2(X; \mathbb{Q}) \longrightarrow H^4(X; \mathbb{Q}) \qquad (2.1.18)$$

---

[9]A nonzero section of the canonical bundle defines an isomorphism $\Omega_X^{2n-p} \cong (\Omega_X^p)^\vee = \bigwedge^p T_X$ and the symplectic form defines an isomorphism $T_X \cong \Omega_X$ and hence $\bigwedge^p T_X \cong \Omega_X^p$.

induced by cup-product is an isomorphism.

Proof. Let $b_i := b_i(X)$. Salamon's equation (2.1.17) for X reads

$$b_4 = 46 + 10b_2 - b_3. \tag{2.1.19}$$

By Verbitsky's theorem — see (2.1.16) — we have

$$\binom{b_2 + 1}{2} \leq b_4. \tag{2.1.20}$$

Replacing b_4 by the right-hand side of (2.1.19) we get that

$$b_2^2 + b_2 \leq 92 + 20b_2 - 2b_3 \leq 92 + 20b_2. \tag{2.1.21}$$

It follows that $b_2 \leq 23$ and that if equality holds then $b_3 = 0$. Suppose that $b_2 = 23$: then $b_4 = 276$ by (2.1.19) and hence (2.1.18) follows from Verbitsky's theorem (Theorem 2.3). $\qquad\square$

Guan [2001] has obtained other restrictions on $b_2(X)$ for a HK four-fold X: for example, $8 < b_2(X) < 23$ is "forbidden".

2.2. The Kähler cone. Let X be a HK manifold of dimension $2n$. The convex cone $\mathcal{K}_X \subset H_{\mathbb{R}}^{1,1}(X)$ of Kähler classes is the *Kähler cone of X*. The inequality in (2.1.2) together with Remark 2.1(3) gives that the restriction of q_X to $H_{\mathbb{R}}^{1,1}(X)$ is nondegenerate of signature $(1, b_2(X) - 3)$; it follows that the cone

$$\{\alpha \in H_{\mathbb{R}}^{1,1}(X) \mid q_X(\alpha) > 0\} \tag{2.2.1}$$

has two connected components. By Remark 2.1(2) $\mathcal{K}_X$ is contained in (2.2.1). Since $\mathcal{K}_X$ is convex it is contained in a single connected component of (2.2.1); that component is the *positive cone* $\mathcal{C}_X$.

Theorem 2.6 [Huybrechts 2003a]. *Let X be a HK manifold. Let $\mathcal{X} \to T$ be a representative of* $\mathrm{Def}(X)$ *with T irreducible. If $t \in T$ is very general (i.e., outside a countable union of proper analytic subsets of T) then*

$$\mathcal{K}_{X_t} = \mathcal{C}_{X_t}. \tag{2.2.2}$$

Proof. Let $0 \in T$ be the point such that $X_0 \cong X$ and the induced map of germs $(T, 0) \to \mathrm{Def}(X)$ is an isomorphism.[10] By shrinking T around 0 if necessary we may assume that T is simply connected and that $\mathcal{X} \to T$ represents $\mathrm{Def}(X_t)$ for every $t \in T$. In particular the Gauss–Manin connection gives an isomorphism $P_t : H^{\bullet}(X; \mathbb{Z}) \xrightarrow{\sim} H^{\bullet}(X_t; \mathbb{Z})$ for every $t \in T$. Given $\gamma \in H^{2p}(X; \mathbb{Z})$ we let

$$T_\gamma := \{t \in T \mid P_t(\gamma) \text{ is of type } (p, p)\}. \tag{2.2.3}$$

---

[10]The map $(T, 0) \to \mathrm{Def}(X)$ depends on the choice of an isomorphism $f : X_0 \xrightarrow{\sim} X$ but whether it is an isomorphism or not is independent of f.

Let

$$t \in \left(T \setminus \bigcup_{T_\gamma \neq T} T_\gamma \right) \tag{2.2.4}$$

and $Z \subset X_t$ be a closed analytic subset of codimension p; we claim that

$$\int_Z \alpha^{2n-p} > 0 \quad \text{if } q_{X_t}(\alpha) > 0. \tag{2.2.5}$$

In fact let $\gamma \in H_{\mathbb{R}}^{p,p}(X_t)$ be the Poincaré dual of Z. By (2.2.4) γ remains of type (p, p) for every deformation of X_t; by Proposition 2.2 p is even and moreover there exists $c_\gamma \in \mathbb{R}$ such that

$$\int_Z \alpha^{2n-p} = c_\gamma q_X(\alpha)^{n-p/2} \quad \text{for all } \alpha \in H^2(X_t). \tag{2.2.6}$$

Let ω be a Kähler class. Since $0 < \int_Z \omega^{2n-p}$ and $0 < q_X(\omega)$ we get that $c_\gamma > 0$; thus (2.2.5) follows from (2.2.6). Now apply Demailly and Paun's version of the Nakai–Moishezon ampleness criterion [Demailly and Paun 2004]: $\mathcal{K}_{X_t}$ is a connected component of the set $P(X_t) \subset H_{\mathbb{R}}^{1,1}(X_t)$ of classes α such that $\int_Z \alpha^{2n-p} > 0$ for all closed analytic subsets $Z \subset X_t$ (here $p = \text{cod}(Z, X_t)$). Let t be as in (2.2.4). By (2.2.5) $P(X_t) = \mathcal{C}_{X_t} \bigsqcup (-\mathcal{C}_{X_t})$; since $\mathcal{K}_{X_t} \subset \mathcal{C}_{X_t}$ we get the proposition. $\qquad\qquad\square$

Theorem 2.6 leads to this *projectivity criterion*:

Theorem 2.7 [Huybrechts 1999]. *A HK manifold X is projective if and only if there exists a (holomorphic) line bundle L on X such that $q_X(c_1(L)) > 0$.*

Boucksom, elaborating on ideas of Huybrechts, gave the following characterization of $\mathcal{K}_X$ for arbitrary X:

Theorem 2.8 [Boucksom 2001]. *Let X be a HK manifold. A class $\alpha \in H_{\mathbb{R}}^{1,1}(X)$ is Kähler if and only if it belongs to the positive cone $\mathcal{C}_X$ and moreover $\int_C \alpha > 0$ for every rational curve C.*[11]

One would like to have a numerical description of the Kähler (or ample) cone as in the 2-dimensional case. There is this result:

Theorem 2.9 [Hassett and Tschinkel 2009b]. *Let X be a HK variety deformation equivalent to $K3^{[2]}$ and L_0 an ample line bundle on X. Let L be a line bundle on X such that $c_1(L) \in \mathcal{C}_X$. Suppose that $(c_1(L), \alpha)_X > 0$ for all $\alpha \in H_{\mathbb{Z}}^{1,1}(X)$ such that $(c_1(L_0), \alpha)_X > 0$ and*

(a) $q_X(\alpha) = -2$ *or*

(b) $q_X(\alpha) = -10$ *and* $(\alpha, H^2(X; \mathbb{Z}))_X = 2\mathbb{Z}$.

Then L is ample.

---

[11] A curve is rational if it is irreducible and its normalization is rational.

Hassett and Tschinkel [2001] conjectured that the converse of this theorem holds, in the sense that its conditions are also necessary for L to be ample. We explain the appearance of the conditions in the theorem and why one expects that the converse holds. We start with (a). Let X be a HK manifold deformation equivalent to K3$^{[2]}$ and L a line bundle on X: Hirzebruch–Riemann–Roch for X reads

$$\chi(L) = \tfrac{1}{8}(q(L)+4)(q(L)+6).\tag{2.2.7}$$

(We let $q = q_X$.) It follows that $\chi(L) = 1$ if and only if $q(L) = -2$ or $q(L) = -8$.

Conjecture 2.10 (Folk). Let X be a HK manifold deformation equivalent to K3$^{[2]}$. Let L be a line bundle on X such that $q_X(L) = -2$.

(1) If $(c_1(L), H^2(X; \mathbb{Z}))_X = \mathbb{Z}$ then either L or L^{-1} has a nonzero section.

(2) If $(c_1(L), H^2(X; \mathbb{Z}))_X = 2\mathbb{Z}$ then either L^2 or L^{-2} has a nonzero section. (Notice that $q_X(L^{\pm 2}) = -8$.)

If this conjecture holds then given $\alpha \in H^{1,1}_{\mathbb{Z}}(X)$ with $q_X(\alpha) = -2$ we have that either $(\alpha, \cdot)_X$ is strictly positive or strictly negative on $\mathcal{K}_X$; in particular the condition corresponding to Theorem 2.9(a) is necessary for a line bundle to be ample. Below are examples of line bundles satisfying the items (1) and (2) in the conjecture.

Example. Let S be a K3 containing a smooth rational curve C and $X = S^{[2]}$. Let

$$D := \{[Z] \in S^{[2]} \mid Z \cap C \neq \varnothing\}.\tag{2.2.8}$$

Let $L := \mathcal{O}_X(D)$; then $c_1(L) = \tilde{\mu}_2(c_1(\mathcal{O}_S(C)))$, where $\tilde{\mu}_2$ is given by (1.1.6). Since $\tilde{\mu}_2$ is an isometry we have $q_X(L) = C \cdot C = -2$; moreover $(c_1(L), H^2(X; \mathbb{Z}))_X = \mathbb{Z}$. For another example see Remark 3.3(5).

Example. Let S be a K3 and $X = S^{[2]}$. Let L_2 be the square root of $\mathcal{O}_X(\Delta_2)$ where $\Delta_2 \subset S^{[2]}$ is the divisor parametrizing nonreduced subschemes — thus $c_1(L_2) = \xi_2$. Then $q(L_2) = -2$ and L_2^2 has "the" nonzero section vanishing on Δ_2. Notice that neither L_2 nor L_2^{-1} has a nonzero section.

Summarizing: line bundles of square -2 on a HK deformation of K3$^{[2]}$ should be similar to (-2)-classes on a K3. (Recall that if L is a line bundle on a K3 with $c_1(L)^2 = -2$ then by Hirzebruch–Riemann–Roch and Serre duality either L or L^{-1} has a nonzero section.) Next we explain Theorem 2.9(b). Suppose that X is a HK deformation of K3$^{[2]}$ and that $Z \subset X$ is a closed submanifold isomorphic to $\mathbb{P}^2$ — see Section 1.3. Let $C \subset Z$ be a line. Since $(\cdot, \cdot)_X$ is nondegenerate (but not unimodular!) there exists $\beta \in H^2(X; \mathbb{Q})$ such that

$$\int_C \gamma = (\beta, \gamma)_X \quad \text{for all } \gamma \in H^2(X).\tag{2.2.9}$$

One proves that

$$q_X(\beta) = -\tfrac{5}{2}. \tag{2.2.10}$$

This follows from the isomorphism (2.1.18) and the good properties of deformations of HK manifolds; see [Hassett and Tschinkel 2009b, Section 4]. Since $(\beta, H^2(X; \mathbb{Z}))_X = \mathbb{Z}$ and the discriminant of $(\,\cdot\,, \cdot\,)_X$ is 2 we have $2\beta \in H^2(X; \mathbb{Z})$; thus $\alpha := 2\beta$ is as in Theorem 2.9(b) and if L is ample then $0 < \int_C c_1(L) = \tfrac{1}{2}(c_1(L), \alpha)_X$.

Remark 2.11. Hassett and Tschinkel [2009a] stated conjectures that extend Theorem 2.9 and its converse to general HK varieties; in particular they have given a conjectural numerical description of the effective cone of a HK variety. The papers [Boucksom 2004; Druel 2011] contain key results in this circle of ideas. Markman [2009, Section 1.4] formulated a conjecture on HK manifolds deformation equivalent to $(K3)^{[n]}$ which generalizes Conjecture 2.10 and provided a proof relying on Verbitsky's global Torelli.

We close the section by stating a beautiful result of Huybrechts [2003c] — the proof is based on results on the Kähler cone and uses in an essential way the existence of the twistor family.

Theorem 2.12. *Let X and Y be bimeromorphic HK manifolds. Then X and Y are deformation equivalent.*

3. Complete families of HK varieties

A pair (X, L), where X is a HK variety and L is a primitive[12] ample line bundle on X with $q_X(L) = d$, is a *HK variety of degree d*; an isomorphism $(X, L) \xrightarrow{\sim} (X', L')$ between HK's of degree d consists of an isomorphism $f : X \xrightarrow{\sim} X'$ such that $f^* L' \cong L$. A family of HK varieties of degree d is a pair

$$(f : \mathscr{X} \to T, \mathscr{L}) \tag{3.0.1}$$

where $\mathscr{X} \to T$ is a family of HK varieties deformation equivalent to a fixed HK manifold X and $\mathscr{L}$ is a line bundle such that (X_t, L_t) is a HK variety of degree d for every $t \in T$ (here $X_t := f^{-1}(t)$ and $L_t := \mathscr{L}|_{X_t}$) — we say that it is a family of HK varieties if we are not interested in the value of $q_X(L_t)$. The deformation space of (X, L) is a codimension-1 smooth subgerm $\mathrm{Def}(X, L) \subset \mathrm{Def}(X)$ with tangent space the kernel of the map (0.0.4) with $\alpha = c_1(L)$. The family (3.0.1) is *locally complete* if given any $t_0 \in T$ the map of germs $(T, t_0) \to \mathrm{Def}(X_{t_0}, L_{t_0})$ is surjective, it is *globally complete* if given any HK variety (Y, L) of degree d with Y deformation equivalent to X there exists $t_0 \in T$ such that $(Y, L) \cong (X_{t_0}, L_{t_0})$. In dimension 2 — that is, for K3 surfaces — one has explicit globally complete

---

[12]That is, $c_1(L)$ is indivisible in $H^2(X; \mathbb{Z})$.

families of low degree: If $d = 2$ the family of double covers $S \to \mathbb{P}^2$ branched over a smooth sextic will do,[13] if $d = 4$ we may consider the family of smooth quartic surfaces $S \subset \mathbb{P}^3$ with the addition of certain "limit" surfaces (double covers of smooth quadrics and certain elliptic K3's) corresponding to degenerate quartics (double quadrics and the surface swept out by tangents to a rational normal cubic curve respectively). The list goes on for quite a few values of d, see [Mukai 1988; 2006] and then it necessarily stops — at least in this form — because moduli spaces of high-degree K3's are not unirational [Gritsenko et al. 2007]. We remark that in low degree one shows "by hand" that there exists a globally complete family which is irreducible; the same is true in arbitrary degree but I know of no elementary proof, the most direct argument is via global Torelli. What is the picture in dimensions higher than two? Four distinct (modulo obvious equivalence) locally complete families of higher-dimensional HK varieties have been constructed — they are all deformations of K3[2]. The families are the following:

(1) In [O'Grady 2006] we constructed the family of double covers of certain special sextic hypersurfaces in $\mathbb{P}^5$ that we named EPW-sextics (they had been introduced by Eisenbud, Popescu, and Walter [2001]). The polarization is the pull-back of $\mathcal{O}_{\mathbb{P}^5}(1)$; its degree is 2.

(2) Let $Z \subset \mathbb{P}^5$ be a smooth cubic hypersurface; Beauville and Donagi [1985] proved that the variety parametrizing lines on Z is a deformation of K3[2]. The polarization is given by the Plücker embedding: it has degree 6.

(3) Let σ be a generic 3-form on $\mathbb{C}^{10}$; Debarre and Voisin [Debarre and Voisin 2010] proved that the set $Y_\sigma \subset Gr(6, \mathbb{C}^{10})$ parametrizing subspaces on which σ vanishes is a deformation of K3[2]. The polarization is given by the Plücker embedding: it has degree 22.

(4) Let $Z \subset \mathbb{P}^5$ be a generic cubic hypersurface; Iliev and Ranestad [2001; 2007] have proved that the variety of sums of powers $VSP(Z, 10)^{14}$ is a deformation of K3[2]. For the polarization we refer to [Iliev and Ranestad 2007]; the degree is 38 (unpublished computation by Iliev, Ranestad and van Geemen).

For each of these families — more precisely for the family obtained by adding "limits" — one might ask whether it is globally complete for HK varieties of the given degree which are deformations of K3[2]. As formulated the answer is negative with the possible exception of our family, for a trivial reason: in the

---

[13] In order to get a global family we must go to a suitable double cover of the parameter space of sextic curves.

[14] $VSP(Z, 10)$ parametrizes 9-dimensional linear spaces of $|\mathcal{O}_{\mathbb{P}^5}(3)|$ which contain Z and are 10-secant to the Veronese $\{[L^3] \mid L \in (H^0(\mathcal{O}_{\mathbb{P}^5}(1)) \setminus \{0\})\}$.

lattice $L := H^2(\mathrm{K3}^{[2]}; \mathbb{Z})$ the orbit of a primitive vector v under the action of $O(L)$ is determined by the value of the B-B form $q(v)$ plus the extra information on whether

$$(v, L) = \begin{cases} \mathbb{Z} & or \\ 2\mathbb{Z} \end{cases} \tag{3.0.2}$$

In the first case one says that the *divisibility of v* is 1, in the second case that it is 2; if the latter occurs then $q(v) \equiv 6 \pmod 8$. Thus the divisibility of the polarization in family (1) above equals 1; on the other hand it equals 2 for families (2)–(4). The correct question regarding global completeness is the following. Let X be a HK deformation of $\mathrm{K3}^{[2]}$ with an ample line bundle L such that either $q(L) = 2$ or $q(L) \in \{6, 22, 38\}$ and the divisibility of $c_1(L)$ is equal to 2: does there exist a variety Y parametrized by one of the families above — or a limit of such — and an isomorphism $(X, L) \cong (Y, \mathcal{O}_Y(1))$? Yes, by Verbitsky's global Torelli and Markmans' monodromy computations.

None of the families above is as easy to construct as are the families of low-degree K3 surfaces. There is the following Hodge-theoretic explanation. In order to get a locally complete family of varieties one usually constructs complete intersections (or sections of ample vector bundles) in homogeneous varieties: by Lefschetz's hyperplane theorem such a construction will never produce a higher-dimensional HK. On the other hand the families (1), (2), and (3) are related to complete intersections as follows (I do not know whether one may view the Iliev–Ranestad family from a similar perspective). First if $f : X \to Y$ is a double EPW-sextic (family (1) above) then f is the quotient map of an involution $X \to X$ which has one-dimensional $(+1)$-eigenspace on $H^2(X)$ — in particular it kills $H^{2,0}$ — and "allows" the quotient to be a hypersurface. Regarding family (2): let $Z \subset \mathbb{P}^5$ be a smooth cubic hypersurface and X the variety of lines on Z, the incidence correspondence in $Z \times X$ induces an isomorphism of the primitive Hodge structures $H^4(Z)_{pr} \xrightarrow{\sim} H^2(X)_{pr}$. Thus a Tate twist of $H^2(X)_{pr}$ has become the primitive intermediate cohomology of a hypersurface. A similar comment applies to the Debarre–Voisin family (and there is a similar incidence-type construction of double EPW-sextics given in [Iliev and Manivel 2009]).

In this section we will describe in some detail the family of double EPW-sextics and we will say a few words about analogies with the Beauville–Donagi family.

3.1. *Double EPW-sextics, I.*
We start by giving the definition of EPW-sextic [Eisenbud et al. 2001]. Let V be a 6-dimensional complex vector space. We choose a volume form $vol : \bigwedge^6 V \xrightarrow{\sim} \mathbb{C}$ and we equip $\bigwedge^3 V$ with the symplectic form

$$(\alpha, \beta)_V := vol(\alpha \wedge \beta). \tag{3.1.1}$$

Let $\mathbb{LG}(\bigwedge^3 V)$ be the symplectic Grassmannian parametrizing Lagrangian subspaces of $\bigwedge^3 V$ — notice that $\mathbb{LG}(\bigwedge^3 V)$ is independent of the chosen volume form vol. Given a nonzero $v \in V$ we let

$$F_v := \{\alpha \in \bigwedge^3 V \mid v \wedge \alpha = 0\}. \qquad (3.1.2)$$

Notice that $(\cdot, \cdot)_V$ is zero on F_v and $\dim(F_v) = 10$, i.e., $F_v \in \mathbb{LG}(\bigwedge^3 V)$. Let

$$F \subset \bigwedge^3 V \otimes \mathcal{O}_{\mathbb{P}(V)} \qquad (3.1.3)$$

be the vector subbundle with fiber F_v over $[v] \in \mathbb{P}(V)$. Given $A \in \mathbb{LG}(\bigwedge^3 V)$ we let

$$Y_A = \{[v] \in \mathbb{P}(V) \mid F_v \cap A \neq \{0\}\}. \qquad (3.1.4)$$

Thus Y_A is the degeneracy locus of the map

$$F \xrightarrow{\lambda_A} (\bigwedge^3 V/A) \otimes \mathcal{O}_{\mathbb{P}(V)} \qquad (3.1.5)$$

where λ_A is given by Inclusion (3.1.3) followed by the quotient map $\bigwedge^3 V \otimes \mathcal{O}_{\mathbb{P}(V)} \to (\bigwedge^3 V/A) \otimes \mathcal{O}_{\mathbb{P}(V)}$. Since the vector bundles appearing in (3.1.5) have equal rank Y_A is the zero-locus of $\det \lambda_A \in H^0(\det F^{\vee})$ — in particular it has a natural structure of closed subscheme of $\mathbb{P}(V)$. A straightforward computation gives that $\det F \cong \mathcal{O}_{\mathbb{P}(V)}(-6)$ and hence Y_A is a sextic hypersurface unless it equals $\mathbb{P}(V)$;[15] if the former holds we say that Y_A is an *EPW-sextic*. What do EPW-sextics look like? The main point is that locally they are the degeneracy locus of a symmetric map of vector bundles (they were introduced by Eisenbud, Popescu and Walter to give examples of a "quadratic sheaf", namely $coker(\lambda_A)$, which can not be expressed **globally** as the cokernel of a symmetric map of vector bundles on $\mathbb{P}^5$). More precisely given $B \in \mathbb{LG}(\bigwedge^3 V)$ we let $\mathcal{U}_B \subset \mathbb{P}(V)$ be the open subset defined by

$$\mathcal{U}_B := \{[v] \in \mathbb{P}(V) \mid F_v \cap B = \{0\}\}. \qquad (3.1.6)$$

Now choose B transversal to A. We have a direct-sum decomposition $\bigwedge^3 V = A \oplus B$; since A is lagrangian the symplectic form $(\cdot, \cdot)_V$ defines an isomorphism $B \cong A^{\vee}$. Let $[v] \in \mathcal{U}_B$: since F_v is transversal to B it is the graph of a map

$$\tau_A^B([v]): A \to B \cong A^{\vee}, \qquad [v] \in \mathcal{U}_B. \qquad (3.1.7)$$

The map $\tau_A^B([v])$ is symmetric because A, B and F_v are lagrangians.

---

[15]Given $[v] \in \mathbb{P}(V)$ there exists $A \in \mathbb{LG}(\bigwedge^3 V)$ such that $A \cap F_v = \{0\}$ and hence $[v] \notin Y_A$; thus Y_A is a sextic hypersurface for generic $A \in \mathbb{LG}(\bigwedge^3 V)$. On the other hand if $A = F_w$ for some $[w] \in \mathbb{P}(V)$ then $Y_A = \mathbb{P}(V)$.

Remark 3.1. There is one choice of B which produces a "classical" description of Y_A, namely $B = \bigwedge^3 V_0$ where $V_0 \subset V$ is a codimension-1 subspace.[16] With such a choice of B we have $\mathcal{U}_B = (\mathbb{P}(V) \setminus \mathbb{P}(V_0))$; we identify it with V_0 by choosing $v_0 \in (V \setminus V_0)$ and mapping

$$V_0 \xrightarrow{\sim} \mathbb{P}(V) \setminus \mathbb{P}(V_0), \quad v \mapsto [v_0 + v]. \tag{3.1.8}$$

The direct-sum decomposition $\bigwedge^3 V = F_{v_0} \oplus \bigwedge^3 V_0$ and transversality $A \pitchfork \bigwedge^3 V_0$ allows us to view A as the graph of a (symmetric) map $\widetilde{q}_A : F_{v_0} \to \bigwedge^3 V_0$. Identifying $\bigwedge^2 V_0$ with F_{v_0} via the isomorphism

$$\bigwedge^2 V_0 \xrightarrow{\sim} F_{v_0}, \quad \alpha \mapsto v_0 \wedge \alpha, \tag{3.1.9}$$

we may view $\widetilde{q}_A$ as a symmetric map

$$\bigwedge^2 V_0 \longrightarrow \bigwedge^3 V_0 = \bigwedge^2 V_0^\vee. \tag{3.1.10}$$

We let $q_A \in \mathrm{Sym}^2(\bigwedge^2 V_0^\vee)$ be the quadratic form corresponding to $\widetilde{q}_A$. Given $v \in V_0$ let $q_v \in \mathrm{Sym}^2(\bigwedge^2 V_0^\vee)$ be the Plücker quadratic form $q_v(\alpha) := vol(v_0 \wedge v \wedge \alpha \wedge \alpha)$. Modulo the identification (3.1.8) we have

$$Y_A \cap (\mathbb{P}(V) \setminus \mathbb{P}(V_0)) = V(\det(q_A + q_v)). \tag{3.1.11}$$

Equivalently let

$$Z_A := V(q_A) \cap \mathrm{Gr}(2, V_0) \subset \mathbb{P}(\bigwedge^2 V_0) \cong \mathbb{P}^9. \tag{3.1.12}$$

Then we have an isomorphism

$$\mathbb{P}(V) \xrightarrow{\sim} |\mathscr{I}_{Z_A}(2)|, \quad [\lambda v_0 + \mu v] \mapsto V(\lambda q_A + \mu q_v). \tag{3.1.13}$$

(Here $\lambda, \mu \in \mathbb{C}$ and $v \in V_0$.) Let $D_A \subset |\mathscr{I}_{Z_A}(2)|$ be the discriminant locus; modulo the identification above we have

$$Y_A \cap (\mathbb{P}(V) \setminus \mathbb{P}(V_0)) = D_A \cap (|\mathscr{I}_{Z_A}(2)| \setminus |\mathscr{I}_{\mathrm{Gr}(2, V_0)}(2)|). \tag{3.1.14}$$

Notice that $|\mathscr{I}_{\mathrm{Gr}(2, V_0)}(2)|$ is a hyperplane contained in D_A with multiplicity 4; that explains why $\deg Y_A = 6$ while $\deg D_A = 10$.

We go back to general considerations regarding Y_A. The symmetric map τ_A^B of (3.1.7) allows us to give a structure of scheme to the degeneracy locus

$$Y_A[k] = \{[v] \in \mathbb{P}(V) \mid \dim(A \cap F_v) \geq k\} \tag{3.1.15}$$

by declaring that $Y_A[k] \cap \mathcal{U}_B = V(\bigwedge^{(11-k)} \tau_A^B)$. By a standard dimension count we expect that the following holds for generic $A \in \mathbb{LG}(\bigwedge^3 V)$: $Y_A[3] = \varnothing$,

---

[16]It might happen that there is no V_0 such that $\bigwedge^3 V_0$ is transversal to A: in that case A is unstable for the natural $PGL(V)$-action on $\mathbb{LG}(\bigwedge^3 V)$ and hence we may forget about it.

$Y_A[2] = \text{sing } Y_A$ and $Y_A[2]$ is a smooth surface (of degree 40 by (6.7) of [Fulton and Pragacz 1998]), in particular Y_A should be a very special sextic hypersurface. This is indeed the case; in order to be less "generic" let

$$\Delta := \{A \in \mathbb{LG}(\wedge^3 V) \mid Y_A[3] \neq \varnothing\}, \qquad (3.1.16)$$

$$\Sigma := \{A \in \mathbb{LG}(\wedge^3 V) \mid \exists W \in Gr(3, V) \text{ s.t. } \wedge^3 W \subset A\}. \qquad (3.1.17)$$

A straightforward computation shows that Σ and Δ are distinct closed irreducible codimension-1 subsets of $\mathbb{LG}(\wedge^3 V)$. Let

$$\mathbb{LG}(\wedge^3 V)^0 := \mathbb{LG}(\wedge^3 V) \setminus \Sigma \setminus \Delta. \qquad (3.1.18)$$

Then Y_A has the generic behavior described above if and only if A belongs to $\mathbb{LG}(\wedge^3 V)^0$. Next let $A \in \mathbb{LG}(\wedge^3 V)$ and suppose that $Y_A \neq \mathbb{P}(V)$: then Y_A comes equipped with a natural double cover $f_A : X_A \to Y_A$ defined as follows. Let $i : Y_A \hookrightarrow \mathbb{P}(V)$ be the inclusion map: since $coker(\lambda_A)$ is annihilated by a local generator of $\det \lambda_A$ we have $coker(\lambda_A) = i_* \zeta_A$ for a sheaf ζ_A on Y_A. Choose $B \in \mathbb{LG}(\wedge^3 V)$ transversal to A; the direct-sum decomposition $\wedge^3 V = A \oplus B$ defines a projection map $\wedge^3 V \to A$; thus we get a map $\mu_{A,B} : F \to A \otimes \mathcal{O}_{\mathbb{P}(V)}$. We claim that there is a commutative diagram with exact rows

$$
\begin{array}{ccccccccc}
0 & \to & F & \xrightarrow{\lambda_A} & A^\vee \otimes \mathcal{O}_{\mathbb{P}(V)} & \longrightarrow & i_* \zeta_A & \to & 0 \\
& & \downarrow{\mu_{A,B}} & & \downarrow{\mu_{A,B}^t} & & \downarrow{\beta_A} & & \\
0 & \to & A \otimes \mathcal{O}_{\mathbb{P}(V)} & \xrightarrow{\lambda_A^t} & F^\vee & \longrightarrow & \text{Ext}^1(i_* \zeta_A, \mathcal{O}_{\mathbb{P}(V)}) & \to & 0
\end{array} \qquad (3.1.19)
$$

(Since A is Lagrangian the symplectic form defines a canonical isomorphism $(\wedge^3 V/A) \cong A^\vee$; that is why we may write λ_A as above.) In fact the second row is obtained by applying the $\text{Hom}(\cdot, \mathcal{O}_{\mathbb{P}(V)})$-functor to the first row and the equality $\mu_{A,B}^t \circ \lambda_A = \lambda_A^t \circ \mu_{A,B}$ holds because F is a Lagrangian subbundle of $\wedge^3 V \otimes \mathcal{O}_{\mathbb{P}(V)}$. Lastly β_A is defined to be the unique map making the diagram commutative; as suggested by notation it is independent of B. Next by applying the $\text{Hom}(i_* \zeta_A, \cdot)$-functor to the exact sequence

$$0 \to \mathcal{O}_{\mathbb{P}(V)} \longrightarrow \mathcal{O}_{\mathbb{P}(V)}(6) \longrightarrow \mathcal{O}_{Y_A}(6) \longrightarrow 0 \qquad (3.1.20)$$

we get the exact sequence

$$0 \longrightarrow i_* \text{Hom}(\zeta_A, \mathcal{O}_{Y_A}(6)) \xrightarrow{\partial} \text{Ext}^1(i_* \zeta_A, \mathcal{O}_{\mathbb{P}(V)}) \xrightarrow{n} \text{Ext}^1(i_* \zeta_A, \mathcal{O}_{\mathbb{P}(V)}(6)) \qquad (3.1.21)$$

where n is locally equal to multiplication by $\det \lambda_A$. Since the second row of (3.1.19) is exact a local generator of $\det \lambda_A$ annihilates $\text{Ext}^1(i_* \zeta_A, \mathcal{O}_{\mathbb{P}(V)})$; thus

$n = 0$ and hence we get a canonical isomorphism

$$\partial^{-1} : \text{Ext}^1(i_*\zeta_A, \mathcal{O}_{\mathbb{P}(V)}) \xrightarrow{\sim} i_* \text{Hom}(\zeta_A, \mathcal{O}_{Y_A}(6)). \tag{3.1.22}$$

Let

$$\zeta_A \times \zeta_A \xrightarrow{\widetilde{m}_A} \mathcal{O}_{Y_A}(6), \quad (\sigma_1, \sigma_2) \mapsto (\partial^{-1} \circ \beta_A(\sigma_1))(\sigma_2). \tag{3.1.23}$$

Let $\xi_A := \zeta_A(-3)$; tensoring both sides of (3.1.23) by $\mathcal{O}_{Y_A}(-6)$ we get a multiplication map

$$m_A : \xi_A \times \xi_A \to \mathcal{O}_{Y_A}. \tag{3.1.24}$$

This multiplication map equips $\mathcal{O}_{Y_A} \oplus \xi_A$ with the structure of a commutative and associative $\mathcal{O}_{Y_A}$-algebra. We let

$$X_A := Spec(\mathcal{O}_{Y_A} \oplus \xi_A), \qquad f_A : X_A \to Y_A. \tag{3.1.25}$$

Then X_A is a *double EPW-sextic*. Let $\mathcal{U}_B$ be as in (3.1.6): we may describe $f_A^{-1}(Y_A \cap \mathcal{U}_B)$ as follows. Let M be the symmetric matrix associated to (3.1.7) by a choice of basis of A and M^c be the matrix of cofactors of M. Let $Z = (z_1, \ldots, z_{10})^t$ be the coordinates on A associated to the given basis; then $f_A^{-1}(Y_A \cap \mathcal{U}_B) \subset \mathcal{U}_B \times \mathbb{A}_Z^{10}$ and its ideal is generated by the entries of the matrices

$$M \cdot Z, \quad Z \cdot Z^t - M^c. \tag{3.1.26}$$

(The "missing" equation $\det M = 0$ follows by Cramer's rule.) One may reduce the size of M in a neighborhood of $[v_0] \in \mathcal{U}_B$ as follows. The kernel of the symmetric map $\tau_A^B([v_0])$ equals $A \cap F_{v_0}$; let $J \subset A$ be complementary to $A \cap F_{v_0}$. Diagonalizing the restriction of τ_A^B to J we may assume that

$$M([v]) = \begin{pmatrix} M_0([v]) & 0 \\ 0 & 1_{10-k} \end{pmatrix} \tag{3.1.27}$$

where $k := \dim(A \cap F_{v_0})$ and M_0 is a symmetric $k \times k$ matrix. It follows at once that f_A is étale over $(Y_A \setminus Y_A[2])$. We also get the following description of f_A over a point $[v_0] \in (Y_A[2] \setminus Y_A[3])$ under the hypothesis that there is no $0 \neq v_0 \wedge v_1 \wedge v_2 \in A$. First $f_A^{-1}([v_0])$ is a single point p_0, secondly X_A is smooth at p_0 and there exists an involution ϕ on (X_A, p_0) with 2-dimensional fixed-point set such that f_A is identified with the quotient map $(X_A, p_0) \to (X_A, p_0)/\langle \phi \rangle$. It follows that X_A is smooth if $A \in \mathbb{LG}(\wedge^3 V)^0$. We may fit together all smooth double EPW-sextics by going to a suitable double cover $\rho : \mathbb{LG}(\wedge^3 V)^* \to \mathbb{LG}(\wedge^3 V)^0$; there exist a family of HK four-folds $\mathscr{X} \to \mathbb{LG}(\wedge^3 V)^*$ and a relatively ample line bundle $\mathscr{L}$ over $\mathscr{X}$ such that for all $t \in \mathbb{LG}(\wedge^3 V)^*$ we have $(X_t, L_t) \cong (X_{A_t}, f_{A_t}^* \mathcal{O}_{Y_{A_t}}(1))$ where

$$X_t := \rho^{-1}(t), \quad L_t = \mathscr{L}|_{X_t}, \quad A_t := \rho(t). \tag{3.1.28}$$

Theorem 3.2 [O'Grady 2006]. *Let $A \in \mathbb{LG}(\bigwedge^3 V)^0$. Then X_A is a HK four-fold deformation equivalent to K3$^{[2]}$. Moreover $\mathcal{X} \to \mathbb{LG}(\bigwedge^3 V)^\star$ is a locally complete family of HK varieties of degree 2.*

Sketch of proof. The main issue is to prove that X_A is a HK deformation of K3$^{[2]}$. In fact once this is known the equality

$$\int_{X_A} f_A^* c_1(\mathcal{O}_{Y_A}(1))^4 = 2 \cdot 6 = 12 \tag{3.1.29}$$

together with (2.1.2) gives that $q(f_A^* c_1(\mathcal{O}_{Y_A}(1))) = 2$ and moreover the family $\mathcal{X} \to \mathbb{LG}(\bigwedge^3 V)^\star$ is locally complete by the following argument. First Kodaira vanishing and Formula (2.2.7) give that

$$h^0(f_A^* \mathcal{O}_{Y_A}(1)) = \chi(f_A^* \mathcal{O}_{Y_A}(1)) = 6 \tag{3.1.30}$$

and hence the map

$$X_A \xrightarrow{f_A} Y_A \hookrightarrow \mathbb{P}(V) \tag{3.1.31}$$

may be identified with the map $X_A \to |f_A^* \mathcal{O}_{Y_A}(1)|^\vee$. From this one gets that the natural map $(\mathbb{LG}(\bigwedge^3 V)^0 // PGL(V), [A]) \to \mathrm{Def}(X_A, f_A^* \mathcal{O}_{Y_A}(1))$ is injective. One concludes that $\mathcal{X} \to \mathbb{LG}(\bigwedge^3 V)^\star$ is locally complete by a dimension count:

$$\dim(\mathbb{LG}(\bigwedge^3 V)^0 // PGL(V)) = 20 = \dim \mathrm{Def}(X_A, f_A^* \mathcal{O}_{Y_A}(1)). \tag{3.1.32}$$

Thus we are left with the task of proving that X_A is a HK deformation of K3$^{[2]}$ if $A \in \mathbb{LG}(\bigwedge^3 V)^0$. We do this by analyzing X_A for

$$A \in (\Delta \setminus \Sigma). \tag{3.1.33}$$

By definition $Y_A[3]$ is nonempty; one shows that it is finite, that sing $X_A = f_A^{-1} Y_A[3]$ and that $f_A^{-1}[v_i]$ is a single point for each $[v_i] \in Y_A[3]$. There exists a small resolution

$$\pi_A : \widehat{X}_A \longrightarrow X_A, \quad (f_A \circ \pi_A)^{-1}([v_i]) \cong \mathbb{P}^2 \quad \forall [v_i] \in Y_A. \tag{3.1.34}$$

In fact one gets that locally over the points of sing X_A the above resolution is identified with the contraction c (or $c^\vee$) appearing in (1.3.?) — in particular $\widehat{X}_A$ is not unique, in fact there are $2^{|Y_A[3]|}$ choices involved in the construction of $\widehat{X}_A$. The resolution $\widehat{X}_A$ fits into a simultaneous resolution; i.e., given a sufficiently small open (in the classical topology) $A \in U \subset (\mathbb{LG}(\bigwedge^3 V) \setminus \Sigma)$ we have proper maps π, ψ

$$\widehat{\mathcal{X}}_U \xrightarrow{\pi} \mathcal{X}_U \xrightarrow{\psi} U \tag{3.1.35}$$

where ψ is a tautological family of double EPW-sextics over U, i.e., $\psi^{-1} A \cong X_A$ and $(\psi \circ \pi)^{-1} A \to \psi^{-1} A = X_A$ is a small resolution as above if $A \in U \cap \Delta$ while

$\pi^{-1}A \cong X_A$ if $A \in (U \setminus \Delta)$. Thus it suffices to prove that there exist $A \in (\Delta \setminus \Sigma)$ such that $\widehat{X}_A$ is a HK deformation of $K3^{[2]}$. Let $[v_i] \in Y_A[3]$; we define a K3 surface $S_A(v_i)$ as follows. There exists a codimension-1 subspace $V_0 \subset V$ not containing v_i and such that $\bigwedge^3 V_0$ is transversal to A. Thus Y_A can be described as in Remark 3.1: we adopt notation introduced in that remark, in particular we have the quadric $Q_A := V(q_A) \subset \mathbb{P}(\bigwedge^2 V_0)$. The singular locus of Q_A is $\mathbb{P}(A \cap F_{v_i})$ — we recall the identification (3.1.9). By hypothesis $\mathbb{P}(A \cap F_{v_i}) \cap Gr(2, V_0) = \varnothing$; it follows that $\dim \mathbb{P}(A \cap F_{v_i}) = 2$ (by hypothesis $\dim \mathbb{P}(A \cap F_{v_i}) \geq 2$). Let

$$S_A(v_i) := Q_A^{\vee} \cap Gr(2, V_0^{\vee}) \subset \mathbb{P}(\bigwedge^2 V_0^{\vee}). \qquad (3.1.36)$$

Then $S_A(v_i) \subset \mathbb{P}(Ann(A \cap F_{v_i})) \cong \mathbb{P}^6$ is the transverse intersection of a smooth quadric and the Fano 3-fold of index 2 and degree 5, i.e., the generic K3 of genus 6. There is a natural degree-2 rational map

$$g_i : S_A(v_i)^{[2]} \dashrightarrow |\mathcal{I}_{S_A(v_i)}(2)|^{\vee} \qquad (3.1.37)$$

which associates to $[Z]$ the set of quadrics in $|\mathcal{I}_{S_A(v_i)}(2)|$ which contain the line spanned by Z — thus g_i is regular if $S_A(v_i)$ contains no lines. One proves that $\operatorname{Im}(g_i)$ may be identified with Y_A; it follows that there exists a birational map

$$h_i : S_A(v_i)^{[2]} \dashrightarrow \widehat{X}_A \qquad (3.1.38)$$

Moreover if $S_A(v_i)$ contains no lines (that is true for generic $A \in (\Delta \setminus \Sigma)$) there is a choice of small resolution $\widehat{X}_A$ such that h_i is regular and hence an isomorphism — in particular $\widehat{X}_A$ is projective.[17] This proves that X_A is a HK deformation of $K3^{[2]}$ for $A \in \mathbb{LG}(\bigwedge^3 V)^0$. $\qquad \square$

Remark 3.3. This proof of Theorem 3.2 provides a description of X_A for $A \in (\Delta \setminus \Sigma)$; what about X_A for $A \in \Sigma$? One proves that if $A \in \Sigma$ is generic — in particular there is a unique $W \in Gr(3, V)$ such that $\bigwedge^3 W \subset A$ — then the following hold:

(1) $C_{W,A} := \{[v] \in \mathbb{P}(W) \mid \dim(A \cap F_v) \geq 2\}$ is a smooth sextic curve.

(2) $\operatorname{sing} X_A = f_A^{-1}\mathbb{P}(W)$ and the restriction of f_A to $\operatorname{sing} X_A$ is the double cover of $\mathbb{P}(W)$ branched over $C_{W,A}$, i.e., a K3 surface of degree 2.

(3) If $p \in \operatorname{sing} X_A$ the germ (X_A, p) (in the classical topology) is isomorphic to the product of a smooth 2-dimensional germ and an A_1 singularity; thus the blow-up $\widetilde{X}_A \to X_A$ resolves the singularities of X_A.

(4) Let $U \subset \mathbb{LG}(\bigwedge^3 V)$ be a small open (classical topology) subset containing A. After a base change $\widetilde{U} \to U$ of order 2 branched over $U \cap \Sigma$ there is a

---

[17]There is no reason a priori why $\widehat{X}_A$ should be Kähler, in fact one should expect it to be non-Kähler for some A and some choice of small resolution.

simultaneous resolution of singularities of the tautological family of double EPW's parametrized by $\tilde{U}$. It follows that $\tilde{X}_A$ is a HK deformation of K3$^{[2]}$.

(5) Let E_A be the exceptional divisor of the blow-up $\tilde{X}_A \to X_A$ and $e_A \in H^2(\tilde{X}_A; \mathbb{Z})$ be its Poincaré dual; then $q(e_A) = -2$ and $(e_A, H^2(\tilde{X}_A; \mathbb{Z})) = \mathbb{Z}$.

3.2. The Beauville–Donagi family. Let $\mathcal{D}, \mathcal{P} \subset |\mathcal{O}_{\mathbb{P}^5}(3)|$ be the prime divisors parametrizing singular cubics and cubics containing a plane respectively. We recall that if $Z \in (|\mathcal{O}_{\mathbb{P}^5}(3)| \setminus \mathcal{D})$ then

$$X = F(Z) := \{L \in \mathbb{G}r(1, \mathbb{P}^5) \mid L \subset X\} \tag{3.2.1}$$

is a HK four-fold deformation equivalent to K3$^{[2]}$. Let H be the Plücker ample divisor on X and $h = c_1(\mathcal{O}_X(H))$; then

$$q(h) = 6, \quad (h, H^2(X; \mathbb{Z}))_X = 2\mathbb{Z}. \tag{3.2.2}$$

These results are proved in [Beauville and Donagi 1985] by considering the codimension-1 locus of Pfaffian cubics; they show that if Z is a generic such Pfaffian cubic then X is isomorphic to $S^{[2]}$ where S is a K3 of genus 8 that one associates to Z, moreover the class h is identified with $2\tilde{\mu}(D) - 5\xi_2$ where D is the class of the (genus 8) hyperplane class of S. Here we will stress the similarities between the HK four-folds parametrized by $\mathcal{D}, \mathcal{P}$ and those parametrized by the loci $\Delta, \Sigma \subset \mathbb{LG}(\bigwedge^3 V)$ described in the previous subsection. Let $Z \in \mathcal{D}$ be generic. Then Z has a unique singular point p and it is ordinary quadratic, moreover the set of lines in Z containing p is a K3 surface S of genus 4. The variety $X = F(Z)$ parametrizing lines in Z is birational to $S^{[2]}$; the birational map is given by

$$S^{[2]} \dashrightarrow F(Z), \quad \{L_1, L_2\} \mapsto R, \tag{3.2.3}$$

where $L_1 + L_2 + R = \langle L_1, L_2 \rangle \cdot Z$. Moreover $F(Z)$ is singular with singular locus equal to S. Thus from this point of view $\mathcal{D}$ is similar to Δ. On the other hand let $Z_0 \in (|\mathcal{O}_{\mathbb{P}^5}(3)| \setminus \mathcal{D})$ be "close" to Z; the monodromy action on $H^2(F(Z_0))$ of a loop in $(|\mathcal{O}_{\mathbb{P}^5}(3)| \setminus \mathcal{D})$ which goes once around $\mathcal{D}$ has order 2 and hence as far as monodromy is concerned $\mathcal{D}$ is similar to Σ. (Let $U \subset |\mathcal{O}_{\mathbb{P}^5}(3)|$ be a small open (classical topology) set containing Z; it is natural to expect that after a base change $\pi : \tilde{U} \to U$ of order 2 ramified over $\mathcal{D}$ the family of $F(Z_u)$ for $u \in (\tilde{U} \setminus \pi^{-1}\mathcal{D})$ can be completed over points of $\pi^{-1}\mathcal{D}$ with HK four-folds birational (isomorphic?) to $S^{[2]}$.) Now let $Z \in \mathcal{P}$ be generic, in particular it contains a unique plane P. Let $T \cong \mathbb{P}^2$ parametrize 3-dimensional linear subspaces of $\mathbb{P}^5$ containing P; given $t \in T$ and L_t the corresponding 3-space the intersection $L_t \cdot Z$ decomposes as $P + Q_t$ where Q_t is a quadric surface. Let

$E \subset X = F(Z)$ be the set defined by

$$E := \{L \in F(Z) \mid \exists t \in T \text{ such that } L \subset Q_t\}. \qquad (3.2.4)$$

For Z generic we have a well-defined map $E \to T$ obtained by associating to L the unique t such that $L \subset Q_t$; the Stein factorization of $E \to T$ is $E \to S \to T$ where $S \to T$ is the double cover ramified over the curve $B \subset T$ parametrizing singular quadrics. The locus B is a smooth sextic curve and hence S is a K3 surface of genus 2. The picture is: E is a conic bundle over the K3 surface S and we have

$$q(E) = -2, \qquad (e, H^2(X; \mathbb{Z})) = \mathbb{Z}, \qquad e := c_1(\mathcal{O}_X(E)). \qquad (3.2.5)$$

Thus from this point of view $\mathscr{P}$ is similar to Σ — of course if we look at monodromy the analogy fails.

4. Numerical Hilbert squares

A *numerical Hilbert square* is a HK four-fold X such that c_X is equal to the Fujiki constant of K3$^{[2]}$ and the lattice $H^2(X; \mathbb{Z})$ is isometric to $H^2(\text{K3}^{[2]}; \mathbb{Z})$; by (2.1.8), (2.1.9) this holds if and only if

$$H^2(X; \mathbb{Z}) \cong U^3 \widehat{\oplus} E_8(-1) \widehat{\oplus} \langle -2 \rangle, \qquad c_X = 1. \qquad (4.0.1)$$

We will present a program which aims to prove that a numerical Hilbert square is a deformation of K3$^{[2]}$, i.e., an analogue of Kodaira's theorem that any two K3's are deformation equivalent. First we recall how Kodaira [1964] proved that K3 surfaces form a single deformation class. Let X_0 be a K3. Let $\mathscr{X} \to T$ be a representative of the deformation space Def(X_0). The image of the local period map $\pi : T \to \mathbb{P}(H^2(X_0))$ contains an open (classical topology) subset of the quadric $\mathcal{Q} := V(q_{X_0})$. The set $\mathcal{Q}(\mathbb{Q})$ of rational points of $\mathcal{Q}$ is dense (classical topology) in the set of real points $\mathcal{Q}(\mathbb{R})$; it follows that the image $\pi(T)$ contains a point $[\sigma]$ such that $\sigma^{\perp} \cap H^2(X_0; \mathbb{Q})$ is generated by a nonzero α such that $q_X(\alpha) = 0$. Let $t \in T$ such that $\pi(t) = [\sigma]$ and set $X := X_t$; by the Lefschetz $(1, 1)$ theorem we have

$$H^{1,1}_{\mathbb{Z}}(X) = \mathbb{Z}c_1(L), \qquad q_X(c_1(L)) = 0, \qquad (4.0.2)$$

where L is a holomorphic line bundle on X. By Hirzebruch–Riemann–Roch and Serre duality we get that $h^0(L) + h^0(L^{-1}) \geq 2$. Thus we may assume that $h^0(L) \geq 2$. It follows that L is globally generated, $h^0(L) = 2$ and the map $\phi_L : X \to |L| \cong \mathbb{P}^1$ is an elliptic fibration. Kodaira then proved that any two elliptic K3's are deformation equivalent. J. Sawon [2003] has launched a similar program with the goal of classifying deformation classes of higher-dimensional HK manifolds[18] by deforming them to Lagrangian fibrations — we notice that Matsushita [1999; 2001; 2005] has proved quite a few results on HK manifolds which have nontrivial fibrations. The program is quite ambitious; it runs immediately into the problem of proving that if L is a nontrivial line bundle on a HK manifold X with $q_X(c_1(L)) = 0$ then $h^0(L) + h^0(L^{-1}) > 0$.[19] On the other hand Kodaira's theorem on K3's can be proved (see [Le Potier 1985]) by deforming X_0 to a K3 surface X such that $H_{\mathbb{Z}}^{1,1}(X) = \mathbb{Z}c_1(L)$ where L is a holomorphic line bundle such that $q_X(L)$ is a small positive integer, say 2. By Hirzebruch–Riemann–Roch and Serre duality $h^0(L) + h^0(L^{-1}) \geq 3$ and hence we may assume that $h^0(L) \geq 3$; it follows easily that L is globally generated, $h^0(L) = 3$ and the map $\phi_L : X \to |L|^\vee \cong \mathbb{P}^2$ is a double cover ramified over a smooth sextic curve. Thus every K3 is deformation equivalent to a double cover of $\mathbb{P}^2$ ramified over a sextic; since the parameter space for smooth sextics is connected it follows that any two K3 surfaces are deformation equivalent. Our idea is to adapt this proof to the case of numerical Hilbert squares. In short the plan is as follows. Let X_0 be a numerical Hilbert square. First we deform X_0 to a HK four-fold X such that

$$H_{\mathbb{Z}}^{1,1}(X) = \mathbb{Z}c_1(L), \qquad q_X(c_1(L)) = 2 \qquad (4.0.3)$$

and the Hodge structure of X is very generic given the constraint (4.0.3), see Section 4.1 for the precise conditions. By Huybrechts' Projectivity Criterion (Theorem 2.7) we may assume that L is ample and then Hirzebruch–Riemann–Roch together with Kodaira vanishing gives that $h^0(L) = 6$. Thus we must study the map $f : X \dashrightarrow |L|^\vee \cong \mathbb{P}^5$. We prove that either f is the natural double cover of an EPW-sextic or else it is birational onto its image (a hypersurface of degree at most 12). We conjecture that the latter never holds; if the conjecture is true then any numerical Hilbert square is a deformation of a double EPW-sextic and hence is a deformation of K3$^{[2]}$.

4.1. _The deformation._ We recall Huybrechts' theorem on surjectivity of the global period map for HK manifolds. Let X_0 be a HK manifold. Let L be a lattice isomorphic to the lattice $H^2(X_0; \mathbb{Z})$; we denote by $(\,\cdot\,,\,\cdot\,)_L$ the extension to $L \otimes \mathbb{C}$ of the bilinear symmetric form on L. The period domain $\Omega_L \subset \mathbb{P}(L \otimes \mathbb{C})$ is given by

$$\Omega_L := \{[\sigma] \in \mathbb{P}(L \otimes \mathbb{C}) \mid (\sigma, \sigma)_L = 0, \quad (\sigma, \overline{\sigma})_L > 0\}. \qquad (4.1.1)$$

---

[18]One should assume that $b_2 \geq 5$ in order to ensure that the set of rational points in $V(q_X)$ is nonempty (and hence dense in the set of real points).

[19]Let dim $X = 2n$. Hirzebruch–Riemann–Roch gives that $\chi(L) = n + 1$, one would like to show that $h^q(L) = 0$ for $0 < q < 2n$.

A HK manifold X deformation equivalent to X_0 is *marked* if it is equipped with an isometry of lattices $\psi : L \xrightarrow{\sim} H^2(X; \mathbb{Z})$. Two pairs (X, ψ) and (X', ψ') are equivalent if there exists an isomorphism $f : X \to X'$ such that $H^2(f) \circ \psi' = \pm \psi$. The moduli space $\mathcal{M}_{X_0}$ of marked HK manifolds deformation equivalent to X_0 is the set of equivalence classes of pairs as above. If $t \in \mathcal{M}_{X_0}$ we let (X_t, ψ_t) be a representative of t. Choosing a representative $\mathcal{X} \to T$ of the deformation space of X_t with T contractible we may put a natural structure of (nonseparated) complex analytic manifold on $\mathcal{M}_{X_0}$; see for example Theorem (2.4) of [Looijenga and Peters 1980/81]. The period map is given by

$$\mathcal{M}_{X_0} \xrightarrow{\mathscr{P}} \Omega_L, \quad (X, \psi) \mapsto \psi^{-1} H^{2,0}(X). \tag{4.1.2}$$

(We denote by the same symbol both the isometry $L \xrightarrow{\sim} H^2(X; \mathbb{Z})$ and its linear extension $L \otimes \mathbb{C} \to H^2(X; \mathbb{C})$.) The map $\mathscr{P}$ is locally an isomorphism by infinitesimal Torelli and local surjectivity of the period map. The following result is proved in [Huybrechts 1999]; the proof is an adaptation of Todorov's proof of surjectivity for K3 surfaces [Todorov 1980].

Theorem 4.1 (Todorov, Huybrechts). *Keep notation as above and let $\mathcal{M}_{X_0}^0$ be a connected component of $\mathcal{M}_{X_0}$. The restriction of $\mathscr{P}$ to $\mathcal{M}_{X_0}^0$ is surjective.*

Let

$$\Lambda := U^3 \widehat{\oplus} E_8 \langle -1 \rangle^2 \widehat{\oplus} \langle -2 \rangle \tag{4.1.3}$$

be the Hilbert square lattice; see (2.1.8). Thus Ω_Λ is the period space for numerical Hilbert squares. A straightforward computation gives the following result; see Lemma 3.5 of [O'Grady 2008].

Lemma 4.2. *Suppose that $\alpha_1, \alpha_2 \in \Lambda$ satisfy*

$$(\alpha_1, \alpha_1)_\Lambda = (\alpha_2, \alpha_2)_\Lambda = 2, \qquad (\alpha_1, \alpha_2)_\Lambda \equiv 1 \mod 2. \tag{4.1.4}$$

Let X_0 be a numerical Hilbert square. Let $\mathcal{M}_{X_0}^0$ be a connected component of the moduli space of marked HK four-folds deformation equivalent to X_0. There exists $1 \le i \le 2$ such that for every $t \in \mathcal{M}_{X_0}^0$ the class of $\psi_t(\alpha_i)^2$ in $H^4(X_t; \mathbb{Z})/$Tors is indivisible.

Notice that Λ contains (many) pairs α_1, α_2 which satisfy (4.1.4); it follows that there exists $\alpha \in \Lambda$ such that for every $t \in \mathcal{M}_{X_0}^0$ the class of $\psi_t(\alpha)^2$ in $H^4(X_t; \mathbb{Z})/$Tors is indivisible. There exists $[\sigma] \in \Omega_\Lambda$ such that

$$\sigma^\perp \cap \Lambda = \mathbb{Z}\alpha. \tag{4.1.5}$$

By Theorem 4.1 there exists $t \in \mathcal{M}_{X_0}$ such that $\mathscr{P}(t) = [\sigma]$. Equality (4.1.5) gives

$$H_{\mathbb{Z}}^{1,1}(X_t) = \mathbb{Z}\alpha. \tag{4.1.6}$$

Since $q(\psi_t(\alpha)) = 2 > 0$ the HK manifold X_t is projective by Theorem 2.7; by (4.1.6) either $\psi_t(\alpha)$ or $\psi_t(-\alpha)$ is ample and hence we may assume that $\psi_t(\alpha)$ is ample. Let $X' := X_t$ and H' be the divisor class such that $c_1(\mathcal{O}_{X'}(H')) = \psi_t(\alpha)$; X' is a first approximation to the deformation of X_0 that we will consider. The reason for requiring that $\psi_t(\alpha)^2$ be indivisible in $H^4(X_t; \mathbb{Z})/\,\mathrm{Tors}$ will become apparent in the sketch of the proof of Theorem 4.5.

Remark 4.3. If X is a deformation of $\mathrm{K3}^{[2]}$ and $\alpha \in H^2(X; \mathbb{Z})$ is an arbitrary class such that $q(\alpha) = 2$ then the class of α^2 in $H^4(X; \mathbb{Z})/\,\mathrm{Tors}$ is not divisible; see Proposition 3.6 of [O'Grady 2008].

Let $\pi : \mathscr{X} \to S$ be a representative of the deformation space $\mathrm{Def}(X', H')$. Thus letting $X_s := \pi^{-1}(s)$ there exist $0 \in S$ and a given isomorphism $X_0 \xrightarrow{\sim} X'$ and moreover there is a divisor-class $\mathscr{H}$ on $\mathscr{X}$ which restricts to H' on X_0; we let $H_s := \mathscr{H}|_{X_s}$. We will replace (X', H') by (X_s, H_s) for s very general in S in order to ensure that $H^4(X_s)$ has the simplest possible Hodge structure. First we describe the Hodge substructures of $H^4(X_s)$ that are forced by the Beauville–Bogomolov quadratic form and the integral $(1, 1)$ class $\psi_t(\alpha)$. Let X be a HK manifold. The Beauville–Bogomolov quadratic form q_X provides us with a nontrivial class $q_X^\vee \in H_\mathbb{Q}^{2,2}(X)$. In fact since q_X is nondegenerate it defines an isomorphism

$$L_X : H^2(X) \xrightarrow{\sim} H^2(X)^\vee. \qquad (4.1.7)$$

Viewing q_X as a symmetric tensor in $H^2(X)^\vee \otimes H^2(X)^\vee$ and applying L_X^{-1} we get a class

$$(L_X^{-1} \otimes L_X^{-1})(q_X) \in H^2(X) \otimes H^2(X);$$

applying the cup-product map $H^2(X) \otimes H^2(X) \to H^4(X)$ to $(L_X^{-1} \otimes L_X^{-1})(q_X)$ we get an element $q_X^\vee \in H^4(X; \mathbb{Q})$ which is of type $(2, 2)$ by (2.1.10). Now we assume that X is a numerical Hilbert square and that H is a divisor class such that $q(H) = 2$. Let $h := c_1(\mathcal{O}_X(H))$. We have an orthogonal (with respect to q_X) direct sum decomposition

$$H^2(X) = \mathbb{C}h \widehat{\oplus} h^\perp \qquad (4.1.8)$$

into Hodge substructures of levels 0 and 2 respectively. Since $b_2(X) = 23$ we get by Corollary 2.5 that cup-product defines an isomorphism

$$\mathrm{Sym}^2 H^2(X) \xrightarrow{\sim} H^4(X). \qquad (4.1.9)$$

Because of (4.1.9) we will identify $H^4(X)$ with $\mathrm{Sym}^2 H^2(X)$. Thus (4.1.8) gives a direct sum decomposition

$$H^4(X) = \mathbb{C}h^2 \oplus (\mathbb{C}h \otimes h^\perp) \oplus \mathrm{Sym}^2(h^\perp) \qquad (4.1.10)$$

into Hodge substructures of levels 0, 2 and 4 respectively. As is easily checked $q_X^\vee \in (\mathbb{C}h^2 \oplus \mathrm{Sym}^2(h^\perp)$. Let

$$W(h) := (q^\vee)^\perp \cap \mathrm{Sym}^2(h^\perp). \tag{4.1.11}$$

(To avoid misunderstandings: the first orthogonality is with respect to the inter-section form on $H^4(X)$, the second one is with respect to q_X.) One proves easily (see Claim 3.1 of [O'Grady 2008]) that $W(h)$ is a codimension-1 rational sub Hodge structure of $\mathrm{Sym}^2(h^\perp)$, and that we have a direct sum decomposition

$$\mathbb{C}h^2 \oplus \mathrm{Sym}^2(h^\perp) = \mathbb{C}h^2 \oplus \mathbb{C}q^\vee \oplus W(h). \tag{4.1.12}$$

Thus we have the decomposition

$$H^4(X; \mathbb{C}) = (\mathbb{C}h^2 \oplus \mathbb{C}q^\vee) \oplus (\mathbb{C}h \otimes h^\perp) \oplus W(h) \tag{4.1.13}$$

into sub Hodge structures of levels 0, 2 and 4 respectively.

Claim 4.4 [O'Grady 2008, Proposition 3.2]. *Keep notation as above. Let $s \in S$ be very general, i.e., outside a countable union of proper analytic subsets of S. Then:*

(1) $H_\mathbb{Z}^{1,1}(X_s) = \mathbb{Z}h_s$ *where* $h_s = c_1(\mathcal{O}_{X_s}(H_s))$.

(2) *Let* $\Sigma \in Z_1(X_s)$ *be an integral algebraic 1-cycle on* X_s *and* $cl(\Sigma) \in H_\mathbb{Q}^{3,3}(X_3)$ *be its Poincaré dual. Then* $cl(\Sigma) = mh_s^3/6$ *for some* $m \in \mathbb{Z}$.

(3) *If* $V \subset H^4(X_s)$ *is a rational sub Hodge structure then* $V = V_1 \oplus V_2 \oplus V_3$ *where* $V_1 \subset (\mathbb{C}h_s^2 \oplus \mathbb{C}q_{X_s}^\vee)$, V_2 *is either* 0 *or equal to* $\mathbb{C}h_s \otimes h_s^\perp$, *and* V_3 *is either* 0 *or equal to* $W(h_s)$.

(4) *The image of* h_s^2 *in* $H^4(X_s; \mathbb{Z})/\mathrm{Tors}$ *is indivisible.*

(5) $H_\mathbb{Z}^{2,2}(X_s)/\mathrm{Tors} \subset \mathbb{Z}(h_s^2/2) \oplus \mathbb{Z}(q_{X_s}^\vee/5)$.

Let $s \in S$ be such that the five conclusions of Claim 4.4 hold. Let $X := X_s$, $H := H_s$ and $h := c_1(\mathcal{O}_X(H))$. Since H is in the positive cone and h generates $H_\mathbb{Z}^{1,1}(X)$ we get that H is ample. By construction X is a deformation of our given numerical Hilbert square. The goal is to analyze the linear system $|H|$. First we compute its dimension. A computation (see pp. 564-565 of [O'Grady 2008]) gives that $c_2(X) = 6q_X^\vee/5$; it follows that (2.2.7) holds for numerical Hilbert squares. Thus $\chi(\mathcal{O}_X(H)) = 6$. By Kodaira vanishing we get $h^0(\mathcal{O}_X(H)) = 6$. Thus we have the map

$$f : X \dashrightarrow |H|^\vee \cong \mathbb{P}^5. \tag{4.1.14}$$

Theorem 4.5 [O'Grady 2008]. *Let (X, H) be as above. One of the following holds:*

(a) *The line bundle $\mathcal{O}_X(H)$ is globally generated and there exist an antisymplectic involution $\phi : X \to X$ and an inclusion $X/\langle \phi \rangle \hookrightarrow |H|^\vee$ such that the map f of (4.1.14) is identified with the composition*

$$X \xrightarrow{\rho} X/\langle \phi \rangle \hookrightarrow |H|^\vee \qquad (4.1.15)$$

where ρ is the quotient map.

(b) *The map f of (4.1.14) is birational onto its image (a hypersurface of degree between 6 and 12).*

Sketch of proof. We use the following result, which follows from conclusions (4) and (5) of Claim 4.4 plus a straightforward computation; see Proposition 4.1 of [O'Grady 2008].

Claim 4.6. *If $D_1, D_2 \in |H|$ are distinct then $D_1 \cap D_2$ is a reduced irreducible surface.*

In fact we chose h such that h^2 is not divisible in $H^4(X; \mathbb{Z})/$ Tors precisely to ensure that this claim holds. Let $Y \subset \mathbb{P}^5$ be the image of f (to be precise the closure of the image by f of its regular points). Thus (abusing notation) we have $f : X \dashrightarrow Y$. Of course $\dim Y \le 4$. Suppose that $\dim Y = 4$ and that $\deg f = 2$. Then there exists a nontrivial rational involution $\phi : X \dashrightarrow X$ commuting with f. Since $\operatorname{Pic}(X) = \mathbb{Z}[H]$ we get that $\phi^* H \sim H$; since $K_X \sim 0$ it follows that ϕ is regular; it follows easily that (a) holds. Thus it suffices to reach a contradiction assuming that $\dim Y < 4$ or $\dim Y = 4$ and $\deg f > 2$. One goes through a (painful) case-by-case analysis. In each case, with the exception of Y a quartic 4-fold, one invokes either Claim 4.6 or Claim 4.4(3). We give two "baby" cases. First suppose that Y is a quadric 4-fold. Let Y_0 be an open dense subset containing the image by f of its regular points. There exists a 3-dimensional linear space $L \subset \mathbb{P}^5$ such that $L \cap Y_0$ is a reducible surface. Now L corresponds to the intersection of two distinct $D_1, D_2 \in |H|$ and since $L \cap Y_0$ is reducible so is $D_1 \cap D_2$; this contradicts Claim 4.6. As a second example we suppose that Y is a smooth cubic 4-fold and f is regular. Notice that

$$H \cdot H \cdot H \cdot H = 12 \qquad (4.1.16)$$

by (2.1.3) and hence $\deg f = 4$. Let $H^4(Y)_{pr} \subset H^4(Y)$ be the primitive cohomology. By Claim 4.4(3) we must have $f^* H^4(Y)_{pr} \subset \mathbb{C}h \otimes h^\perp$. The restriction to $f^* H^4(Y; \mathbb{Q})_{pr}$ of the intersection form on $H^4(X)$ equals the intersection form on $H^4(Y; \mathbb{Q})_{pr}$ multiplied by 4 because $\deg f = 4$; one gets a contradiction by comparing discriminants. $\qquad \square$

Conjecture 4.7. Item (b) of Theorem 4.5 does not occur.

As we will explain in the next subsection, Conjecture 4.7 implies that a numerical Hilbert square is in fact a deformation of $K3^{[2]}$. The following question arose in connection with the proof of Theorem 4.5.

Question 4.8. Let X be a HK 4-fold and H an ample divisor on X. Is $\mathcal{O}_X(2H)$ globally generated?

The analogous question in dimension 2 has a positive answer; see for example [Mayer 1972]. We notice that if X is a 4-fold with trivial canonical bundle and H is ample on X then $\mathcal{O}_X(5H)$ is globally generated, by [Kawamata 1997]. The relation between Question 4.8 and Theorem 4.5 is the following.

Claim 4.9. *Suppose that the answer to Question 4.8 is positive. Let X be a numerical Hilbert square equipped with an ample divisor H such that $q_X(H) = 2$. Let $Y \subset |H|^\vee$ be the closure of the image of the set of regular points of the rational map $X \dashrightarrow |H|^\vee$. Then one of the following holds:*

(1) $\mathcal{O}_X(H)$ *is globally generated.*

(2) Y *is contained in a quadric.*

Proof. Suppose that alternative (2) does not hold. Then multiplication of sections defines an injection $\mathrm{Sym}^2 H^0(\mathcal{O}_X(H)) \hookrightarrow H^0(\mathcal{O}_X(2H))$; on the other hand we have

$$\dim \mathrm{Sym}^2 H^0(\mathcal{O}_X(H)) = 21 = \dim H^0(\mathcal{O}_X(2H)). \qquad (4.1.17)$$

(The last equation holds by (2.2.7), which is valid for numerical Hilbert squares as noticed above.) Since $\mathcal{O}_X(2H)$ is globally generated it follows that $\mathcal{O}_X(H)$ is globally generated as well, and alternative (1) holds. $\square$

We remark that alternatives (1) and (2) of the claim are not mutually exclusive. In fact let $S \subset \mathbb{P}^3$ be a smooth quartic surface (a K3) not containing lines. We have a finite map

$$S^{[2]} \xrightarrow{f} \mathrm{Gr}(1, \mathbb{P}^3) \subset \mathbb{P}^5, \qquad [Z] \mapsto \langle Z \rangle \qquad (4.1.18)$$

with image the Plücker quadric in $\mathbb{P}^5$. Let $H := f^* \mathcal{O}_{\mathbb{P}^5}(1)$; since f is finite H is ample. Moreover $q(H) = 2$ because $H \cdot H \cdot H \cdot H = 12$; thus (4.1.18) may be identified with the map associated to the complete linear system $|H|$.

4.2. *Double EPW-sextics, II.* Let (X, H) be as in Theorem 4.5(a). In [O'Grady 2006] we proved that there exists $A \in \mathbb{L}G(\bigwedge^3 \mathbb{C}^6)^0$ such that $Y_A = f(X)$ and the double cover $X \to f(X)$ may be identified with the canonical double cover $X_A \to Y_A$. Since X_A is a deformation of $K3^{[2]}$ it follows that if Conjecture 4.7 holds then numerical Hilbert squares are deformations of $K3^{[2]}$. The precise result is this:

Theorem 4.10 [O'Grady 2006]. *Let X be a numerical Hilbert square. Suppose that H is an ample divisor class on X such that the following hold*:

(1) $q_X(H) = 2$ *(and hence* $\dim |H| = 5$).

(2) $\mathcal{O}_X(H)$ *is globally generated.*

(3) *There exist an antisymplectic involution* $\phi : X \to X$ *and an inclusion* $X/\langle\phi\rangle \hookrightarrow |H|^\vee$ *such that the map* $X \to |H|^\vee$ *is identified with the composition*

$$X \xrightarrow{\rho} X/\langle\phi\rangle \hookrightarrow |H|^\vee \qquad (4.2.1)$$

where ρ is the quotient map.

Then there exists $A \in \mathbb{L}G(\bigwedge^3\mathbb{C}^6)^0$ such that $Y_A = Y$ and the double cover $X \to f(X)$ may be identified with the canonical double cover $X_A \to Y_A$.

Proof. Step I. Let $Y := f(X)$; abusing notation we let $f : X \to Y$ be the double cover which is identified with the quotient map for the action of $\langle\phi\rangle$. We have the decomposition $f_*\mathcal{O}_X = \mathcal{O}_Y \oplus \eta$ where η is the (-1)-eigensheaf for the action of ϕ on $\mathcal{O}_X$. One proves that $\zeta := \eta \otimes \mathcal{O}_Y(3)$ is globally generated — an intermediate step is the proof that $3H$ is very ample. Thus we have an exact sequence

$$0 \to G \longrightarrow H^0(\zeta) \otimes \mathcal{O}_{|H|^\vee} \longrightarrow i_*\zeta \to 0. \qquad (4.2.2)$$

where $i : Y \hookrightarrow |H|^\vee$ is inclusion.

Step II. One computes $h^0(\zeta)$ as follows. First $H^0(\zeta)$ is equal to $H^0(\mathcal{O}_X(3H))^-$, the space of ϕ-anti-invariant sections of $\mathcal{O}_X(3H)$. Using Equation (2.2.7) one gets that $h^0(\zeta) = 10$. A local computation shows that G is locally free. By invoking Beilinson's spectral sequence for vector bundles on projective spaces one gets that $G \cong \Omega^3_{|H|^\vee}(3)$. On the other hand one checks easily (Euler sequence) that the vector bundle F of (3.1.3) is isomorphic to $\Omega^3_{\mathbb{P}(V)}(3)$. Hence if we identify $\mathbb{P}(V)$ with $|H|^\vee$ then F is isomorphic to the sheaf G appearing in (4.2.2). In other words (4.2.2) starts looking like the top horizontal sequence of (3.1.19).

Step III. The multiplication map $\eta \otimes \eta \to \mathcal{O}_Y$ defines an isomorphism $\beta : i_*\zeta \xrightarrow{\sim} \mathrm{Ext}^1(i_*\zeta, \mathcal{O}_{|H|^\vee})$. Applying general results of Eisenbud, Popescu, and Walter [Eisenbud et al. 2001] (alternatively see the proof of Claim (2.1) of [Casnati and Catanese 1997]) one gets that β fits into a commutative diagram

$$
\begin{array}{ccccccccc}
0 & \longrightarrow & \Omega^3_{|H|^\vee}(3) & \xrightarrow{\kappa} & H^0(\theta) \otimes \mathcal{O}_{|H|^\vee} & \longrightarrow & i_*\zeta & \longrightarrow & 0 \\
& & \downarrow{\scriptstyle s^t} & & \downarrow{\scriptstyle s} & & \downarrow{\scriptstyle \beta} & & \\
0 & \longrightarrow & H^0(\theta)^\vee \otimes \mathcal{O}_{|H|^\vee} & \xrightarrow{\kappa^t} & \Theta^3_{|H|^\vee}(-3) & \xrightarrow{\partial} & \mathrm{Ext}^1(i_*\zeta, \mathcal{O}_{|H|^\vee}) & \longrightarrow & 0
\end{array}
$$

$$(4.2.3)$$

where the second row is obtained from the first one by applying $\mathrm{Hom}(\,\cdot\,, \mathcal{O}_{|H|^\vee})$.

Step IV. One checks that

$$\Omega^3_{|H|^\vee}(3) \xrightarrow{(\kappa, s^t)} \left(H^0(\zeta) \oplus H^0(\zeta)^\vee \right) \otimes \mathcal{O}_{|H|^\vee} \tag{4.2.4}$$

is an injection of vector bundles. The transpose of the map above induces an isomorphism $\left(H^0(\zeta)^\vee \oplus H^0(\zeta) \right) \xrightarrow{\sim} H^0(\Omega^3_{|H|^\vee}(3)^\vee)$. The same argument shows that the transpose of (3.1.3) induces an isomorphism $\bigwedge^3 V^\vee \xrightarrow{\sim} H^0(F^\vee)$. Since F is isomorphic to $\Omega^3_{|H|^\vee}(3)$ we get an isomorphism $\rho : H^0(\zeta) \oplus H^0(\zeta)^\vee \xrightarrow{\sim} \bigwedge^3 V$ such that (abusing notation) $\rho(\Omega^3_{|H|^\vee}(3)) = F$. Lastly one checks that the standard symplectic form on $(H^0(\zeta) \oplus H^0(\zeta)^\vee)$ is identified (up to a multiple) via ρ with the symplectic form $(\cdot, \cdot)_V$ of (3.1.1). Now let $A = \rho(H^0(\zeta)^\vee)$; then (4.2.3) is identified with (3.1.19). This ends the proof of Theorem 4.10. $\square$

References

[Beauville 1983] A. Beauville, "Variétés Kähleriennes dont la première classe de Chern est nulle", *J. Differential Geom.* **18**:4 (1983), 755–782.

[Beauville and Donagi 1985] A. Beauville and R. Donagi, "La variété des droites d'une hypersurface cubique de dimension 4", *C. R. Acad. Sci. Paris Sér. I Math.* **301**:14 (1985), 703–706.

[Bogomolov 1978] F. A. Bogomolov, "Hamiltonian Kählerian manifolds", *Dokl. Akad. Nauk SSSR* **243**:5 (1978), 1101–1104. In Russian; translated in *Soviet Math. Dokl.* **19** (1978), 1462–1465.

[Bogomolov 1996] F. A. Bogomolov, "On the cohomology ring of a simple hyper-Kähler manifold (on the results of Verbitsky)", *Geom. Funct. Anal.* **6**:4 (1996), 612–618.

[Boucksom 2001] S. Boucksom, "Le cône kählérien d'une variété hyperkählérienne", *C. R. Acad. Sci. Paris Sér. I Math.* **333**:10 (2001), 935–938.

[Boucksom 2004] S. Boucksom, "Divisorial Zariski decompositions on compact complex manifolds", *Ann. Sci. École Norm. Sup.* (4) **37**:1 (2004), 45–76.

[Burns and Rapoport 1975] D. Burns, Jr. and M. Rapoport, "On the Torelli problem for kählerian K-3 surfaces", *Ann. Sci. École Norm. Sup.* (4) **8**:2 (1975), 235–273.

[Casnati and Catanese 1997] G. Casnati and F. Catanese, "Even sets of nodes are bundle symmetric", *J. Differential Geom.* **47**:2 (1997), 237–256.

[Debarre 1984] O. Debarre, "Un contre-exemple au théorème de Torelli pour les variétés symplectiques irréductibles", *C. R. Acad. Sci. Paris Sér. I Math.* **299**:14 (1984), 681–684.

[Debarre and Voisin 2010] O. Debarre and C. Voisin, "Hyper-Kähler fourfolds and Grassmann geometry", *J. Reine Angew. Math.* **649** (2010), 63–87.

[Demailly and Paun 2004] J.-P. Demailly and M. Paun, "Numerical characterization of the Kähler cone of a compact Kähler manifold", *Ann. of Math.* (2) **159**:3 (2004), 1247–1274.

[Druel 2011] S. Druel, "Quelques remarques sur la décomposition de Zariski divisorielle sur les variétés dont la première classe de Chern est nulle", *Math. Z.* **267**:1-2 (2011), 413–423.

[Eisenbud et al. 2001] D. Eisenbud, S. Popescu, and C. Walter, "Lagrangian subbundles and codimension 3 subcanonical subschemes", *Duke Math. J.* **107**:3 (2001), 427–467.

[Fantechi 1995] B. Fantechi, "Deformation of Hilbert schemes of points on a surface", *Compositio Math.* **98**:2 (1995), 205–217.

[Fujiki 1983] A. Fujiki, "On primitively symplectic compact Kähler V-manifolds of dimension four", pp. 71–250 in *Classification of algebraic and analytic manifolds* (Katata, 1982), Progr. Math. **39**, Birkhäuser, Boston, 1983.

[Fujiki 1987] A. Fujiki, "On the de Rham cohomology group of a compact Kähler symplectic manifold", pp. 105–165 in *Algebraic geometry* (Sendai, 1985), Adv. Stud. Pure Math. **10**, North-Holland, Amsterdam, 1987.

[Fulton and Pragacz 1998] W. Fulton and P. Pragacz, *Schubert varieties and degeneracy loci*, Lecture Notes in Mathematics **1689**, Springer, Berlin, 1998.

[Göttsche 1994] L. Göttsche, *Hilbert schemes of zero-dimensional subschemes of smooth varieties*, Lecture Notes in Mathematics **1572**, Springer, Berlin, 1994.

[Göttsche and Huybrechts 1996] L. Göttsche and D. Huybrechts, "Hodge numbers of moduli spaces of stable bundles on K3 surfaces", *Internat. J. Math.* **7**:3 (1996), 359–372.

[Gritsenko et al. 2007] V. A. Gritsenko, K. Hulek, and G. K. Sankaran, "The Kodaira dimension of the moduli of K3 surfaces", *Invent. Math.* **169**:3 (2007), 519–567.

[Guan 2001] D. Guan, "On the Betti numbers of irreducible compact hyperkähler manifolds of complex dimension four", *Math. Res. Lett.* **8**:5-6 (2001), 663–669.

[Hassett and Tschinkel 2001] B. Hassett and Y. Tschinkel, "Rational curves on holomorphic symplectic fourfolds", *Geom. Funct. Anal.* **11**:6 (2001), 1201–1228.

[Hassett and Tschinkel 2009a] B. Hassett and Y. Tschinkel, "Intersection numbers of extremal rays on holomorphic symplectic four-folds", preprint, 2009. arXiv 0909.4745

[Hassett and Tschinkel 2009b] B. Hassett and Y. Tschinkel, "Moving and ample cones of holomorphic symplectic fourfolds", *Geom. Funct. Anal.* **19**:4 (2009), 1065–1080.

[Huybrechts 1999] D. Huybrechts, "Compact hyper-Kähler manifolds: basic results", *Invent. Math.* **135**:1 (1999), 63–113. Erratum in [Huybrechts 2003a].

[Huybrechts 2003a] D. Huybrechts, "Erratum to 'Compact hyper-Kähler manifolds: basic results'", *Invent. Math.* **152**:1 (2003), 209–212.

[Huybrechts 2003b] D. Huybrechts, "Finiteness results for compact hyperkähler manifolds", *J. Reine Angew. Math.* **558** (2003), 15–22.

[Huybrechts 2003c] D. Huybrechts, "The Kähler cone of a compact hyperkähler manifold", *Math. Ann.* **326**:3 (2003), 499–513.

[Huybrechts 2011] D. Huybrechts, "A global Torelli theorem for hyperkaehler manifolds (after Verbitsky)", preprint, 2011. arXiv 1106.5573

[Iliev and Manivel 2009] A. Iliev and L. Manivel, "Fano manifolds of degree ten and EPW sextics", preprint, 2009. arXiv 0907.2781

[Iliev and Ranestad 2001] A. Iliev and K. Ranestad, "K3 surfaces of genus 8 and varieties of sums of powers of cubic fourfolds", *Trans. Amer. Math. Soc.* **353**:4 (2001), 1455–1468.

[Iliev and Ranestad 2007] A. Iliev and K. Ranestad, "Addendum to "K3 surfaces of genus 8 and varieties of sums of powers of cubic fourfolds" [Trans. Amer. Math. Soc. **353** (2001), no. 4, 1455–1468; MR1806733]", *C. R. Acad. Bulgare Sci.* **60**:12 (2007), 1265–1270.

[Kaledin et al. 2006] D. Kaledin, M. Lehn, and C. Sorger, "Singular symplectic moduli spaces", *Invent. Math.* **164**:3 (2006), 591–614.

[Kawamata 1997] Y. Kawamata, "On Fujita's freeness conjecture for 3-folds and 4-folds", *Math. Ann.* **308**:3 (1997), 491–505.

[Kodaira 1964] K. Kodaira, "On the structure of compact complex analytic surfaces, I", *Amer. J. Math.* **86** (1964), 751–798.

[Le Potier 1985] J. Le Potier, "Simple connexité des surfaces K3", pp. 79–89 in *Geometry of K3 surfaces: moduli and periods* (Palaiseau, 1981/1982), Astérisque **126**, Soc. Math. de France, 1985.

[Lehn and Sorger 2006] M. Lehn and C. Sorger, "La singularité de O'Grady", *J. Algebraic Geom.* **15**:4 (2006), 753–770.

[Looijenga and Peters 1980/81] E. Looijenga and C. Peters, "Torelli theorems for Kähler K3 surfaces", *Compositio Math.* **42**:2 (1980/81), 145–186.

[Markman 2001] E. Markman, "Brill-Noether duality for moduli spaces of sheaves on K3 surfaces", *J. Algebraic Geom.* **10**:4 (2001), 623–694.

[Markman 2009] E. Markman, "Prime exceptional divisors on holomorphic symplectic varieties and modromy-reflections", preprint, 2009. arXiv 0912.4981

[Markman 2010] E. Markman, "Integral constraints on the monodromy group of the hyperKähler resolution of a symmetric product of a K3 surface", *Internat. J. Math.* **21**:2 (2010), 169–223.

[Matsushita 1999] D. Matsushita, "On fibre space structures of a projective irreducible symplectic manifold", *Topology* **38**:1 (1999), 79–83.

[Matsushita 2001] D. Matsushita, "Addendum: "On fibre space structures of a projective irreducible symplectic manifold" [Topology **38** (1999), no. 1, 79–83; MR1644091 (99f:14054)]", *Topology* **40**:2 (2001), 431–432.

[Matsushita 2005] D. Matsushita, "Higher direct images of dualizing sheaves of Lagrangian fibrations", *Amer. J. Math.* **127**:2 (2005), 243–259.

[Mayer 1972] A. L. Mayer, "Families of $K - 3$ surfaces", *Nagoya Math. J.* **48** (1972), 1–17.

[Mukai 1984] S. Mukai, "Symplectic structure of the moduli space of sheaves on an abelian or K3 surface", *Invent. Math.* **77**:1 (1984), 101–116.

[Mukai 1987a] S. Mukai, "Moduli of vector bundles on K3 surfaces and symplectic manifolds", *Sūgaku* **39**:3 (1987), 216–235. Reprinted in *Sugaku Expositions* **1**:2 (1988), 139–174.

[Mukai 1987b] S. Mukai, "On the moduli space of bundles on K3 surfaces, I", pp. 341–413 in *Vector bundles on algebraic varieties*, Tata Inst. Fund. Res. Stud. Math. **11**, Tata Inst. Fund. Res., Bombay, 1987.

[Mukai 1988] S. Mukai, "Curves, K3 surfaces and Fano 3-folds of genus ≤ 10", pp. 357–377 in *Algebraic geometry and commutative algebra*, vol. I, Kinokuniya, Tokyo, 1988.

[Mukai 2006] S. Mukai, "Polarized K3 surfaces of genus thirteen", pp. 315–326 in *Moduli spaces and arithmetic geometry*, Adv. Stud. Pure Math. **45**, Math. Soc. Japan, Tokyo, 2006.

[Namikawa 2002] Y. Namikawa, "Counter-example to global Torelli problem for irreducible symplectic manifolds", *Math. Ann.* **324**:4 (2002), 841–845.

[O'Grady 1997] K. G. O'Grady, "The weight-two Hodge structure of moduli spaces of sheaves on a K3 surface", *J. Algebraic Geom.* **6**:4 (1997), 599–644.

[O'Grady 1999] K. G. O'Grady, "Desingularized moduli spaces of sheaves on a K3", *J. Reine Angew. Math.* **512** (1999), 49–117.

[O'Grady 2003] K. G. O'Grady, "A new six-dimensional irreducible symplectic variety", *J. Algebraic Geom.* **12**:3 (2003), 435–505.

[O'Grady 2006] K. G. O'Grady, "Irreducible symplectic 4-folds and Eisenbud-Popescu-Walter sextics", *Duke Math. J.* **134**:1 (2006), 99–137.

[O'Grady 2008] K. G. O'Grady, "Irreducible symplectic 4-folds numerically equivalent to $(K3)^{[2]}$", *Commun. Contemp. Math.* **10**:4 (2008), 553–608.

[Perego and Rapagnetta 2010] A. Perego and A. Rapagnetta, "Deformation of the O'Grady moduli spaces", preprint, 2010. arXiv 1008.0190

[Piatetski-Shapiro and Shafarevich 1971] I. Piatetski-Shapiro and I. R. Shafarevich, "A Torelli theorem for algebraic surfaces of type K3", *Math. USSR Izvestija* **5** (1971), 547–588.

[Rapagnetta 2007] A. Rapagnetta, "Topological invariants of O'Grady's six dimensional irreducible symplectic variety", *Math. Z.* **256**:1 (2007), 1–34.

[Rapagnetta 2008] A. Rapagnetta, "On the Beauville form of the known irreducible symplectic varieties", *Math. Ann.* **340**:1 (2008), 77–95.

[Salamon 1996] S. M. Salamon, "On the cohomology of Kähler and hyper-Kähler manifolds", *Topology* **35**:1 (1996), 137–155.

[Sawon 2003] J. Sawon, "Abelian fibred holomorphic symplectic manifolds", *Turkish J. Math.* **27**:1 (2003), 197–230.

[Todorov 1980] A. N. Todorov, "Applications of the Kähler–Einstein–Calabi–Yau metric to moduli of K3 surfaces", *Invent. Math.* **61**:3 (1980), 251–265.

[Tyurin 1987] A. N. Tyurin, "Cycles, curves and vector bundles on an algebraic surface", *Duke Math. J.* **54**:1 (1987), 1–26.

[Verbitsky 1996] M. Verbitsky, "Cohomology of compact hyper-Kähler manifolds and its applications", *Geom. Funct. Anal.* **6**:4 (1996), 601–611.

[Verbitsky 2009] M. Verbitsky, "Mapping class group and a global Torelli theorem for hyper-Kähler manifolds", preprint, 2009. arXiv 0908.4121v5

[Wierzba and Wiśniewski 2003] J. Wierzba and J. A. Wiśniewski, "Small contractions of symplectic 4-folds", *Duke Math. J.* **120**:1 (2003), 65–95.

[Yoshioka 1999] K. Yoshioka, "Some examples of Mukai's reflections on K3 surfaces", *J. Reine Angew. Math.* **515** (1999), 97–123.

[Yoshioka 2001] K. Yoshioka, "Moduli spaces of stable sheaves on abelian surfaces", *Math. Ann.* **321**:4 (2001), 817–884.

ogrady@mat.uniroma1.it   *Dipartimento di Matematica, Università di Roma "La Sapienza", Piazzale A. Moro 5, 00185 Roma, Italy*

Current Developments in Algebraic Geometry
MSRI Publications
Volume 59, 2011

Compactifications of moduli of abelian varieties: an introduction

MARTIN OLSSON

We survey the various approaches to compactifying moduli stacks of polarized abelian varieties. To motivate the different approaches to compactifying, we first discuss three different points of view of the moduli stacks themselves. Then we explain how each point of view leads to a different compactification. Throughout we emphasize maximal degenerations which capture much of the essence of the theory without many of the technicalities.

1. Introduction

A central theme in modern algebraic geometry is to study the degenerations of algebraic varieties, and its relationship with compactifications of moduli stacks. The standard example considered in this context is the moduli stack $\mathcal{M}_g$ of genus g curves (where $g \geq 2$) and the Deligne–Mumford compactification $\mathcal{M}_g \subset \overline{\mathcal{M}}_g$ [Deligne and Mumford 1969]. The stack $\overline{\mathcal{M}}_g$ has many wonderful properties:

(1) It has a moduli interpretation as the moduli stack of stable genus g curves.

(2) The stack $\overline{\mathcal{M}}_g$ is smooth.

(3) The inclusion $\mathcal{M}_g \hookrightarrow \overline{\mathcal{M}}_g$ is a dense open immersion and $\overline{\mathcal{M}}_g \setminus \mathcal{M}_g$ is a divisor with normal crossings in $\overline{\mathcal{M}}_g$.

Unfortunately the story of the compactification $\mathcal{M}_g \subset \overline{\mathcal{M}}_g$ is not reflective of the general situation. There are very few known instances where one has a moduli stack $\mathcal{M}$ classifying some kind of algebraic varieties and a compactification $\mathcal{M} \subset \overline{\mathcal{M}}$ with the three properties above.

After studying moduli of curves, perhaps to next natural example to consider is the moduli stack $\mathcal{A}_g$ of principally polarized abelian varieties of a fixed dimension g. Already here the story becomes much more complicated, though work of several people has led to a compactification $\mathcal{A}_g \subset \overline{\mathcal{A}}_g$ which enjoys the following properties:

(1) The stack $\overline{\mathcal{A}}_g$ is the solution to a natural moduli problem.

(2′) The stack $\overline{\mathcal{A}}_g$ has only toric singularities.

Partially supported by NSF grant DMS-0714086, NSF CAREER grant DMS-0748718, and an Alfred P. Sloan Research Fellowship.

(3′) The inclusion $\mathscr{A}_g \hookrightarrow \overline{\mathscr{A}}_g$ is a dense open immersion, and the complement $\overline{\mathscr{A}}_g \setminus \mathscr{A}_g$ defines an fs-log structure $M_{\overline{\mathscr{A}}_g}$ (in the sense of Fontaine and Illusie [Kato 1989]) on $\overline{\mathscr{A}}_g$ such that $(\overline{\mathscr{A}}_g, M_{\overline{\mathscr{A}}_g})$ is log smooth over $\mathrm{Spec}(\mathbb{Z})$.

Our aim in this paper is to give an overview of the various approaches to compactifying $\mathscr{A}_g$, and to outline the story of the canonical compactification $\mathscr{A}_g \hookrightarrow \overline{\mathscr{A}}_g$. In addition, we also consider higher degree polarizations.

What one considers a 'natural' compactification of $\mathscr{A}_g$ depends to a large extent on one's view of $\mathscr{A}_g$ itself. There are three basic points of view of this moduli stack (which of course are all closely related):

(*The standard approach*). Here one views $\mathscr{A}_g$ as classifying pairs (A, λ), where A is an abelian variety of dimension g and $\lambda : A \to A^t$ is an isomorphism between A and its dual (a *principal polarization*), such that λ is equal to the map defined by an ample line bundle, but one does not fix such a line bundle. This point of view is the algebraic approach most closely tied to Hodge theory.

(*Moduli of pairs approach*). This is the point of view taken in Alexeev's work [2002]. Here one encodes the ambiguity of the choice of line bundle defining λ into a torsor under A. So $\mathscr{A}_g$ is viewed as classifying collections of data (A, P, L, θ), where A is an abelian variety of dimension g, P is an A-torsor, L is an ample line bundle on P defining a principal polarization on A (see 2.2.3), and $\theta \in \Gamma(P, L)$ is a nonzero global section.

(*Theta group approach*). This point of view comes out of Mumford's theory [1966; 1967] of the theta group, combined with Alexeev's approach via torsors. Here one considers triples (A, P, L), where A is an abelian variety of dimension g, P is an A-torsor, and L is an ample line bundle on P defining a principal polarization on A (but one does not fix a section of L). This gives a stack $\mathscr{T}_g$ which is a gerbe over $\mathscr{A}_g$ bound by $\mathbb{G}_m$. Using a standard stack-theoretic construction called *rigidification* one can then construct $\mathscr{A}_g$ from $\mathscr{T}_g$, but in the theta group approach the stack $\mathscr{T}_g$ is the more basic object.

In Section 2 we discuss each of these three points of view of the moduli of principally polarized abelian varieties (and moduli of abelian varieties with higher degree polarization). Then in sections 3 and 4 we discuss how each of these three approaches leads to different compactifications (toroidal, Alexeev, and $\overline{\mathscr{A}}_g$ respectively). We discuss in some detail in the maximally degenerate case the relationship between degenerating abelian varieties and quadratic forms. This relationship is at the heart of all of the different approaches to compactification. We do not discuss the case of partial degenerations where one has to introduce the theory of biextensions (for this the reader should consult [Faltings and Chai 1990]), since most of the main ideas can already be seen in the maximally degenerate case.

Finally in Section 5 we give an overview of how the canonical compactification can be used to compactify moduli stacks for abelian varieties with level structure and higher degree polarizations, using the theta group approach.

Our aim here is not to give a complete treatment, but rather to give the reader an indication of some of the basic ideas involved. Much of our focus is on the local structure of these moduli stacks at points of maximal degeneration in the boundary of the various compactifications (i.e., points where the degeneration of the abelian scheme is a torus). This is because the local structure of the moduli stacks can be seen more clearly here, and because the case of partial degeneration introduces many more technicalities (in particular, in this paper we do not discuss the theory of biextensions). We hardly touch upon the issues involved in going from the local study to the global. The interested reader should consult the original sources [Alexeev 2002; Faltings and Chai 1990; Olsson 2008].

Perhaps preceding the entire discussion of this paper is the theory of the Sataka/Baily–Borel/minimal compactification of $\mathcal{A}_g$, and the connection with modular forms. We should also remark that a beautiful modular interpretation of the toroidal compactifications using log abelian varieties has been developed by Kajiwara, Kato, and Nakayama [Kajiwara et al. 2008a; 2008b]. We do not, however, discuss either of these topics here.

Acknowledgements. The aim of this article is to give a survey of known results, and there are no new theorems. The results discuss here are the fruits of work of many people. We won't try to make an exhaustive list, but let us at least mention two basic sources: [Faltings and Chai 1990] and [Alexeev 2002], from which we learned the bulk of the material on toroidal compactifications and Alexeev's compactification, respectively. We thank the referee for helpful comments on the first version of the paper.

Prerequisites and conventions. We assume that the reader is familiar with the basic theory of abelian varieties as developed for example in [Mumford 1970]. We also assume the reader is familiar with stacks at the level of [Laumon and Moret-Bailly 2000]. Finally knowledge of logarithmic geometry in the sense of Fontaine and Illusie [Kato 1989] will be assumed for sections 4.5 and 5.

Our conventions about algebraic stacks are those of [Laumon and Moret-Bailly 2000].

2. Three perspectives on $\mathcal{A}_g$

2.1. *The standard definition.*

2.1.1. Let k be an algebraically closed field, and let A/k be an abelian variety. Let A^t denote the dual abelian variety of A (see [Mumford 1970, Chapter III,

§13]). Recall that A^t is the connected component of the identity in the Picard variety $\underline{\mathrm{Pic}}_{A/k}$ of A. If L is a line bundle on A, then we obtain a map

$$\lambda_L : A \to A^t, \quad x \mapsto [t_x^* L \otimes L^{-1}],$$

where $t_x : A \to A$ denotes translation by the point x. If L is ample then λ_L is finite and the kernel is a finite group scheme over k (by [Mumford 1970, Application 1 on p. 60]) whose rank is a square by [Mumford 1970, Riemann–Roch theorem, p. 150]. The *degree* of an ample line bundle L is defined to be the positive integer d for which the rank of $\mathrm{Ker}(\lambda_L)$ is d^2. The degree d can also be characterized as the dimension of the k-space $\Gamma(A, L)$ (loc. cit.).

Definition 2.1.2. Let $d \geq 1$ be an integer. A *polarization of degree d* on an abelian variety A/k is a morphism $\lambda : A \to A^t$ of degree d^2, which is equal to λ_L for some ample line bundle L on A. A *principal polarization* is a polarization of degree 1.

Remark 2.1.3. If L and L' are two ample line bundles on an abelian variety A/k, then $\lambda_L = \lambda_{L'}$ if and only if $L' \simeq t_x^* L$ for some point $x \in A(k)$. Indeed $\lambda_L = \lambda_{L'}$ if and only if

$$\lambda_{L' \otimes L^{-1}} = \{e\} \quad \text{(constant map)},$$

which by the definition of the dual abelian variety (see for example [Mumford 1970, p. 125]) is equivalent to the statement that the line bundle $L' \otimes L^{-1}$ defines a point of A^t. Since λ_L is surjective, this in turn is equivalent to the statement that there exists a point $x \in A(k)$ such that

$$t_x^* L \otimes L^{-1} \simeq L' \otimes L^{-1},$$

or equivalently that $t_x^* L \simeq L'$. The same argument shows that if L and L' are line bundles such that $\lambda_L = \lambda_{L'}$ then L is ample if and only if L' is ample.

2.1.4. These definitions extend naturally to families. Recall [Mumford 1965, Definition 6.1] that if S is a scheme then an *abelian scheme over S* is a smooth proper group scheme A/S with geometrically connected fibers. As in the case of abelian varieties, the group scheme structure on A is determined by the zero section [Mumford 1965, Corollary 6.6].

For an abelian scheme A/S, one can define the dual abelian scheme A^t/S as a certain subgroup scheme of the relative Picard scheme $\underline{\mathrm{Pic}}_{A/S}$ (see [Mumford 1965, Corollary 6.8] for more details). As in the case of a field, any line bundle L on A defines a homomorphism

$$\lambda_L : A \to A^t.$$

If L is relatively ample then λ_L is finite and flat, and the kernel $\mathrm{Ker}(\lambda_L)$ has rank d^2 for some locally constant positive integer-valued function d on S. If $\pi : A \to S$ denotes the structure morphism, then we have

$$R^i \pi_* L = 0, \quad i > 0,$$

and $\pi_* L$ is a locally free sheaf of rank d on S whose formation commutes with arbitrary base change $S' \to S$ (this follows from the vanishing theorem for higher cohomology over fields [Mumford 1970, p. 150] and cohomology and base change).

Definition 2.1.5. Let $d \geq 1$ be an integer. A *polarization of degree d* on an abelian scheme A/S is a homomorphism $\lambda : A \to A^t$ such that for every geometric point $\bar{s} \to S$ the map on geometric fibers $A_{\bar{s}} \to A_{\bar{s}}^t$ is a polarization of degree d in the sense of 2.1.2.

Remark 2.1.6. By a similar argument as in 2.1.3, if A/S is an abelian scheme over a base S, and if L and L' are two relatively ample line bundles on A, then $\lambda_L = \lambda_{L'}$ if and only if there exists a point $x \in A(S)$ such that L' and $t_x^* L$ differ by the pullback of a line bundle on S.

2.1.7. If (A, λ) and (A', λ') are two abelian schemes over a scheme S with polarizations of degree d, then an isomorphism $(A, \lambda) \to (A', \lambda')$ is an isomorphism of abelian schemes

$$f : A \to A'$$

such that the diagram

$$
\begin{array}{ccc}
A & \xrightarrow{\ f\ } & A' \\
\downarrow{\scriptstyle \lambda} & & \downarrow{\scriptstyle \lambda'} \\
A^t & \xleftarrow{\ f^t\ } & A'^t
\end{array}
$$

commutes, where f^t denotes the isomorphism of dual abelian schemes induced by f.

Lemma 2.1.8. *Let A/S be an abelian scheme and $\lambda : A \to A^t$ a homomorphism. Suppose $s \in S$ is a point such that the restriction $\lambda_s : A_s \to A_s^t$ of λ to the fiber at s is equal to λ_{L_s} for some ample line bundle L_s on A_s. Then after replacing S by an étale neighborhood of s, there exists a relatively ample line bundle L on A such that $\lambda = \lambda_L$.*

Proof. By a standard limit argument, it suffices to consider the case when S is of finite type over an excellent Dedekind ring. By the Artin approximation theorem [1969, 2.2] applied to the functor

$$F : (S\text{-schemes})^{\mathrm{op}} \to \mathrm{Sets}$$

sending an S-scheme T to the set of isomorphism classes of line bundles L on A_T such that $\lambda = \lambda_L$, it suffices to consider the case when $S = \mathrm{Spec}(R)$ is the spectrum of a complete noetherian local ring. In this case it follows from [Oort 1971, 2.3.2 and its proof] that there exists a line bundle L on A whose fiber over the closed point s is isomorphic to L_s. Now note that the two maps

$$\lambda_L, \lambda : A \to A^t$$

are equal by [Mumford 1965, Chapter 6, Corollary 6.2]. $\qquad\qquad\qquad\qquad\square$

Lemma 2.1.9. *Let A/S be an abelian scheme over a scheme S, and let $\lambda : A \to A^t$ be a polarization. Then fppf-locally on S there exists a relatively ample line bundle L on A such that $\lambda = \lambda_L$. If 2 is invertible on S, then there exists such a line bundle étale locally on S.*

Proof. Consider first the case when $S = \mathrm{Spec}(k)$, for some field k. In this case, there exists by [Mumford 1965, Chapter 6, Proposition 6.10] a line bundle M on A such that $\lambda_M = 2\lambda$. Let Z denote the fiber product of the diagram

$$\begin{array}{c}
\mathrm{Spec}(k) \\
\downarrow {\scriptstyle [M]} \\
\underline{\mathrm{Pic}}_{A/k} \xrightarrow{\ \cdot 2\ } \underline{\mathrm{Pic}}_{A/k}.
\end{array}$$

The scheme Z represents the fppf-sheaf associated to the presheaf which to any k-scheme T associates the set of isomorphism classes of line bundles L for which $L^{\otimes 2} \simeq M$.

By assumption, there exists a field extension $k \to K$ and a line bundle L on A_K such that $\lambda|_{A_K} = \lambda_L$. Then

$$\lambda_{L^{\otimes 2}} = 2\lambda = \lambda_M,$$

so by 2.1.3 there exists, after possibly replacing K by an even bigger field extension, a point $x \in A(K)$ such that $t_x^*(L^{\otimes 2}) \simeq M$. It follows that $t_x^* L$ defines a point of $Z(K)$. Note also that if L is a line bundle on A such that $L^{\otimes 2} \simeq M$ then for any other line bundle R on A the product $L \otimes R$ defines a point of Z if and only if the class of the line bundle R is a point of $A^t[2]$. From this we conclude that Z is a torsor under $A^t[2]$. In particular, Z is étale if 2 is invertible in k, whence in this case there exists étale locally a section of Z.

To conclude the proof in the case of a field, note that if L is a line bundle on A with $L^{\otimes 2} \simeq M$, then

$$\lambda_L - \lambda : A \to A^t$$

has image in $A^t[2]$ since $2\lambda_L = 2\lambda$, and since $A^t[2]$ is affine the map $\lambda_L - \lambda$ must be the trivial homomorphism.

For the general case, let $s \in S$ be a point. Then we can find a finite field extension $k(s) \to K$ and a line bundle L on A_K such that $\lambda_L = \lambda|_{A_K}$. By the above we can further assume $k(s) \to K$ is separable if 2 is invertible in S. Now by [EGA 1961, chapitre 0, proposition 10.3.1, p. 20] there exists a quasifinite flat morphism $S' \to S$ and a point $s' \in S'$ such that the induced extension

$$k(s) \to k(s')$$

is isomorphic to $k(s) \to K$. If $k(s) \to K$ is separable then we can even choose $S' \to S$ to be étale. Now we obtain the result from 2.1.8 applied to $s' \in S'$. $\qquad\square$

2.1.10. For integers $d, g \geq 1$, let $\mathcal{A}_{g,d}$ denote the fibered category over the category of schemes, whose fiber over a scheme S is the groupoid of pairs $(A/S, \lambda)$, where A is an abelian scheme of dimension g and $\lambda : A \to A^t$ is a polarization of degree d. We denote $\mathcal{A}_{g,1}$ simply by $\mathcal{A}_g$.

The basic result on the fibered category $\mathcal{A}_{g,d}$ is the following:

Theorem 2.1.11. *The fibered category $\mathcal{A}_{g,d}$ is a Deligne–Mumford stack over $\mathbb{Z}$, with quasiprojective coarse moduli space $A_{g,d}$. Over $\mathbb{Z}[1/d]$ the stack $\mathcal{A}_{g,d}$ is smooth.*

Proof. For the convenience of the reader we indicate how to obtain this theorem from the results of [Mumford 1965], which does not use the language of stacks.

Recall that if S is a scheme and A/S is an abelian scheme, then for any integer n invertible on S the kernel of multiplication by n on A

$$A[n] := \mathrm{Ker}(\cdot n : A \to A)$$

is a finite étale group scheme over S of rank n^{2g}, étale locally isomorphic to $(\mathbb{Z}/(n))^{2g}$. Define a *full level-n-structure on A/S* to be an isomorphism

$$\sigma : (\mathbb{Z}/(n))^{2g} \simeq A[n],$$

and let $\mathcal{A}_{g,d,n}$ be the fibered category over $\mathbb{Z}[1/n]$ whose fiber over a $\mathbb{Z}[1/n]$-scheme S is the groupoid of triples (A, λ, σ), where $(A, \lambda) \in \mathcal{A}_{g,d}(S)$ and σ is a full level-n-structure on A. Here an isomorphism between two objects

$$(A, \lambda, \sigma), \quad (A', \lambda', \sigma') \in \mathcal{A}_{g,d,n}(S)$$

is an isomorphism $f : (A, \lambda) \to (A', \lambda')$ in $\mathcal{A}_{g,d}(S)$ such that the diagram

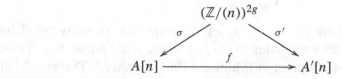

commutes. By [Mumford 1965, Chapter 7, Theorem 7.9 and remark following its proof], if $n \geq 3$ then $\mathscr{A}_{g,d,n}$ is equivalent to the functor represented by a quasiprojective $\mathbb{Z}[1/n]$-scheme. Let us also write $\mathscr{A}_{g,d,n}$ for this scheme. There is a natural action of $\mathrm{GL}_{2g}(\mathbb{Z}/(n))$ on $\mathscr{A}_{g,d,n}$ for which $g \in \mathrm{GL}_{2g}(\mathbb{Z}/(n))$ sends (A, λ, σ) to $(A, \lambda, \sigma \circ g)$. Furthermore, we have an isomorphism

$$\mathscr{A}_{g,d}|_{\mathbb{Z}[1/n]} \simeq [\mathscr{A}_{g,d,n}/\mathrm{GL}_{2g}(\mathbb{Z}/(n))].$$

Now choose two integer $n, n' \geq 3$ such that n and n' are relatively prime. We then get a covering

$$\mathscr{A}_{g,d} \simeq [\mathscr{A}_{g,d,n}/\mathrm{GL}_{2g}(\mathbb{Z}/(n))] \cup [\mathscr{A}_{g,d,n'}/\mathrm{GL}_{2g}(\mathbb{Z}/(n'))]$$

of $\mathscr{A}_{g,d}$ by open substacks which are Deligne–Mumford stacks, whence $\mathscr{A}_{g,d}$ is also a Deligne–Mumford stack.

By [Keel and Mori 1997, 1.3] the stack $\mathscr{A}_{g,d}$ has a coarse moduli space, which we denote by $A_{g,d}$. A priori $A_{g,d}$ is an algebraic space, but we show that $A_{g,d}$ is a quasiprojective scheme as follows.

Recall from [Mumford 1965, Chapter 6, Propositon 6.10], that to any object $(A, \lambda) \in \mathscr{A}_{g,d}(S)$ over some scheme S, there is a canonically associated relatively ample line bundle M on A which is rigidified at the zero section of A and such that $\lambda_M = 2\lambda$. By [Mumford 1970, theorem on p. 163] and cohomology and base change, the sheaf $M^{\otimes 3}$ is relatively very ample on A/S, and if $f : A \to S$ denotes the structure morphism then $f_*(M^{\otimes 3})$ is a locally free sheaf on S whose formation commutes with arbitrary base change $S' \to S$ and whose rank N is independent of (A, λ).

Let

$$f : \mathscr{X} \to \mathscr{A}_{g,d}$$

denote the universal abelian scheme, and let $\mathcal{M}$ denote the invertible sheaf on $\mathscr{X}$ given by the association

$$(A, \lambda, \sigma) \mapsto M.$$

For $r \geq 1$, let $\mathscr{E}_r$ denote the vector bundle on $\mathscr{A}_{g,d}$ given by $f_*(\mathcal{M}^{\otimes 3r})$, and let $\mathscr{L}_r$ denote the top exterior power of $\mathscr{E}_r$. We claim that for suitable choices of r and s the line bundle $\mathscr{L}_{mr}^{\otimes ms}$ descends to an ample line bundle on $A_{g,d}$ for any $m \geq 1$. Note that if this is the case, then the descended line bundle is unique up to unique isomorphism, for if R is any line bundle on $A_{g,d}$ then the adjunction map

$$R \to \pi_*\pi^*R$$

is an isomorphism, where $\pi : \mathscr{A}_{g,d} \to A_{g,d}$ is the projection. To verify this claim it suffices to verify it after restricting to $\mathbb{Z}[1/p]$, where p is a prime. In this case the claim follows from the proof of [Mumford 1965, Chapter 7, Theorem 7.10].

Finally the statement that $\mathscr{A}_{g,d}$ is smooth over $\mathbb{Z}[1/d]$ follows from [Oort 1971, 2.4.1]. $\qquad\qquad\square$

2.2. Moduli of pairs.

2.2.1. In [Alexeev 2002], Alexeev introduced a different perspective on $\mathscr{A}_g$. The key point is to encode into a torsor the ambiguity in the choice of line bundle for a given polarization. To make this precise let us first introduce some basic results about torsors under abelian varieties.

2.2.2. Let S be a scheme and A/S an abelian scheme. An A-*torsor* is a smooth scheme $f : P \to S$ with an action of A on P over S such that the graph of the action map

$$A \times_S P \to P \times_S P, \qquad (a, p) \mapsto (p, a * p)$$

is an isomorphism. This implies that if we have a section $s : S \to P$ of f then the induced map

$$A \to P, \qquad a \mapsto a * s$$

is an isomorphism of schemes compatible with the A-action, where A acts on itself by left translation. In particular, f is a proper morphism.

2.2.3. If A/S is an abelian scheme, and P/S is an A-torsor, then any line bundle L on P defines a homomorphism

$$\lambda_L : A \to A^t.$$

Namely, since $P \to S$ is smooth, there exists étale locally a section $s : S \to P$ which defines an isomorphism $\iota_s : A \to P$. In this situation we define λ_L to be the map

$$\lambda_{\iota_s^* L} : A \to A^t, \qquad a \mapsto t_a^*(\iota_s^* L) \otimes \iota_s^* L^{-1}.$$

We claim that this is independent of the choice of section s. To see this let $s' : S \to P$ be another section. Since P is an A-torsor there exists a unique point $b \in A(S)$ such that $s' = b * s$. It follows that $\iota_{s'}^* L \simeq t_b^* \iota_s^* L$, so the claim follows from [Alexeev 2002, 4.1.12]. It follows that even when there is no section of P/S, we can define the map λ_L by descent theory using local sections.

2.2.4. With notation as in the preceding paragraph, suppose L is an ample line bundle on P, and let $f : P \to S$ be the structure morphism. Then:

(1) $f_* L$ is a locally free sheaf of finite rank on S whose formation commutes with arbitrary base change on S.

(2) If d denotes the rank of $f_* L$, then the kernel of $\lambda_L : A \to A^t$ is a finite flat group scheme over S of rank d^2.

Indeed both these assertions are local on S in the étale topology, so to prove them it suffices to consider the case when P admits a section, in which case they follow from the corresponding statements for ample line bundles on abelian schemes.

2.2.5. The most important example of torsors for this paper is the following. Let S be a scheme and let A/S be an abelian scheme with a principal polarization $\lambda : A \to A^t$. Consider the functor

$$P : (S\text{-schemes}) \to \text{Sets}$$

which to any S-scheme T associates the set of isomorphism classes of pairs (L, ϵ), where L is a line bundle on A_T such that $\lambda_L = \lambda|_{A_T}$ and $\epsilon : \mathcal{O}_T \to e^*L$ is an isomorphism of $\mathcal{O}_T$-modules. Note that two objects (L, ϵ) and (L', ϵ') are isomorphic if and only if the line bundles L and L' are isomorphic, in which case there exists a unique isomorphism $\sigma : L \to L'$ such that the induced diagram

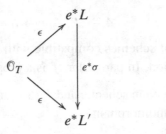

commutes.

There is an action of A on P defined as follows. Given an S-scheme T, a T-valued point $x \in A(T)$, and an element $(L, \epsilon) \in P(T)$, define $x * (L, \epsilon)$ to be the line bundle

$$t_x^*L \otimes_{\mathcal{O}_S} x^*L^{-1} \otimes_{\mathcal{O}_S} e^*L$$

on A_T, where $t_x : A_T \to A_T$ is the translation, and let $x * \epsilon$ be the isomorphism obtained from ϵ and the canonical isomorphism

$$e^*(t_x^*L \otimes_{\mathcal{O}_S} x^*L^{-1} \otimes_{\mathcal{O}_S} e^*L) \simeq x^*L \otimes x^*L^{-1} \otimes e^*L \simeq e^*L.$$

Then the functor P is representable, and the action of A makes P an A-torsor. Note that we can also think of P is the sheaf (with respect to the étale topology) associated to the presheaf which to any S-scheme T associates the set of isomorphism classes of line bundles L on A_T such that $\lambda_L = \lambda|_T$.

On P there is a tautological line bundle $\mathcal{L}$ together with a global section $\theta \in \Gamma(P, \mathcal{L})$. Indeed giving such a line bundle and section is equivalent to giving for every scheme-valued point $p \in P(T)$ a line bundle $\mathcal{L}_p$ on T together with a section $\theta_p \in \Gamma(T, \mathcal{L}_p)$. We obtain such a pair by noting that since P is a torsor, the point p corresponds to a pair (L_p, ϵ_p) on A_T, and we define $\mathcal{L}_p$ to be e^*L_p

with the section θ_p being the image of 1 under the map

$$\epsilon_p : \mathcal{O}_T \to e^* L_p.$$

2.2.6. Define $\mathscr{A}_g^{\mathrm{Alex}}$ to be the fibered category over the category of schemes, whose fiber over a scheme S is the groupoid of quadruples (A, P, L, θ) as follows:

(1) A/S is an abelian scheme of relative dimension g.

(2) P is an A-torsor. Let $f : P \to S$ denote the structure morphism.

(3) L is an ample line bundle on P such that $\lambda_L : A \to A^t$ is an isomorphism.

(4) $\theta : \mathcal{O}_S \to f_* L$ is an isomorphism of line bundles on S.

Note that for any $(A, P, L, \theta) \in \mathscr{A}_g^{\mathrm{Alex}}(S)$ the pair (A, λ_L) is an object of $\mathscr{A}_g(S)$. We therefore get a morphism of fibered categories

$$F : \mathscr{A}_g^{\mathrm{Alex}} \to \mathscr{A}_g \tag{1}$$

Proposition 2.2.7. *The morphism* (1) *is an equivalence.*

Proof. The construction in 2.2.5 defines another functor

$$G : \mathscr{A}_g \to \mathscr{A}_g^{\mathrm{Alex}}$$

which we claim is a quasi-inverse to F.

For this note that given a quadruple $(A, P, L, \theta) \in \mathscr{A}_g^{\mathrm{Alex}}(S)$ over some scheme S, and if (A, P', L', θ') denote the object obtained by applying $G \circ F$, then there is a natural map of A-torsors

$$\rho : P \to P'$$

obtained by associating to any S-scheme-valued point $p \in P(T)$ the class of the line bundle $\iota_p^* L$, where

$$\iota_p : A \to P$$

is the A-equivariant isomorphism obtained by sending $e \in A$ to p (here we think of P' as the sheaf associated to the presheaf of isomorphism classes of line bundles on A defining λ). By construction the isomorphism ρ has the property that $\rho^* L'$ and L are locally on S isomorphic. Since the automorphism group scheme of any line bundle on P is isomorphic to $\mathbb{G}_m$, we see that there exists a unique isomorphism

$$\tilde{\rho} : \rho^* L' \to L$$

sending θ' to θ. We therefore obtain a natural isomorphism

$$(A, P, L, \theta) \simeq (A, P', L', \theta')$$

in $\mathscr{A}_g^{\mathrm{Alex}}$. This construction defines an isomorphism of functors $\mathrm{id} \to G \circ F$.

To construct an isomorphism id $\to F \circ G$, it suffices to show that if (A, λ) belongs to $\mathcal{A}_g(S)$ for some scheme S, and if (A, P, L, θ) denotes $G(A, P, L, \theta)$, then $\lambda_L = \lambda$, which is immediate from the construction in 2.2.5. $\square$

Remark 2.2.8. In what follows we will usually not use the notation $\mathcal{A}_g^{\mathrm{Alex}}$.

Remark 2.2.9. While we find the language of line bundles with sections most convenient, note that giving the pair (L, θ) is equivalent to giving the corresponding Cartier divisor $D \hookrightarrow P$.

2.3. *Approach via theta group.* The third approach to the moduli stacks $\mathcal{A}_{g,d}$ is through a study of theta groups of line bundles. Before explaining this we first need a general stack theoretic construction that will be needed. The notion of *rigidification* we describe below has been discussed in various level of generality in many papers (see for example [Abramovich et al. 2003, Theorem 5.1.5]).

2.3.1. Let $\mathcal{X}$ be an algebraic stack, and let $\mathcal{I}_{\mathcal{X}} \to \mathcal{X}$ be its inertia stack. By definition, the stack $\mathcal{I}_{\mathcal{X}}$ has fiber over a scheme S the groupoid of pairs (x, α), where $x \in \mathcal{X}(S)$ and $\alpha : x \to x$ is an automorphism of x. In particular, $\mathcal{I}_{\mathcal{X}}$ is a relative group space over $\mathcal{X}$. The stack $\mathcal{I}_{\mathcal{X}}$ can also be described as the fiber product of the diagram

$$
\begin{array}{ccc}
 & & \mathcal{X} \\
 & & \downarrow{\scriptstyle \Delta} \\
\mathcal{X} & \xrightarrow{\ \Delta\ } & \mathcal{X} \times \mathcal{X}.
\end{array}
$$

Suppose further given a closed substack $\mathcal{G} \subset \mathcal{I}_{\mathcal{X}}$ such that the following hold:

(i) For every $x : S \to \mathcal{X}$ with S a scheme, the base change $\mathcal{G}_S \hookrightarrow \mathcal{I}_S$ is a normal subgroup space of the group space $\mathcal{G}_S$.

(ii) The structure map $\mathcal{G} \to \mathcal{X}$ is flat.

Then one can construct a new stack $\bar{\mathcal{X}}$, called the *rigidification of $\mathcal{X}$ with respect to $\mathcal{G}$*, together with a map

$$\pi : \mathcal{X} \to \bar{\mathcal{X}}$$

such that the following hold:

(i) The morphism on inertia stacks

$$\mathcal{I}_{\mathcal{X}} \to \mathcal{I}_{\bar{\mathcal{X}}}$$

sends $\mathcal{G}$ to the identity in $\mathcal{I}_{\bar{\mathcal{X}}}$.

(ii) The morphism π is universal with respect to this property: If $\mathcal{Y}$ is any algebraic stack, then

$$\pi^* : \mathrm{HOM}(\bar{\mathcal{X}}, \mathcal{Y}) \to \mathrm{HOM}(\mathcal{X}, \mathcal{Y})$$

identifies the category $\mathrm{HOM}(\bar{\mathfrak{X}}, \mathfrak{Y})$ with the full subcategory of $\mathrm{HOM}(\mathfrak{X}, \mathfrak{Y})$ of morphisms $f : \mathfrak{X} \to \mathfrak{Y}$ for which the induced morphism of inertia stacks

$$\mathcal{I}_{\mathfrak{X}} \to \mathcal{I}_{\mathfrak{Y}}$$

sends $\mathcal{G}$ to the identity.

(iii) The map π is faithfully flat, and $\mathfrak{X}$ is a gerbe over $\bar{\mathfrak{X}}$.

2.3.2. The stack $\bar{\mathfrak{X}}$ is obtained as the stack associated to the prestack $\bar{\mathfrak{X}}^{\mathrm{ps}}$ whose objects are the same as those of $\mathfrak{X}$ but whose morphisms between two objects $x, x' \in \mathfrak{X}(S)$ over a scheme S is given by the quotient of $\mathrm{Hom}_{\mathfrak{X}(S)}(x, x')$ by the natural action of $\mathcal{G}(S, x)$ (a subgroup scheme of the scheme of automorphisms of x). One checks (see for example [Olsson 2008, §1.5]) that the composition law for morphisms in $\mathfrak{X}$ descends to a composition law for morphisms modulo the action of $\mathcal{G}$.

Remark 2.3.3. The faithful flatness of the map π implies that one can frequently descend objects from $\mathfrak{X}$ to $\bar{\mathfrak{X}}$. Let us explain this in the case of quasicoherent sheaves, but the same argument applies in many other contexts (in particular to finite flat group schemes and logarithmic structures, which will be considered later). For an object $x \in \mathfrak{X}(S)$ over a scheme S, let $\mathcal{G}_x$ denote the pullback of $\mathcal{G}$, so $\mathcal{G}_x$ is a flat group scheme over S. If $\mathcal{F}$ is a quasicoherent sheaf on $\mathfrak{X}$ then pullback by x also defines a quasicoherent sheaf $\mathcal{F}_x$ on S, and there is an action of the group $\mathcal{G}_x(S)$ on $\mathcal{F}_x$. It is immediate that if $\mathcal{F}$ is of the form $\pi^* \bar{\mathcal{F}}$ for some quasicoherent sheaf $\bar{\mathcal{F}}$ on $\bar{\mathfrak{X}}$, then these actions of $\mathcal{G}_x(S)$ on the $\mathcal{F}_x$ are trivial. An exercise in descent theory, which we leave to the reader, shows that in fact π^* induces an equivalence of categories between quasicoherent sheaves on $\bar{\mathfrak{X}}$ and the category of quasicoherent sheaves $\mathcal{F}$ on $\mathfrak{X}$ such that for every object $x \in \mathfrak{X}(S)$ the action of $\mathcal{G}_x(S)$ on $\mathcal{F}_x$ is trivial.

2.3.4. We will apply this rigidification construction to get another view on $\mathcal{A}_{g,d}$.

Consider first the case of $\mathcal{A}_g$. Let $\mathcal{T}_g$ denote the fibered category over the category of schemes whose fiber over a scheme S is the groupoid of triples (A, P, L), where A/S is an abelian scheme of relative dimension g, P is an A-torsor, and L is a relatively ample line bundle on P such that the induced map

$$\lambda_L : A \to A^t$$

is an isomorphism.

Note that for any such triple, there is a natural inclusion

$$\mathbb{G}_m \hookrightarrow \underline{\mathrm{Aut}}_{\mathcal{T}_g}(A, P, L) \tag{2}$$

given by sending $u \in \mathbb{G}_m$ to the automorphism which is the identity on A and P and multiplication by u on L.

Proposition 2.3.5. *The stack $\mathcal{T}_g$ is algebraic, and the map*

$$\mathcal{T}_g \to \mathcal{A}_g, \quad (A, P, L) \mapsto (A, \lambda_L) \tag{3}$$

identifies $\mathcal{A}_g$ with the rigidification of $\mathcal{T}_g$ with respect to the subgroup space $\mathcal{G} \hookrightarrow \mathcal{I}_{\mathcal{T}_g}$ defined by the inclusions (2).

Proof. Since any object of $\mathcal{A}_g$ is locally in the image of (3), it suffices to show that for any scheme S and two objects (A, P, L) and (A', P', L') in $\mathcal{T}_g(S)$, the map sheaves on S-schemes (with the étale topology)

$$\underline{\mathrm{Hom}}_{\mathcal{T}_g}((A, P, L), (A', P', L')) \to \underline{\mathrm{Hom}}_{\mathcal{A}_g}((A, \lambda_L), (A', \lambda_{L'}))$$

provides an identification between $\underline{\mathrm{Hom}}_{\mathcal{A}_g}((A, \lambda_L), (A', \lambda_{L'}))$ and the sheaf quotient of $\underline{\mathrm{Hom}}_{\mathcal{T}_g}((A, P, L), (A', P', L'))$ by the natural action of $\mathbb{G}_m$. To verify this we may work étale locally on S, and hence may assume that P and P' are trivial torsors. Fix trivializations of these torsors, and view L and L' as line bundles on A and A' respectively.

In this case we need to show that for any isomorphism $\sigma : A \to A'$ such that the diagram

$$\begin{array}{ccc} A & \xrightarrow{\;\sigma\;} & A' \\ \downarrow{\scriptstyle \lambda_L} & & \downarrow{\scriptstyle \lambda_{L'}} \\ A^t & \xleftarrow{\;\sigma^*\;} & A'^t \end{array}$$

commutes, there exists a unique point $a \in A(S)$ such that the two line bundles

$$L, \quad t_a^* \sigma^* L'$$

are locally on S isomorphic. This follows from 2.1.6 applied to the two line bundles L and $\sigma^* L'$ which define the same principal polarization on A. $\qquad\square$

2.3.6. For any object $(A, P, L) \in \mathcal{T}_g(S)$ over a scheme S, we have a line bundle $\mathcal{W}_{(A,P,L)}$ on S given by f_*L, where $f : P \to S$ is the structure morphisms, and the formation of this line bundle commutes with arbitrary base change $S' \to S$. It follows that we get a line bundle $\mathcal{W}$ on the stack $\mathcal{T}_g$. Let

$$\mathcal{V} \to \mathcal{T}_g$$

denote the $\mathbb{G}_m$-torsor corresponding to $\mathcal{W}$. As a stack, $\mathcal{V}$ classifies quadruples (A, P, L, θ), where $(A, P, L) \in \mathcal{T}_g$ and $\theta \in \mathcal{W}_{(A,P,L)}$ is a nowhere vanishing section. From this and 2.2.7 we conclude that the composite map

$$\mathcal{V} \to \mathcal{T}_g \to \mathcal{A}_g$$

is an isomorphism, and therefore defines a section

$$s : \mathscr{A}_g \to \mathscr{T}_g.$$

Since $\mathscr{T}_g$ is a $\mathbb{G}_m$-gerbe over $\mathscr{A}_g$, we conclude that in fact

$$\mathscr{T}_g \simeq \mathscr{A}_g \times B\mathbb{G}_m.$$

2.3.7. The description of $\mathscr{A}_g$ in 2.3.5 can be generalized to higher degree polarizations as follows.

Let S be a scheme and consider a triple (A, P, L), where A/S is an abelian scheme, P is an A-torsor, and L is a line bundle on P. Define the *theta group* of (A, P, L), denoted $\mathscr{G}_{(A,P,L)}$ to be the functor on S-schemes which to any S'/S associates the group of pairs (α, ι), where $\alpha : P_{S'} \to P_{S'}$ is a morphism of $A_{S'}$-torsors, and $\iota : \alpha^* L_{S'} \to L_{S'}$ is an isomorphism of line bundles. Here $P_{S'}$, $A_{S'}$, and $L_{S'}$ denote the base changes to S'. Note that α is equal to translation by a, for a unique point $a \in A(S')$.

It follows that there is a natural map

$$\mathscr{G}_{(A,P,L)} \to A. \tag{4}$$

Its image consists of scheme-valued points $b \in A$ for which $t_b^* L$ and L are locally isomorphic. This is precisely the kernel of λ_L. Note also that there is a natural central inclusion

$$\mathbb{G}_m \hookrightarrow \mathscr{G}_{(A,P,L)}$$

given by sending a unit u to (id_P, u). This is in fact the kernel of (4) so we have an exact sequence of functors

$$1 \to \mathbb{G}_m \to \mathscr{G}_{(A,P,L)} \to K_{(A,P,L)} \to 1,$$

where

$$K_{(A,P,L)} := \mathrm{Ker}(\lambda_L).$$

In particular, if L is ample then $K_{(A,P,L)}$ is a finite flat group scheme over S, which also implies that $\mathscr{G}_{(A,P,L)}$ is a group scheme flat over S.

2.3.8. Suppose now that L is relatively ample on P, so that $K_{(A,P,L)}$ is a finite flat group scheme over S. We then get a skew-symmetric pairing

$$e : K_{(A,P,L)} \times K_{(A,P,L)} \to \mathbb{G}_m,$$

defined by sending sections $x, y \in K_{(A,P,L)}$ to the commutator

$$e(x, y) := \tilde{x}\tilde{y}\tilde{x}^{-1}\tilde{y}^{-1},$$

where $\tilde{x}, \tilde{y} \in \mathscr{G}_{(A,P,L)}$ are local liftings of x and y respectively. Note that this is well-defined (in particular independent of the choices of liftings) since $\mathbb{G}_m$ is central in $\mathscr{G}_{(A,P,L)}$.

The pairing e is called the *Weil pairing* and is nondegenerate. Indeed, this can be verified étale locally on S, so it suffices to consider the case when P is a trivial torsor in which case the result is [Mumford 1970, Corollary 2, p. 234].

2.3.9. Fix integers $g, d \geq 1$, and let $\mathcal{T}_{g,d}$ be the stack over the category of schemes whose fiber over a scheme S is the groupoid of triples (A, P, L), where A is an abelian scheme of relative dimension g over S, P is an A-torsor, and L is a relatively ample line bundle on P of degree d.

Proposition 2.3.10. *The stack $\mathcal{T}_{g,d}$ is an algebraic stack. If $\mathcal{G} \subset \mathcal{I}_{\mathcal{T}_{g,d}}$ denotes the subgroup of the inertia stack defined by the theta groups, then $\mathcal{G}$ is flat over $\mathcal{T}_{g,d}$ and the rigidification of $\mathcal{T}_{g,d}$ with respect to $\mathcal{G}$ is canonically isomorphic to $\mathcal{A}_{g,d}$.*

Proof. This follows from an argument similar to the proof of 2.3.5, which we leave to the reader. □

2.3.11. The stacks $\mathcal{A}_{g,d}$ arise naturally when considering level structures, even if one is only interested in principally polarized abelian varieties. Namely, suppose $d' = d \cdot k$ is a second integer. Then there is a natural map

$$\mathcal{A}_{g,d} \to \mathcal{A}_{g,d'}, \quad (A, \lambda) \mapsto (A, k \cdot \lambda). \tag{5}$$

This map is obtained by passing to rigidifications from the map

$$\mathcal{T}_{g,d} \to \mathcal{T}_{g,d'}, \quad (A, P, L) \mapsto (A, P, L^{\otimes k}).$$

Proposition 2.3.12. *Over $\mathbb{Z}[1/d]$, the map (5) is an open and closed immersion.*

Proof. See [Olsson 2008, 6.2.3]. □

2.3.13. As we discuss in Section 5 below, this result can be used to study moduli of principally polarized abelian varieties with level structure using moduli stacks for abelian varieties with higher degree polarizations.

3. Degenerations

3.1. *Semiabelian schemes.*

3.1.1. By a *torus* over a scheme S, we mean a commutative group scheme T/S which étale locally on S is isomorphic to $\mathbb{G}_m^r$, for some integer $r \geq 0$. For such a group scheme T, let

$$X_T := \underline{\mathrm{Hom}}(T, \mathbb{G}_m)$$

be the sheaf on the big étale site of S classifying homomorphisms $T \to \mathbb{G}_m$. Then X_T is a locally constant sheaf of free finitely generated abelian groups (indeed this can be verified étale locally where it follows from the fact that $\mathrm{Hom}(\mathbb{G}_m^r, \mathbb{G}_m) \simeq \mathbb{Z}^r$), and the natural map

$$T \to \underline{\mathrm{Hom}}(X_T, \mathbb{G}_m), \quad u \mapsto (\chi \mapsto \chi(u))$$

is an isomorphism of group schemes (again to verify this it suffices to consider the case when $T = \mathbb{G}_m^r$). The sheaf X_T is called the *sheaf of characters* of T.

One can also consider the *sheaf of cocharacters* of T defined to be the sheaf

$$Y_T := \underline{\mathrm{Hom}}(\mathbb{G}_m, T)$$

of homomorphisms $\mathbb{G}_m \to T$. Again this is a locally constant sheaf of finitely generated free abelian groups and the natural map

$$X_T \times Y_T \to \underline{\mathrm{Hom}}(\mathbb{G}_m, \mathbb{G}_m) \simeq \mathbb{Z}, \quad (\chi, \rho) \mapsto \chi \circ \rho$$

identifies Y_T with $\underline{\mathrm{Hom}}(X_T, \mathbb{Z})$. Furthermore, the natural map

$$Y_T \otimes_{\mathbb{Z}} \mathbb{G}_m \to T, \quad \rho \otimes u \mapsto \rho(u)$$

is an isomorphism (where both sides are viewed as sheaves on the big étale site of S).

3.1.2. A *semiabelian variety* over a field k is a commutative group scheme G/k which fits into an exact sequence

$$1 \to T \to G \to A \to 1,$$

where T is a torus and A is an abelian variety over k.

Lemma 3.1.3. *For any scheme S and abelian scheme A/S there are no nonconstant homomorphisms*

$$\mathbb{G}_{m,S} \to A$$

over S.

Proof. Consider first the case when $S = \mathrm{Spec}(k)$ is the spectrum of a field k. If $f : \mathbb{G}_m \to A$ is a homomorphism, then since A is proper f extends to a $\mathbb{G}_m$-equivariant morphism

$$\mathbb{P}^1 \to A,$$

where $\mathbb{G}_m$ acts on A through f. Since $0, \infty \in \mathbb{P}^1(k)$ are fixed points for the $\mathbb{G}_m$-action, their images in A must also be fixed points of the $\mathbb{G}_m$-action, which implies that f is constant.

For the general case, note first that by a standard limit argument it suffices to consider the case when S is noetherian. Furthermore, to verify that a morphism $f : \mathbb{G}_{m,S} \to A$ is constant we may pass to the local rings of S at geometric points, and may therefore assume that S is strictly henselian local. Reducing modulo powers of the maximal ideal, we are then reduced to the case when S is the spectrum of an artinian local ring R with algebraically closed residue field k. Let

$$f : \mathbb{G}_{m,R} \to A$$

be a morphism. Then the reduction of f modulo the maximal ideal of R is a constant morphism by the case of a field. It follows that for each integer n invertible in k the restriction of f to $\mu_{n,R} \subset \mathbb{G}_{m,R}$ is constant, as $\mu_{n,R}$ is étale over R and must have image in the étale group scheme $A[n]$ of n-torsion points of A (and a map of étale schemes over R is determined by its reduction modulo the maximal ideal). It follows that the preimage of the identity $f^{-1}(e) \subset \mathbb{G}_{m,R}$ is a closed subscheme which contains all the subgroup schemes $\mu_{n,R}$ for n invertible in k. From this it follows that $f^{-1}(e) = \mathbb{G}_{m,R}$. $\qquad\square$

3.1.4. In particular, in the setting of 3.1.2 any homomorphism $\mathbb{G}_m \to G$ factors through the subtorus $T \subset G$. This implies that the subtorus $T \subset G$ is canonically defined. Indeed if Y denotes the sheaf

$$\underline{\mathrm{Hom}}(\mathbb{G}_m, G),$$

then from above we conclude that Y is a locally constant sheaf of finitely generated abelian groups, and the natural map

$$Y \otimes_{\mathbb{Z}} \mathbb{G}_m \to G, \quad \rho \otimes u \mapsto \rho(u)$$

is a closed immersion with image T.

Note that this implies in particular that if G/k is a smooth group scheme such that the base change $G_{\bar{k}}$ to an algebraic closure is a semiabelian variety, then G is also a semiabelian variety as the subtorus $T_{\bar{k}} \subset G_{\bar{k}}$ descends to G.

3.1.5. For a general base scheme S, we define a *semiabelian scheme over S* to be a smooth commutative group scheme G/S all of whose fibers are semiabelian varieties. Semiabelian schemes arise as degenerations of abelian varieties. The basic theorem in this regard is the following:

Theorem 3.1.6 (Semistable reduction theorem [SGA 1972, IX.3.6]). *Let V be a regular noetherian local ring of dimension 1, with field of fractions K, and let A_K be an abelian scheme over K. Then there exists a finite extension K'/K such that the base change $A_{K'}$ of K' extends to a semiabelian scheme G over the integral closure V' of V in K'.*

3.2. *Fourier expansions and quadratic forms.* The key to understanding degenerations of abelian varieties and how it relates to moduli, is the connection with quadratic forms. This connection was originally established in the algebraic context by Mumford in [Mumford 1972], and then developed more fully for

partial degenerations in [Faltings and Chai 1990]. In this section we explain from
the algebraic point of view the basic idea of why quadratic forms are related to
degenerations.

3.2.1. First we need some facts about line bundles on tori. Let R be a complete
noetherian local ring with maximal ideal $\mathfrak{m} \subset R$ and reside field k. Let G/R be a
smooth commutative group scheme such that the reduction G_k is a torus. Assume
further that the character group sheaf X of G_k is constant (so G_k is isomorphic
to $\mathbb{G}_m^g$ for some g), and write also X for the free abelian group $\Gamma(\mathrm{Spec}(k), X)$.
For every integer n, let G_n denote the reduction of G modulo $\mathfrak{m}^{n+1}$, and let T_n
denote the torus over $R_n := R/\mathfrak{m}^{n+1}$ defined by the group X. By [SGA 1970,
chapitre IX, théorème 3.6] there exists for every $n \geq 0$ a unique isomorphism of
group schemes

$$\sigma_n : T_n \to G_n$$

restricting to the identity over k.

Suppose now that L_n is a line bundle on T_n. Then L_n is a trivial line bundle.
Indeed since T_0 has trivial Picard group and T_n is affine, there exists a global
section $s \in \Gamma(T_n, L_n)$ whose pullback to T_0 is a basis. By Nakayama's lemma
this implies that s defines an isomorphism $\mathcal{O}_{T_n} \simeq L_n$.

In particular, the line bundle L_n admits a T_n-linearization. Recall that such a
linearization is given by an isomorphism

$$\alpha : m^* L_n \to \mathrm{pr}_1^* L_n$$

over $T_n \times_{\mathrm{Spec}(R_n)} T_n$, where

$$m : T_n \times_{\mathrm{Spec}(R_n)} T_n \to T_n$$

is the group law, and such that over

$$T_n \times_{\mathrm{Spec}(R_n)} T_n \times_{\mathrm{Spec}(R_n)} T_n$$

the diagram

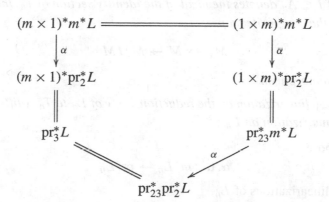

commutes, where we write

$$\mathrm{pr}_{23} : T_n \times_{\mathrm{Spec}(R_n)} T_n \times_{\mathrm{Spec}(R_n)} T_n \to T_n \times_{\mathrm{Spec}(R_n)} T_n$$

for the projection onto the second two components etc.

Since T_n is affine a T_n-linearization can also be described as follows. Let M_n denote $\Gamma(T_n, L_n)$ which is a module over $A_n := \Gamma(T_n, \mathcal{O}_{T_n}) \simeq R_n[X]$ (the group ring on X). Note that since A_n is canonically identified with the group ring on X we have a grading

$$A_n = \bigoplus_{\chi \in X} A_{n,\chi}.$$

Then giving a T_n-linearization on L_n is equivalent to giving a decomposition

$$M = \bigoplus_{\chi \in X} M_\chi$$

of M into submodules indexed by X which is compatible with the X-grading on A_n in the sense that for every $\chi, \eta \in X$ the map

$$A_{n,\chi} \otimes M_\eta \to M$$

has image in $M_{\chi+\eta}$.

Note that if $\chi_0 \in X$ is a fixed element, then we obtain a new T_n-linearization

$$M = \bigoplus_{\chi \in X} (M^{(\chi_0)})_\chi,$$

by setting

$$(M^{(\chi_0)})_\chi := M_{\chi+\chi_0}.$$

We call this new T_n-linearization the χ_0-*translate* of the original one.

Lemma 3.2.2.   (i) *Translation by elements of X gives the set of T_n-linearizations on L_n the structure of an X-torsor.*

(ii) *For any T_n-linearization of L_n corresponding to a decomposition $M = \bigoplus_\chi M_\chi$ each of the modules M_χ is a free module over R_n of rank 1. Moreover, if $I \subset A_n$ denotes the ideal of the identity section of T_n, then for every $\chi \in X$ the composite map*

$$M_\chi \hookrightarrow M \to M/IM$$

is an isomorphism.

(iii) *Any T_{n-1}-linearization on the reduction L_{n-1} of L_n to T_{n-1} lifts uniquely to a T_n-linearization on L_n.*

Proof. Suppose

$$\alpha, \alpha' : m^* L_n \to \mathrm{pr}_1^* L_n$$

are two T_n-linearizations of L_n.

For any R_n-scheme S let $T_{n,S}$ denote the base change of T_n to S, and let $L_{n,S}$ denote the pullback of L_n to $T_{n,S}$. For any point $u \in T_n(S)$, let

$$t_u : T_{n,S} \to T_{n,S}$$

denote translation by u, and let

$$\alpha_u, \alpha_u' : t_u^* L_{n,S} \to L_{n,S}$$

be the two isomorphisms obtained by pulling back α and α' along the map

$$T_{n,S} = T_n \times_{\mathrm{Spec}(R_n)} S \xrightarrow{\ \mathrm{id} \times u\ } T_n \times_{\mathrm{Spec}(R_n)} T_n.$$

The map $\alpha_u' \circ \alpha_u^{-1}$ is then an automorphism of $L_{n,S}$ over $T_{n,S}$, and hence is specified by a global section

$$s_u \in \Gamma(T_{n,S}, \mathcal{O}_{T_{n,S}}^*) = \mathbb{G}_m(S) \times X.$$

By sending $u \in T(S)$ to s_u we therefore obtain a natural transformation of functors

$$s : T_n \to \mathbb{G}_m \times X,$$

or equivalently by Yoneda's lemma a morphism of schemes. Since T_n is connected this map has connected image, and since the identity in T_n goes to the identity in $\mathbb{G}_m \times X$, the map s in fact has image in

$$\mathbb{G}_m \hookrightarrow \mathbb{G}_m \times X, \quad u \mapsto (u, 0).$$

Now the fact that α and α' are compatible with composition implies that the map

$$s : T_n \to \mathbb{G}_m$$

is a homomorphism, whence given by a character $\chi_0 \in X$. From this and the correspondence between T_n-linearizations and gradings on M, we get that α' is obtained from α by translation by χ_0.

This shows that the translation action of X on the set of T_n-linearizations of L_n is transitive. In particular, to verify (ii) it suffices to verify it for a single choice of T_n-linearization, as the validity of (ii) is clearly invariant under translation by elements of X. To verify (ii) it therefore suffices to consider $L_n = \mathcal{O}_{T_n}$ with the standard linearization, where the result is immediate.

Now once we know that each M_χ has rank 1, then it also follows that the action in (i) is simply transitive, as the character χ_0 is determined by the image of M_0.

Finally (iii) follows immediately from (i). $\qquad\square$

3.2.3. Consider again the setting of 3.2.1, and let L be a line bundle on G. For every $n \geq 0$ we then get by reduction (and using the isomorphisms σ_n) compatible line bundles L_n on T_n. Fix the following data:

A. A trivialization $t : R \simeq e^*L$, where $e : \mathrm{Spec}(R) \to G$ is the identity section.

B. A T_0-linearization α_0 of L_0.

By 3.2.2 (iii) the T_0-linearization α_0 lifts uniquely to a compatible system of T_n-linearizations $\{\alpha_n\}$. For every $n \geq 0$ and $\chi \in X$, we then get by 3.2.2 (ii) an isomorphism

$$\Gamma(T_n, L_n)_\chi \simeq e^*L_n \simeq R_n$$

where the second isomorphism is given by t. We therefore obtain a compatible system of basis elements $f_{n,\chi} \in \Gamma(T_n, L_n)_\chi$ defining an isomorphism

$$\Gamma(T_n, L_n) \simeq \bigoplus_{\chi \in X} R_n \cdot f_{n,\chi}.$$

Passing to the inverse limit we get an isomorphism

$$\varprojlim_n \Gamma(T_n, L_n) \simeq \prod'_{\chi \in X} R \cdot f_\chi,$$

where

$$\prod'_{\chi \in X} R \cdot f_\chi \subset \prod_{\chi \in X} R \cdot f_\chi$$

denotes the submodule of elements $(g_\chi \cdot f_\chi)_{\chi \in X}$ such that for every $n \geq 0$ almost all $g_\chi \in \mathfrak{m}^{n+1}$.

For any $\mu \in X$, we get by composing the natural map $\Gamma(G, L) \to \varprojlim_n \Gamma(T_n, L_n)$ with the projection

$$\prod'_{\chi \in X} R \cdot f_\chi \to R \cdot f_\mu$$

a map

$$\sigma_\mu : \Gamma(G, L) \to R.$$

If $m \in \Gamma(G, L)$ then we write

$$m = \sum_\chi \sigma_\chi(m) \cdot f_\chi$$

for the resulting expression in $\prod'_\chi R \cdot f_\chi$. We call this the *Fourier expansion of m*.

If R is an integral domain with field of fractions K, then we can tensor the maps σ_μ with K to get maps

$$\Gamma(G_K, L_K) \to K,$$

which we again denote by σ_μ. Note that for any $m \in \Gamma(G_K, L_K)$ the elements $\sigma_\mu(m)$ have bounded denominators in the sense that for any $n \geq 0$ we have $\sigma_\mu(\dot{m}) \in \mathfrak{m}^{n+1}$ for all but finitely many μ.

3.2.4. Suppose

$$t' : R \simeq e^*L$$

is a second choice of trivialization, and α'_0 is a second T_0-linearization of L_0. Let

$$\sigma'_\mu : \Gamma(G, L) \to R$$

be the maps obtained using this second choice. Suppose

$$t'(-) = vt(-),$$

for some unit $v \in R^*$ and that α'_0 is the χ_0-translate of α_0 for some $\chi_0 \in X$. Then the collections $\{\sigma_\mu\}$ and $\{\sigma'_\mu\}$ are related by the formula

$$\sigma'_\mu(-) = v\sigma_{\mu+\chi_0}(-).$$

3.2.5. Suppose now that our complete noetherian local ring R is also normal, and let K be the field of fractions. Let G/R be a semiabelian scheme whose generic fiber G_K is an abelian variety, and assume as above that the closed fiber G_k is a split torus. As before let X denote the character group of G_k.

Assume given an ample line bundle L_K on G_K, and let

$$\lambda_K : G_K \to G^t_K$$

be the induced polarization, where G^t_K denotes the dual abelian variety of G_K. As explained in [Faltings and Chai 1990, Chapter II, §2], the abelian scheme G^t_K extends uniquely to a semiabelian scheme G^t/R, and the map λ_K extends uniquely to a homomorphism

$$\lambda : G \to G^t.$$

Moreover, the closed fiber G^t_k is also a split torus, say Y is the character group of G^t_k. The map λ defines an inclusion

$$\phi : Y \hookrightarrow X.$$

Since G/R is smooth, the line bundle L_K extends to a line bundle L on G, unique up to isomorphism. Fix a trivialization

$$t : R \simeq e^*L$$

and a T_0-linearization α_0 on L_0, so we get maps

$$\sigma_\mu(-) : \Gamma(G_K, L_K) \to K.$$

Theorem 3.2.6 [Faltings and Chai 1990, Chapter II, 4.1]. *There exist unique functions*

$$a : Y \to K^*, \quad b : Y \times X \to K^*$$

such that the following hold:

(i) *The map b is bilinear.*

(ii) *For any $\mu \in X$ and $y \in Y$ we have*

$$\sigma_{\mu+\phi(y)}(-) = a(y)b(y,\mu)\sigma_\mu(-).$$

(iii) *For any $y, y' \in Y$ we have*

$$b(y, \phi(y')) = b(y', \phi(y)).$$

(iv) *For $y, y' \in Y$ we have*

$$a(y+y') = b(y, \phi(y'))a(y)a(y').$$

(v) *For every nonzero $y \in Y$ we have $b(y, \phi(y)) \in \mathfrak{m}$, and for every $n \geq 0$ we have $a(y) \in \mathfrak{m}^n$ for all but finitely many $y \in Y$.*

Remark 3.2.7. If we choose a different trivialization t' of e^*L and a different T_0-linearization α'_0, then we get new functions a' and b', which differ from a and b as follows. By 3.2.4 there exists a unit $v \in R^*$ and an element χ_0 such that

$$\sigma'_\mu(-) = v\sigma_{\mu+\chi_0}(-)$$

for all $\mu \in X$. From this we get that for any $\mu \in X$ and $y \in Y$ we have

$$\sigma'_{\mu+\phi(y)}(-) = a(y)b(y, \mu + \chi_0)\sigma'_\mu(-).$$

Since b is bilinear we have

$$b(y, \mu + \chi_0) = b(y, \mu)b(y, \chi_0).$$

It follows that

$$a'(y) = a(y)b(y, \chi_0), \quad b'(y, x) = b(y, x).$$

3.2.8. In particular, if R is a discrete valuation ring, then we also have a valuation map

$$v : K^* \to \mathbb{Z}.$$

Let A (resp. B) denote the composite of a (resp. b) with v, so we have functions

$$A : Y \to \mathbb{Z}, \quad B : Y \times X \to \mathbb{Z}.$$

If we fix a uniformizer $\pi \in R$ then we also get functions

$$\alpha : Y \to R^*, \quad \beta : Y \times X \to R^*$$

such that

$$a(y) = \alpha(y)\pi^{A(y)}, \quad b(y, x) = \beta(y, x)\pi^{B(y,x)}.$$

Now observe that since G and G^t have the same dimension, the map ϕ induces an isomorphism upon tensoring with $\mathbb{Q}$, so B induces a map

$$B_{\mathbb{Q}} : X_{\mathbb{Q}} \times X_{\mathbb{Q}} \to \mathbb{Q}$$

which is a positive definite quadratic form by 3.2.6(v). Note also that the difference

$$L : Y \to \mathbb{Q}, \qquad y \mapsto A(y) - \tfrac{1}{2}B(y, \phi(y))$$

is a linear form on Y, and that B can be recovered from A by the formula

$$B(y, \phi(y')) = A(y+y') - A(y) - A(y').$$

Note that by 3.2.7 the functions B is independent of the choice of (t, α_0), and for different choices of (t, α_0) the corresponding A-functions differ by a linear form.

3.2.9. The situation when G is not totally degenerate (i.e., the closed fiber G_k has an abelian part) is more complicated, and the functions a and b in the above get replaced with data involving the theory of biextensions. We will not go through that here (the interested reader should consult [Faltings and Chai 1990, Chapter II, §5] and [Olsson 2008, proof of 4.7.2]). One important thing to know about this, however, is that even in this case one obtains a positive semidefinite quadratic form

$$B : X_{\mathbb{Q}} \times X_{\mathbb{Q}} \to \mathbb{Q}$$

on the character group X of the maximal torus in G_k. We will use this in what follows.

4. Compactifications

4.1. *Toroidal.* The toroidal compactifications of $\mathcal{A}_g$ defined in [Faltings and Chai 1990] depend on some auxiliary choice of data, which we now explain.

4.1.1. Let X be a free finitely generated abelian group of rank g. For $A = \mathbb{Z}, \mathbb{Q}$ or $\mathbb{R}$, let $B(X_A)$ denote the space of A-valued quadratic forms on X

$$B(X_A) := \operatorname{Hom}(S^2 X, A).$$

For a bilinear form $b \in B(X_{\mathbb{R}})$ the *radical of b*, denoted $\operatorname{rad}(b)$, is defined to be the kernel of the map

$$X_{\mathbb{R}} \to \operatorname{Hom}(X_{\mathbb{R}}, \mathbb{R}), \qquad y \mapsto b(y, -).$$

Let $C(X) \subset B(X_{\mathbb{R}})$ denote the subset of positive semidefinite bilinear forms b such that $\operatorname{rad}(b)$ is defined over $\mathbb{Q}$. Then $C(X)$ is a convex cone in the real vector space $B(X_\mathbb{R})$, and its interior $C(X)^\circ \subset C(X)$ is the set of positive definite forms b.

Note also that there is an action of $\mathrm{GL}(X)$ on $B(X_\mathbb{R})$ induced by the action on X, and $C(X)$ and $C(X)^\circ$ are invariant under this action.

4.1.2. Degenerations of abelian varieties give subsets of $C(X)$ as follows.

Let S be an irreducible normal scheme with generic point η. Let G/S be a semiabelian scheme, and assume that the generic fiber G_η of G is an abelian scheme of dimension g. Suppose further given a principal polarization λ_η on G_η. Then for any complete discrete valuation ring V with algebraically closed residue field and morphism

$$\rho : \mathrm{Spec}(V) \to S \qquad (6)$$

sending the generic point of $\mathrm{Spec}(V)$ to η, we can pull back G to get a semiabelian scheme G_ρ/V whose generic fiber is a principally polarized abelian variety. As mentioned in 3.2.9, we therefore get a quadratic form

$$B_\rho \in C(X_{s,\mathbb{Q}}),$$

where X_s denotes the character group of the torus part of the closed fiber G_s of G_ρ. Choosing any surjection $X \to X_s$ we get an element $B'_\rho \in C(X_\mathbb{Q})$, well-defined up to the natural $\mathrm{GL}(X)$-action on $C(X_\mathbb{Q})$.

4.1.3. An *admissible cone decomposition of* $C(X)$ is a collection $\Sigma = \{\sigma_\alpha\}_{\alpha \in J}$ (where J is some indexing set) as follows:

(1) Each σ_α is a subcone of $C(X)$ of the form

$$\sigma_\alpha = \mathbb{R}_{>0} \cdot v_1 + \cdots + \mathbb{R}_{>0} \cdot v_r$$

for some elements $v_1, \ldots, v_r \in B(X_\mathbb{Q})$, and such that σ_α does not contain any line.

(2) $C(X)$ is equal to the disjoint union of the σ_α, and the closure of each σ_α is a disjoint union of σ_β's.

(3) For any $g \in \mathrm{GL}(X)$ and $\alpha \in J$ we have $g(\sigma_\alpha) = \sigma_\beta$ for some $\beta \in J$, and the quotient $J/\mathrm{GL}(X)$ of the set of cones J by the induced action of $\mathrm{GL}(X)$ is finite.

4.1.4. An admissible cone decomposition Σ of $C(X)$ is called *smooth* if for every $\sigma_\alpha \in \Sigma$ we can write

$$\sigma_\alpha = \mathbb{R}_{>0} \cdot v_1 + \cdots + \mathbb{R}_{>0} \cdot v_r$$

where $v_1, \ldots, v_r \in B(X_\mathbb{Z})$ can be extended to a $\mathbb{Z}$-basis for $B(X_\mathbb{Z})$.

4.1.5. Let Σ be an admissible cone decomposition of $C(X)$ and let B be a regular scheme (the case of interest is when B is the spectrum of a field or $\mathbb{Z}$). A *toroidal compactification of $\mathscr{A}_g$ with respect to Σ over B* is a Deligne–Mumford stack $\mathscr{A}_{g,\Sigma}$ over B together with a dense open immersion $j : \mathscr{A}_{g,B} \hookrightarrow \mathscr{A}_{g,\Sigma}$ over B such that the following hold:

(1) $\mathscr{A}_{g,\Sigma}$ is an irreducible normal algebraic stack, which is smooth over B if Σ is smooth.

(2) The universal abelian scheme $X \to \mathscr{A}_{g,B}$ extends to a semiabelian scheme $X_\Sigma \to \mathscr{A}_{g,\Sigma}$.

(3) Let S be an irreducible normal B-scheme and let G/S be a semiabelian scheme of relative dimension g whose generic fiber G_η is abelian with a principal polarization λ_η. Let $U \subset S$ be a dense open subset such that (G_η, λ_η) defines a morphism

$$f_U : U \to \mathscr{A}_g.$$

Then f_U extends to a (necessarily unique) morphism $f : S \to \mathscr{A}_{g,\Sigma}$ if and only if the following condition holds: For any point $s \in S$ there exists $\alpha \in J$ and a surjection $X \to X_{\bar{s}}$ such that for any morphism (6) sending the closed point of $\mathrm{Spec}(V)$ to s the element $B'_\rho \in C(X_{\mathbb{Q}})$ lies in σ_α.

Remark 4.1.6. The extension X_Σ of X in (2) is unique up to unique isomorphism by [Faltings and Chai 1990, I.2.7].

Remark 4.1.7. Properties (1), (2), and (3) characterize the stack $\mathscr{A}_{g,\Sigma}$ up to unique isomorphism. Indeed suppose we have another irreducible normal algebraic stack $\mathscr{A}'_g$ over B (this stack could be just an Artin stack, and doesn't have to be Deligne–Mumford) together with a dense open immersion $j' : \mathscr{A}_{g,B} \hookrightarrow \mathscr{A}'_g$ and an extension $X' \to \mathscr{A}'_g$ of the universal abelian scheme $X/\mathscr{A}_{g,B}$ to a semiabelian scheme over $\mathscr{A}'_g$. Suppose further that for any smooth morphism $g : W \to \mathscr{A}'_g$ the pullback $X_W \to W$ of X' to W satisfies the condition in (3). We then get a unique extension

$$\tilde{f} : W \to \mathscr{A}_{g,\Sigma}$$

of the map induced by X_W over the preimage of $\mathscr{A}_{g,B}$. Moreover, the two arrows

$$W \times_{\mathscr{A}'_g} W \to \mathscr{A}_{g,\Sigma}$$

obtained by composing the two projections with $\tilde{f}$ are canonically isomorphic by the uniqueness part of (3). In addition, the usual cocycle condition over $W \times_{\mathscr{A}'_g} W \times_{\mathscr{A}'_g} W$ holds again by the uniqueness. The map $\tilde{f}$ therefore descends to a unique morphism

$$f : \mathscr{A}'_g \to \mathscr{A}_{g,\Sigma}$$

compatible with the inclusions of $\mathcal{A}_{g,B}$. In particular, if $\mathcal{A}'_g$ is also a Deligne–Mumford stack satisfying (1), (2), and (3) then we also get a map

$$g : \mathcal{A}_{g,\Sigma} \to \mathcal{A}'_g$$

such that $f \circ g = \mathrm{id}_{\mathcal{A}_{g,\Sigma}}$ and $g \circ f = \mathrm{id}_{\mathcal{A}'_g}$.

One of the main results of [Faltings and Chai 1990] is then the following:

Theorem 4.1.8 [Faltings and Chai 1990, IV.5.7]. *For any smooth admissible cone decomposition Σ of $C(X)$, there exists a toroidal compactification of $\mathcal{A}_g$ with respect to Σ over $\mathrm{Spec}(\mathbb{Z})$. Moreover, for any regular scheme B, the base change $\mathcal{A}_{g,\Sigma,B}$ of $\mathcal{A}_{g,\Sigma}$ to B is a toroidal compactification of $\mathcal{A}_g$ with respect to Σ over B.*

Over $\mathbb{C}$, the smoothness assumption on the cone Σ can be omitted. This follows from Mumford's theory of toroidal embeddings [Ash et al. 1975]. A more accessible discussion in the case of $\mathcal{A}_g$ can be found in [Namikawa 1980].

Theorem 4.1.9. *For any admissible cone decomposition Σ of $C(X)$, there exists a toroidal compactification of $\mathcal{A}_g$ with respect to Σ over $\mathrm{Spec}(\mathbb{C})$.*

Remark 4.1.10. It seems widely believed that for any admissible cone decomposition Σ of $C(X)$ there exists a toroidal compactification of $\mathcal{A}_g$ with respect to Σ over $\mathrm{Spec}(\mathbb{Z})$, and it should have the property that for any regular scheme B the base change $\mathcal{A}_{g,\Sigma,B}$ is again a toroidal compactification of $\mathcal{A}_{g,B}$ with respect to Σ over B. However, no proof seems to be available in the literature.

4.2. Alexeev's compactification $\overline{\mathcal{A}}_g^{Alex}$.

4.2.1. Alexeev's compactification of $\mathcal{A}_g$ arises from considering $\mathcal{A}_g$ as the moduli stack of quadruples (A, P, L, θ), where A is an abelian variety, P is an A-torsor, L is an ample line bundle on P defining a principal polarization, and θ is a nonzero global section of L (see 2.2.7).

4.2.2. To get a sense for Alexeev's compactification let us consider a 1-parameter degeneration, and explain how the quadratic form obtained in 3.2.8 defines a degeneration of the whole quadruple (A, P, L, θ). So let V be a complete discrete valuation ring, let S denote $\mathrm{Spec}(V)$, and let η (resp. s) denote the generic (resp. closed) point of S. Let G/S be a semiabelian scheme with G_η an abelian variety and G_s a split torus. Assume further given a line bundle L on G whose restriction L_η to G_η is ample and defines a principal polarization. Let X denote the character group of G_s and let T denote the torus over V defined by X. Fix a trivialization $t : V \simeq e^*L$ (where $e \in G(V)$ is the identity section) and a T-linearization of L_s (the pullback of L to G_s). Finally let $\theta_\eta \in \Gamma(G_\eta, L_\eta)$ be a global section.

Let P_η denote G_η viewed as a trivial G_η-torsor. We can then construct a degeneration of the quadruple $(G_\eta, P_\eta, L_\eta, \theta_\eta)$ as follows.

4.2.3. Let

$$A : X \to \mathbb{Z}$$

be the quadratic function defined as in 3.2.8 (and using the identification $Y \simeq X$ defined by ϕ, which is an isomorphism since L_η is a principal polarization). Let

$$S := \{(x, A(x)) | x \in X\} \subset X_{\mathbb{R}} \oplus \mathbb{R}$$

be the graph of A, and let $S_{\mathbb{R}} \subset X_{\mathbb{R}} \oplus \mathbb{R}$ denote the convex hull of the set S. Then the projection

$$S_{\mathbb{R}} \to X_{\mathbb{R}}$$

is a bijection, and therefore $S_{\mathbb{R}}$ is the graph of a unique function

$$g : X_{\mathbb{R}} \to \mathbb{R}.$$

This function is piece-wise linear in the sense that there exists a unique collection $\Sigma = \{\omega\}$ of polytopes $\omega \subset X_{\mathbb{R}}$ such that the following hold:

(1) For any two elements $\omega, \eta \in \Sigma$ the intersection $\omega \cap \eta$ is also in Σ.

(2) Any face of a polytope $\omega \in \Sigma$ is also in Σ.

(3) $X_{\mathbb{R}} = \cup_{\omega \in \Sigma} \omega$ and for any two distinct elements $\omega, \eta \in \Sigma$ the interiors of ω and η are disjoint.

(4) For any bounded subset $W \subset X_{\mathbb{R}}$ there are only finitely many $\omega \in \Sigma$ with $\omega \cap W \neq \varnothing$.

(5) The top-dimensional polytopes $\omega \in \Sigma$ are precisely the domains of linearity of the function g.

A decomposition Σ of $X_{\mathbb{R}}$ into polytopes which arises from a quadratic function $A : X \to \mathbb{Z}$ by the construction above is called an *integral regular paving* of $X_{\mathbb{R}}$.

Note that the paving Σ is invariant under the action of elements of X acting by translation on $X_{\mathbb{R}}$. Indeed for $x, y \in X$ we have

$$A(x + y) = A(x) + A(y) + B(x, y) \qquad (7)$$

so if $t_y : X_{\mathbb{R}} \to X_{\mathbb{R}}$ denotes translation by y, then the composite function

$$X_{\mathbb{R}} \xrightarrow{t_y} X_{\mathbb{R}} \xrightarrow{g} \mathbb{R}$$

is equal to

$$x \mapsto g(x) + B(x, y) + A(y),$$

which differs from g by the linear function $B(-, y) + A(y)$.

Remark 4.2.4. Note that any positive definite quadratic form

$$B : S^2 X \to \mathbb{Q}$$

defines an X-invariant paving of $X_{\mathbb{R}}$ by the construction above. If more generally we allow also infinite polytopes in the definition of paving then we can also consider the pavings associated to positive semidefinite quadratic forms.

4.2.5. We use the function g to define a graded V-subalgebra

$$\mathfrak{R} \subset K[X \oplus \mathbb{N}].$$

For $\omega \in \Sigma$ let $C_\omega \subset X \oplus \mathbb{N}$ be the integral points of the cone over $\omega \times \{1\} \subset X_{\mathbb{R}} \oplus \mathbb{R}$, so C_ω is the set of elements $(x, d) \in X \oplus \mathbb{N}$ such that the element $(1/d) \cdot x \in X_{\mathbb{Q}}$ lies in ω. Since g is a linear function on ω it extends uniquely to an additive function

$$g_\omega : C_\omega \to \mathbb{Q}, \quad (x, d) \mapsto d \cdot g((1/d) \cdot x).$$

These functions define a function

$$\tilde{g} : X \oplus \mathbb{N}_{>0} \to \mathbb{Q}$$

by sending (x, d) to $g_\omega(x, d)$ for any $\omega \in \Sigma$ such that $(x, d) \in C_\omega$ (note that this is independent of the choice of ω).

Let $C'_\omega \subset C_\omega$ be the submonoid generated by degree 1 elements. Then $C'^{\mathrm{gp}}_\omega \subset C^{\mathrm{gp}}_\omega$ has finite index, say N_ω. Now using property (4) and the translation invariance of the paving, we see that there exists an integer N such that for every $\omega \in \Sigma$ the index of C'^{gp}_ω in C^{gp}_ω divides N. In particular, the function g_ω has image in $(1/N) \cdot \mathbb{Z}$ for all ω.

Also observe that making a base change $V \to V'$ with ramification e in the construction above has the effect of multiplying the function g by e. Therefore, after possibly replacing V by a ramified extension, we may assume that all the g_ω's, and hence also $\tilde{g}$, are integer valued.

Let

$$\mathfrak{R} \subset K[X \oplus \mathbb{N}]$$

be the graded V-subalgebra generated by the elements

$$\xi^{(x,d)} := \pi^{\tilde{g}(x,d)} e^{(x,d)},$$

where we write $e^{(x,d)} \in K[X \oplus \mathbb{N}]$ for the element corresponding to $(x, d) \in X \oplus \mathbb{N}$.

Then $\mathfrak{R}$ is a graded V-algebra and we can consider the V-scheme

$$\widetilde{P} := \mathrm{Proj}(\mathfrak{R}).$$

This scheme comes equipped with a line bundle $L_{\widetilde{P}}$, and we usually consider the pair $(\widetilde{P}, L_{\widetilde{P}})$.

4.2.6. There is a natural action of T on $(\widetilde{P}, L_{\widetilde{P}})$ induced by the X-grading on $\mathcal{R}$.

4.2.7. There is an action of X on $(\widetilde{P}, L_{\widetilde{P}})$ defined as follows. Let

$$\alpha : X \to V^*, \qquad \beta : X \times X \to V^*$$

be the maps defined in 3.2.8. Recall that for $x, y \in X$ we have

$$\alpha(x + y) = \beta(x, y)\alpha(x)\alpha(y).$$

The action of $y \in X$ is then given by

$$\xi^{(x,d)} \mapsto \alpha(y)^d \beta(y, x)\xi^{(x+dy,d)}.$$

Note that the actions of T and X on $\widetilde{P}$ commute, but that if $\chi \in T$ (a scheme-valued point) and $y \in X$ then the induced automorphism of $L_{\widetilde{P}}$

$$(T_y \circ S_\chi)^{-1} \circ (S_\chi \circ T_y)$$

is equal to multiplication by $\chi(y)$.

4.2.8. The generic fiber of $\widetilde{P}$ is isomorphic to T_K with the standard action of T_K and trivial action of X.

4.2.9. The closed fiber $\widetilde{P}_0$ of $\widetilde{P}$ has the following description. Note first of all that for any $(x, d), (y, e) \in X \oplus \mathbb{N}_{>0}$ we have

$$\tilde{g}(x + y, d + e) - \tilde{g}(x, d) - \tilde{g}(y, e) < 0$$

unless (x, d) and (y, e) lie in the same C_ω for some $\omega \in \Sigma$. Therefore

$$\xi^{(x,d)} \cdot \xi^{(y,e)} \equiv 0 \pmod{\pi}$$

if (x, d) and (y, e) lie in different cones. We therefore get a map

$$\mathcal{R} \otimes_V k \to k[C_\omega]$$

by sending $\xi^{(d,e)}$ to 0 unless $(d, e) \in C_\omega$ in which case we send $\xi^{(d,e)}$ to the element $e^{(d,e)}$. In this case we get a closed immersion

$$P_\omega := \text{Proj}(k[C_\omega]) \hookrightarrow \widetilde{P}_0,$$

and it follows from the construction that $\widetilde{P}_0$ is equal to the union of the P_ω's glued along the natural inclusions $P_\eta \hookrightarrow P_\omega$, whenever η is a face of ω. Moreover, the T-action on P_ω is given by the natural T-action on each P_ω, and the translation action of $y \in X$ is given by the isomorphisms

$$P_\omega \to P_{\omega+y}$$

given by the natural identification of C_ω and $C_{\omega+y}$ given by the translation invariance of the paving.

Remark 4.2.10. Similarly, for every integer s and $\omega \in \Sigma$, there exists only finitely many cones $\eta \in \Sigma$ such that there exists $(x, d) \in C_\omega$ and $(y, e) \in C_\eta$ with the property

$$\xi^{(x,d)} \cdot \xi^{(\eta,e)} \neq 0 \ (\mathrm{mod} \ \pi^s).$$

4.2.11. This description of the closed fiber $\widetilde{P}_0$ implies in particular that the action of X on $\widetilde{P}_0$, and hence also the action on $\widetilde{P}_n := \widetilde{P} \otimes (V/\pi^{n+1})$, is properly discontinuous. We can therefore take the quotient

$$P_n := \widetilde{P}_n / X,$$

which is a finite type $V/(\pi^{n+1})$-scheme. The X-action on $L_{\widetilde{P}}$ gives descent data for the line bundles $L_{\widetilde{P}_n} := L_{\widetilde{P}}|_{\widetilde{P}_n}$, so we get a compatible collection of line bundles L_{P_n} on the schemes P_n. One can show that the line bundles L_{P_n} are in fact ample, so by the Grothendieck existence theorem [EGA 1961, chapitre III, corollaire 5.1.8, p. 151] the projective schemes $\{P_n\}$ are induced by a unique projective scheme P/V with a line bundle L_P inducing the L_{P_n}.

4.2.12. Since the action of T on $\widetilde{P}_n$ commutes with the action of X, there is an action of T on each of the P_n which is compatible with the reduction maps. One can show that there is a unique action of G on P inducing these compatible actions of T on the P_n's (recall that there is a canonical identification $G_n \simeq T$). This is one of the most subtle aspects of the construction. A detailed discussion in this special case can be found in [Mumford 1972, §3].

4.2.13. There is a compatible set of global sections $\theta_n \in \Gamma(P_n, L_{P_n})$ defined as follows. First of all note that since the map

$$\pi_n : \widetilde{P}_n \to P_n$$

is an X-torsor, we have a canonical isomorphism

$$\Gamma(P_n, L_{P_n}) \simeq \Gamma(\widetilde{P}_n, L_{\widetilde{P}_n})^X.$$

It therefore suffices to construct an X-invariant section

$$\widetilde{\theta}_n \in \Gamma(\widetilde{P}_n, L_{\widetilde{P}_n}).$$

For $x \in X$ let $D(x)_n \subset \widetilde{P}_n$ denote the open subset defined by $\xi^{(x,1)}$, so

$$D(x)_n = \mathrm{Spec}(\mathcal{R}_{n,\xi^{(x,1)}})_0,$$

where $(\mathcal{R}_{n,\xi^{(x,1)}})_0$ denotes the degree 0 elements in $\mathcal{R}_{n,\xi^{(x,1)}}$. Then the $D(x)_n$ cover $\widetilde{P}_n$. Now for every x, all but finitely many $\xi^{(1,y)}$ map to zero in $\mathcal{R}_{n,\xi^{(x,1)}}$

by 4.2.10. Therefore the sum in $R_{n,\xi^{(x,1)}}$

$$\sum_{y \in Y} \alpha(y)\xi^{(y,1)}$$

is finite and defines a section $\tilde{\theta}_n \in \Gamma(D(x), L_{\tilde{P}_n})$. These sections clearly glue to define the section $\tilde{\theta}_n \in \Gamma(\tilde{P}_n, L_{\tilde{P}_n})$. The relation

$$\alpha(x + y) = \alpha(x)\alpha(y)\beta(y, x), \quad x, y \in X$$

and the definition of the X-action on $(\tilde{P}, L_{\tilde{P}})$ implies that the section $\tilde{\theta}_n$ is X-invariant and therefore defines the section $\theta_n \in \Gamma(P_n, L_{P_n})$.

Finally since

$$\Gamma(P, L_P) = \varprojlim_n \Gamma(P_n, L_{P_n})$$

the sections $\{\theta_n\}$ are induced by a unique section $\theta \in \Gamma(P, L)$.

4.2.14. Summarizing the preceding discussion, we started with the quadruple $(G_\eta, P_\eta, L_\eta, \theta_\eta)$ over the fraction field K of V, and ended up with a quadruple (G, P, L, θ) as follows:

(1) G is a semiabelian scheme over V;

(2) P is a proper V-scheme with action of G;

(3) L is an ample line bundle on P;

(4) $\theta \in \Gamma(P, L)$ is a global section.

It follows from [Faltings and Chai 1990, Chapter III, 6.4] that the restriction of this quadruple to $\mathrm{Spec}(K)$ is canonically isomorphic to the original quadruple $(G_\eta, P_\eta, L_\eta, \theta_\eta)$. The collection (G, P, L, θ) should be viewed as the degeneration of $(G_\eta, P_\eta, L_\eta, \theta_\eta)$.

4.2.15. A careful investigation of this construction, as well as its generalization to the case when G is not totally degenerate, is the starting point for the definition of Alexeev's moduli problem which gives his compactification $\overline{\mathscr{A}}_g^{\mathrm{Alex}}$ of $\mathscr{A}_g$. The end result of this investigation is the following.

4.2.16. Following [Alexeev 2002, 1.1.3.2], define a *stable semiabelic variety* over an algebraically closed field k to be a proper scheme P/k with an action of a semiabelian variety G/k such that the following hold:

(1) The dimension of each irreducible component of P is equal to the dimension of G.

(2) There are only finitely many orbits for the G-action.

(3) The stabilizer group scheme of every point of P is connected, reduced, and lies in the toric part of G.

(4) The scheme P is seminormal (recall that this means that the following property holds: If $f : P' \to P$ is a proper bijective morphism with P' reduced and with the property that for any $p' \in P'$ the map on residue fields $k(f(p')) \to k(k)$ is an isomorphism, then f is an isomorphism).

A *stable semiabelic pair* is a stable semiabelic variety P and a pair (L, θ), where L is an ample line bundle on P and $\theta \in H^0(P, L)$ is a global section whose zero locus does not contain any G-orbits.

Remark 4.2.17. If G is an abelian variety, then condition (3) implies that P is a disjoint union of G-torsors. If, moreover, we have a stable semiabelic pair (G, P, L, θ) with G abelian and $H^0(P, L)$ of dimension 1, then P must be connected so P is a G-torsor.

4.2.18. If S is a general base scheme, we define a *stable semiabelic pair over S* to be a quadruple (G, P, L, θ), where

(1) G/S is a semiabelian scheme.

(2) $f : P \to S$ is a projective flat morphism and G acts on P over S.

(3) L is a relatively ample invertible sheaf on P.

(4) $\theta \in H^0(P, L)$ is a global section.

(5) For every geometric point $\bar{s} \to S$, the geometric fiber $(G_{\bar{s}}, P_{\bar{s}}, L_{\bar{s}}, \theta_{\bar{s}})$ of this data is a stable semiabelic pair over the field $k(\bar{s})$.

Remark 4.2.19. It follows from cohomology and base change and [Alexeev 2002, 5.2.6] that if (G, P, L, θ) is a stable semiabelic pair over a scheme S as above, then f_*L is a locally free sheaf of finite rank on S whose formation commutes with arbitrary base change $S' \to S$. We define the *degree* of L to be the the rank of f_*L (a locally constant function on S).

Definition 4.2.20. Let $\mathscr{A}_g^{\text{Alex}}$ be the stack over the category of schemes, whose fiber over a scheme S is the groupoid of semiabelic pairs (G, P, L, θ) over S with G of dimension g and L of degree 1.

4.2.21. By 2.2.7, there is a morphism of stacks

$$j : \mathscr{A}_g \to \mathscr{A}_g^{\text{Alex}}$$

identifying $\mathscr{A}_g$ with the substack of semiabelic pairs (G, P, L, θ) with G an abelian scheme.

Theorem 4.2.22 [Alexeev 2002, 5.10.1]. *The stack $\mathscr{A}_g^{\text{Alex}}$ is an Artin stack of finite type over $\mathbb{Z}$ with finite diagonal, and the map j is an open immersion.*

Example 4.3. The quadruple (G, P, L, θ) constructed starting in 4.2.2 is a semiabelic pair of degree 1 over $\mathrm{Spec}(V)$ (i.e., a V-point of $\mathscr{A}_g^{\mathrm{Alex}}$). Indeed note that the closed fiber P_0 of P can be described as follows.

Let

$$\widetilde{P}_0 \to P_0$$

be the X-torsor which is the reduction of the scheme $\widetilde{P}$, so as in 4.2.9 the scheme $\widetilde{P}_0$ is equal to a union of the toric varieties P_ω ($\omega \in \Sigma$). Since $\widetilde{P}_0$ is reduced so is P_0, and the irreducible components of $\widetilde{P}_0$ are the subschemes P_ω with ω top dimensional. From this it follows that each irreducible component of P has dimension equal to the dimension of $G_0 = T$. Also note that the orbits for the T-action on P are in bijection with Σ/X, and hence is finite. To compute the stabilizer group schemes, note that if $\tilde{x} \in \widetilde{P}_0$ is a point in $\widetilde{P}_0$ with image $x \in P_0$, then the stabilizer group scheme of $\tilde{x}$ is equal to the stabilizer group scheme of x. Since each P_ω is a toric variety it follows that the stabilizer of any point of P_0 is a subtorus of T.

That the scheme P_0 is seminormal can be seen as follows. Let $f : Q \to P_0$ be a proper bijective morphism with Q reduced and the property that for any $q \in Q$ the map on residue fields $k(f(q)) \to k(q)$ is an isomorphism, and let $\mathscr{A}$ be the coherent sheaf of $\mathcal{O}_{P_0}$-algebras corresponding to Q. Since P_0 is reduced the map

$$\mathcal{O}_{P_0} \to \mathscr{A}$$

is injective, and we must show that it is also surjective. Let $\widetilde{\mathscr{A}}$ be the pullback of $\mathscr{A}$ to $\widetilde{P}_0$. Then $\widetilde{\mathscr{A}}$ is a coherent sheaf of $\mathcal{O}_{\widetilde{P}_0}$-algebras with an X-action lifting the X-action on $\widetilde{P}_0$. For each $\omega \in \Sigma$, let

$$j_\omega : P_\omega \hookrightarrow \widetilde{P}_0$$

be the inclusion. We construct an X-invariant morphism $s : \widetilde{\mathscr{A}} \to \mathcal{O}_{\widetilde{P}_0}$ such that the composite map

$$\mathcal{O}_{\widetilde{P}_0} \longrightarrow \widetilde{\mathscr{A}} \xrightarrow{s} \mathcal{O}_{\widetilde{P}_0}$$

is the identity. This will prove the seminormality of P_0, for by the X-invariance the map s descends to a morphism of algebras

$$\bar{s} : \mathscr{A} \to \mathcal{O}_{P_0}.$$

The kernel of this homomorphism is an ideal $\mathscr{I} \subset \mathscr{A}$ which is nilpotent since the map $Q \to P_0$ is bijective. Since Q is assumed reduced this implies that $\mathscr{I}$ is the zero ideal.

To construct the map s, proceed as follows. For each $\omega \in \Sigma$ let

$$i_\omega : P_\omega \hookrightarrow \widetilde{P}_0$$

be the inclusion. Let $\mathcal{S} \subset \Sigma$ be the subset of top-dimensional simplices, and choose an ordering of $\mathcal{S}$. We then have a map

$$\partial : \prod_{\omega \in \mathcal{S}} i_{\omega *} \mathcal{O}_{P_\omega} \to \prod_{\substack{\omega < \omega' \\ \omega, \omega' \in \mathcal{S}}} i_{\omega \cap \omega' *} \mathcal{O}_{P_{\omega \cap \omega'}},$$

defined by sending a local section $(\xi_\omega)_{\omega \in \mathcal{S}}$ to the section of the product whose image in the factor corresponding to $\omega < \omega'$ is

$$\xi_{\omega'}|_{P_{\omega \cap \omega'}} - \xi_\omega|_{P_{\omega \cap \omega'}}.$$

Then a straightforward verification, using the grading on the ring $\mathcal{R}$, shows that the natural map

$$\mathcal{O}_{\widetilde{P}_0} \to \operatorname{Ker}(\partial)$$

is an isomorphism of rings. To construct the map s it therefore suffices to construct compatible maps from $\widetilde{\mathcal{A}}$ to the $i_{\omega *} \mathcal{O}_{P_\omega}$. To construct these maps, note that since P_ω is normal the composite map

$$\operatorname{Spec}(i_\omega^* \widetilde{\mathcal{A}})_{\mathrm{red}} \hookrightarrow \operatorname{Spec}(i_\omega^* \widetilde{\mathcal{A}}) \to P_\omega$$

is an isomorphism, and hence we get maps

$$i_\omega^* \widetilde{\mathcal{A}} \to \mathcal{O}_{P_\omega}$$

which define maps

$$\widetilde{\mathcal{A}} \to i_{\omega *} \mathcal{O}_{P_\omega}$$

which are clearly compatible.

Finally we need to verify that the zero locus of the section $\theta_0 \in \Gamma(P_0, L_0)$ does not contain any T-orbit. For this let L_ω be the pullback of L_0 to P_ω and let $\theta_\omega \in \Gamma(P_\omega, L_\omega)$ be the pullback of θ. Then it suffices to show that the zero locus of θ_ω in P_ω does not contain any T-orbits. For this recall that we have

$$P_\omega = \operatorname{Proj}(k[C_\omega]),$$

and L_ω is equal to $\mathcal{O}_{P_\omega}(1)$. It follows that

$$\Gamma(P_\omega, L_\omega)$$

is isomorphic to the k-vector space with basis $\xi^{(x,1)}$, with $x \in \omega$. In terms of this basis the section θ_ω is by construction given by the sum of the elements $\alpha(x) \xi^{(x,1)}$. From this it follows immediately that the restriction of θ_ω to any T-invariant subset of P_ω is nonzero.

Remark 4.3.1. The stack $\mathcal{A}_g^{\mathrm{Alex}}$ is not irreducible. Explicit examples illustrating this is given in [Alexeev 2001]. In [Olsson 2008] we gave a modular interpretation of the closure of $\mathcal{A}_g$ in $\mathcal{A}_g^{\mathrm{Alex}}$ which we will describe in Section 4.5.

4.4. Canonical compactification $\mathscr{A}_g \subset \overline{\mathscr{A}}_g$ and the second Voronoi compactification.

4.4.1. Let $\overline{\mathscr{A}}_g$ denote the normalization of the closure of $\mathscr{A}_g$ in $\mathscr{A}_g^{\text{Alex}}$. We call $\overline{\mathscr{A}}_g$ the *canonical compactification of* $\mathscr{A}_g$ (in Section 4.5 below we discuss a modular interpretation of $\overline{\mathscr{A}}_g$)).

4.4.2. Consider again the lattice X of rank g, and the integral regular paving Σ defined in 4.2.3. View Σ as a category whose objects are the polytopes $\omega \in \Sigma$ and in which the set of morphisms $\omega \to \eta$ is the unital set if $\omega \subset \eta$ and the empty set otherwise. We have a functor

$$P_. : \Sigma \to \text{Monoids}$$

sending ω to the monoid C_ω. Taking the associated group we also obtain a functor

$$P_.^{\text{gp}} : \Sigma \to \text{Abelian groups}$$

by sending ω to C_ω^{gp}. Consider the inductive limit

$$\varinjlim P_.^{\text{gp}}.$$

For every $\omega \in \Sigma$ define

$$\rho_\omega : C_\omega \to \varinjlim P_.^{\text{gp}}$$

to be the composite map

$$C_\omega \hookrightarrow C_\omega^{\text{gp}} \to \varinjlim P_.^{\text{gp}}.$$

Note that if $\eta \subset \omega$ then the diagram

$$\begin{array}{ccc} C_\eta & \hookrightarrow & C_\omega \\ & \searrow{\scriptstyle\rho_\eta} & \downarrow{\scriptstyle\rho_\omega} \\ & & \varinjlim P_.^{\text{gp}} \end{array}$$

commutes. In particular, the $\{\rho_\omega\}$ define a set map

$$\rho : P \to \varinjlim P_.^{\text{gp}},$$

where P denotes the integral points of the cone

$$\text{Cone}(1, X_{\mathbb{R}}) \subset \mathbb{R} \oplus X_{\mathbb{R}}.$$

Define

$$\widetilde{H}_\Sigma \subset \varinjlim P_.^{\text{gp}}$$

to be the submonoid generated by elements of the form

$$\rho(p) + \rho(q) - \rho(p+q), \quad p, q \in P.$$

4.4.3. There is a natural action of X on $\mathbb{R} \oplus X_{\mathbb{R}}$ given by

$$y * (a, x) := (a, ay + x).$$

Since the paving Σ is X-invariant, this action induces actions of X on $\varinjlim P^{gp}$, P, and $\widetilde{H}_{\Sigma}$.

Let H_{Σ} denote the quotient (in the category of integral monoids)

$$H_{\Sigma} := \widetilde{H}_{\Sigma}/X,$$

and let

$$\pi : \widetilde{H}_{\Sigma} \to H_{\Sigma}$$

be the projection. For elements $p, q \in P$ define

$$p * q := \pi(\rho(p) + \rho(q) - \rho(p+q)).$$

By [Olsson 2008, 4.1.6] the monoid H_{Σ} is finitely generated.

4.4.4. We have a monoid

$$P \rtimes H_{\Sigma}$$

defined as follows. As a set, $P \rtimes H_{\Sigma}$ is equal to the product $P \times H_{\Sigma}$, but the monoid law is given by

$$(p, \alpha) + (q, \beta) := (p+q, \alpha + \beta + p*q).$$

With this definition we get a commutative integral monoid $P \rtimes H_{\Sigma}$.

There is a natural projection

$$P \rtimes H_{\Sigma} \to P, \quad (p, \alpha) \mapsto p,$$

and therefore we get a grading on $P \rtimes H_{\Sigma}$ from the $\mathbb{N}$-grading on P. The scheme

$$\widetilde{\mathscr{P}} := \mathrm{Proj}(\mathbb{Z}[P \rtimes H_{\Sigma}])$$

over $\mathrm{Spec}(\mathbb{Z}[H_{\Sigma}])$ generalizes the scheme $\widetilde{P}$ in 4.2.5.

Lemma 4.4.5. *There exists a morphism of monoids*

$$h : H_{\Sigma} \to \mathbb{N}$$

sending all nonzero elements of H_{Σ} to strictly positive numbers. In particular, the monoid H_{Σ} is unit-free.

Proof. Let

$$\tilde{g} : \mathbb{N}_{>0} \oplus X \to \mathbb{Q}$$

be the function defined in 4.2.5. The function $\tilde{g}$ is linear on each C_ω, and therefore induces a function

$$\tilde{h} : \varinjlim P^{\mathrm{gp}} \to \mathbb{Q}.$$

This function has the property that whenever $p, q \in P$ lies in different cones of Σ then we have

$$\tilde{h}(\rho(p) + \rho(q) - \rho(p+q)) > 0.$$

In particular, we get a morphism of monoids

$$\tilde{h} : \tilde{H}_\Sigma \to \mathbb{Q}_{\geq 0}$$

sending all nonzero generators, and hence also all nonzero elements, to $\mathbb{Q}_{>0}$. Now observe that if $p = (d, x)$ and $q = (e, y)$ are two elements of P, and if $z \in X$ is an element, then an exercise using (7), which we leave to the reader, shows that

$$\tilde{h}(\rho(d, x + dz) + \rho(e, y + ez) - \rho(d + e, x + y - (d + e)z))$$
$$= \tilde{h}(\rho(d, x) + \rho(e, y) - \rho(d + e, x + y)).$$

The map $\tilde{h}$ therefore descends to a homomorphism

$$h : H_\Sigma \to \mathbb{Q}_{\geq 0}.$$

Now since H_Σ is finitely generated, we can by replacing h with Nh for suitable N assume that this has image in $\mathbb{N}$, which gives the desired morphism of monoids. $\qquad\square$

4.4.6. In particular, there is a closed immersion

$$\mathrm{Spec}(\mathbb{Z}) \to \mathrm{Spec}(\mathbb{Z}[H_\Sigma]) \qquad\qquad (8)$$

induced by the map

$$\mathbb{Z}[H_\Sigma] \to \mathbb{Z}$$

sending all nonzero elements of H_Σ to 0. Let $\mathbb{Z}[[H_\Sigma]]$ be the completion of $\mathbb{Z}[[H_\Sigma]]$ with respect to the ideal $J \subset \mathbb{Z}[H_\Sigma]$ defining this closed immersion. Let $\mathcal{V}$ denote the spectrum of $\mathbb{Z}[[H_\Sigma]]$, and for $n \geq 0$ let $\mathcal{V}_n$ denote the closed subscheme of $\mathcal{V}$ defined by J^{n+1}.

As before let T denote the torus associated to X. We define a compatible family of projective schemes with T-action

$$(\mathscr{P}_n, L_{\mathscr{P}_n})$$

over the schemes $\mathcal{V}_n$ as follows. Let $\tilde{\mathcal{P}}_n$ denote the pullback of $\tilde{P}$ to $\mathcal{V}_n$, and let $L_{\tilde{P}_n}$ denote the pullback of $\mathcal{O}_{\tilde{P}}(1)$. Note that the scheme $\tilde{\mathcal{P}}_0$ over $\mathrm{Spec}(\mathbb{Z})$ can be described as in 4.2.9 as the union of the toric varieties $\mathrm{Spec}(\mathbb{Z}[C_\omega])$ for $\omega \in \Sigma$, glued along the natural closed immersions

$$\mathrm{Spec}(\mathbb{Z}[C_\eta]) \hookrightarrow \mathrm{Spec}(\mathbb{Z}[C_\omega])$$

for $\eta \subset \omega$. This implies in particular that the natural X-action on $\tilde{\mathcal{P}}_n$ is free, and hence we can form the quotient of $(\tilde{P}_n, L_{\tilde{\mathcal{P}}_n})$ to get a compatible system of projective schemes $\{(\mathcal{P}_n, L_{\mathcal{P}_n})\}$ over the $\mathcal{V}_n$.

There is a T-action on $\tilde{\mathcal{P}}$ defined as follows. For this note that the inclusion

$$P \hookrightarrow \mathbb{Z} \oplus X$$

induces an isomorphism $P^{\mathrm{gp}} \simeq \mathbb{Z} \oplus X$, so the projection $P \to \mathbb{N}$ defines a morphism of monoids

$$P \rtimes H_\Sigma \to \mathbb{Z} \oplus X.$$

This defines an action of $\mathbb{G}_m \times T$ on the affine scheme

$$\mathrm{Spec}(\mathbb{Z}[P \rtimes H_\Sigma]).$$

Since

$$\tilde{\mathcal{P}} = (\mathrm{Spec}(\mathbb{Z}[P \rtimes H_\Sigma]) - \{\text{zero section}\})/\mathbb{G}_m$$

we therefore get an action of T on $\tilde{\mathcal{P}}$. By construction this action commutes with the X-action, and hence we get also compatible actions of T on the $\mathcal{P}_n$.

Each of the line bundles $L_{\mathcal{P}_n}$ is ample on $\mathcal{P}_n$, so by the Grothendieck existence theorem the compatible system $\{(\mathcal{P}_n, L_{\mathcal{P}_n})\}$ is induced by a unique projective scheme $\mathcal{P}/\mathcal{V}$ with ample line bundle $L_{\mathcal{P}}$.

If $f : \mathcal{P} \to \mathcal{V}$ is the structure morphism, then $f_* L_{\mathcal{P}}$ is a locally free sheaf of rank 1 on $\mathcal{V}$ whose formation commutes with arbitrary base change (this follows from cohomology and base change and [Alexeev and Nakamura 1999, 4.4]). If we choose a nonzero global section $\theta \in f_* L_{\mathcal{P}}$, we then get a compatible family of objects

$$(T_{\mathcal{V}_n}, \mathcal{P}_n, L_{\mathcal{P}_n}, \theta_n) \in \mathcal{A}_g^{\mathrm{Alex}}(\mathcal{V}_n),$$

which induce a morphism

$$\mathrm{Spec}(\mathcal{V}) \to \mathcal{A}_g^{\mathrm{Alex}}. \tag{9}$$

We conclude that there exists a semiabelian scheme $G/\mathcal{V}$ with abelian generic fiber and closed fiber T which acts on $\mathcal{P}$ such that

$$(G, \mathcal{P}, L_{\mathcal{P}}, \theta)$$

defines a point of $\mathcal{A}_g^{\mathrm{Alex}}(\mathcal{V})$.

Remark 4.4.7. The discussion above is a bit circular, and it would be better to construct G using the theory of degenerations discussed in [Faltings and Chai 1990, Chapters II and III]. In fact, this theory enters into the construction of $\mathscr{A}_g^{\text{Alex}}$.

4.4.8. Let H_Σ^{sat} denote the saturation of the monoid H_Σ, and let $\mathscr{V}^{\text{sat}}$ denote the fiber product

$$\mathscr{V}^{\text{sat}} := \mathscr{V} \times_{\text{Spec}(\mathbb{Z}[H_\Sigma])} \text{Spec}(\mathbb{Z}[H_\Sigma^{\text{sat}}]).$$

Note that the map

$$\mathbb{Z}[H_\Sigma] \to \mathbb{Z}[H_\Sigma^{\text{sat}}]$$

is finite so the coordinate ring of the affine scheme $\mathscr{V}^{\text{sat}}$ is J-adically complete.

Let $\bar{\mathscr{A}}_g$ denote the normalization of the scheme-theoretic closure of $\mathscr{A}_g$ in $\mathscr{A}_g^{\text{Alex}}$ (below we shall give a modular interpretation of this stack). Then the map (9) induces a map

$$\mathscr{V}^{\text{sat}} \to \bar{\mathscr{A}}_g, \tag{10}$$

since $\mathscr{V}^{\text{sat}}$ is normal and the restriction of $(G, \mathscr{P}, L_\mathscr{P}, \theta)$ to the generic fiber of $\mathscr{V}^{\text{sat}}$ defines a point of $\mathscr{A}_g$.

This map (10) is étale (a more general result is given in [Olsson 2008, 4.5.20]).

4.4.9. The relationship between H_Σ and quadratic forms is the following. Consider the exact sequence

$$0 \to \widetilde{H}_\Sigma^{\text{gp}} \to (P \rtimes \widetilde{H}_\Sigma)^{\text{gp}} \to P^{\text{gp}} \to 0. \tag{11}$$

Now by the universal property of the group associated to a monoid, we have

$$H_\Sigma^{\text{gp}} = (\widetilde{H}_\Sigma / X)^{\text{gp}} = (\widetilde{H}_\Sigma^{\text{gp}})/X.$$

In particular, the long exact sequence of group homology arising from (11) defines a morphism

$$H_1(X, P^{\text{gp}}) \to H_0(X, \widetilde{H}_\Sigma^{\text{gp}}) = H_\Sigma^{\text{gp}}. \tag{12}$$

Now we have a short exact sequence of groups with X-action

$$0 \to X \to P^{\text{gp}} \to \mathbb{Z} \to 0,$$

where the inclusion $X \hookrightarrow P^{\text{gp}}$ is the identification of X with the degree 0 elements of P^{gp}, and the X-action on X and $\mathbb{Z}$ is trivial. We therefore obtain a map

$$H_1(X, \mathbb{Z}) \otimes X \to H_1(X, P^{\text{gp}}),$$

and hence by composing with (12) a map

$$H_1(X, \mathbb{Z}) \otimes X \to H_\Sigma^{\text{gp}}.$$

Now as explained in [Olsson 2008, 5.8.4] there is a natural identification of $H_1(X, \mathbb{Z})$ with X, and hence we get a map

$$X \otimes X \to H_\Sigma^{\text{gp}}.$$

As explained in [Olsson 2008, 5.8.8] this map is equal to the map sending $x \otimes y \in X \otimes X$ to

$$(1, x + y) * (1, 0) - (1, x) * (1, y).$$

In particular, the map is symmetric and therefore defines a map

$$\tau : S^2 X \to H_\Sigma^{\text{gp}}.$$

By [Olsson 2008, 5.8.15] this map induces an isomorphism after tensoring with $\mathbb{Q}$.

4.4.10. In particular we get an inclusion

$$\text{Hom}(H_\Sigma, \mathbb{Q}_{\geq 0}) \hookrightarrow \text{Hom}(S^2 X, \mathbb{Q})$$

of the rational dual of H_Σ into the space of quadratic forms on X. By [Olsson 2008, 5.8.16] this identifies the cone $\text{Hom}(H_\Sigma, \mathbb{Q}_{\geq 0})$ with the cone

$$U(\Sigma) \subset \text{Hom}(S^2 X, \mathbb{Q})$$

of positive semidefinite quadratic forms whose associated paving is coarser than the paving Σ.

4.4.11. As we now discuss, this description of H_Σ leads naturally to the *second Voronoi decomposition* of the space of quadratic forms. As explained in [Namikawa 1976, 2.3] there exists a unique admissible cone decomposition Σ^{Vor} of $C(X)$ (notation as in 4.1.1), called the *second Voronoi decomposition*, such that two quadratic forms $B, B' \in C(X)$ lie in the same $\sigma \in \Sigma^{\text{Vor}}$ if and only if the pavings of $X_\mathbb{R}$ defined by B and B' as in 4.2.3 are equal. This paving is known to be smooth if $g \leq 4$, but for $g > 4$ is not smooth (see [Alexeev and Nakamura 1999, 1.14]). Let

$$\mathscr{A}_g^{\text{Vor}}$$

denote the corresponding toroidal compactification over $\mathbb{C}$.

4.4.12. If V is a complete discrete valuation ring and

$$\rho : \text{Spec}(V) \to \mathscr{V}$$

is a morphism sending the closed point of $\text{Spec}(V)$ to a point in $\mathscr{V}_0$ and the generic point to the open subset of $\mathscr{V}$ over which G is an abelian scheme, then the pullback of G to V defines by the discussion in 3.2.8 a quadratic form

$$B_\rho : S^2 X \to \mathbb{Q}.$$

It follows the construction that this quadratic form is equal to the composite map

$$S^2 X \xrightarrow{\ \tau\ } H_\Sigma^{gp} \xrightarrow{\ \rho^*\ } K^* \xrightarrow{\ \text{val}\ } \mathbb{Z}.$$

In particular, it follows from 4.1.7 that the inclusion

$$\mathscr{A}_{g,\mathbb{C}} \hookrightarrow \mathscr{A}_g^{Vor}$$

extends to some neighborhood of the image of $\mathscr{V}_\mathbb{C}^{sat}$ in $\overline{\mathscr{A}}_{g,\mathbb{C}}$.

A similar description of the versal deformation space of partial degenerations (as discussed in [Olsson 2008, §4.5]), and again using 4.1.7, shows that in fact the inclusion $\mathscr{A}_{g,\mathbb{C}} \subset \mathscr{A}_g^{Vor}$ extends to a morphism of stacks

$$\pi : \overline{\mathscr{A}}_{g,\mathbb{C}} \to \mathscr{A}_g^{Vor}. \tag{13}$$

4.4.13. The local description of the map π is the following.

Let $\mathscr{V}_\mathbb{C}$ denote the spectrum of the completion of $\mathbb{C}[H_\Sigma]$ with respect to the morphism to $\mathbb{C}$ defined by (8), and let $\mathscr{V}_\mathbb{C}^{sat}$ denote the base change

$$\mathscr{V}_\mathbb{C} \times_{\text{Spec}(\mathbb{C}[H_\Sigma])} \text{Spec}(\mathbb{C}[H_\Sigma^{sat}]).$$

Consider the composite map

$$\mathscr{V}_\mathbb{C}^{sat} \longrightarrow \overline{\mathscr{A}}_{g,\mathbb{C}} \longrightarrow \mathscr{A}_g^{Vor}.$$

Let $Q \subset S^2 X$ be the cone of elements $q \in S^2 X$ such that for every $B \in U(\Sigma)$ we have

$$B(q) \geq 0.$$

Note that by 4.4.10 we have a natural inclusion

$$Q \hookrightarrow H_\Sigma^{sat}.$$

Let $\mathscr{W}$ denote the spectrum of the completion of $\mathbb{C}[Q]$ with respect to the kernel of the composite map

$$\mathbb{C}[Q] \to \mathbb{C}[H_\Sigma^{sat}] \to \mathbb{C}[H_\Sigma^{sat}]/J^{sat},$$

where $J^{sat} \subset \mathbb{C}[H_\Sigma^{sat}]$ is the ideal induced by $J \subset \mathbb{C}[H_\Sigma]$. The inclusion $Q \hookrightarrow H_\Sigma^{sat}$ induces a map

$$\lambda : \mathscr{V}_\mathbb{C}^{sat} \to \mathscr{W}.$$

4.4.14. By construction of the toroidal compactification $\mathcal{A}_g^{\mathrm{Vor}}$ we then have a formally étale map

$$\mathcal{W} \to \mathcal{A}_g^{\mathrm{Vor}},$$

and it follows from the construction of the toroidal compactification (see [Faltings and Chai 1990, Chapter IV, §3]) that the resulting diagram

$$
\begin{array}{ccc}
\mathcal{V}_{\mathbb{C}}^{\mathrm{sat}} & \xrightarrow{\ \lambda\ } & \mathcal{W} \\
\downarrow & & \downarrow \\
\overline{\mathcal{A}}_{g,\mathbb{C}} & \longrightarrow & \mathcal{A}_g^{\mathrm{Vor}}
\end{array}
$$

commutes.

4.4.15. This implies in particular that in a neighborhood of any totally degenerate point of $\overline{\mathcal{A}}_{g,\mathbb{C}}$ the map (13) is étale locally quasifinite, whence quasifinite. A suitable generalization of the preceding discussion to the partially degenerate case, gives that in fact that map (13) is a quasifinite morphism. This together with the fact that $\mathcal{A}_g^{\mathrm{Vor}}$ is normal implies that the map (13) identifies $\mathcal{A}_g^{\mathrm{Vor}}$ with the relative coarse moduli space of the morphism (13), in the sense of [Abramovich et al. 2011, §3].

This implies in particular that the map (13) induces an isomorphism on coarse moduli spaces.

The map (13) is not, however, in general an isomorphism. This can be seen from the fact that the map $Q \hookrightarrow H_{\Sigma}^{\mathrm{sat}}$ is not in general an isomorphism. The stack $\overline{\mathcal{A}}_g$ has some additional "stacky structure" at the boundary.

4.4.16. Granting that one has also a toroidal compactification of $\mathcal{A}_g$ over $\mathbb{Z}$ with respect to the second Voronoi decomposition over $\mathbb{Z}$ (this is known if $g \leq 4$), the preceding discussion applies verbatim over $\mathbb{Z}$ as well. Here one can see the difference between $\overline{\mathcal{A}}_g$ and $\mathcal{A}_g^{\mathrm{Vor}}$ even more clearly, for while $\mathcal{A}_g^{\mathrm{Vor}}$ is a Deligne–Mumford stack, the stack $\overline{\mathcal{A}}_g$ is only an Artin stack with finite diagonal, as the stabilizer group schemes in positive characteristic may have a diagonalizable local component.

4.5. *Modular interpretation of* $\overline{\mathcal{A}}_g$.

4.5.1. The key to giving $\overline{\mathcal{A}}_g$ a modular interpretation is to systematically use the toric nature of the construction in 4.4.6 using logarithmic geometry. We will assume in this section that the reader is familiar with the basic language of logarithmic geometry (the basic reference is [Kato 1989]).

4.5.2. Consider again the family

$$\tilde{f} : \widetilde{\mathcal{P}} \to \mathcal{V}$$

constructed in 4.4.6. The natural map

$$H_\Sigma \to \mathcal{O}_\mathcal{V}$$

defines a fine log structure $M_\mathcal{V}$ on M. Moreover, there is a fine log structure $M_{\widetilde{\mathcal{P}}}$ on $\widetilde{\mathcal{P}}$ and a morphism

$$f^b : f^* M_\mathcal{V} \to M_{\widetilde{\mathcal{P}}}$$

such that the induced morphism of fine log schemes

$$(f, f^b) : (\widetilde{\mathcal{P}}, M_{\widetilde{\mathcal{P}}}) \to (\mathcal{V}, M_\mathcal{V})$$

is log smooth. Moreover, the T-action on $\widetilde{\mathcal{P}}$ extends naturally to a T-action on the log scheme $(\widetilde{\mathcal{P}}, M_{\widetilde{\mathcal{P}}})$ over $(\mathcal{V}, M_\mathcal{V})$.

This log structure $M_{\widetilde{\mathcal{P}}}$ can be constructed as follows. The scheme $\widetilde{\mathcal{P}}$ is equal to the quotient of

$$\mathrm{Spec}(\mathbb{Z}[P \rtimes H_\Sigma]) - \{\text{zero section}\}$$

by the action of $\mathbb{G}_m$ defined by the $\mathbb{N}$-grading on $P \rtimes H_\Sigma$. The action of $\mathbb{G}_m$ extends naturally to an action on the log scheme

$$(\mathrm{Spec}(\mathbb{Z}[P \rtimes H_\Sigma]), \text{log structure associated to } P \rtimes H_\Sigma \to \mathbb{Z}[P \rtimes H_\Sigma])$$

over the log scheme

$$(\mathrm{Spec}(\mathbb{Z}[H_\Sigma]), \text{log structure associated to } H_\Sigma \to \mathbb{Z}[H_\Sigma]).$$

Passing to the quotient by this $\mathbb{G}_m$-action and base changing to $\mathcal{V}$, we therefore get the map

$$(f, f^b) : (\widetilde{\mathcal{P}}, M_{\widetilde{\mathcal{P}}}) \to (\mathcal{V}, M_\mathcal{V}).$$

Note that the X-action on $\widetilde{\mathcal{P}}$ extends naturally to an action of X on the log scheme $(\widetilde{\mathcal{P}}, M_{\widetilde{\mathcal{P}}})$ over $(\mathcal{V}, M_\mathcal{V})$. In particular, base changing to $\mathcal{V}_n$ and passing to the quotient by the X-action we get the log structure $M_{\mathcal{P}_n}$ on $\mathcal{P}_n$ and a morphism of log schemes

$$(\mathcal{P}_n, M_{\mathcal{P}_n}) \to (\mathcal{V}_n, M_{\mathcal{V}_n}). \tag{14}$$

4.5.3. Let H_Σ^{sat} be the saturation of H_Σ, and let $\mathcal{V}^{\mathrm{sat}}$ be as in 4.4.8. Define $M_{\mathcal{V}^{\mathrm{sat}}}$ to be the log structure on $\mathcal{V}^{\mathrm{sat}}$ defined by the natural map

$$H_\Sigma^{\mathrm{sat}} \to \mathcal{O}_{\mathcal{V}^{\mathrm{sat}}}$$

so we have a morphism of log schemes

$$(\mathcal{V}^{\mathrm{sat}}, M_{\mathcal{V}^{\mathrm{sat}}}) \to (\mathcal{V}, M_\mathcal{V}).$$

If $\mathcal{V}_n^{\mathrm{sat}}$ denotes $\mathcal{V}^{\mathrm{sat}} \times_{\mathcal{V}} \mathcal{V}_n$, then we get by base change a compatible collection of morphisms

$$(\mathcal{P}_n^{\mathrm{sat}}, M_{\mathcal{P}_n^{\mathrm{sat}}}) \to (\mathcal{V}_n^{\mathrm{sat}}, M_{\mathcal{V}_n^{\mathrm{sat}}})$$

from the collection (14).

4.5.4. If k is a field, define a *totally degenerate standard family* over k to be a collection of data

$$\big(M_k, T, f : (P, M_P) \to (\mathrm{Spec}(k), M_k), L_P\big)$$

as follows:

(1) M_k is a fine saturated log structure on $\mathrm{Spec}(k)$;

(2) T is a torus over k of dimension g;

(3) $f : (P, M_P) \to (\mathrm{Spec}(k), M_k)$ is a log smooth morphism with P/k proper, together with a T-action on (P, M_P) over $(\mathrm{Spec}(k), M_k)$.

(4) L_P is an ample line bundle on P such that $H^0(P, L_P)$ has dimension 1.

(5) The data is isomorphic to the collection obtained from the closed fiber of the family constructed in 4.5.3.

More generally, as explained in [Olsson 2008, §4.1] given a semiabelian scheme G/k with toric part X, a paving of X corresponding to a quadratic form etc., there is a generalization of the preceding construction which gives a fine saturated log structure M_k on $\mathrm{Spec}(k)$ and a log smooth morphism

$$f : (P, M_P) \to (\mathrm{Spec}(k), M_k),$$

where P/k is proper, and G acts on (P, M_P) over $(\mathrm{Spec}(k), M_k)$. Moreover, the construction gives a line bundle L_P on P which is ample and such that $H^0(P, L_P)$ has dimension 1. We define a *standard family* over k to be a collection of data

$$(M_k, G, f : (P, M_P) \to (\mathrm{Spec}(k), M_k), L_P)$$

obtained in this way (so the G-action on (P, M_P) is part of the data of a standard family).

For an arbitrary scheme S define $\overline{\mathcal{T}}_g(S)$ as the groupoid of collections of data

$$\big(M_S, G, f : (P, M_P) \to (S, M_S), L_P\big) \tag{15}$$

as follows:

(1) M_S is a fine saturated log structure on S.

(2) G/S is a semiabelian scheme of dimension g.

(3) $f : (P, M_P) \to (S, M_S)$ is a log smooth morphism with P/S proper.

(4) L_P is a relatively ample invertible sheaf on P.

(5) For every geometric point $\bar{s} \to S$, the collection of data over $\bar{s}$

$$(M_{\bar{s}}, G_{\bar{s}}, f_{\bar{s}} : (P_{\bar{s}}, M_{P_{\bar{s}}}) \to (\bar{s}, M_{\bar{s}}), L_{P_{\bar{s}}})$$

obtained by pullback, is a standard family in the preceding sense.

By definition a morphism

$$\big(M_S, G, f : (P, M_P) \to (S, M_S), L_P\big)$$
$$\to \big(M'_S, G', f' : (P', M_{P'}) \to (S, M'_S), L_{P'}\big)$$

between two objects of $\overline{\mathcal{T}}_g$ consists of the following data:

(1) An isomorphism $\sigma : M'_S \to M_S$ of log structures on S.

(2) An isomorphism of fine log schemes

$$\tilde{\sigma} : (P, M_P) \to (P', M_{P'})$$

such that the square

$$
\begin{array}{ccc}
(P, M_P) & \xrightarrow{\tilde{\sigma}} & (P', M_{P'}) \\
\downarrow{\scriptstyle f} & & \downarrow{\scriptstyle f'} \\
(S, M_S) & \xrightarrow{(\mathrm{id}, \sigma)} & (S, M'_S)
\end{array}
$$

commutes.

(3) An isomorphism $\tau : G \to G'$ of semiabelian group schemes over S such that the diagram

$$
\begin{array}{ccc}
G \times_S (P, M_P) & \xrightarrow{\text{action}} & (P, M_P) \\
\downarrow{\scriptstyle \tau \times \tilde{\sigma}} & & \downarrow{\scriptstyle \tilde{\sigma}} \\
G' \times_S (P', M_{P'}) & \xrightarrow{\text{action}} & (P', M_{P'})
\end{array}
$$

commutes.

(4) $\lambda : \tilde{\sigma}^* L_{P'} \to L_P$ is an isomorphism of line bundles on P.

In particular, for any object (15) of $\overline{\mathcal{T}}_g(S)$ and element $u \in \mathbb{G}_m(S)$ we get an automorphism of (15) by taking $\sigma = \mathrm{id}$, $\tilde{\sigma} = \mathrm{id}$, $\tau = \mathrm{id}$, and λ equal to multiplication by u.

With the natural notion of pullback we then get a stack $\overline{\mathcal{T}}_g$ over the category of schemes, together with an inclusion

$$\mathbb{G}_m \hookrightarrow \mathcal{I}_{\overline{\mathcal{T}}_g}$$

of $\mathbb{G}_m$ into the inertia stack of $\overline{\mathcal{T}}_g$.

Theorem 4.5.5 [Olsson 2008, 4.6.2]. *The stack $\overline{\mathcal{T}}_g$ is algebraic and there is a natural map $\overline{\mathcal{T}}_g \to \overline{\mathcal{A}}_g$ identifying $\overline{\mathcal{A}}_g$ with the rigidification of $\overline{\mathcal{T}}_g$ with respect to the subgroup $\mathbb{G}_m$ of the inertia stack.*

4.5.6. In fact the map $\overline{\mathcal{T}}_g \to \overline{\mathcal{A}}_g$ has a section. Consider the stack $\overline{\mathcal{A}}'_g$ whose fiber over a scheme S is the groupoid of data

$$(M_S, G, f : (P, M_P) \to (S, M_S), L_P, \theta),$$

where

$$(M_S, G, f : (P, M_P) \to (S, M_S), L_P) \in \overline{\mathcal{T}}_g(S)$$

is an object and $\theta \in f_* L_P$ is a section which is nonzero in every fiber. So $\overline{\mathcal{A}}'_g$ is the total space of the $\mathbb{G}_m$-torsor over $\overline{\mathcal{T}}_g$ corresponding to the line bundle defined by the sheaves $f_* L_P$ (which are locally free of rank 1 and whose formation commutes with arbitrary base change). Then it follows, by an argument similar to the one proving 2.3.5, that the composite map

$$\overline{\mathcal{A}}'_g \to \overline{\mathcal{T}}_g \to \overline{\mathcal{A}}_g$$

is an isomorphism. So $\overline{\mathcal{A}}_g$ can be viewed as the stack whose fiber over a scheme S is the groupoid of collections of data

$$(M_S, G, f : (P, M_P) \to (S, M_S), L_P, \theta)$$

as above. In particular, from the log structures M_S in this collection, we get a natural log structure $M_{\overline{\mathcal{A}}_g}$ on $\overline{\mathcal{A}}_g$, whose open locus of triviality is the stack $\mathcal{A}_g$.

5. Higher degree polarizations

5.0.7. One advantage of the approach to $\overline{\mathcal{A}}_g$ using $\overline{\mathcal{T}}_g$ and rigidification is that it generalizes well to higher degree polarizations and moduli spaces for abelian varieties with level structure.

Fix an integer $d \geq 1$, and let $\mathcal{A}_{g,d}$ be the stack of abelian schemes of dimension g with polarization of degree d. Let $\mathcal{T}_{g,d}$ be the stack defined in 2.3.9, so that $\mathcal{A}_{g,d}$ is the rigidification of $\mathcal{A}_{g,d}$ with respect to the universal theta group $\mathcal{G}$ over $\mathcal{T}_{g,d}$. To compactify $\mathcal{A}_{g,d}$, we first construct a dense open immersion $\mathcal{T}_{g,d} \hookrightarrow \overline{\mathcal{T}}_{g,d}$ and an extension of the universal theta group over $\mathcal{T}_{g,d}$ to a subgroup $\overline{\mathcal{G}} \subset \mathcal{I}_{\overline{\mathcal{T}}_{g,d}}$, and then $\overline{\mathcal{A}}_{g,d}$ will be obtained as the rigidification of $\overline{\mathcal{T}}_{g,d}$ with respect to $\overline{\mathcal{G}}$. Though the stack $\overline{\mathcal{T}}_{g,d}$ is not separated, it should be viewed as a compactification of $\mathcal{T}_{g,d}$ as it gives a proper stack $\overline{\mathcal{A}}_{g,d}$ after rigidifying.

5.1. *Standard families.*

5.1.1. To get a sense for the boundary points of $\overline{\mathcal{T}}_{g,d}$, let us again consider the case of maximal degeneration. Let V be a complete discrete valuation ring, $S = \mathrm{Spec}(V)$, and let G/V be a semiabelian scheme over V whose generic fiber G_η is an abelian variety and whose closed fiber is a split torus T. Let X denote the character group of T. Assume further given a polarization

$$\lambda : G_\eta \to G_\eta^t$$

of degree d. In this case we again get by 3.2.8 a quadratic form on $X_{\mathbb{Q}} = Y_{\mathbb{Q}}$, where Y is as in 3.2.5. Note that in this case we only get a quadratic function

$$A : X \to \mathbb{Q},$$

but after making a suitable base change of V we may assume that this function actually takes values in $\mathbb{Z}$. We then get a paving Σ of $X_{\mathbb{R}}$ by considering the convex hull of the set of points

$$\{(x, A(x)) | x \in X\} \subset X_{\mathbb{R}} \oplus \mathbb{R}.$$

Just as before we get a paving Σ of $X_{\mathbb{R}}$ and we can consider the scheme

$$\widetilde{P} \to \mathrm{Spec}(V)$$

defined in the same way as in 4.2.5. The main difference is that now we get an action of Y on $\widetilde{P}$ as opposed to an action of X. Taking the quotient of the reductions of $\widetilde{Y}$ by this Y-action, and algebraizing as before we end up with a projective V-scheme P/V with G-action and an ample line bundle L_P on P, such that the generic fiber P_η is a torsor under G_η, and the map

$$G_\eta \to G_\eta^t$$

defined by the line bundle L_P is equal to λ.

5.1.2. The construction of the logarithmic structures in 4.5.2 also generalizes to the case of higher degree polarization by the same construction. From the construction in 4.5.2 we therefore obtain candidates for the boundary points of $\overline{\mathcal{T}}_{g,d}$ over an algebraically closed field k as collections of data

$$(M_k, G, f : (P, M_P) \to (\mathrm{Spec}(k), M_k), L_P),$$

where

(1) M_k is a fine saturated log structure on $\mathrm{Spec}(k)$.

(2) f is a log smooth morphism of fine saturated log schemes such that P/k is proper.

(3) G is a semiabelian variety over k which acts on (P, M_P) over $(\text{Spec}(k), M_k)$.

(4) L_P is an ample line bundle on P.

(5) This data is required to be isomorphic to the data arising from a paving Σ of $X_{\mathbb{R}}$ coming from a quadratic form as above.

We call such a collection of data over k a *standard family*. More generally, there is a notion of standard family in the case when G is not totally degenerate (see [Olsson 2008, §5.2] for the precise definition).

Over a general base scheme S we define $\overline{\mathscr{T}}_{g,d}(S)$ to be the groupoid of collections of data

$$(M_S, G, f : (P, M_P) \to (S, M_S), L_P),$$

where

(1) M_S is a fine saturated log structure on S.

(2) f is a log smooth morphism whose underlying morphism $P \to S$ is proper.

(3) G/S is a semiabelian scheme which acts on (P, M_P) over (S, M_S).

(4) L_P is a relatively ample invertible sheaf on P.

(5) For every geometric point $\bar{s} \to S$ the pullback

$$(M_{\bar{s}}, G_{\bar{s}}, f_{\bar{s}} : (P_{\bar{s}}, M_{P_{\bar{s}}}) \to (\bar{s}, M_{\bar{s}}), L_{P_{\bar{s}}})$$

is a standard family over $\bar{s}$.

With the natural notion of pullback we get a stack $\overline{\mathscr{T}}_{g,d}$ over S.

Theorem 5.1.3 [Olsson 2008, 5.10.3]. *The stack $\overline{\mathscr{T}}_{g,d}$ is an algebraic stack of finite type. If $M_{\overline{\mathscr{T}}_{g,d}}$ denotes the natural log structure on $\overline{\mathscr{T}}_{g,d}$, then the restriction of $(\overline{\mathscr{T}}_{g,d}, M_{\overline{\mathscr{T}}_{g,d}})$ to $\mathbb{Z}[1/d]$ is log smooth.*

5.2. The theta group.

5.2.1. The stack $\overline{\mathscr{T}}_{g,d}$ is not separated, but it does have an extension of the theta group. Namely, for any objects

$$\mathscr{S} = (M_S, G, f : (P, M_P) \to (S, M_S), L_P) \in \overline{\mathscr{T}}_{g,d}(S)$$

over some scheme S, define

$$\mathscr{G}_{\mathscr{S}} : (S\text{-schemes})^{\text{op}} \to (\text{Groups})$$

to be the functor which to any S'/S associates the group of pairs

$$(\rho, \iota),$$

where

$$\rho : (P_{S'}, M_{P_{S'}}) \to (P_{S'}, M_{P_{S'}})$$

is an automorphism of log schemes over $(S', M_{S'})$ (where $M_{S'}$ is the pullback of M_S to $M_{S'}$), and

$$\iota : \rho^* L_{P_{S'}} \to L_{P_{S'}}$$

is an isomorphism of line bundles. We call $\mathcal{G}_{\mathcal{G}}$ the *theta group of* $\mathcal{G}$.

Note that there is a natural inclusion

$$i : \mathbb{G}_m \hookrightarrow \mathcal{G}_S$$

sending a unit u to the automorphism with $\rho = \mathrm{id}$ and ι multiplication by u.

Theorem 5.2.2 [Olsson 2008, 5.4.2]. *The functor* $\mathcal{G}_{\mathcal{G}}$ *is representable by a flat group scheme over* S, *which we again denote by* $\mathcal{G}_{\mathcal{G}}$. *The quotient of* $\mathcal{G}_{\mathcal{G}}$ *by* $\mathbb{G}_m$ *is a finite flat commutative group scheme* $H_{\mathcal{G}}$ *of rank* d^2.

5.2.3. So we have a central extension

$$1 \to \mathbb{G}_m \to \mathcal{G}_S \to H_{\mathcal{G}} \to 1,$$

with $H_{\mathcal{G}}$ commutative. We can then define a skew symmetric pairing

$$e : H_{\mathcal{G}} \times H_{\mathcal{G}} \to \mathbb{G}_m$$

by setting

$$e(x, y) := \tilde{x}\tilde{y}\tilde{x}^{-1}\tilde{y}^{-1} \in \mathbb{G}_m,$$

where $\tilde{x}, \tilde{e} \in \mathcal{G}_{\mathcal{G}}$ are local lifts of x and y respectively. We call this pairing on $H_{\mathcal{G}}$ the *Weil pairing*. It is shown in [Olsson 2008, 5.4.2] that this pairing is nondegenerate.

5.3. *The stack* $\bar{\mathcal{A}}_{g,d}$.

5.3.1. The theta groups of objects of $\overline{\mathcal{T}}_{g,d}$ define a flat subgroup scheme

$$\mathcal{G} \hookrightarrow \mathcal{I}_{\overline{\mathcal{T}}_{g,d}}$$

of the inertia stack of $\overline{\mathcal{T}}_{g,d}$, and we define

$$\bar{\mathcal{A}}_{g,d}$$

to be the rigidification of $\overline{\mathcal{T}}_{g,d}$ with respect to $\mathcal{G}$.

Theorem 5.3.2 [Olsson 2008, §5.11]. (i) *The stack* $\bar{\mathcal{A}}_{g,d}$ *is a proper algebraic stack over* $\mathbb{Z}$.

(ii) *The log structure* $M_{\overline{\mathcal{T}}_{g,d}}$ *on* $\overline{\mathcal{T}}_{g,d}$ *descends uniquely to a log structure* $M_{\bar{\mathcal{A}}_{g,d}}$ *on* $\bar{\mathcal{A}}_{g,d}$. *The restriction of* $(\bar{\mathcal{A}}_{g,d}, M_{\bar{\mathcal{A}}_{g,d}})$ *to* $\mathbb{Z}[1/d]$ *is log smooth.*

(iii) *The natural inclusion $\mathscr{A}_{g,d} \hookrightarrow \overline{\mathscr{A}}_{g,d}$ is a dense open immersion and identifies $\mathscr{A}_{g,d}$ with the open substack of $\overline{\mathscr{A}}_{g,d}$ where $M_{\overline{\mathscr{A}}_{g,d}}$ is trivial.*

(iv) *The finite flat group scheme $\mathscr{H} := \mathscr{G}/\mathbb{G}_m$ with its Weil pairing e descends to a finite flat group scheme with perfect pairing (still denoted $(\mathscr{H}, e)$) on $\overline{\mathscr{A}}_{g,d}$. The restriction of $\mathscr{H}$ to $\mathscr{A}_{g,d}$ is the kernel of the universal polarization*

$$\lambda : X \to X^t$$

on the universal abelian scheme $X/\mathscr{A}_{g,d}$.

5.4. Moduli spaces for abelian varieties with level structure. Theorem 5.3.2 enables one to give compactifications for moduli spaces of abelian varieties with level structure. We illustrate this with an example.

5.4.1. Let $g \geq 1$ be an integer, let p be a prime, and let $\mathscr{A}_g(p)$ denote the stack over $\mathbb{Z}[1/p]$ which to any scheme S associates the groupoid of pairs

$$(A, \lambda, x : S \to A),$$

where (A, λ) is a principally polarized abelian variety of dimension g, and $x \in A(S)$ is a point of exact order p.

Note that if

$$X[p] \to \mathscr{A}_g$$

denotes the p-torsion subgroup of the universal principally polarized abelian scheme over $\mathscr{A}_g$, then $\mathscr{A}_g(p)$ is equal to the restriction to $\mathbb{Z}[1/p]$ of the complement of the zero section of $X[p]$ (which is finite over $\mathscr{A}_g[1/p]$ since the restriction of $X[p]$ to $\mathscr{A}_g[1/p]$ is finite étale). So we can view the problem of compactifying $\mathscr{A}_g(p)$ as a problem of compactifying the universal p-torsion subgroup scheme over $\mathscr{A}_g$.

5.4.2. For this note first that if (A, λ) is a principally polarized abelian scheme over a scheme S, then the p-torsion subgroup $A[p]$ is the kernel of

$$p\lambda : A \to A^t.$$

Let

$$j : \mathscr{A}_g[1/p] \to \mathscr{A}_{g,p^g}[1/p]$$

be the map sending (A, λ) to $(A, p\lambda)$. By 2.3.12 this map is an open and closed immersion, and if

$$\eta : X \to X^t$$

denotes the universal polarization over $\mathscr{A}_{g,p^g}[1/p]$ then the universal p-torsion subgroup over $\mathscr{A}_g[1/p]$ is the restriction of the finite étale group scheme

$$\mathrm{Ker}(\eta) \to \mathscr{A}_{g,p^g}[1/p].$$

5.4.3. Let

$$\mathcal{H} \to \overline{\mathcal{A}}_{g,p^g}$$

be the finite flat group scheme discussed in 5.3.2(iii). The rank of $\mathcal{H}$ is p^{2g}, so its restriction $\mathcal{H}[1/p]$ to $\overline{\mathcal{A}}_{g,p^g}[1/p]$ is a finite étale group scheme of rank p^{2g}, whose restriction to $\mathcal{A}_{g,p^g}[1/p]$ is $\mathrm{Ker}(\eta)$. We then get a compactification of $\mathcal{A}_g(p)$ by taking the closure of $\mathcal{A}_g(p)$ in the complement of the identity section in $\mathcal{H}[1/p]$. Since $\mathcal{H}[1/p]$ is finite étale over $\overline{\mathcal{A}}_{g,p^g}$ the resulting space $\overline{\mathcal{A}_g(p)}$ is finite étale over $\overline{\mathcal{A}}_{g,p^g}[1/p]$, and in particular is proper over $\mathbb{Z}[1/p]$ with toric singularities.

References

[Abramovich et al. 2003] D. Abramovich, A. Corti, and A. Vistoli, "Twisted bundles and admissible covers", *Comm. Algebra* **31**:8 (2003), 3547–3618.

[Abramovich et al. 2011] D. Abramovich, M. Olsson, and A. Vistoli, "Twisted stable maps to tame Artin stacks", *J. Algebraic Geom.* **20**:3 (2011), 399–477.

[Alexeev 2001] V. Alexeev, "On extra components in the functorial compactification of A_g", pp. 1–9 in *Moduli of abelian varieties* (Texel, 1999), Progr. Math. **195**, Birkhäuser, Basel, 2001.

[Alexeev 2002] V. Alexeev, "Complete moduli in the presence of semiabelian group action", *Ann. of Math.* (2) **155**:3 (2002), 611–708.

[Alexeev and Nakamura 1999] V. Alexeev and I. Nakamura, "On Mumford's construction of degenerating abelian varieties", *Tohoku Math. J.* (2) **51**:3 (1999), 399–420.

[Artin 1969] M. Artin, "Algebraic approximation of structures over complete local rings", *Inst. Hautes Études Sci. Publ. Math.* 36 (1969), 23–58.

[Ash et al. 1975] A. Ash, D. Mumford, M. Rapoport, and Y. Tai, *Smooth compactification of locally symmetric varieties*, Lie Groups: History, Frontiers and Applications **4**, Math. Sci. Press, Brookline, MA, 1975.

[Deligne and Mumford 1969] P. Deligne and D. Mumford, "The irreducibility of the space of curves of given genus", *Inst. Hautes Études Sci. Publ. Math.* 36 (1969), 75–109.

[EGA 1961] A. Grothendieck, "Éléments de géométrie algébrique, III: étude cohomologique des faisceaux cohérents (première partie)", *Inst. Hautes Études Sci. Publ. Math.* **11** (1961), 5–167. Zbl 0153.22301

[Faltings and Chai 1990] G. Faltings and C.-L. Chai, *Degeneration of abelian varieties*, Ergebnisse der Mathematik und ihrer Grenzgebiete (3) **22**, Springer, Berlin, 1990.

[Kajiwara et al. 2008a] T. Kajiwara, K. Kato, and C. Nakayama, "Logarithmic abelian varieties, I: complex analytic theory", *J. Math. Sci. Univ. Tokyo* **15**:1 (2008), 69–193.

[Kajiwara et al. 2008b] T. Kajiwara, K. Kato, and C. Nakayama, "Logarithmic abelian varieties", *Nagoya Math. J.* **189** (2008), 63–138.

[Kato 1989] K. Kato, "Logarithmic structures of Fontaine–Illusie", pp. 191–224 in *Algebraic analysis, geometry, and number theory* (Baltimore, 1988), Johns Hopkins Univ. Press, Baltimore, MD, 1989.

[Keel and Mori 1997] S. Keel and S. Mori, "Quotients by groupoids", *Ann. of Math.* (2) **145**:1 (1997), 193–213.

[Laumon and Moret-Bailly 2000] G. Laumon and L. Moret-Bailly, *Champs algébriques*, Ergebnisse der Mathematik und ihrer Grenzgebiete (3) **39**, Springer, Berlin, 2000.

[Mumford 1965] D. Mumford, *Geometric invariant theory*, Ergebnisse der Mathematik und ihrer Grenzgebiete, N.F. **34**, Springer, Berlin, 1965.

[Mumford 1966] D. Mumford, "On the equations defining abelian varieties, I", *Invent. Math.* **1** (1966), 287–354.

[Mumford 1967] D. Mumford, "On the equations defining abelian varieties, II and III", *Invent. Math.* **3** (1967), 71–135 and 215–244.

[Mumford 1970] D. Mumford, *Abelian varieties*, Tata Studies in Mathematics **5**, Tata Institute of Fundamental Research, Bombay, 1970.

[Mumford 1972] D. Mumford, "An analytic construction of degenerating abelian varieties over complete rings", *Compositio Math.* **24** (1972), 239–272.

[Namikawa 1976] Y. Namikawa, "A new compactification of the Siegel space and degeneration of Abelian varieties, I", *Math. Ann.* **221**:2 (1976), 97–141.

[Namikawa 1980] Y. Namikawa, *Toroidal compactification of Siegel spaces*, Lecture Notes in Mathematics **812**, Springer, Berlin, 1980.

[Olsson 2008] M. C. Olsson, *Compactifying moduli spaces for abelian varieties*, Lecture Notes in Mathematics **1958**, Springer, Berlin, 2008.

[Oort 1971] F. Oort, "Finite group schemes, local moduli for abelian varieties, and lifting problems", *Compositio Math.* **23** (1971), 265–296.

[SGA 1970] M. Demazure and A. Grothendieck (editors), *Schémas en groupes, II: Groupes de type multiplicatif, et structure des schémas en groupes généraux* (Séminaire de Géométrie Algébrique du Bois Marie 1962/64 = SGA 3 II), Lecture Notes in Math. **152**, Springer, Berlin, 1970.

[SGA 1972] A. Grothendieck (editor), *Groupes de monodromie en géométrie algébrique, I* (Séminaire de Géométrie Algébrique du Bois Marie 1967/69 = SGA 7 I), Lecture Notes in Mathematics **288**, Springer, Berlin, 1972.

molsson@math.berkeley.edu   *Department of Mathematics, University of California, Berkeley, CA 94720, United States*

Current Developments in Algebraic Geometry
MSRI Publications
Volume 59, 2011

The geography of irregular surfaces

MARGARIDA MENDES LOPES AND RITA PARDINI

We give an overview of irregular complex surfaces of general type, discussing in particular the distribution of the numerical invariants K^2 and χ for minimal ones.

1. Introduction

Let S be a minimal surface of general type and let K^2, χ be its main numerical invariants (Section 2.3). For every pair of positive integers a, b the surfaces with $K^2 = a$, $\chi = b$ belong to finitely many irreducible families, so that in principle their classification is possible. In practice, the much weaker *geographical problem*, i.e., determining the pairs a, b for which there exists a minimal surface of general type with $K^2 = a$ and $\chi = b$, is quite hard.

In the past, the main focus in the study of both the geographical problem and the fine classification of surfaces of general type has been on *regular* surfaces, namely surfaces that have no global 1-forms, or, equivalently, whose first Betti number is 0. The reason for this is twofold: on the one hand, the canonical map of regular surfaces is easier to understand, on the other hand complex surfaces are the main source of examples of differentiable 4-manifolds, hence the simply connected ones are considered especially interesting from that point of view.

So, while, for instance, the geographical problem is by now almost settled and the fine classification of some families of regular surfaces is accomplished

The first author is a member of the Center for Mathematical Analysis, Geometry and Dynamical Systems, IST/UTL and the second author is a member of G.N.S.A.G.A.-I.N.d.A.M. This research was partially supported by the Fundação para a Ciência e Tecnologia (FCT Portugal).
This is an expanded version of the talk given by the second author at the workshop "Classical Algebraic Geometry Today", MSRI, January 26–30, 2009.
MSC2000: 14J29.

[Barth et al. 1984, §9–11, Chapter VII], little is known about irregular surfaces of general type. In recent years, however, the use of new methods and the revisiting of old ones have produced several new results.

Here we give an overview of these results, with special emphasis on the geographical problem. In addition, we give several examples and discuss some open questions and possible generalizations to higher dimensions (e.g., Theorem 5.2.2).

Notation and conventions. We work over the complex numbers. All varieties are projective algebraic and, unless otherwise specified, smooth. We denote by $J(C)$ the Jacobian of a curve C.

Given varieties X_i and sheaves $\mathcal{F}_i$ on X_i, $i = 1, 2$, we denote by $\mathcal{F}_1 \boxtimes \mathcal{F}_2$ the sheaf $p_1^* \mathcal{F}_1 \otimes p_2^* \mathcal{F}_2$ on $X_1 \times X_2$, where $p_i : X_1 \times X_2 \to X_i$ is the projection.

2. Irregular surfaces of general type

Unless otherwise specified, a surface is a smooth projective complex surface.

A surface S is *of general type* if the canonical divisor K_S is big. Every surface of general type has a birational morphism onto a unique minimal model, which is characterized by the fact that K_S is nef. A surface (or more generally a variety) S is called *irregular* if the *irregularity* $q(S) := h^0(\Omega_S^1) = h^1(\mathcal{O}_S)$ is > 0.

2.1. *Irrational pencils.* A *pencil* on a surface S is a morphism with connected fibers $f : S \to B$, where B is a smooth curve. A map $\psi : S \to X$, X a variety, is *composed with a pencil* if there exists a pencil $f : S \to B$ and a map $\overline{\psi} : B \to X$ such that $\psi = \overline{\psi} \circ f$. The *genus* of the pencil f is by definition the genus b of B. The pencil f is *irrational* if $b > 0$. Since pullback of forms induces an injective map $H^0(\Omega_B^1) \to H^0(\Omega_S^1)$, a surface with an irrational pencil is irregular. Clearly, the converse is not true (Remark 2.2.1).

In addition, if the pencil f has genus ≥ 2, by pulling back two independent 1-forms of B one gets independent 1-forms α and β on S such that $\alpha \wedge \beta = 0$. The following classical result (see [Beauville 1996] for a proof) states that this condition is equivalent to the existence of an irrational pencil of genus ≥ 2:

Theorem 2.1.1 (Castelnuovo and de Franchis). *Let $\alpha, \beta \in H^0(\Omega_S^1)$ be linearly independent forms such that $\alpha \wedge \beta = 0$. Then there exists a pencil $f : S \to B$ of genus ≥ 2 and $\alpha_0, \beta_0 \in H^0(\omega_B)$ such that $\alpha = f^* \alpha_0$, $\beta = f^* \beta_0$.*

Let $\alpha_1, \dots, \alpha_k, \beta \in H^0(\Omega_S^1)$ be linearly independent forms such that $\alpha_1, \dots, \alpha_k$ are pullbacks from a curve via a pencil $f : S \to B$ and $\beta \wedge \alpha_j = 0$ (notice that it is the same to require this for one index j or for all $j = 1, \dots, k$). By Theorem 2.1.1, there exists a pencil $h : S \to D$ such that, say, α_1 and β are pull backs of independent 1-forms of D. Let $\psi := f \times h : S \to B \times D$. Then the forms $\alpha_j \wedge \beta$ are pullbacks of nonzero 2-forms of $B \times D$. Since $\alpha_j \wedge \beta = 0$, it follows that the image of ψ is a curve C and that $\alpha_1, \ldots, \alpha_k, \beta$ are pullbacks of 1-forms of C. This shows that for every $b \geq 2$ there is one-to-one correspondence between pencils of S of genus b and subspaces $W \subset H^0(\Omega_S^1)$ of dimension b such that $\wedge^2 W = 0$ and W is maximal with this property.

The existence of a subspace W as above can be interpreted in terms of the cohomology of S with complex coefficients, thus showing that the existence of a pencil of genus ≥ 2 is a topological property [Catanese 1991, Theorem 1.10]. On the contrary, the existence of pencils of genus 1 is not detected by the topology (Remark 2.2.1).

A classical result of Severi (see [Severi 1932; Samuel 1966]) states that a surface of general type has finitely many pencils of genus ≥ 2. However a surface of general type can have infinitely many pencils of genus 1, as it is shown by the following example:

Example 2.1.2. Let E be a curve of genus 1, $O \in E$ a point, $A := E \times E$ and $L := \mathbb{O}_A(\{O\} \times E + E \times \{O\})$. For every $n \geq 1$, the map $h_n : A \to E$ defined by $(x, y) \mapsto x + ny$ is a pencil of genus 1. Let now $S \to A$ be a double cover branched on a smooth ample curve $D \in |2dL|$ (see Section 2.4(d) for a quick review of double covers). The surface S is minimal of general type, and for every n the map h_n induces a pencil $f_n : S \to E$. The general fiber of f_n is a double cover of an elliptic curve isomorphic to E, branched on $2d(n^2 + 1)$ points, hence it has genus $dn^2 + d + 1$. It follows that the pencils f_n are all distinct.

2.2. The Albanese map. The *Albanese variety* of S is defined as $\mathrm{Alb}(S) := H^0(\Omega_S^1)^\vee / H_1(S, \mathbb{Z})$. By Hodge theory, $\mathrm{Alb}(S)$ is a compact complex torus and, in addition, it can be embedded in projective space, namely it is an abelian variety. For a fixed base point $x_0 \in S$ one defines the Albanese morphism $a_{x_0} : S \to \mathrm{Alb}(S)$ by $x \mapsto \int_{x_0}^x -$; see [Beauville 1996, Chapter V]. Choosing a different base point in S, the Albanese morphism just changes by a translation of $\mathrm{Alb}(S)$, so we often ignore the base point and just write a. By construction, the map $a_* : H_1(S, \mathbb{Z}) \to H_1(\mathrm{Alb}(S), \mathbb{Z})$ is surjective, with kernel equal to the torsion subgroup of $H_1(S, \mathbb{Z})$, and the map $a^* : H^0(\Omega_{\mathrm{Alb}(S)}^1) \to H^0(\Omega_S^1)$ is an isomorphism, so if $q(S) > 0$ it follows immediately that a is nonconstant. The dimension of $a(S)$ is called the *Albanese dimension* of S and it is denoted by $\mathrm{Albdim}(S)$. S is of *Albanese general type* if $\mathrm{Albdim}(S) = 2$ and $q(S) > 2$.

The morphism $a : S \to \mathrm{Alb}(S)$ is characterized up to unique isomorphism by the following universal property: for every morphism $S \to T$, with T a complex torus, there exists a unique factorization $S \xrightarrow{a} \mathrm{Alb}(S) \to T$. It follows immediately that the image of a generates $\mathrm{Alb}(S)$, namely that $a(S)$ is not contained in any proper subtorus of $\mathrm{Alb}(S)$. Using the Stein factorization and the fact that for a smooth curve B the Abel–Jacobi map $B \to J(B)$ is an embedding, one can show that if $\mathrm{Albdim}(S) = 1$, then $B := a(S)$ is a smooth curve of genus $q(S)$ and the map $a : S \to B$ has connected fibers. In this case, the map $a : S \to B$ is called the *Albanese pencil* of S. By the analysis of Section 2.1, $\mathrm{Albdim}(S) = 1$ ("a is composed with a pencil") if and only if $\alpha \wedge \beta = 0$ for every pair of 1-forms $\alpha, \beta \in H^0(\Omega_S^1)$ if and only if there exists a surjective map $f : S \to B$ where B is a smooth curve of genus $q(S)$. In this case, if $q(S) > 1$ then f coincides with the Albanese pencil.

Since, as we recalled in Section 2.1, the existence of a pencil of given genus ≥ 2 is a topological property of S, the Albanese dimension of a surface is a topological property.

Remark 2.2.1. Let $f : S \to B$ be an irrational pencil. By the universal property of the Albanese variety, there is a morphism of tori $\mathrm{Alb}(S) \to J(B)$. The differential of this morphism at 0 is dual to $f^* : H^0(\omega_B) \to H^0(\Omega_S^1)$, hence it is surjective and $\mathrm{Alb}(S) \to J(B)$ is surjective, too. So if $\mathrm{Albdim}(S) = 2$ and $\mathrm{Alb}(S)$ is simple, S has no irrational pencil. It is easy to produce examples of this situation, for instance by considering surfaces that are complete intersections inside a simple abelian variety; see Section 2.4(c).

If $\mathrm{Albdim}(S) = q(S) = 2$, S has an irrational pencil if and only if there exists a surjective map $\mathrm{Alb}(S) \to E$, where E is an elliptic curve. So, by taking double covers of principally polarized abelian surfaces branched on a smooth ample curve — see Section 2.4(d) — one can construct a family of minimal surfaces of general type such that the general surface in the family has no irrational pencil but some special ones have.

If $\mathrm{Albdim}(S) = 2$, then a contracts finitely many irreducible curves. By Grauert's criterion ([Barth et al. 1984, Theorem III.2.1]), the intersection matrix of the set of curves contracted by a is negative definite. An irreducible curve C of S is contracted by a if and only if the restriction map $H^0(\Omega_S^1) \to H^0(\omega_{C^\nu})$ is trivial, where $C^\nu \to C$ is the normalization. So, every rational curve of S is contracted by a. More generally, by the universal property the map $C^\nu \to \mathrm{Alb}(S)$ factorizes through $J(C^\nu) = \mathrm{Alb}(C^\nu)$, hence $a(C)$ spans an abelian subvariety of $\mathrm{Alb}(C)$ of dimension at most $g(C^\nu)$. So, for instance, if $\mathrm{Alb}(S)$ is simple every curve of S of geometric genus $< q(S)$ is contracted by a.

An irreducible curve C of S such that $a(C)$ spans a proper abelian subvariety $T \subset \mathrm{Alb}(S)$ has $C^2 \leq 0$. In particular, if C has geometric genus $< q(S)$ then $C^2 \leq 0$. Indeed, consider the nonconstant map $\bar{a} : S \to A/T$ induced by a. If the image of $\bar{a}$ is a surface, then C is contracted to a point, hence $C^2 < 0$. If the image of $\bar{a}$ is a curve, then C is contained in a fiber, hence by Zariski's lemma ([Barth et al. 1984, Lem.III.8.9]) $C^2 \leq 0$ and $C^2 = 0$ if and only if C moves in an irrational pencil. In view of the fact that S has finitely many irrational pencils of genus ≥ 2, the irreducible curves of S whose image via a spans an abelian subvariety of $\mathrm{Alb}(S)$ of codimension > 1 belong to finitely many numerical equivalence classes.

We close this section by giving an example that shows that the degree of the Albanese map of a surface with $\mathrm{Albdim} = 2$ is not a topological invariant.

Example 2.2.2. We describe an irreducible family of smooth minimal surfaces of general type such that the Albanese map of the general element of the family is generically injective but for some special elements the Albanese map has degree 2 onto its image. The examples are constructed as divisors in a double cover $p : V \to A$ of an abelian threefold A.

Let A be an abelian threefold and let L be an ample line bundle of A such that $|2L|$ contains a smooth divisor D. There is a double cover $p : V \to A$ branched on D and such that $p_* \mathcal{O}_V = \mathcal{O}_A \oplus L^{-1}$; see Section 2.4(d). The variety V is smooth and, arguing as in Section 2.4(d), one shows that the Albanese map of V coincides with p. Notice that this implies that the map $p_* : H_1(V, \mathbb{Z}) \to H_1(A, \mathbb{Z})$ is an isomorphism up to torsion (see Section 2.2).

Let now $Y \subset A$ be a very ample divisor such that $h^0(\mathcal{O}_A(Y) \otimes L^{-1}) > 0$ and set $X := p^*Y$. If Y is general, then both X and Y are smooth. Let now $X' \in |X|$ be a smooth element. By the adjunction formula, X' is smooth of general type. By the Lefschetz theorem on hyperplane sections, the inclusion $X' \to V$ induces an isomorphism $\pi_1(X') \simeq \pi_1(V)$, which in turn gives an isomorphism $H_1(X', \mathbb{Z}) \simeq H_1(V, \mathbb{Z})$. Composing with $p_* : H_1(V, \mathbb{Z}) \to H_1(A, \mathbb{Z})$ we get an isomorphism (up to torsion) $H_1(X', \mathbb{Z}) \to H_1(A, \mathbb{Z})$, which is induced by the map $p|_{X'} : X' \to A$. Hence $p|_{X'} : X' \to A$ is the Albanese map of X'. Now the map $p|_{X'}$ has degree 2 onto its image if X' is invariant under the involution σ associated to p, and it is generically injective otherwise. By the projection formula for double covers, the general element of $|X|$ is not invariant under σ if and only if $h^0(\mathcal{O}_A(Y) \otimes L^{-1}) > 0$, hence we have the required example.

2.3. Numerical invariants and geography. To a minimal complex surface S of general type, one can attach several integer invariants, besides the irregularity $q(S) = h^0(\Omega_S^1) > 0$:

- the self intersection K_S^2 of the canonical class,

- the *geometric genus* $p_g(S) := h^0(K_S) = h^2(\mathcal{O}_S)$,

- the *holomorphic Euler–Poincaré characteristic*, $\chi(S) := h^0(\mathcal{O}_S) - h^1(\mathcal{O}_S) + h^2(\mathcal{O}_S) = 1 - q(S) + p_g(S)$,

- the second Chern class $c_2(S)$ of the tangent bundle, which coincides with the topological Euler characteristic of S.

All these invariants are determined by the topology of S plus the orientation induced by the complex structure. Indeed (see [Barth et al. 1984, I.1.5]), by Noether's formula we have:

$$K_S^2 + c_2(S) = 12\chi(S), \qquad (2.3.1)$$

and by the Thom–Hirzebruch index theorem:

$$\tau(S) = 2(K_S^2 - 8\chi(S)), \qquad (2.3.2)$$

where $\tau(S)$ denotes the index of the intersection form on $H^2(S, \mathbb{C})$, namely the difference between the number of positive and negative eigenvalues. So K_S^2 and $\chi(S)$ are determined by the (oriented) topological invariants $c_2(S)$ and $\tau(S)$. By Hodge theory the irregularity $q(S)$ is equal to $\frac{1}{2}b_1(S)$, where $b_1(S)$ is the first Betti number. So $q(S)$ is also determined by the topology of S and the same is true for $p_g(S) = \chi(S) + q(S) - 1$.

It is apparent from the definition that these invariants are not independent. So it is usual to take K_S^2, $\chi(S)$ (or, equivalently, K_S^2, $c_2(S)$) as the main numerical invariants. These determine the Hilbert polynomial of the n-canonical image of S for $n \geq 2$, and by a classical result [Gieseker 1977] the coarse moduli space $\mathcal{M}_{a,b}$ of surfaces of general type with $K^2 = a$, $\chi = b$ is a quasiprojective variety. Roughly speaking, this means that surfaces with fixed K^2 and χ are parametrized by a finite number of irreducible varieties, hence in principle they can be classified. In practice, however, the much more basic *geographical question*, i.e., "for what values of a, b is $\mathcal{M}_{a,b}$ nonempty?" is already nontrivial.

The invariants K^2, χ are subject to the following restrictions:

- $K^2, \chi > 0$,
- $K^2 \geq 2\chi - 6$ (Noether's inequality),
- $K^2 \leq 9\chi$ (Bogomolov–Miyaoka–Yau inequality).

All these inequalities are sharp and it is known that for "almost all" a, b in the admissible range the space $\mathcal{M}_{a,b}$ is nonempty. (The possible exceptions seem to be due to the method of proof and not to the existence of special areas in the admissible region for the invariants of surfaces of general type).

In this note we focus on the geographical question for irregular surfaces. More precisely, we address the following questions:

"for what values of a, b does there exist a minimal surface of general type S with $K^2 = a$, $\chi = b$ such that:

- $q(S) > 0$?"
- S has an irrational pencil?"
- Albdim$(S) = 2$?"
- $q(S) > 0$ and S has no irrational pencil?"

Remark 2.3.1. If $S' \to S$ is an étale cover of degree d and S is minimal of general type, S' is also minimal of general type with invariants $K^2_{S'} = d K^2_S$ and $\chi(S') = d\chi(S)$. The first three properties listed above are stable under étale covers. Since the first Betti number of a surface is equal to $2q$, an irregular surface has étale covers of degree d for any $d > 0$. Hence, if for some a, b the answer to one of these three questions is affirmative, the same is true for all the pairs da, db, with d a positive integer.

The main known inequalities for the invariants of irregular surfaces of general type are illustrated in the following sections. Here we only point out the following simple consequence of Noether's inequality:

Proposition 2.3.2. *Let S be a minimal irregular surface of general type. Then*:

$$K^2_S \geq 2\chi(S).$$

Proof. Assume for contradiction that $K^2_S < 2\chi(S)$. Then an étale cover S' of degree $d \geq 7$ has $K^2_{S'} < 2\chi(S') - 6$, violating Noether's inequality. $\square$

More generally, the following inequality holds for minimal irregular surfaces of general type [Debarre 1982]:

$$K^2 \geq \max\{2p_g, 2p_g + 2(q - 4)\}. \tag{2.3.3}$$

The inequality (2.3.3) implies that irregular surfaces with $K^2 = 2\chi$ have $q = 1$. These surfaces are described in Section 2.5(b).

2.4. Basic constructions. Some constructions of irregular surfaces of general type have already been presented in the previous sections. We list and describe briefly the most standard ones:

(a) <u>Products of curves.</u> Take $S := C_1 \times C_2$, with C_i a curve of genus $g_i \geq 2$. S has invariants

$$K^2 = 8(g_1 - 1)(g_2 - 1), \quad \chi = (g_1 - 1)(g_2 - 1), \quad q = g_1 + g_2, \quad p_g = g_1 g_2.$$

In particular these surfaces satisfy $K^2 = 8\chi$. The Albanese variety is the product $J(C_1) \times J(C_2)$ and the Albanese map induces an isomorphism onto its image. The two projections $S \to C_i$ are pencils of genus $g_i \geq 2$.

(b) <u>Symmetric products.</u> Take $S := S^2 C$, where C is a smooth curve of genus $g \geq 3$. Consider the natural map $p : C \times C \to S$, which is the quotient map by the involution ι that exchanges the two factors of $C \times C$. The ramification divisor of p is the diagonal $\Delta \subset C \times C$, hence we have:

$$K_{C \times C} = p^* K_S + \Delta.$$

Computing intersections on $C \times C$ we get

$$K_S^2 = (g-1)(4g-9).$$

Global 1 and 2-forms on $S^2 C$ correspond to forms on $C \times C$ that are invariant under ι. Writing down the action of ι on $H^0(\Omega_{C \times C}^i)$, one obtains canonical identifications:

$$H^0(\omega_S) = \wedge^2 H^0(\omega_C), \quad H^0(\Omega_S^1) = H^0(\omega_C). \tag{2.4.1}$$

Thus we have:

$$p_g = g(g-1)/2, \quad q = g, \quad \chi = g(g-3)/2 + 1.$$

Since $p_g(S) > 0$ and $K_S^2 > 0$, it follows that S is of general type. Notice that by Theorem 2.1.1 S has no irrational pencil of genus ≥ 2, since by (2.4.1) the natural map $\wedge^2 H^0(\Omega_S^1) \to H^0(\omega_S)$ is injective.

The points of S can be identified with the effective divisors of degree 2 of C. If C is hyperelliptic, then the g_2^1 of C gives a smooth rational curve Γ of S such that $\Gamma^2 = 1 - g$. Let $(P_0, Q_0) \in C \times C$ be a point: the map $C \times C \to J(C) \times J(C)$ defined by $(P, Q) \mapsto (P - P_0, Q - Q_0)$ is the Albanese map of $C \times C$ with base point (P_0, Q_0). Composing with the addition map, one obtains a map $C \times C \to J(C)$ that is invariant for the action of ι and therefore induces a map $a : S \to J(C)$, which can be written explicitly as $P + Q \mapsto P + Q - P_0 - Q_0$. Using the universal property, one shows that a is the Albanese map of S. By the Riemann–Roch theorem, if C is not hyperelliptic a is injective, while if C is hyperelliptic a contracts Γ to a point and is injective on $S \setminus \Gamma$. Since $H^0(\omega_S)$ is the pullback of $H^0(\Omega_{J(C)}^2) = \wedge^2 H^0(\omega_C)$ via the Albanese map a, the points of S where the differential of a fails to be injective are precisely the base points of $|K_S|$. So, if C is not hyperelliptic then a is an isomorphism of S with its image and if C is hyperelliptic, then a gives an isomorphism of $S \setminus \Gamma$ with its image.

Notice that as g goes to infinity, the ratio $K_S^2 / \chi(S)$ approaches 8 from below.

(c) Complete intersections. Let V be an irregular variety of dimension $k+2 \geq 3$. For instance, one can take as V an abelian variety or a product of curves not all rational. Given $|D_1|, \ldots, |D_k|$ free and ample linear systems on V such that $K_V + D_1 + \cdots + D_k$ is nef and big, we take

$$S = D_1 \cap \cdots \cap D_k,$$

with $D_i \in |D_i|$ general, so that S is smooth. By the adjunction formula, K_S is the restriction to S of $K_V + D_1 + \cdots + D_k$, hence S is minimal of general type. Since the D_i are ample, the Lefschetz Theorem for hyperplane sections gives an isomorphism $H_1(S, \mathbb{Z}) \cong H_1(V, \mathbb{Z})$. Hence the Albanese map of S is just the restriction of the Albanese map of V.

The numerical invariants of S can be computed by means of standard exact sequences on V. If $k = 1$ and $D_1 \in |rH|$, where H is a fixed ample divisor, one has:

$$K_S^2 = r^3 H^3 + O(r^2), \quad \chi(S) = r^3 H^3/6 + O(r^2),$$

so the ratio $K_S^2/\chi(S)$ tends to 6 as r goes to infinity. Similarly, for $k = 2$ and $D_1, D_2 \in |rH|$, one has:

$$K_S^2 = 4r^4 H^4 + O(r^3), \quad \chi(S) = 7r^4 H^4/12 + O(r^3),$$

and $K_S^2/\chi(S)$ tends to 48/7 as r goes to infinity.

(d) <u>Double covers.</u> If Y is an irregular surface, any surface S that dominates Y is irregular, too, and $\mathrm{Albdim}(S) \geq \mathrm{Albdim}(Y)$. The simplest instance of this situation in which the map $S \to Y$ is not birational is that of a double cover. A smooth double cover of a variety Y is determined uniquely by a line bundle L on Y and a smooth divisor $D \in |2L|$. Set $\mathcal{E} := \mathcal{O}_Y \oplus L^{-1}$ and let $\mathbb{Z}_2$ act on $\mathcal{E}$ as multiplication by 1 on $\mathcal{O}_Y$ and multiplication by -1 on L^{-1}. To define on $\mathcal{E}$ an $\mathcal{O}_Y$-algebra structure compatible with this $\mathbb{Z}_2$-action it suffices to give a map $\mu : L^{-2} \to \mathcal{O}_Y$: we take μ to be a section whose zero locus is D, set $S := \mathrm{Spec}\,\mathcal{E}$ and let $\pi : S \to Y$ be the natural map. S is easily seen to be smooth if and only if D is. By construction, one has:

$$H^i(\mathcal{O}_S) = H^i(\mathcal{O}_Y) \oplus H^i(L^{-1}).$$

In particular, if L is nef and big then by Kawamata–Viehweg vanishing $H^1(\mathcal{O}_S) = H^1(\mathcal{O}_Y)$, hence the induced map $\mathrm{Alb}(S) \to \mathrm{Alb}(Y)$ is an isogeny. It is actually an isomorphism: since $H^0(\Omega_S^1) = H^0(\Omega_Y^1)$, the induced $\mathbb{Z}_2$ action on $\mathrm{Alb}(S)$ is trivial. Since $D = 2L$ is nef and big and effective, it is nonempty and therefore we may choose a base point $x_0 \in S$ that is fixed by $\mathbb{Z}_2$. The Albanese map with base point x_0 is $\mathbb{Z}_2$-equivariant, hence it descends to a map $Y \to \mathrm{Alb}(S)$. So by the universal property there is a morphism $\mathrm{Alb}(Y) \to \mathrm{Alb}(S)$ which is the inverse of the morphism $\mathrm{Alb}(S) \to \mathrm{Alb}(Y)$ induced by π.

A local computation gives the following pullback formula for the canonical divisor:

$$K_S = \pi^*(K_Y + L).$$

By this formula, if $K_Y + L$ is nef and big the surface S is minimal of general type. The numerical invariants of S are:

$$K_S^2 = 2(K_Y + L)^2, \quad \chi(S) = 2\chi(Y) + L(K_Y + L)/2.$$

Hence for "large" L, the ratio $K_S^2/\chi(S)$ tends to 4.

If Y is an abelian surface and L is ample, one has $\mathrm{Albdim}(S) = 2$ and $K_S^2 = 4\chi(S)$.

2.5. *Examples.* The constructions of irregular surfaces of Section 2.4 can be combined to produce more sophisticated examples; see Example 2.2.2, for instance. However the computations of the numerical invariants suggest that in infinite families of examples the ratio K^2/χ converges, so that it does not seem easy to fill by these methods large areas of the admissible region for the invariants K^2, χ.

Here we collect some existence results for irregular surfaces.

(a) $\underline{\chi = 1}$. As explained in Section 2.3, $\chi = 1$ is the smallest possible value for a surface of general type. Since $K^2 \leq 9\chi$ (Section 2.3), in this case we have $K^2 \leq 9$, hence by (2.3.3) $q = p_g \leq 4$. To our knowledge, the only known example of an irregular surface of general type with $K^2 = 9$, $\chi = 1$ was recently constructed by Donald Cartwright and Tim Steiger (unpublished). It has $q = 1$.

We recall briefly what is known about the classification of these surfaces for the possible values of q.

$q = 4$: S is the product of two curves of genus 2 by Theorem 3.0.4, due to Beauville. So $K^2 = 8$ in this case.

$q = 3$: By [Hacon and Pardini 2002] and [Pirola 2002] (see also [Catanese et al. 1998]) these surfaces belong to two families. They are either the symmetric product S^2C of a curve C of genus 3 ($K^2 = 6$) or free $\mathbb{Z}_2$-quotients of a product of curves $C_1 \times C_2$ where $g(C_1) = 2$, $g(C_2) = 3$ ($K^2 = 8$).

$q = 2$: Surfaces with $p_g = q = 2$ having an irrational pencil (hence in particular those with Albdim$(S) = 1$) are classified in [Zucconi 2003]. They have either $K^2 = 4$ or $K^2 = 8$.

Let (A, Θ) be a principally polarized abelian surface A. A double cover $S \to A$ branched on a smooth curve of $|2\Theta|$ is a minimal surface of general type with $K^2 = 4$, $p_g = q = 2$ and it has no irrational pencil if and only if A is simple; see Section 2.4(d). In [Ciliberto and Mendes Lopes 2002] it is proven that this is the only surface with $p_g = q = 2$ and nonbirational bicanonical map that has no pencil of curves of genus 2. An example with $p_g = q = 2$, $K^2 = 5$ and no irrational pencil is constructed in [Chen and Hacon 2006].

$q = 1$: For S a minimal surface of general type with $p_g = q = 1$, we denote by E the Albanese curve of S and by g the genus of the general fiber of the Albanese pencil $a : S \to E$.

The case $K^2 = 2$ is classified in [Catanese 1981]. These surfaces are constructed as follows. Let E be an elliptic curve with origin O. The map $E \times E \to E$ defined by $(P, Q) \mapsto P + Q$ descends to a map $S^2E \to E$ whose fibers are smooth rational curves. We denote by F the algebraic equivalence class of a fiber of $S^2E \to E$. The curves $\{P\} \times E$ and $E \times \{P\}$ map to curves

$D_P \subset S^2 E$ such that $D_P F = D_P^2 = 1$. The curves D_P, $P \in A$, are algebraically equivalent and $h^0(D_P) = 1$. We denote by D the algebraic equivalence class of D_P. Clearly D and F generate the Néron–Severi group of $S^2 E$. All the surfaces S are (minimal desingularizations of) double covers of $S^2 E$ branched on a divisor B numerically equivalent to $6D - 2F$ and with at most simple singularities. The composite map $S \to S^2 E \to E$ is the Albanese pencil of X and its general fiber has genus 2.

The case $K^2 = 3$ is studied in [Catanese and Ciliberto 1991; 1993]. One has either $g = 2$ or $g = 3$. If $g = 2$, then S is birationally a double cover of $S^2 E$, while if $g = 3$ S is birational to a divisor in $S^3 E$. For $K^2 = 4$, several components of the moduli space are constructed in [Pignatelli 2009] (all these examples have $g = 2$). Rito [2007; 2010b; 2010a] gave examples with $K^2 = 2, \ldots, 8$. The case in which S is birational to a quotient $(C \times F)/G$, where C and F are curves and G is a finite group is considered in [Carnovale and Polizzi 2009; Mistretta and Polizzi 2010; Polizzi 2008; 2009]: when $(C \times F)/G$ has at most canonical singularities the surface S has $K^2 = 8$, but there are also examples with $K^2 = 2, 3, 5$.

(b) The line $K^2 = 2\chi$. As pointed out in Proposition 2.3.2, for irregular surfaces the lower bound for the ratio K^2/χ is 2. Irregular surfaces attaining this lower bound were studied in [Horikawa 1977; 1981]. Their structure is fairly simple: they have $q = 1$, the fibers of the Albanese pencil $a : S \to E$ have genus 2 (compare Proposition 4.1.4) and the quotient of the canonical model of S by the hyperelliptic involution is a $\mathbb{P}^1$-bundle over E. The moduli space of these surfaces is studied in [Horikawa 1981].

We just show here that for every integer $d > 0$ there exists a minimal irregular surface of general type with $K^2 = 2\chi$ and $\chi = d$. In (a) above we have sketched the construction of such a surface S with $K_S^2 = 2$, $\chi(S) = 1$. Let $a : S \to E$ be the Albanese pencil and let $E' \to E$ be an unramified cover of degree d. Then the map $S' \to S$ obtained from $E' \to E$ by taking base change with $S \to E$ is a connected étale cover and S' is minimal of general type with $K^2 = 2d$, $\chi = d$. By construction S' maps onto E', hence $q(S') > 0$. By Proposition 4.1.4 we have $q(S') = 1$, hence $S' \to E'$, having connected fibers, is the Albanese pencil of S'.

Alternatively, here is a direct construction for χ even. Let $Y = \mathbb{P}^1 \times E$, with E an elliptic curve and let $L := \mathcal{O}_{\mathbb{P}^1}(3) \boxtimes \mathcal{O}_E(kO)$, where $k \geq 1$ is an integer and $O \in E$ is a point. Let $D \in |2L|$ be a smooth curve and let $\pi : S \to Y$ be the double cover given by the relation $2L \equiv D$; see Section 2.4(d). The surface is smooth, since D is smooth, and it is minimal of general type since $K_S = \pi^*(K_Y + L) = \pi^*(\mathcal{O}_{\mathbb{P}^1}(1) \boxtimes \mathcal{O}_E(kO))$ is ample. The invariants are

$$K^2 = 4k, \chi = 2k.$$

One shows as above that $q(S) = 1$ and $S \to E$ is the Albanese pencil.

(c) Surfaces with an irrational pencil with general fiber of genus g. We use the same construction as in the previous case. Let $g \geq 2$. Take $Y = \mathbb{P}^1 \times E$ with E an elliptic curve, Δ a divisor of positive degree of E and set $L :=$ $\mathcal{O}_{\mathbb{P}^1}(g+1) \boxtimes \mathcal{O}_E(\Delta)$. Let $D \in |2L|$ be a smooth curve and $\pi : S \to Y$ the double cover given by the relation $2L \equiv D$. S is smooth minimal of general type, with invariants

$$K_S^2 = 4(g-1)\deg\Delta, \quad \chi(S) = g \deg\Delta.$$

The projection $Y \to E$ lifts to a pencil $S \to E$ of hyperelliptic curves of genus g. Here $K_S^2/\chi(S)$ is equal to $4(g-1)/g$, which is the lowest possible value by Theorem 4.1.3.

(d) The line $K^2 = 9\chi$. By [Miyaoka 1984, Theorem 2.1] minimal surfaces of general type with $K^2 = 9\chi$ have ample canonical class. By Yau's results [1977], surfaces with $K^2 = 9\chi$ and ample canonical class are quotients of the unit ball in $\mathbb{C}^2$ by a discrete subgroup. The existence of several examples has been shown using this description [Barth et al. 1984, §9, Chapter VII].

Three examples have been constructed in [Hirzebruch 1983] as Galois covers of the plane branched on an arrangement of lines.

For later use, we sketch here one of these constructions. Let $P_1, \ldots, P_4 \in \mathbb{P}^2$ be points in general positions and let $L_1, \ldots, L_6$ be equations for the lines through $P_1, \ldots, P_4$. Let $X \to \mathbb{P}^2$ be the normal finite cover corresponding to the field inclusion

$$\mathbb{C}(\mathbb{P}^2) \subset \mathbb{C}(\mathbb{P}^2)((L_1/L_6)^{\frac{1}{5}}, \ldots, (L_5/L_6)^{\frac{1}{5}}).$$

The cover $X \to \mathbb{P}^2$ is abelian with Galois group $\mathbb{Z}_5^5$ and one can show, for instance by the methods of [Pardini 1991], that X is singular over $P_1, \ldots, P_4$ and that the cover $S \to \hat{\mathbb{P}}^2$ obtained by blowing up $P_1, \ldots, P_4$ and taking base change and normalization is smooth. The cover $S \to \hat{\mathbb{P}}^2$ is branched of order 5 on the union B of the exceptional curves and of the strict transforms of the lines L_j. Hence the canonical class K_S is numerically equivalent to the pull back of $\frac{1}{5}(9L - 3(E_1 + \cdots + E_4))$, where L is the pullback on $\hat{\mathbb{P}}^2$ class of a line in $\mathbb{P}^2$ and the E_i are the exceptional curves of $\hat{\mathbb{P}}^2 \to \mathbb{P}^2$. It follows that K_S is ample and $K_S^2 = 3^2 \cdot 5^4$. The divisor B has 15 singular points, that are precisely the points of $\hat{\mathbb{P}}^2$ whose preimage consists of 5^3 points. Hence, denoting by e the topological Euler characteristic of a variety, we have

$$c_2(S) = e(S) = 5^5[e(\hat{\mathbb{P}}^2) - e(B)] + 5^4[e(B) - 15] + 5^3 \cdot 15 = 3 \cdot 5^4.$$

Thus S satisfies $K^2 = 3c_2$ or, equivalently, $K^2 = 9\chi$. In [Ishida 1983] it is shown that the irregularity $q(S)$ is equal to 30. To prove that Albdim $S = 2$, by the discussion in Section 2.1 it is enough to show that S has more than one irrational pencil. The surface $\hat{\mathbb{P}}^2$ has 5 pencils of smooth rational curves, induced by the systems h_i of lines through each of the P_i, $i = 1, \ldots, 4$, and by the system h_5 of conics through $P_1, \ldots, P_4$. For $i = 1, \ldots, 5$, denote by $f_i : S \to B_i$ the pencil induced by h_i and denote by F_i the general fiber of f_i. For $i = 1, \ldots, 5$, the subgroup $H_i < \mathbb{Z}_5^5$ that maps F_i to itself has order 5^3 and the restricted cover $F_i \to \mathbb{P}^1$ is branched at 4 points. So F_i has genus 76 by the Hurwitz formula. There is a commutative diagram

$$
\begin{array}{ccc}
S & \longrightarrow & \hat{\mathbb{P}}^2 \\
\Big\downarrow{f_i} & & \Big\downarrow \\
B_i & \longrightarrow & \mathbb{P}^1
\end{array}
\qquad (2.5.1)
$$

where the map $B_i \to \mathbb{P}^1$ is an abelian cover with Galois group $\mathbb{Z}_5^5 / H_i \cong \mathbb{Z}_5^2$. The branch points of $B_i \to \mathbb{P}^1$ correspond to the multiple fibers of f_i, hence there are 3 of them and B_i has genus 6 by the Hurwitz formula. One computes $F_i F_j = 5$ for $i \neq j$, hence the pencils F_i are all distinct.

Since the group $H_i \cap H_j$ acts faithfully on the set $F_i \cap F_j$ for F_i, F_j general, it follows that $H_i \cap H_j$ has order 5 and $H_i + H_j = \mathbb{Z}_5^5$. We use this remark to show that $H^0(\Omega_S^1) = \oplus_{i=1}^5 V_i$, where $V_i := f_i^* H^0(\omega_{B_i})$, and therefore that Alb$(S)$ is isogenous to $J(B_1) \times \cdots \times J(B_5)$. Since $q(S) = 30$ and dim $V_i = 6$, it is enough to show that the V_i are in direct sum in $H^0(\Omega_S^1)$. Each subspace V_i decomposes under the action of $\mathbb{Z}_5^5$ as a direct sum of eigenspaces relative to some subset of the group of characters Hom$(\mathbb{Z}_5^5, \mathbb{C}^*)$. Notice that the trivial character never occurs in the decomposition since $B_i / \mathbb{Z}_5^5 = \mathbb{P}^1$. The diagram (2.5.1) shows that the characters occurring in the decomposition of V_i belong to $H_i^\perp$. Since $H_i + H_j = \mathbb{Z}_5^5$ for $i \neq j$, it follows that each character occurs in the decompositions of at most one the V_i, and therefore that there is no linear relation among the V_i.

(e) The ratio K^2/χ. Sommese [1984] showed that the ratios $K^2(S)/\chi(S)$, for S a minimal surface of general type, form a dense set in the admissible interval $[2, 9]$. His construction can be used to prove:

Proposition 2.5.1. (i) *The ratios $K_S^2/\chi(S)$, as S ranges among surfaces with* Albdim$(S) = 1$, *are dense in the interval* $[2, 8]$.

(ii) *The ratios $K_S^2/\chi(S)$, as S ranges among surfaces with* Albdim$(S) = 2$, *are dense in the interval* $[4, 9]$.

Proof. Let X be a minimal surface of general type and let $f : X \to B$ be an irrational pencil. Denote by $g \geq 2$ the genus of a general fiber of f and write K^2, χ for K_X^2, $\chi(X)$. Given positive integers d, k, we construct a surface $S_{d,k}$ as follows:

(1) We take an unramified degree d cover $B' \to B$ and let $Y_d \to X$ be the cover obtained by taking base change with $f : X \to B$.

(2) We take a double cover $B'' \to B'$ branched on $2k > 0$ general points and let $S_{d,k} \to Y_d$ be the cover obtained from $B'' \to B'$ by base change.

The étale cover $Y_d \to X$ is connected, hence Y_d is a minimal surface of general type with $K_{Y_d}^2 = dK^2$ and $\chi(Y_d) = d\chi$. By Section 2.4(d), the surface $S_{d,k}$ is smooth, since the branch points of $B'' \to B'$ are general, and it is minimal of general type since $K_{S_{d,k}}$ is numerically the pullback of $K_{Y_d} + kF$, where F is a fiber of $Y_d \to B'$ (F is the same as the general fiber of $X \to B$). By the formulae for double covers we have

$$\frac{K_{S_{d,k}}^2}{\chi(S_{d,k})} = \frac{2dK^2 + 8k(g-1)}{2d\chi + k(g-1)} = \frac{K^2}{\chi} + \left(8 - \frac{K^2}{\chi}\right)\frac{k(g-1)}{2d\chi + k(g-1)}. \quad (2.5.2)$$

This formula shows that the ratio $K_{S_{d,k}}^2/\chi(S_{d,k})$ is in the interval $[8, K^2/\chi]$ if $K^2 \geq 8\chi$ and it is in $[K^2/\chi, 8]$ otherwise. It is not difficult to show that as d, k vary one obtains a dense set in the appropriate interval [Sommese 1984].

Now to prove the statement it is enough to apply the construction to suitable surfaces. If one takes X to be the surface with $K^2 = 9\chi$ described in (c) and $f : X \to B$ one of the 5 irrational pencils of X, then the surfaces $S_{d,k}$ have Albanese dimension 2 and the ratios of their numerical invariants are dense in $[8, 9]$.

If one takes X to be a double cover of $E \times E$ branched on a smooth ample curve as in Section 2.4(d) and $f : X \to E$ one of the induced pencil, then the surfaces $S_{d,k}$ have Albanese dimension 2 and the ratio of their numerical invariants are dense in $[4, 8]$.

Finally, we take X an irregular surface with $K^2 = 2\chi$. Since $q(X) = 1$ (see (c) above), we can take $f : X \to B$ to be the Albanese pencil. In this case the ratios of the numerical invariants of the surfaces $S_{d,k}$ are dense in the interval $[2, 8]$. To complete the proof we show that the surfaces $S_{d,k}$ have Albanese dimension 1. The surfaces Y_d satisfy $K^2 = 2\chi$, hence they also have $q = 1$. The induced pencil $S_{d,k} \to B''$ has genus $k + 1$, so we need to show that the irregularity of $S_{d,k}$ is equal to $k + 1$. Denote by L the line bundle of Y_d associated to the double cover $S_{d,k} \to Y_d$. By construction $L = \mathcal{O}_{Y_d}(F_1 + \cdots + F_k)$, where the F_i are fibers of the Albanese pencil, and if $k > 1$ we can take the F_i smooth and distinct. We have $q(S_{d,k}) = q(Y_d) + h^1(L^{-1}) = 1 + h^1(L^{-1})$. Finally, $h^1(L^{-1}) = k$ can be proven using the restriction sequence

$$0 \to L^{-1} \to \mathcal{O}_{Y_d} \to \mathcal{O}_{F_1 + \cdots + F_k} \to 0.$$

(Notice that the map $H^1(\mathcal{O}_{Y_d}) \to H^1(\mathcal{O}_{F_1 + \cdots + F_k})$ is 0, since the curves F_i are contracted by the Albanese map). □

Remark 2.5.2. Due to the method of proof, all the surfaces constructed in the proof of Proposition 2.5.1 have an irrational pencil. The examples in Section 2.4 show that, for instance, $4, 6, \frac{48}{7}, 8$ are accumulation points for the ratio K^2/χ of irregular surfaces without irrational pencils. We have no further information on the distribution of the ratios K^2/χ for these surfaces.

3. The Castelnuovo–de Franchis inequality

Let S be an irregular minimal surface of general type. Set $V := H^0(\Omega_S^1)$ and denote by $w : \bigwedge^2 V \to H^0(\omega_S)$ the natural map. If $f : S \to B$ is a pencil of genus $b \geq 2$, then $f^* H^0(\Omega_B^1)$ is a subspace of V such that $\bigwedge^2 f^* H^0(\Omega_B^1)$ is contained in $\ker w$ (Section 2.1). Conversely if $p_g = h^0(\omega_S) < 2q - 3$, the intersection of $\ker w$ with the cone of decomposable elements is nonzero and by 2.1.1, S has a pencil of genus $b \geq 2$. Thus:

Theorem 3.0.3 (Castelnuovo–de Franchis inequality). *Let S be an irregular surface of general type having no irrational pencil of genus $b \geq 2$. Then $p_g \geq 2q - 3$.*

In fact, using this theorem and positivity properties of the relative canonical bundle of a fibration (Theorem 4.1.1), Beauville showed:

Theorem 3.0.4 [Beauville 1982]. *Let S be a minimal surface of general type. Then $p_g \geq 2q - 4$ and if equality holds then S is the product of a curve of genus 2 and a curve of genus $q - 2 \geq 2$.*

So surfaces satisfying $p_g = 2q - 4$ have a particularly simple structure and it is natural to ask what are the irregular surfaces satisfying $p_g = 2q - 3$. Those having an irrational pencil have again a simple structure, as explained in the following theorem, which was proven in [Catanese et al. 1998] for $q = 3$, in [Barja et al. 2007] for $q = 4$ and for $q \geq 5$ in [Mendes Lopes and Pardini 2010].

Theorem 3.0.5. *Let S be a minimal surface of general type satisfying $p_g = 2q - 3$. If S has an irrational pencil of genus $b \geq 2$, then there are the following possibilities for S:*

(i) $S = (C \times F)/\mathbb{Z}_2$, *where C and F are genus 3 curves with a free involution ($q = 4$).*

(ii) S *is the product of two curves of genus 3 ($q = 6$).*

(iii) $S = (C \times F)/\mathbb{Z}_2$, where C is a curve of genus $2q - 3$ with a free action of $\mathbb{Z}_2$, F is a curve of genus 2 with a $\mathbb{Z}_2$-action such that $F/\mathbb{Z}_2$ has genus 1 and $\mathbb{Z}_2$ acts diagonally on $C \times F$ ($q \geq 3$).

In particular, $K_S^2 = 8\chi$.

So the main issue is to study the case when S has no irrational pencil. Various numerical restrictions on the invariants have been obtained. For instance if $q \geq 5$, we have (see [Mendes Lopes and Pardini 2010]):

- $K_S^2 \geq 7\chi(S) - 1$.
- If $K_S^2 < 8\chi(S)$, then $|K_S|$ has fixed components and the degree of the canonical map is 1 or 3.
- If $q(S) \geq 7$ and $K^2 < 8\chi(S) - 6$, then the canonical map is birational.

However, it is hard to say whether these results are sharp, since the only known example of a surface with $p_g = 2q - 3$ and no irrational pencil is the symmetric product of a general curve of genus 3. For low values of q we have:

- if $q = 3$, then S is the symmetric product of a curve of genus 3 — see Section 2.5(a);
- if $q = 4$, then $K^2 = 16, 17$ [Barja et al. 2007; Causin and Pirola 2006];
- if $q = 5$, there exists no such surface [Lopes et al. 2012].

Theorem 3.0.3 has been generalized to the case of Kähler manifolds of arbitrary dimension:

Theorem 3.0.6 [Pareschi and Popa 2009]. *Let X be a compact Kähler manifold with $\dim X = \text{Albdim } X = n$. If there is no surjective morphism $X \to Z$ with Z a normal analytic variety such that $0 < \dim Z = \text{Albdim } Z < \min\{n, q(Z)\}$, then*

$$\chi(\omega_X) \geq q(X) - n.$$

4. The slope inequality

4.1. *Relative canonical class and slope of a fibration.* In this section we consider a fibration ("pencil") $f : S \to B$ with S a smooth projective surface and B a smooth curve of genus $b \geq 0$. Recall that a fibration is *smooth* if and only if all its fibers are smooth, and it is *isotrivial* if and only if all the smooth fibers of f are isomorphic, or, equivalently, if the fibers over a nonempty open set of B are isomorphic. Isotrivial fibrations are also said to have "constant moduli".

We assume that the general fiber F of f has genus $g \geq 2$ and that f is *relatively minimal*, namely that there is no -1-curve contained in the fibers of f. Notice that these assumptions are always satisfied if S is minimal of general type. Notice also that given a nonminimal fibration f it is always possible to pass to a minimal one by blowing down the -1-curves in the fibers.

The *relative canonical class* is defined by $K_f := K_S - f^* K_B$. We also write ω_f for the corresponding line bundle $\mathcal{O}_S(K_f) = \omega_S \otimes f^* \omega_B^{-1}$.

K_f has the following positivity properties:

Theorem 4.1.1 (Arakelov; cf. [Beauville 1982]). *Let f a relatively minimal fibration of genus $g \geq 2$.*

(i) K_f *is nef. Hence,* $K_f^2 = K_S^2 - 8(g-1)(b-1) \geq 0$.

(ii) *If f is not isotrivial, then:*

    (a) $K_f^2 > 0$

    (b) $K_f C = 0$ *for an irreducible curve C of S if and only if C is a -2-curve contained in a fiber of f.*

Let $f : S \to B$ be relatively minimal and let $\bar{S}$ be the surface obtained by contracting the -2 curves contained in the fibers of f. There is an induced fibration $\bar{f} : \bar{S} \to B$ and, since $\bar{S}$ has canonical singularities, K_f is the pullback of $K_{\bar{f}} := K_{\bar{S}} - \bar{f}^* K_B$. By the Nakai criterion for ampleness, Theorem 4.1.1(ii) can be restated by saying that if $\bar{f}$ (equivalently, f) is not isotrivial then $K_{\bar{f}}$ is ample on $\bar{S}$.

Given a pencil f with general fiber of genus g, the push forward $f_* \omega_f$ is a rank g vector bundle on B of degree $\chi_f = \chi(S) - (b-1)(g-1)$. Recall that a vector bundle E on a variety X is said to be *nef* if the tautological line bundle $\mathbb{P}(E)$ is nef.

Theorem 4.1.2 [Fujita 1978] (see also [Beauville 1982]). *Let $f : S \to B$ be a relatively minimal fibration with general fiber of genus $g \geq 2$. Then:*

(i) $f_* \omega_f$ *is nef. In particular,* $\mathcal{O}_{\mathbb{P}(f_* \omega_f)}(1)^g = \chi_f \geq 0$;

(ii) $\chi_f = 0$ *if and only if f is smooth and isotrivial.*

From now one we assume that the relatively minimal fibration $f : S \to B$ is not isotrivial. The *slope* of f is defined as

$$\lambda(f) := \frac{K_f^2}{\chi_f} = \frac{K_S^2 - 8(b-1)(g-1)}{\chi(S) - (b-1)(g-1)}. \qquad (4.1.1)$$

By Theorems 4.1.1 and 4.1.2, $\lambda(f)$ is well defined and > 0.

Theorem 4.1.3 (Slope inequality). *Let $f : S \to B$ be a relatively minimal fibration with fibers of genus $g \geq 2$. If f is not smooth and isotrivial, then*

$$\frac{4(g-1)}{g} \leq \lambda(f) \leq 12.$$

The inequality $\lambda(f) \leq 12$ follows from Noether's formula $12\chi(S) = K_S^2 + c_2(S)$ and from the well-known formula for the Euler characteristic of a fibered surface [Barth et al. 1984, Proposition II.11.24]:

$$c_2(S) = 4(b-1)(g-1) + \sum_{P \in T} e(F_P) - e(F), \qquad (4.1.2)$$

where e denotes the topological Euler characteristic, T is the set of critical values of f, F_p is the fiber of f over the point P and F is a general fiber. For any point $P \in B$ one has $e(F_P) \geq e(F)$, with equality holding only if F_P is smooth (loc. cit.). Hence, the main content of Theorem 4.1.3 is the lower bound $\lambda(f) \geq 4(g-1)/g$.

We have seen (Proposition 2.5.1 and Remark 2.5.2) that the ratios K^2/χ for surfaces with an irrational pencil are dense in the interval $[2, 9]$. The slope inequality gives a lower bound for this ratio in terms of the genus g of the general fiber of the pencil.

Proposition 4.1.4. *Let S be a minimal surface of general type that has an irrational pencil $f : S \to B$ with general fiber of genus g. Then*

$$K_S^2 \geq \frac{4(g-1)}{g} \chi(S) \geq 2\chi(S).$$

In particular, if $K_S^2 = 2\chi(S)$, then $g = 2$ and B has genus 1.

Proof. Assume that the pencil S is not smooth and isotrivial. If $K_S^2 \geq 8\chi(S)$ then of course the statement holds. If $K^2 < 8\chi$, then $K_S^2/\chi(S) \geq \lambda(f)$ and the statement follows by the slope inequality.

If f is smooth and isotrivial, then by [Serrano 1996, §1], S is a quotient $(C \times D)/G$ where C and D are curves of genus ≥ 2 and G is a finite group that acts freely. In particular, $K_S^2 = 8\chi(S)$ and the inequality is satisfied also in this case.

If $K_S^2 = 2\chi(S)$, then by the previous remarks f is not smooth and isotrivial. We have

$$2 = \frac{K_S^2}{\chi(S)} \geq \lambda(f) \geq \frac{4(g-1)}{g}.$$

It follows immediately that $g = 2$ and B has genus 1. □

Further applications of the slope inequality are given in Section 5.

4.2. *History and proofs.* Theorem 4.1.3 was stated and proved first in the case of hyperelliptic fibrations [Persson 1981; Horikawa 1981].

The general statement was proved in [Xiao 1987a] and, independently, in [Cornalba and Harris 1988] under the extra assumption that the fibers of f are semistable curves, i.e., nodal curves. The proof by Cornalba and Harris has been recently generalized in [Stoppino 2008] to cover the general case. Yet another proof was given in [Moriwaki 1996]. One can also find in [Ashikaga and Konno 2002] a nice account of Xiao's proof and of Moriwaki's proof. Hence it seems superfluous to include the various proofs here.

We just wish to point out that the three methods of proof are different. Xiao's proof uses the Harder–Narasimhan filtration of the vector bundle $f_*\omega_f$ and Clifford's Lemma.

The Cornalba–Harris proof uses GIT and relies on the fact that a canonically embedded curve of genus $g \geq 3$ is stable (fibrations whose general fiber is hyperelliptic are treated separately).

Moriwaki's proof is based on Bogomolov's instability theorem for vector bundles on surfaces.

4.3. Refinements and generalizations.

(a) Fibrations attaining the lower or the upper bound for the slope. The examples constructed in Section 2.5(c) show that the lower bound for $\lambda(f)$ given in Theorem 4.1.3 is sharp. By construction, the general fiber in all these examples is hyperelliptic. This is not a coincidence: in [Konno 1993] it is proven that the general fiber of a fibration attaining the minimum possible value of the slope is hyperelliptic. On the other hand, if $\lambda(f) = 12$, then by Noether's formula one has $c_2(S) = 4(g-1)(b-1)$. Hence by (4.1.2), this happens if and only if all the fibers of f are smooth.

(b) Nonhyperelliptic fibrations. Since, as explained in (a), the minimum value of the slope is attained only by hyperelliptic fibrations, it is natural to look for a better bound for nonhyperelliptic fibrations. In [Konno 1993], such a lower bound is established for $g \leq 5$. In [Konno 1999], it is shown that the inequality

$$\lambda(f) \geq \frac{6(g-1)}{g+1}$$

holds if: (1) g is odd, (2) the general fiber of f has maximal Clifford index, (3) Green's conjecture is true for curves of genus g.

Konno's result actually holds under assumption (1) and (2), since Green's conjecture has been proved for curves of odd genus and maximal Clifford index [Voisin 2005; Hirschowitz and Ramanan 1998].

The influence of the Clifford index and of the *relative irregularity* $q_f :=$ $q(S) - b$ has been studied also in [Barja and Stoppino 2008].

(c) Fibrations with general fiber of special type. Refinements of the slope inequality have been obtained also under the assumption that the general fiber has some special geometrical property.

Konno [1996b] showed that if the general fiber of f is trigonal and $g \geq 6$, then $\lambda(f) \geq 14(g - 1)/(3g + 1)$. Barja and Stoppino [2009] showed that the better bound $\lambda(f) \geq (5g - 6)/g$ holds if $g \geq 6$ is even and the general fiber F of f has Maroni invariant 0. (This means that the intersection of all the quadrics containing the canonical image of F is a surface of minimal degree isomorphic to $\mathbb{P}^1 \times \mathbb{P}^1$. For the definition of the Maroni invariant see [Barja and Stoppino 2009, Remark 3.1], for instance).

The case in which the general fiber of f has an involution with quotient a curve of genus γ has been considered in [Barja 2001; Barja and Zucconi 2001; Cornalba and Stoppino 2008]: the most general result in this direction, proved in the last of these papers, is the inequality

$$\lambda(f) \geq \frac{4(g - 1)}{g - \gamma} \quad \text{for } g > 4\gamma + 1.$$

(d) Generalizations to higher dimensions. Let $f : X \to B$ be a fibration, where X is an n-dimensional $\mathbb{Q}$-Gorenstein variety and B is a smooth curve. As in the case of surfaces, one can consider the relative canonical divisor $K_f := K_X - f^* K_B$ and define the slope of f as

$$\lambda(f) := \frac{K_f^n}{\deg f_*(\mathcal{O}_X(K_f))}.$$

This situation is studied in [Barja and Stoppino 2009], where some lower bounds are obtained under quite restrictive assumptions on the fibers.

The relative numerical invariants of threefolds fibered over a curve have also been studied in [Ohno 1992] and [Barja 2000].

5. The Severi inequality

5.1. *History and proofs.* Severi [1932] stated the inequality that bears his name:

Theorem 5.1.1 (Severi's inequality). *If S is a minimal surface of general type with* $\mathrm{Albdim}(S) = 2$, *then*

$$K_S^2 \geq 4\chi(S).$$

Unfortunately, Severi's proof contained a fatal gap, as pointed out by Catanese [1983], who posed the inequality as a conjecture. More or less at the same time, Reid [1979] made the following conjecture, which for irregular surfaces is a consequence of Theorem 5.1.1 (compare Proposition 5.3.3):

Conjecture 5.1.2 (Reid). *If S is a minimal surface of general type such that $K_S^2 < 4\chi(S)$ then either $\pi_1^{\mathrm{alg}}(S)$ is finite or there exists a finite étale cover $S' \to S$ and a pencil $f : S' \to B$ such that the induced surjective map on the algebraic fundamental groups has finite kernel.*

(Recall that for a complex variety X the algebraic fundamental group $\pi_1^{\text{alg}}(X)$ is the profinite completion of the topological fundamental group $\pi_1(X)$).

Motivated by this conjecture, Xiao wrote the foundational paper [Xiao 1987a] on surfaces fibred over a curve, in which he proved both Severi's inequality in the special case of a surface with an irrational pencil and the slope inequality (Section 4).

Building on the results of [Xiao 1987a], Severi's conjecture was proven by Konno ([Konno 1996a]) for even surfaces, namely surfaces such that the canonical class is divisible by 2 in the Picard group. At the end of the 1990's, the conjecture was almost solved by Manetti ([Manetti 2003]), who proved it under the additional assumption that the surface have ample canonical bundle. Finally, the inequality was proven in [Pardini 2005].

The proof given in [Pardini 2005] and the proof given in [Manetti 2003] for K ample are completely different. We sketch briefly both proofs:

(a) Proof for K ample [Manetti 2003]: Let $\pi : \mathbb{P}(\Omega_S^1) \to S$ be the projection and let L be the hyperplane class of $\mathbb{P}(\Omega_S^1)$. Assume for simplicity that $H^0(\Omega_S^1)$ has no base curve. Then two general elements $L_1, L_2 \in |L|$ intersect properly, hence L^2 is represented by the effective 1-cycle $L_1 \cap L_2$. One has

$$L^2(L + \pi^* K_S) = 3(K_S^2 - 4\chi(S)). \tag{5.1.1}$$

If $L + \pi^* K_S$ is nef, then Theorem 5.1.1 follows immediately by (5.1.1). However, this is not the case in general, and one is forced to analyze the cycle $L_1 \cap L_2$ more closely. One can write

$$L_1 \cap L_2 = V + \Gamma_0 + \Gamma_1 + \Gamma_2,$$

where V is a sum of fibers of π, and the Γ_i are sums of sections of π. More precisely, the support of $\pi(\Gamma_0)$ is the union of the curves contracted by the Albanese map a, the support of $\pi(\Gamma_1)$ consists of the curves not contracted by a but on which the differential of a has rank 1 and $\pi(\Gamma_2)$ is supported on curves on which the differential of a is generically nonsingular. The term $\Gamma_0(L + \pi^* K_S)$ can be < 0, but by means of a very careful analysis of the components of Γ_0 and of the multiplicities in the vertical cycle V one can show that

$$L^2(L + \pi^* K_S) = (L_1 \cap L_2)(L + \pi^* K_S) \geq 0.$$

Unfortunately this kind of analysis does not work when $\pi(\Gamma_0)$ contains -2-curves, hence one has to assume K_S ample.

(a) Proof [Pardini 2005]: Set $A := \text{Alb}(S)$ and fix a very ample divisor D on A. For every integer d one constructs a fibered surface $f_d : Y_d \to \mathbb{P}^1$ as follows:

(1) Consider the cartesian diagram

$$
\begin{array}{ccc}
S_d & \xrightarrow{p_d} & S \\
a_d \downarrow & & \downarrow a \\
A & \xrightarrow{\mu_d} & A
\end{array}
\qquad (5.1.2)
$$

where μ_d denotes multiplication by d, and let $H_d := a_d^* D$. The map p_d is a finite connected étale cover, in particular S_d is minimal of general type.

(2) Choose a general pencil $\Lambda_d \subset |H_d|$ and let $f_d : Y_d \to \mathbb{P}^1$ be the relatively minimal fibration obtained by resolving the indeterminacy of the rational map $S_d \to \mathbb{P}^1$ defined by Λ_d.

The key observation of this proof is that as d goes to infinity, the intersection numbers H_d^2 and $K_{S_d} H_d$ grow slower than $K_{S_d}^2$ and $\chi(S_d)$. As a consequence, the slope of f_d converges to the ratio $K_S^2 / \chi(S)$ as d goes to infinity. Since g_d goes to infinity with d, the Severi inequality can be obtained by applying the slope inequality (Theorem 4.1.3) to the fibrations f_d and taking the limit for $d \to \infty$.

5.2. Remarks, refinements and open questions.

(a) <u>Chern numbers of surfaces with Albdim = 2.</u> Severi's inequality is sharp, since double covers of an abelian surface branched on a smooth ample curve satisfy $K^2 = 4\chi$; see Section 2.5(d). Actually, in [Manetti 2003] it is proven that these are the only surfaces with K ample, Albdim = 2 and $K^2 = 4\chi$. Hence it is natural to conjecture that the canonical models of surfaces with Albdim = 2 and $K^2 = 4\chi$ are double covers of abelian surfaces branched on an ample curve with at most simple singularities.

In addition, by Proposition 2.5.1, the ratios $K_S^2 / \chi(S)$ for S a minimal surface with Albdim $S = 2$ are dense in all the admissible interval [4, 9].

(b) <u>Refinements of the inequality.</u> Assume that the differential of the Albanese map $a : S \to A$ is nonsingular outside a finite set. Then the cotangent bundle Ω_S^1 is nef and $L^3 = 2(K_S^2 - 6\chi(S)) \geq 0$. This remark suggests that one may expect an inequality stronger than Theorem 5.1.1 to hold under some assumption on a, e.g., that a is birational. A possible way of obtaining a result of this type would be to apply in the proof of [Pardini 2005] one of the refined versions of the slope inequality; see Section 4.3(b,c). Unfortunately, one has very little control on the general fiber of the fibrations $f_d : Y_d \to \mathbb{P}^1$ constructed in the proof.

Analogously, by the result of Manetti mentioned in (a), it is natural to expect that a better bound holds for surfaces with $q > 2$ [Manetti 2003, §7] for a series of conjectures. A step in this direction is the following:

Theorem 5.2.1 [Mendes Lopes and Pardini 2011]. *Let S be a smooth surface of maximal Albanese dimension and irregularity $q \geq 5$ with K_S ample. Then*

$$K_S^2 \geq 4\chi(S) + \tfrac{10}{3}q - 8.$$

Furthermore, if S has no irrational pencil and the Albanese map $a : S \to A$ is unramified in codimension 1, then

$$K_S^2 \geq 6\chi(S) + 2q - 8.$$

Theorem 5.2.1 is proven by using geometrical arguments to give a lower bound for the term $L\Gamma_2$ in the proof of Severi's inequality for K ample given in [Manetti 2003]; see Section 5.1(a). This is why one needs to assume K ample. In [Mendes Lopes and Pardini 2011], it is also shown that the bounds of Theorem 5.2.1 can be sharpened if one assumes that the Albanese map or the canonical map of S is not birational.

(c) Generalizations to higher dimensions. The proof of Theorem 5.1.1 given in [Pardini 2005] (see Section 5.1) would work in arbitrary dimension n if one had a slope inequality for varieties of dimension $n - 1$. For instance, using the results of [Barja 2000], one can prove:

Theorem 5.2.2. *Let X be a smooth projective threefold such that K_X is nef and big and Albdim $X = 3$. Then:*

(i) $K_X^3 \geq 4\chi(\omega_X)$.

(ii) *If Alb(X) is simple, then $K_X^3 \geq 9\chi(\omega_X)$.*

Proof. We may assume $\chi(\omega_X) > 0$. (Recall $\chi(\omega_X) \geq 0$ by [Green and Lazarsfeld 1987, Corollary to Theorem 1]). Consider a fibered threefold $f : Y \to B$, where B is a smooth curve, denote by F a general fiber and assume $\chi_f := \chi(\omega_X) - \chi(\omega_B)\chi(\omega_F) \neq 0$. Following [Barja 2000] we define

$$\lambda_2(f) := \frac{(K_X - f^*K_B)^3}{\chi_f}.$$

We apply the same construction as in Section 5.1(b) to get for every positive integer d a smooth fibered threefold $f_d : Y_d \to \mathbb{P}^1$ such that $\lambda_2(f_d)$ is defined for $d \gg 0$ and converges to $\frac{K_X^3}{\chi(\omega_X)}$ for $d \to \infty$. Statement (i) now follows by applying [Barja 2000, Theorem 3.1(i)] to f_d and taking the limit for $d \to \infty$.

Statement (ii) requires a little more care. The threefold Y_d is the blow up along a smooth curve of an étale cover $Z_d \to X$. Since $A := \text{Alb}(X)$ is simple, $V^1(X) := \{P \in \text{Pic}^0(X)|h^1(-P) > 0\}$ is a finite set by the Generic Vanishing theorem of [Green and Lazarsfeld 1987]. Then there are infinitely many values of d such that $dP \neq 0$ for every $P \in V^1(X) \setminus \{0\}$. For those values $q(Y_d) =$

$q(Z_d) = q(X)$ and $A := \mathrm{Alb}(X)$ and $\mathrm{Alb}(Y_d) = \mathrm{Alb}(Z_d)$ are isogenous; in particular $\mathrm{Alb}(Z_d)$ is simple. By construction, the general fiber F_d of f_d is a surface of maximal Albanese dimension. In addition, since F_d is isomorphic to an element of the nef and big linear system $|H_d|$ (notation as in Section 5.1(b)), it follows that $q(F_d) = q(X)$ and $\mathrm{Alb}(F_d)$ is isogenous to the simple abelian variety $\mathrm{Alb}(Y_d)$. Hence F_d has no irrational pencil and we get statement (ii) by applying [Barja 2000, Theorem 3.1(ii)] and taking the limit for $d \to \infty$. $\qquad\square$

Remark 5.2.3. In order to keep the proof of Theorem 5.2.2 simple, we have made stronger assumptions than necessary: for instance one can assume that X has terminal singularities and, with some more work, one can eliminate the assumption that $\mathrm{Alb}(X)$ is simple in (ii).

5.3. *Applications.* We use the following result due to Xiao Gang:

Proposition 5.3.1 [Xiao 1987a]. *Let $f : S \to B$ be a relatively minimal fibration with fibers of genus $g \geq 2$. If $\lambda(f) < 4$ and f has no multiple fibers, then there is an exact sequence*

$$1 \to N \to \pi_1^{\mathrm{alg}}(S) \to \pi_1^{\mathrm{alg}}(B) \to 1,$$

where $|N| \leq 2$.

We also need the following consequence of Severi's inequality and of the slope inequality.

Lemma 5.3.2. *Let S be a minimal regular surface of general type S with $K_S^2 < 4\chi(S)$ that has an irregular étale cover $S' \to S$. Then S has a pencil $f : S \to \mathbb{P}^1$ such that*

(i) *f has multiple fibers $m_1 F_1, \ldots, m_k F_k$ with $\displaystyle\sum_j \frac{m_j - 1}{m_j} \geq 2$;*

(ii) *the general fiber of f has genus $g \leq 1 + \dfrac{K_S^2}{4\chi(S) - K_S^2}$.*

Proof. Up to passing to the Galois closure, we may assume that $S' \to S$ is a Galois cover with Galois group G. By Severi's inequality (Theorem 5.1.1) the Albanese map of S' is a pencil $a' : S' \to B$, where B has genus $b > 0$. Clearly G acts on f and on B, inducing a pencil $f : S \to B/G$. Since S is regular, B/G is isomorphic to $\mathbb{P}^1$. Denote by $\overline{G}$ the quotient of G that acts effectively on B. Let $y \in B$ be a ramification point of order ν of the map $B \to B/\overline{G} = \mathbb{P}^1$ and let $H < \overline{G}$ be the stabilizer of y. The group H is cyclic of order ν and it acts freely on the fiber F_y of a' over y. It follows that the fiber of f over the image x of y is a multiple fiber of multiplicity divisible by ν. Let $x_1, \ldots, x_r \in \mathbb{P}^1$ be the critical values of $B \to B/\overline{G}$, let ν_i be the ramification order of x_i and let

$m_1 F_1, \ldots, m_k F_k$ be the multiple fibers of f. Then by the Hurwitz formula we have

$$\sum_j \frac{m_j - 1}{m_j} \geq \sum_i \frac{v_i - 1}{v_i} = \frac{2b - 2}{|\overline{G}|} + 2 \geq 2,$$

hence (i) is proven.

To prove (ii), we observe that the fibers of f are quotients (possibly by a trivial action) of the fibers of a, hence $g \leq \gamma$, where γ is the genus of a general fiber of a. In turn, by the slope inequality one has

$$\frac{K_S^2}{\chi(S)} \geq \lambda(a) \geq \frac{4(\gamma - 1)}{\gamma},$$

which gives the required bound

$$g \leq \gamma \leq 1 + \frac{K_S^2}{4\chi(S) - K_S^2}. \qquad \square$$

(a) <u>Reid's conjecture for irregular surfaces.</u> Severi's inequality implies Reid's Conjecture 5.1.2 for irregular surfaces and, more generally, surfaces that have an irregular étale cover:

Proposition 5.3.3. *Let S be a minimal surface of general type with $K_S^2 < 4\chi(S)$. If S has an irregular étale cover, then there exists a finite étale cover $S' \to S$ and a pencil $f : S' \to B$ that induces an exact sequence*

$$0 \to N \to \pi_1^{\mathrm{alg}}(S') \to \pi_1^{\mathrm{alg}}(B) \to 0,$$

where $|N| \leq 2$.

Proof. Let $X \to S$ be an irregular étale cover. By Theorem 5.1.1 the Albanese map of X is a pencil $a : X \to C$ and, if $S' \to X$ is étale, then the Albanese pencil of S' is obtained by pulling back the Albanese pencil of X and taking the Stein factorization. So, up to taking a suitable base change $B \to C$ and normalizing, we can pass to an étale cover $S' \to S$ such that the Albanese pencil $a' : S' \to B$ has no multiple fiber. The statement now follows by applying Proposition 5.3.1 to a'. $\qquad \square$

Remark 5.3.4. By Proposition 5.3.3, to prove Reid's conjecture it is enough to show:

If S is a surface with $K_S^2 < 4\chi(S)$ that has no irregular étale cover, then $\pi_1^{\mathrm{alg}}(S)$ is finite.

This is known to be true for $K^2 < 3\chi$; see [Mendes Lopes and Pardini 2007] and references therein. In the same reference and in [Ciliberto et al. 2007] the following sharp bound on the order of $\pi_1^{\mathrm{alg}}(S)$ is given:

If $K_S^2 < 3\chi(S)$ and S has no irregular étale cover, then $|\pi_1^{\mathrm{alg}}(S)| \leq 9$. Furthermore, if $|\pi_1^{\mathrm{alg}}(S)| = 8$ or 9 then $K_S^2 = 2$ and $p_g(S) = 0$ (namely, S is a Campedelli surface).

Surfaces with $K^2 = 2$, $p_g = 0$ and $|\pi_1^{\mathrm{alg}}| = 8$ or 9 are classified in [Mendes Lopes et al. 2009] and [Lopes and Pardini 2008], respectively.

However, all the above mentioned results make essential use of Castelnuovo's inequality $K^2 \geq 3\chi - 10$ for surfaces whose canonical map is not two-to-on onto a ruled surface. Hence, different methods are needed to deal with surfaces with $3\chi \leq K^2 < 4\chi$.

(b) Surfaces with "small" K^2. By Proposition 2.3.2, a minimal irregular surface of general type satisfies $K^2 \geq 2\chi$. Irregular surfaces on or near the line $K^2 = 2\chi$ have been studied in [Bombieri 1973; Horikawa 1981; Reid 1979; Xiao 1987a; 1987b]. As an application of Severi's inequality and of the slope inequality, we give quick proofs of some of these results.

Proposition 5.3.5. *Let S be a minimal irregular surface of general type. Then*

(i) *If $K_S^2 = 2\chi(S)$, then $q(S) = 1$ and the fibers of the Albanese pencil $a : S \to B$ have genus 2 (Proposition 4.1.4);*

(ii) *if $K_S^2 < \frac{8}{3}\chi(S)$, then the Albanese map is a pencil of curves of genus 2;*

(iii) *if $K_S^2 < 3\chi(S)$, then the Albanese map is a pencil of hyperelliptic curves of genus ≤ 3.*

Proof. By Severi's inequality, the Albanese map of S is a pencil $a : S \to B$, where B has genus b. By the slope inequality we have

$$\frac{K_S^2}{\chi(S)} \geq \lambda(a) \geq \frac{4(g-1)}{g} \geq 2, \qquad (5.3.1)$$

where, as usual, g denotes the genus of a general fiber of a. If $K_S^2 = 2\chi(S)$, then all the inequalities in (5.3.1) are equalities, hence $g = 2$ and $b = 1$.

If $K_S^2 < \frac{8}{3}\chi(S)$, then (5.3.1) gives $g = 2$.

If $K_S^2 < 3\chi(S)$, then (5.3.1) gives $g \leq 3$. The general fiber of a is hyperelliptic, since otherwise $\lambda(a) \geq 3$ by [Konno 1996b, Lem. 3.1]. $\square$

Next we consider regular surfaces that have an irregular étale cover.

Proposition 5.3.6. *Let S be a minimal regular surface of general type. Then:*

(i) *if $K_S^2 < \frac{8}{3}\chi(S)$, then S has no irregular étale cover;*

(ii) *if $K_S^2 < 3\chi(S)$, S has an irregular étale cover if and only if it has a pencil of hyperelliptic curves of genus 3 with at least 4 double fibers.*

Proof. Assume that $S' \to S$ is an irregular étale cover. Then by Lemma 5.3.2, there is a pencil $f : S \to \mathbb{P}^1$ such that the general fiber of f has genus at most

$$1 + \frac{K_S^2}{4\chi(S) - K_S^2}$$

and there are multiple fibers $m_1 F_1, \ldots, m_k F_k$ such that $\sum_i (m_i - 1)/m_i \geq 2$. Since by the adjunction formula a pencil of curves of genus 2 has no multiple fibers, it follows $g \geq 3$ and $K_S^2 \geq \frac{8}{3}\chi(S)$. This proves (i).

If $K_S^2 < 3\chi(S)$, then $g = 3$ and the general fiber of f is hyperelliptic (compare the proof of Proposition 5.3.5). By the adjunction formula, the multiple fibers of f are double fibers, hence there are at least 4 double fibers.

Conversely, assume that $f : S \to \mathbb{P}^1$ is a pencil and that $y_1, \ldots, y_4 \in \mathbb{P}^1$ are points such that the fiber of f over y_i is double. Let $B \to \mathbb{P}^1$ be the double cover branched on $y_1, \ldots, y_4$ and let $S' \to S$ be obtained from $B \to \mathbb{P}^1$ by taking base change with f and normalizing. The map $S' \to S$ is an étale double cover and by construction S' maps onto the genus 1 curve B, hence $q(S') \geq 1$. $\square$

Remark 5.3.7. With some more work, it can be shown that Proposition 5.3.6(ii) also holds for $K_S^2 = 3\chi(S)$ [Mendes Lopes and Pardini 2007, Theorem 1.1].

References

[Ashikaga and Konno 2002] T. Ashikaga and K. Konno, "Global and local properties of pencils of algebraic curves", pp. 1–49 in *Algebraic geometry 2000* (Azumino), Adv. Stud. Pure Math. **36**, Math. Soc. Japan, Tokyo, 2002.

[Barja 2000] M. A. Barja, "Lower bounds of the slope of fibred threefolds", *Internat. J. Math.* **11**:4 (2000), 461–491.

[Barja 2001] M. A. Barja, "On the slope of bielliptic fibrations", *Proc. Amer. Math. Soc.* **129**:7 (2001), 1899–1906.

[Barja and Stoppino 2008] M. Á. Barja and L. Stoppino, "Linear stability of projected canonical curves with applications to the slope of fibred surfaces", *J. Math. Soc. Japan* **60**:1 (2008), 171–192.

[Barja and Stoppino 2009] M. Á. Barja and L. Stoppino, "Slopes of trigonal fibred surfaces and of higher dimensional fibrations", *Ann. Sc. Norm. Super. Pisa Cl. Sci.* (5) **8**:4 (2009), 647–658.

[Barja and Zucconi 2001] M. Á. Barja and F. Zucconi, "On the slope of fibred surfaces", *Nagoya Math. J.* **164** (2001), 103–131.

[Barja et al. 2007] M. A. Barja, J. C. Naranjo, and G. P. Pirola, "On the topological index of irregular surfaces", *J. Algebraic Geom.* **16**:3 (2007), 435–458.

[Barth et al. 1984] W. Barth, C. Peters, and A. Van de Ven, *Compact complex surfaces*, Ergebnisse der Mathematik und ihrer Grenzgebiete (3) **4**, Springer, Berlin, 1984.

[Beauville 1982] A. Beauville, "L'inégalité $p_g \geq 2q - 4$ pour les surfaces de type général", 1982. Appendix in [Debarre 1982].

[Beauville 1996] A. Beauville, *Complex algebraic surfaces*, 2nd ed., London Mathematical Society Student Texts **34**, Cambridge University Press, Cambridge, 1996.

[Bombieri 1973] E. Bombieri, "Canonical models of surfaces of general type", *Inst. Hautes Études Sci. Publ. Math.* **42** (1973), 171–219.

[Carnovale and Polizzi 2009] G. Carnovale and F. Polizzi, "The classification of surfaces with $p_g = q = 1$ isogenous to a product of curves", *Adv. Geom.* **9**:2 (2009), 233–256.

[Catanese 1981] F. Catanese, "On a class of surfaces of general type", pp. 269–284 in *Algebraic surfaces* (Liguori, Naples, 1981), 1981.

[Catanese 1983] F. Catanese, "Moduli of surfaces of general type", pp. 90–112 in *Algebraic geometry—open problems* (Ravello, 1982), Lecture Notes in Math. **997**, Springer, Berlin, 1983.

[Catanese 1991] F. Catanese, "Moduli and classification of irregular Kaehler manifolds (and algebraic varieties) with Albanese general type fibrations", *Invent. Math.* **104**:2 (1991), 263–289.

[Catanese and Ciliberto 1991] F. Catanese and C. Ciliberto, "Surfaces with $p_g = q = 1$", pp. 49–79 in *Problems in the theory of surfaces and their classification* (Cortona, 1988), Sympos. Math. **32**, Academic Press, London, 1991.

[Catanese and Ciliberto 1993] F. Catanese and C. Ciliberto, "Symmetric products of elliptic curves and surfaces of general type with $p_g = q = 1$", *J. Algebraic Geom.* **2**:3 (1993), 389–411.

[Catanese et al. 1998] F. Catanese, C. Ciliberto, and M. Mendes Lopes, "On the classification of irregular surfaces of general type with nonbirational bicanonical map", *Trans. Amer. Math. Soc.* **350**:1 (1998), 275–308.

[Causin and Pirola 2006] A. Causin and G. P. Pirola, "Hermitian matrices and cohomology of Kähler varieties", *Manuscripta Math.* **121**:2 (2006), 157–168.

[Chen and Hacon 2006] J. A. Chen and C. D. Hacon, "A surface of general type with $p_g = q = 2$ and $K_X^2 = 5$", *Pacific J. Math.* **223**:2 (2006), 219–228.

[Ciliberto and Mendes Lopes 2002] C. Ciliberto and M. Mendes Lopes, "On surfaces with $p_g = q = 2$ and non-birational bicanonical maps", *Adv. Geom.* **2**:3 (2002), 281–300.

[Ciliberto et al. 2007] C. Ciliberto, M. M. Lopes, and R. Pardini, "Surfaces with $K^2 < 3\chi$ and finite fundamental group", *Math. Res. Lett.* **14**:6 (2007), 1069–1086.

[Cornalba and Harris 1988] M. Cornalba and J. Harris, "Divisor classes associated to families of stable varieties, with applications to the moduli space of curves", *Ann. Sci. École Norm. Sup.* (4) **21**:3 (1988), 455–475.

[Cornalba and Stoppino 2008] M. Cornalba and L. Stoppino, "A sharp bound for the slope of double cover fibrations", *Michigan Math. J.* **56**:3 (2008), 551–561.

[Debarre 1982] O. Debarre, "Inégalités numériques pour les surfaces de type général", *Bull. Soc. Math. France* **110**:3 (1982), 319–346.

[Fujita 1978] T. Fujita, "On Kähler fiber spaces over curves", *J. Math. Soc. Japan* **30**:4 (1978), 779–794.

[Gieseker 1977] D. Gieseker, "Global moduli for surfaces of general type", *Invent. Math.* **43**:3 (1977), 233–282.

[Green and Lazarsfeld 1987] M. Green and R. Lazarsfeld, "Deformation theory, generic vanishing theorems, and some conjectures of Enriques, Catanese and Beauville", *Invent. Math.* **90**:2 (1987), 389–407.

[Hacon and Pardini 2002] C. D. Hacon and R. Pardini, "Surfaces with $p_g = q = 3$", *Trans. Amer. Math. Soc.* **354**:7 (2002), 2631–2638.

[Hirschowitz and Ramanan 1998] A. Hirschowitz and S. Ramanan, "New evidence for Green's conjecture on syzygies of canonical curves", *Ann. Sci. École Norm. Sup.* (4) **31**:2 (1998), 145–152.

[Hirzebruch 1983] F. Hirzebruch, "Arrangements of lines and algebraic surfaces", pp. 113–140 in *Arithmetic and geometry*, Progr. Math. **36**, Birkhäuser, Boston, 1983.

[Horikawa 1977] E. Horikawa, "On algebraic surfaces with pencils of curves of genus 2", pp. 79–90 in *Complex analysis and algebraic geometry*, Iwanami Shoten, Tokyo, 1977.

[Horikawa 1981] E. Horikawa, "Algebraic surfaces of general type with small c_1^2. V", *J. Fac. Sci. Univ. Tokyo Sect. IA Math.* **28**:3 (1981), 745–755.

[Ishida 1983] M.-N. Ishida, "The irregularities of Hirzebruch's examples of surfaces of general type with $c_1^2 = 3c_2$", *Math. Ann.* **262**:3 (1983), 407–420.

[Konno 1993] K. Konno, "Nonhyperelliptic fibrations of small genus and certain irregular canonical surfaces", *Ann. Scuola Norm. Sup. Pisa Cl. Sci.* (4) **20**:4 (1993), 575–595.

[Konno 1996a] K. Konno, "Even canonical surfaces with small K^2. III", *Nagoya Math. J.* **143** (1996), 1–11.

[Konno 1996b] K. Konno, "A lower bound of the slope of trigonal fibrations", *Internat. J. Math.* **7**:1 (1996), 19–27.

[Konno 1999] K. Konno, "Clifford index and the slope of fibered surfaces", *J. Algebraic Geom.* **8**:2 (1999), 207–220.

[Lopes and Pardini 2008] M. M. Lopes and R. Pardini, "Numerical Campedelli surfaces with fundamental group of order 9", *J. Eur. Math. Soc. (JEMS)* **10**:2 (2008), 457–476.

[Lopes et al. 2012] M. M. Lopes, R.Pardini, and G. P. Pirola, "On surfaces of general type with $q = 5$", *Ann. Sc. Norm. Super. Pisa Cl. Sci.* (5) (2012). To appear.

[Manetti 2003] M. Manetti, "Surfaces of Albanese general type and the Severi conjecture", *Math. Nachr.* **261/262** (2003), 105–122.

[Mendes Lopes and Pardini 2007] M. Mendes Lopes and R. Pardini, "On the algebraic fundamental group of surfaces with $K^2 \le 3\chi$", *J. Differential Geom.* **77**:2 (2007), 189–199.

[Mendes Lopes and Pardini 2010] M. Mendes Lopes and R. Pardini, "On surfaces with $p_g = 2q - 3$", *Adv. Geom.* **10**:3 (2010), 549–555.

[Mendes Lopes and Pardini 2011] M. Mendes Lopes and R. Pardini, "Severi type inequalities for irregular surfaces with ample canonical class", *Comment. Math. Helv.* **86**:2 (2011), 401–414.

[Mendes Lopes et al. 2009] M. Mendes Lopes, R. Pardini, and M. Reid, "Campedelli surfaces with fundamental group of order 8", *Geom. Dedicata* **139** (2009), 49–55.

[Mistretta and Polizzi 2010] E. Mistretta and F. Polizzi, "Standard isotrivial fibrations with $p_g = q = 1$, II", *J. Pure Appl. Algebra* **214**:4 (2010), 344–369.

[Miyaoka 1984] Y. Miyaoka, "The maximal number of quotient singularities on surfaces with given numerical invariants", *Math. Ann.* **268**:2 (1984), 159–171.

[Moriwaki 1996] A. Moriwaki, "A sharp slope inequality for general stable fibrations of curves", *J. Reine Angew. Math.* **480** (1996), 177–195.

[Ohno 1992] K. Ohno, "Some inequalities for minimal fibrations of surfaces of general type over curves", *J. Math. Soc. Japan* **44**:4 (1992), 643–666.

[Pardini 1991] R. Pardini, "Abelian covers of algebraic varieties", *J. Reine Angew. Math.* **417** (1991), 191–213.

[Pardini 2005] R. Pardini, "The Severi inequality $K^2 \ge 4\chi$ for surfaces of maximal Albanese dimension", *Invent. Math.* **159**:3 (2005), 669–672.

[Pareschi and Popa 2009] G. Pareschi and M. Popa, "Strong generic vanishing and a higher-dimensional Castelnuovo–de Franchis inequality", *Duke Math. J.* **150**:2 (2009), 269–285.

[Persson 1981] U. Persson, "Chern invariants of surfaces of general type", *Compositio Math.* **43**:1 (1981), 3–58.

[Pignatelli 2009] R. Pignatelli, "Some (big) irreducible components of the moduli space of minimal surfaces of general type with $p_g = q = 1$ and $K^2 = 4$", *Atti Accad. Naz. Lincei Cl. Sci. Fis. Mat. Natur. Rend. Lincei* (9) *Mat. Appl.* **20**:3 (2009), 207–226.

[Pirola 2002] G. P. Pirola, "Surfaces with $p_g = q = 3$", *Manuscripta Math.* **108**:2 (2002), 163–170.

[Polizzi 2008] F. Polizzi, "On surfaces of general type with $p_g = q = 1$ isogenous to a product of curves", *Comm. Algebra* **36**:6 (2008), 2023–2053.

[Polizzi 2009] F. Polizzi, "Standard isotrivial fibrations with $p_g = q = 1$", *J. Algebra* **321**:6 (2009), 1600–1631.

[Reid 1979] M. Reid, "π_1 for surfaces with small K^2", pp. 534–544 in *Algebraic geometry* (Copenhagen, 1978), Lecture Notes in Math. **732**, Springer, Berlin, 1979.

[Rito 2007] C. Rito, "On surfaces with $p_g = q = 1$ and non-ruled bicanonial involution", *Ann. Sc. Norm. Super. Pisa Cl. Sci.* (5) **6**:1 (2007), 81–102.

[Rito 2010a] C. Rito, "Involutions on surfaces with $p_g = q = 1$", *Collect. Math.* **61**:1 (2010), 81–106.

[Rito 2010b] C. Rito, "On equations of double planes with $p_g = q = 1$", *Math. Comp.* **79**:270 (2010), 1091–1108.

[Samuel 1966] P. Samuel, "Compléments à un article de Hans Grauert sur la conjecture de Mordell", *Inst. Hautes Études Sci. Publ. Math.* **29** (1966), 55–62.

[Serrano 1996] F. Serrano, "Isotrivial fibred surfaces", *Ann. Mat. Pura Appl.* (4) **171** (1996), 63–81.

[Severi 1932] F. di Severi, "La serie canonica e la teoria delle serie principali di gruppi di punti sopra una superficie algebrica", *Comment. Math. Helv.* **4**:1 (1932), 268–326.

[Sommese 1984] A. J. Sommese, "On the density of ratios of Chern numbers of algebraic surfaces", *Math. Ann.* **268**:2 (1984), 207–221.

[Stoppino 2008] L. Stoppino, "Slope inequalities for fibred surfaces via GIT", *Osaka J. Math.* **45**:4 (2008), 1027–1041.

[Voisin 2005] C. Voisin, "Green's canonical syzygy conjecture for generic curves of odd genus", *Compos. Math.* **141**:5 (2005), 1163–1190.

[Xiao 1987a] G. Xiao, "Fibered algebraic surfaces with low slope", *Math. Ann.* **276**:3 (1987), 449–466.

[Xiao 1987b] G. Xiao, "Hyperelliptic surfaces of general type with $K^2 < 4\chi$", *Manuscripta Math.* **57**:2 (1987), 125–148.

[Yau 1977] S. T. Yau, "Calabi's conjecture and some new results in algebraic geometry", *Proc. Nat. Acad. Sci. U.S.A.* **74**:5 (1977), 1798–1799.

[Zucconi 2003] F. Zucconi, "Surfaces with $p_g = q = 2$ and an irrational pencil", *Canad. J. Math.* **55**:3 (2003), 649–672.

mmlopes@math.ist.utl.pt   *Departamento de Matemática, Instituto Superior Técnico, Universidade Técnica de Lisboa, Av. Rovisco Pais, 1049-001 Lisboa, Portugal*

pardini@dm.unipi.it   *Dipartimento di Matematica Università di Pisa, Largo B. Pontecorvo, 5, 56127 Pisa, Italy*

Current Developments in Algebraic Geometry
MSRI Publications
Volume 59, 2011

Basic results on irregular varieties via Fourier–Mukai methods

GIUSEPPE PARESCHI

Recently Fourier–Mukai methods have proved to be a valuable tool in the study of the geometry of irregular varieties. The purpose of this paper is to illustrate these ideas by revisiting some basic results. In particular, we show a simpler proof of the Chen–Hacon birational characterization of abelian varieties. We also provide a treatment, along the same lines, of previous work of Ein and Lazarsfeld. We complete the exposition by revisiting further results on theta divisors. Two preliminary sections of background material are included.

In recent years the systematic use of the classical Fourier–Mukai transform between dual abelian varieties, and of related integral transforms, has proved to be a valuable tool for investigating the geometry of irregular varieties. An especially interesting point is the interplay between vanishing notions naturally arising in the Fourier–Mukai context, as weak index theorems, and the generic vanishing theorems of Green and Lazarsfeld. This naturally leads to the notion of *generic vanishing sheaves* (GV-sheaves for short). The purpose of this paper is to exemplify these ideas by revisiting some basic results.

To be precise, we focus on the theorem of Chen and Hacon [2001a] characterizing (birationally) abelian varieties by means of the conditions $q(X) = \dim X$ and $h^0(K_X) = h^0(2K_X) = 1$; this is stated as Lemma 4.2 below. We show that the Fourier–Mukai/Generic Vanishing package, in combination with Kollár's theorems on higher direct images of canonical bundles, produces a surprisingly quick and transparent proof of this result. Along the way, we provide a unified Fourier–Mukai treatment of most of the results of [Ein and Lazarsfeld 1997], where both the original and the present proof of the Chen–Hacon theorem find their roots.[1] We complete the exposition with a refinement of Hacon's cohomological characterization of desingularizations of theta divisors, as it fits well in the same framework.

Although many of the results treated here have led to further developments (see, for example, [Chen and Hacon 2002; Hacon and Pardini 2002; Jiang 2011; Debarre and Hacon 2007]), we have not attempted to recover the latter with

---

[1] However, our treatment of the results of Ein and Lazarsfeld differs from the original arguments only in some aspects.

the present approach. However, we hope that the point of view illustrated in this paper will be useful in the further study of irregular varieties with low invariants. In particular, the main lemma used to prove the Chen–Hacon theorem (see Lemma 4.2 and Scholium 4.3) is new, as far as I know, and can be useful in other situations. Moreover the present version of Hacon's characterization of desingularized theta divisors (Theorem 5.1) improves slightly the ones appearing in the literature.

The paper is organized as follows: there are two preliminary sections where the background material is recalled and informally discussed at some length. The first one is about the Fourier–Mukai transform, related integral transforms, and GV-sheaves. The second one is on generic vanishing theorems, including Hacon's generic vanishing theorem for higher direct images of canonical sheaves, which is already one of the most relevant applications of the Fourier–Mukai methods in the present context. The last three sections are devoted respectively to the work of Ein and Lazarsfeld, to the Chen–Hacon characterization of abelian varieties, and to Hacon's characterization of desingularizations of theta divisors.

My view of the material treated in this paper has been largely influenced by my collaboration with Mihnea Popa. I also thank M. A. Barja, J. A. Chen, C. Hacon, M. Lahoz, J. C. Naranjo, and S. Tirabassi.

1. Fourier–Mukai transform, cohomological support loci, and GV-sheaves

Unless otherwise stated, in this paper we will deal with smooth complex projective varieties (but, as it will be pointed out in the sequel, some results work more generally for complex Kähler manifolds and some others for smooth projective varieties on any algebraically closed field). By *sheaf* we will mean always coherent sheaf.

Given a smooth complex projective variety X, its *irregularity* is

$$q(X) := h^0(\Omega_X^1) = h^1(\mathbb{O}_X) = \tfrac{1}{2}b_1(X),$$

and X is said to be *irregular* if $q(X) > 0$. Its *Albanese variety*

$$\mathrm{Alb}\, X := H^0(\Omega_X^1)^* / H_1(X, \mathbb{Z})$$

is a $q(X)$-dimensional complex torus which, since X is assumed to be projective, is an abelian variety. The *Albanese morphism* $\mathrm{alb} : X \to \mathrm{Alb}\, X$ is defined by making sense of $x \to (\omega \mapsto \int_{x_0}^x \omega)$, where x_0 is a fixed point of X. Note that alb is defined up to translation in $\mathrm{Alb}\, X$. The Albanese morphism is a universal morphism of X to abelian varieties (or, more generally, complex tori). The *Albanese dimension* of X is the dimension of the image of its Albanese map. It is easily seen that the Albanese dimension of X is positive as soon as X is irregular (we refer to [Ueno 1975, §9] for a thorough treatment of the Albanese morphism). X is said *of maximal Albanese dimension* if $\dim \mathrm{alb}(X) = \dim X$.

The dual abelian variety of the Albanese variety is

$$\mathrm{Pic}^0 X = H^1(\mathcal{O}_X)/H^1(X, \mathbb{Z}).$$

The exponential sequence shows that $\mathrm{Pic}^0 X$ parametrizes those line bundles on X whose first Chern class vanishes [Griffiths and Harris 1978, p. 313]. Hence $\mathrm{Pic}^0 X$ is a (smooth and compact) space of deformations of the structure sheaf of X. So all sheaves $\mathcal{F}$ on X have a common family of deformations: $\{\mathcal{F} \otimes \alpha\}_{\alpha \in \mathrm{Pic}^0 X}$. Since Riemann it has been natural to consider, rather than the cohomology $H^*(X, \mathcal{F})$ of $\mathcal{F}$, the full family $\{H^*(X, \mathcal{F} \otimes \alpha)\}_{\alpha \in \mathrm{Pic}^0 X}$. For example, a good part of the geometry of curves is captured by the Brill–Noether varieties $W_d^r(C) = \{\alpha \in \mathrm{Pic}^0 C \mid h^0(L \otimes \alpha) \geq r+1\}$, where L is a line bundle on C of degree d [Arbarello et al. 1985]. In fact, it is often convenient to do a related thing. Since $\mathrm{Pic}^0 X$ is a fine moduli space; i.e., $X \times \mathrm{Pic}^0 X$ carries a universal line bundle P, the Poincaré line bundle, one can consider the *integral transform*

$$\mathbf{R}q_*(p^*(\cdot) \otimes P) : \mathbf{D}(X) \to \mathbf{D}(\mathrm{Pic}^0 X),$$

where p and q are respectively the projections on X and $\mathrm{Pic}^0 X$. Given a sheaf $\mathcal{F}$, the cohomology sheaves of $\mathbf{R}q_*(p^*(\mathcal{F}) \otimes P)$ are isomorphic to $R^i q_*(p^*(\mathcal{F}) \otimes P)$. They are naturally related to the family of cohomology groups $H^i(X, \mathcal{F} \otimes \alpha)_{\alpha \in \mathrm{Pic}^0 X}$ via base change (see 1.3 below).

The pullback map $\mathrm{alb}^* : \mathrm{Pic}^0(\mathrm{Alb}\, X) \to \mathrm{Pic}^0 X$ is an isomorphism [Griffiths and Harris 1978, p. 332], and, via this identification, the Poincaré line bundle on $X \times \mathrm{Pic}^0 X$ is the pullback of the Poincaré line bundle on $\mathrm{Alb}\, X \times \mathrm{Pic}^0(\mathrm{Alb}\, X)$. Therefore the above transform should be thought as a tool for studying the part of the geometry of X coming from its Albanese morphism.

Integral transform associated to the Poincaré line bundle, cohomological support loci, GV_{-k}-sheaves. In practice it is convenient to consider an integral transform as above for an arbitrary morphism from X to an abelian variety.

Definition 1.1 (Integral transforms associated to Poincaré line bundles). Let X be a projective variety of dimension d, equipped with a morphism to a q-dimensional abelian variety

$$a : X \to A.$$

Let $\mathcal{P}$ (script) be a Poincaré line bundle on $A \times \mathrm{Pic}^0 A$. We will denote

$$P_a = (a \times \mathrm{id}_{\mathrm{Pic}^0 A})^*(\mathcal{P})$$

and p, q the two projections of $X \times \mathrm{Pic}^0 A$. Given a sheaf $\mathcal{F}$ on X, we define

$$\Phi_{P_a}(\mathcal{F}) = q_*(p^*(\mathcal{F}) \otimes P_a).$$

We consider the derived functor of the functor Φ_{P_a}:

$$\mathbf{R}\Phi_{P_a} : \mathbf{D}(X) \to \mathbf{D}(\mathrm{Pic}^0 A). \qquad (1)$$

Sometimes we will have to consider the analogous derived functor $\mathbf{R}\Phi_{P_a^{-1}} :$ $\mathbf{D}(X) \to \mathbf{D}(\mathrm{Pic}^0 A)$ as well. Since $\mathcal{P}^{-1} \cong (1_A \times (-1)_{\mathrm{Pic}^0 A})^*\mathcal{P}$, there is not much difference between the two:

$$\mathbf{R}\Phi_{P_a^{-1}} = (-1_{\mathrm{Pic}^0 A})^* \mathbf{R}\Phi_{P_a} \qquad (2)$$

Finally, when the map a is the Albanese map of X, denoted by alb : $X \to \mathrm{Alb}\, X$, the map alb* identifies $\mathrm{Pic}^0(\mathrm{Alb}X)$ with $\mathrm{Pic}^0 X$ and the line bundle P_{alb} is identified with the Poincaré line bundle on $X \times \mathrm{Pic}^0 X$. We will simply set $P_{\mathrm{alb}} := P$. When X is an abelian variety then its Albanese morphism is (up to translation) the identity. In this case the transform $\mathbf{R}\Phi_{\mathcal{P}}$ is called the *Fourier–Mukai transform* (see below).

In the sequel, we will adopt the following notation: given a line bundle $\alpha \in \mathrm{Pic}^0 A$, we will denote $[\alpha]$ the point of $\mathrm{Pic}^0 A$ parametrizing α (via the Poincaré line bundle $\mathcal{P}$). In other words $\alpha = \mathcal{P}_{|A \times [\alpha]}$.

Definition 1.2 (Cohomological support loci). Given a coherent sheaf $\mathcal{F}$ on X, its *i-th cohomological support locus with respect to a* is

$$V_a^i(X, \mathcal{F}) = \{[\alpha] \in \mathrm{Pic}^0 A \mid h^i(X, \mathcal{F} \otimes a^*\alpha) > 0\}$$

In the special case when a is the Albanese map of X, we omit the reference to the map, writing

$$V^i(X, \mathcal{F}) = \{[\alpha] \in \mathrm{Pic}^0 X \mid h^i(X, \mathcal{F} \otimes \alpha) > 0\}.$$

As is customary for cohomology groups, when possible we will omit the variety X in the notation, writing simply $V_a^i(\mathcal{F})$ or $V^i(\mathcal{F})$ instead of $V_a^i(X, \mathcal{F})$ and $V^i(X, \mathcal{F})$.

Finally, we will adopt the notation

$$R\Delta(\mathcal{F}) = \mathbf{R}\mathcal{H}om(\mathcal{F}, \omega_X).$$

1.3 (Hyper)cohomology and base change. Given a sheaf, or more generally, a complex of sheaves $\mathcal{G}$ on X, the sheaf $R^i \Phi_{P_a}(\mathcal{G})$ is said to *have the base change property at a given point* $[\alpha]$ *of* $\mathrm{Pic}^0 X$ if the natural map $R^i \Phi_{P_a}(\mathcal{G}) \otimes \mathbb{C}([\alpha]) \to$ $H^i(X, \mathcal{G} \otimes a^*\alpha)$ is an isomorphism, where $\mathbb{C}([\alpha])$ denotes the residue field at the point $[\alpha] \in \mathrm{Pic}^0 X$. We will frequently use the following well-known base-change result (applied to our setting): given a sheaf (or, more generally, a bounded complex) $\mathcal{G}$ on X, if $h^{i+1}(X, \mathcal{G} \otimes a^*\alpha)$ is constant in a neighborhood of $[\alpha]$,[2] then both $R^{i+1}\Phi_{P_a}(\mathcal{G})$ and $R^i\Phi_{P_a}(\mathcal{G})$ have the base-change property in a neighborhood of $[\alpha]$. When $\mathcal{G}$ is a sheaf this is well known; see [Mumford 1970, Corollary 2, p. 52], for instance. For the general case of complexes see [EGA 1963, §7.7, pp. 65–72]. It follows that, if $\mathcal{F}$ is a sheaf, then $R^i\Phi_{P_a}(\mathcal{F})$ and $R^i\Phi_{P_a}(R\Delta\mathcal{F})$ vanish for $i > \dim X$, and both $R^d\Phi_{P_a}(\mathcal{F})$ and $R^d\Phi_{P_a}(R\Delta\mathcal{F})$ have the base change property at all $[\alpha] \in \mathrm{Pic}^0 A$.

The following basic result, whose proof will be outlined in the next page, compares two types of vanishing notions. The first one is *generic vanishing of cohomology groups* i.e., roughly speaking, that certain cohomological support loci $V^i(\mathcal{F})$ are *proper* closed subsets of $\mathrm{Pic}^0 A$. The second one is the *vanishing of cohomology sheaves* of the transform of the derived dual of $\mathcal{F}$:

Theorem 1.4 [Pareschi and Popa 2011a, Theorem A; Pareschi and Popa 2009, Theorem 2.2]. *For $\mathcal{F}$ a sheaf on X and k a nonnegative integer, equivalence holds between*[3]

(a) $$\mathrm{codim}_{\mathrm{Pic}^0 A}\, V_a^i(\mathcal{F}) \geq i - k \quad \text{for all } i \geq 0$$

and

(b) $$R^i\Phi_{P_a}(R\Delta\mathcal{F}) = 0 \quad \text{for all } i \notin [d-k, d].$$

Definition 1.5 (GV_{-k}-sheaves). When one of the two equivalent conditions of Theorem 1.4 holds, the sheaf $\mathcal{F}$ is said to be a GV_{-k}-*sheaf with respect to the morphism a*. When possible, we will omit the reference to the morphism a.

GV-sheaves. We focus on the special case $k = 0$ in Theorem 1.4. For sake of brevity, a GV_0-sheaf will be simply referred to as a GV-sheaf (with respect to the morphism a). Note that in this case it follows from condition (a) of Theorem 1.4 that, for *generic* $\alpha \in \mathrm{Pic}^0 A$, the cohomology groups $H^i(\mathcal{F} \otimes a^*\alpha)$ vanish for all $i > 0$. The second equivalent condition of Theorem 1.4 says that, for a GV-sheaf $\mathcal{F}$, the full transform $\mathbf{R}\Phi_{P_a}(R\Delta(\mathcal{F}))$ is a sheaf concentrated in degree $d - \dim X$:

$$\mathbf{R}\Phi_{P_a}(R\Delta\mathcal{F}) = R^d\Phi_{P_a}(R\Delta\mathcal{F})[-d]$$

(in the terminology of Fourier–Mukai theory, "$R\Delta\mathcal{F}$ satisfies the weak index theorem with index d"). In this situation one usually writes

$$R^d\Phi_{P_a}(R\Delta\mathcal{F}) = \widehat{R\Delta\mathcal{F}}.$$

---

[2]By semicontinuity, this holds if $h^{i+1}(X, \mathcal{G} \otimes a^*\alpha) = 0$.

[3]If $V_a^1(\mathcal{F})$ is empty we declare that its codimension is ∞.

The following proposition provides two basic properties of the sheaf $\widehat{R\Delta\mathcal{F}}$.

Proposition 1.6. *Let $\mathcal{F}$ be a GV-sheaf on X, with respect to a.*

(a) *The rank of $\widehat{R\Delta\mathcal{F}}$ equals $\chi(\mathcal{F})$.*

(b) $\mathcal{E}xt^i_{\mathcal{O}_{\mathrm{Pic}^0 A}}(\widehat{R\Delta\mathcal{F}}, \mathcal{O}_{\mathrm{Pic}^0 A}) \cong (-1_{\mathrm{Pic}^0 A})^* R^i \Phi_{P_a}(\mathcal{F}).$

Proof. (a) By Serre duality and base change, the rank of $\widehat{R\Delta\mathcal{F}}$ at a general point is the generic value of $h^0(\mathcal{F} \otimes a^*\alpha)$, which coincides with $\chi(\mathcal{F} \otimes a^*\alpha)$ (the higher cohomology vanishes for generic $\alpha \in \mathrm{Pic}^0 X$). Then (a) follows from the deformation invariance of the Euler characteristic.

(b) In the context of Definition 1.1, Grothendieck duality says that

$$\mathbf{R}\mathcal{H}om(R\Phi_{P_a}(\mathcal{F}), \mathcal{O}_{\mathrm{Pic}^0 A}) \cong \mathbf{R}\Phi_{P_a^{-1}}(R\Delta\mathcal{F})[d]; \tag{3}$$

see [Pareschi and Popa 2011a, Lemma 2.2]. Therefore Theorem 1.4(b), combined with (2), yields

$$\mathbf{R}\mathcal{H}om(R\Phi_{P_a}(\mathcal{F}), \mathcal{O}_{\mathrm{Pic}^0 A}) \cong (-1_{\mathrm{Pic}^0 X})^* \widehat{R\Delta\mathcal{F}}$$

Since $\mathbf{R}\mathcal{H}om(\,\cdot\,, \mathcal{O}_{\mathrm{Pic}^0 X})$ is an involution on $\mathbf{D}(\mathrm{Pic}^0 X)$, we have also

$$\mathbf{R}\Phi_{P_a}(\mathcal{F}) \cong (-1_{\mathrm{Pic}^0 X})^* \mathbf{R}\mathcal{H}om(\widehat{R\Delta\mathcal{F}}, \mathcal{O}_{\mathrm{Pic}^0 X}) \tag{4}$$

which proves (b). $\qquad\square$

Outline of proof of Theorem 1.4. The implication (b) $\Rightarrow$ (a) of Theorem 1.4 in the case $k = 0$ is proved as follows. Recall that, by the Auslander–Buchsbaum–Serre formula, if $\mathcal{G}$ is a sheaf on $\mathrm{Pic}^0 A$, then the support of $\mathcal{E}xt^i(\mathcal{G}, \mathcal{O}_{\mathrm{Pic}^0 A})$ has codimension $\geq i$ in $\mathrm{Pic}^0 A$ (see [Okonek et al. 1980, Lemma II.1.1.2], for instance). Applying this to the sheaf $\widehat{R\Delta\mathcal{F}}$, from Proposition 1.6(b) (which is a consequence of hypothesis (b) of Theorem 1.4) we get that

$$\mathrm{codim}_{\mathrm{Pic}^0 A} \, \mathrm{Supp} \, R^i \Phi_{P_a}(\mathcal{F}) \geq i \quad \text{for all } i \geq 0. \tag{5}$$

To show that (5) is equivalent to Theorem 1.4(a), we argue by descending induction on i. For $i = d$ this is immediate since $R^d \Phi_{P_a}(\mathcal{F})$ has the base-change property. Now suppose that codim $V_a^{\bar{i}}(\mathcal{F}) < \bar{i}$ for a given $\bar{i} < d$, and let $[\alpha]$ be a general point of a component of $V_a^{\bar{i}}(\mathcal{F})$ achieving the dimension. Because of (5) it must be that $R^{\bar{i}} \Phi_{P_a}(\mathcal{F})$ does not have the base-change property in the neighborhood of $[\alpha]$. Hence, by 1.3, such component has to be contained in $V_a^{\bar{i}+1}(\mathcal{F})$. Therefore codim $V_a^{\bar{i}+1}(\mathcal{F}) < \bar{i}$, violating the inductive hypothesis. Hence codim $V_a^i(\mathcal{F}) \geq i$ for all $i \geq 0$. This proves the implication (b) $\Rightarrow$ (a) for $k = 0$. For arbitrary k one uses the same argument, replacing Proposition 1.6(b) with the spectral sequence arising from (4).

The implication (a) $\Rightarrow$ (b) can be proved, with more effort, using the same ingredients (Grothendieck duality, Auslander–Buchsbaum–Serre formula and base change; see [Pareschi and Popa 2009, Theorem 2.2]). $\qquad\square$

A peculiar property of GV-sheaves is the following:

Lemma 1.7 [Hacon 2004, Corollary 3.2]. *Let $\mathcal{F}$ be a GV-sheaf on X, with respect to a. Then*

$$V_a^d(\mathcal{F}) \subseteq \cdots \subseteq V_a^1(\mathcal{F}) \subseteq V_a^0(\mathcal{F}).$$

Proof. Let $i > 0$ and assume that

$$[\alpha] \in V_a^i(\mathcal{F}) = -V_a^{d-i}(R\Delta(\mathcal{F}))$$

(the equality follows from Serre duality). Since $R^{d-i}\Phi_{P_a}(R\Delta(\mathcal{F})) = 0$, it follows by base change that $[\alpha] \in -V_a^{d-i+1}(R\Delta\mathcal{F}) = V_a^{i-1}(\mathcal{F})$. $\qquad\square$

The usefulness of the concept of GV-sheaf stems from the fact that some features of the cohomology groups $H^i(\mathcal{F} \otimes a^*\alpha)$ and of the cohomological support loci $V_a^i(\mathcal{F})$ can be detected by local and sheaf-theoretic properties of the transform $\widehat{R\Delta\mathcal{F}}$. The following simple example will be repeatedly used in Sections 3 and 4.

Lemma 1.8 [Pareschi and Popa 2011a, Proposition 3.15][4]. *Let $\mathcal{F}$ be a GV-sheaf on X with respect to a, let W be an irreducible component of $V_a^0(\mathcal{F})$, and let $k = \mathrm{codim}_{\mathrm{Pic}^0 A} W$. Then W is also a component of $V_a^k(\mathcal{F})$. Therefore $\dim X \geq k$. In particular, if $[\alpha]$ is an isolated point of $V_a^0(\mathcal{F})$ then $[\alpha]$ is also an isolated point of $V_a^q(\mathcal{F})$ (here $q = \dim A$). Therefore $\dim X \geq \dim A$.*

Proof. Since $\widehat{R\Delta\mathcal{F}}$ has the base-change property, it is supported at $V_a^d(R\Delta\mathcal{F}) = -V_a^0(\mathcal{F})$. Hence $-W$ is a component of the support of $\widehat{R\Delta\mathcal{F}}$. Let $[\alpha]$ be a general point of W. Since

$$R^i\Phi_{P_a}(\mathcal{F}) = (-1_{\mathrm{Pic}^0 A})^*\mathcal{E}xt^i(\widehat{R\Delta\mathcal{F}}, \mathcal{O}_{\mathrm{Pic}^0 A}),$$

from well-known properties of $\mathcal{E}xt$'s it follows that, in a suitable neighborhood in $\mathrm{Pic}^0 A$ of $[\alpha]$, $R^i\Phi_{P_a}(\mathcal{F})$ vanishes for $i < k$ and is supported at W for $i = k$. Therefore, by base change (see 1.3), W is contained in $V_a^k(\mathcal{F})$ (and in fact it is a component since, again by Theorem 1.4, $\mathrm{codim}\, V_a^k(\mathcal{F}) \geq k$). $\qquad\square$

From the previous lemma it follows that, if $\mathcal{F}$ is a GV-sheaf, then either $V_a^0(\mathcal{F}) = \mathrm{Pic}^0 A$ or there is a positive i such that $\mathrm{codim}\, V_a^i(\mathcal{F}) = i$, i.e., such that equality is achieved in condition (a) of Theorem 1.4. This can be rephrased as follows:

---

[4]In that reference, this result appears with an unnecessary hypothesis.

Corollary 1.9. (a) *If $\mathcal{F}$ is a nonzero GV-sheaf, there exists $i \geq 0$ such that* $\operatorname{codim} V_a^i(\mathcal{F}) = i$.

(b) *Let $\mathcal{F}$ be any sheaf on X. Then either $V_a^i(\mathcal{F}) = \varnothing$ for all $i \geq 0$ or there is an $i \geq 0$ such that $\operatorname{codim} V_a^i(\mathcal{F}) \leq i$.*

Proof. (a) follows immediately from the previous lemma.

(b) If not all of the cohomological support loci of $\mathcal{F}$ are empty then either $\mathcal{F}$ is a GV-sheaf, in which case part (a) applies, or there is an i such that $\operatorname{codim} V_a^i(\mathcal{F}) < i$. $\qquad\square$

Lemma 1.8 is a particular instance of a wider and more precise picture. In fact the condition that the cohomological support loci $V_a^0(\mathcal{F})$ is a *proper* subvariety of $\operatorname{Pic}^0 A$ is equivalent, by Serre duality and base change, to the fact that the generic rank of $\widehat{R\Delta\mathcal{F}}$ is zero; i.e., $\widehat{R\Delta\mathcal{F}}$ is a torsion sheaf on $\operatorname{Pic}^0 A$. Under this condition Lemma 1.8 says that there is an $i > 0$ achieving the bound of Theorem 1.4(a), i.e., such that $\operatorname{codim} V^i(\mathcal{F}) = i$. There is the following converse (which will be in use in the proof of Hacon's characterization of theta divisors in Section 5). In what follows we will say that a sheaf *has torsion* if it is not torsion-free.

Theorem 1.10 [Pareschi and Popa 2009, Corollary 3.2; 2011b, Proposition 2.8]. *Let $\mathcal{F}$ be a GV-sheaf on X. The following are equivalent*:

(a) *There is an $i > 0$ such that $\operatorname{codim} V_a^i(\mathcal{F}) = i$.*

(b) *$\widehat{R\Delta\mathcal{F}}$ has torsion.*

Proof. We start by recalling a general commutative algebra result. Let $\mathcal{G}$ be a sheaf on a smooth variety Y. Then $\mathcal{G}$ is torsion-free if and only if

$$\operatorname{codim}_Y \operatorname{Supp}(\mathcal{E}xt_Y^i(\mathcal{G}, \mathcal{O}_Y)) > i \quad \text{for all } i > 0 \tag{6}$$

(see, for example, [Pareschi and Popa 2009, Proposition 6.4] or [Pareschi and Popa 2011b, Lemma 2.9]). We apply this to the sheaf $\widehat{R\Delta\mathcal{F}}$ on the smooth variety $\operatorname{Pic}^0 X$. From (6) and Proposition 1.6(b) it follows that: $\widehat{R\Delta\mathcal{F}}$ has torsion if and only if there exists an $i > 0$ such that

$$\operatorname{codim}_{\operatorname{Pic}^0(X)} \operatorname{Supp}(R^i \Phi_{P_a}\mathcal{F}) = i. \tag{7}$$

By base change (see §1.3), condition (7) implies that there exists a $i > 0$ such that $\operatorname{codim} V_a^i(\mathcal{F}) \leq i$. Hence, since $\mathcal{F}$ is GV, $\operatorname{codim} V_a^i(\mathcal{F}) = i$. Conversely, the same argument as in the indication of proof of Theorem 1.4 proves that if (a) holds, then in any case there is a $j \geq i$ such that $\operatorname{codim} \operatorname{Supp}(R^j \Phi_{P_a}\mathcal{F}) = j$. Therefore, by (6), $\widehat{R\Delta\mathcal{F}}$ has torsion. $\qquad\square$

Mukai's equivalence of derived categories of abelian varieties. Nonvanishing. Assume that X coincides with the abelian variety A (and the map a is the identity). In this special case, according to Notation 1.1, the Poincaré line bundle on $A \times \operatorname{Pic}^0 A$ is denoted by $\mathscr{P}$. A well known theorem of Mukai asserts that $\mathbf{R}\Phi_{\mathscr{P}}$ is an equivalence of categories. More precisely, set $q = \dim A$ and denote the "opposite" functor $\mathbf{R}p_*(q^*(\cdot) \otimes \mathscr{P})$ by

$$\mathbf{R}\Psi_{\mathscr{P}} : \mathbf{D}(\operatorname{Pic}^0 A) \to \mathbf{D}(A).$$

Theorem 1.11 [Mukai 1981, Theorem 2.2]. *Let A be an abelian variety (over any algebraically closed field k). Then*

$$\mathbf{R}\Psi_{\mathscr{P}} \circ \mathbf{R}\Phi_{\mathscr{P}} = (-1_A)^*[-q], \quad \mathbf{R}\Phi_{\mathscr{P}} \circ \mathbf{R}\Psi_{\mathscr{P}} = (-1_{\operatorname{Pic}^0 A})^*[-q].$$

Mukai's theorem can be used to provide nonvanishing criteria for spaces of global sections. Here are some immediate ones:

Lemma 1.12 (nonvanishing). *Let $\mathscr{F}$ be a nonzero sheaf on an abelian variety X.*

(a) *If $\mathscr{F}$ is a GV-sheaf then $V^0(\mathscr{F})$ is nonempty.*

(b) *If $\operatorname{codim} V^i(\mathscr{F}) > i$ for all $i > 0$, then $V^0(\mathscr{F}) = \operatorname{Pic}^0 X$.*

Proof. (a) By base change, $\widehat{R\Delta\mathscr{F}} = R^d \Phi_{\mathscr{P}}(R\Delta\mathscr{F})$ is supported at $-V^0(\mathscr{F})$. Therefore if $V^0(\mathscr{F}) = \varnothing$ then $R^d \Phi_{\mathscr{P}}(R\Delta\mathscr{F})$ is zero, i.e., by Theorem 1.4, $\mathbf{R}\Phi_{\mathscr{P}}(R\Delta\mathscr{F})$ is zero. Then, by Mukai's theorem, $R\Delta\mathscr{F}$ is zero. Therefore $\mathscr{F}$ itself is zero, since $R\Delta$ is an involution on the derived category.

(b) If $V^0(\mathscr{F})$ is a *proper* subvariety of $\operatorname{Pic}^0 X$ then, by Lemma 1.8, there is at least one $i > 0$ such that $\operatorname{codim} V^i(\mathscr{F}) = i$. □

In the context of irregular varieties, Mukai's theorem is frequently used via the following proposition, whose proof is an exercise.

Proposition 1.13. *In the notation of Definition 1.1,*

$$\mathbf{R}\Phi_{P_a} \cong \mathbf{R}\Phi_{\mathscr{P}} \circ \mathbf{R}a_*.$$

Going back to Mukai's Theorem 1.11, the key point of its proof is the verification of the statement for the one-dimensional skyscraper sheaf at the identity point, namely that

$$\mathbf{R}\Phi_{\mathscr{P}}(\mathbf{R}\Psi_{\mathscr{P}}(k(\hat{0}))) = k(\hat{0})[-q].$$

Since $\mathbf{R}\Psi_{\mathscr{P}}(k(\hat{0})) = \mathcal{O}_A$, this amounts to proving that $\mathbf{R}\Phi_{\mathscr{P}}(\mathcal{O}_A) = k(\hat{0})[-q]$, or equivalently

$$R^i \Phi_{\mathscr{P}}(\mathcal{O}_A) = 0 \quad \text{for } i < q \qquad \text{and} \qquad R^q \Phi_{\mathscr{P}}(\mathcal{O}_A) = k(\hat{0}). \tag{8}$$

Since $V^i(\mathcal{O}_A) = \{\hat{0}\}$ for all i such that $0 \le i \le q$, the first part follows from easily from Theorem 1.4. Concerning the second part of (8), it does not follow from base change, and has to be proved with a different argument. Over the complex numbers this can be done easily using the explicit description of the Poincaré line bundle on an abelian variety (see [Kempf 1991, Theorem 3.15] or [Birkenhake and Lange 2004, Corollary 14.1.6]). In arbitrary characteristic it is proved in [Mumford 1970, p. 128]. Another proof can be found in [Huybrechts 2006, p. 202].

Now let X be an irregular variety of dimension d. The next proposition is a generalization of the second part of (8) to any smooth variety and is proved via an argument similar to Mumford's.

Proposition 1.14 [Barja et al. 2012, Proposition 6.1]. *Let X be a smooth projective variety* (*over any algebraically closed field k*), *equipped with a morphism to an abelian variety $a : X \to A$ such that the pullback map $a^* : \mathrm{Pic}^0 A \to \mathrm{Pic}^0 X$ is an embedding. Then*

$$R^d \Phi_{P_a}(\omega_X) \cong k(\hat{0}).$$

Notes 1.15. (1) All results in this section work in any characteristic. They work for compact Kähler manifolds as well.

(2) The implication (b) $\Rightarrow$ (a) of Theorem 1.4 and also Lemma 1.7 were already observed by Hacon [Hacon 2004, Theorem 1.2 and Corollary 3.2]. While the converse implication of Theorem 1.4 makes the picture conceptually more clear — and is also useful in various applications as Proposition 2.4 below — the careful reader will note that in the proof of the Chen–Hacon theorem we are only using the implication (b) $\Rightarrow$ (a).

(3) Theorem 1.10 is a particular case of a much more general statement: *the sheaf $\widehat{R\Delta\mathcal{F}}$ is not a k-syzygy sheaf if and only if there is a $i > 0$ such that* codim $V^i(\mathcal{F}) = i + k$. An example of an application of these ideas is the higher dimensional Castelnuovo–de Franchis inequality [Pareschi and Popa 2009].

(4) The result about k-the syzygy sheaves mentioned in the previous note, together with Theorems 1.4 and 1.10, hold in a much more general setting. In the first place they work not only for sheaves, but also for objects in the bounded derived category of X. Secondly, they are not specific to the transforms (1) but they work for all integral transforms $\Phi_P : \mathbf{D}(X) \to \mathbf{D}(Y)$ whose kernel P is a perfect object of $\mathbf{D}(X \times Y)$, where X is a Cohen–Macaulay equidimensional perfect scheme and Y is a locally noetherian scheme, both defined over a field k. A thorough analysis of the implications at the derived category level of these results is carried out in [Popa 2009].

(5) The hypothesis of Lemma 1.12(b), also called *M-regularity*, has many applications to global generation properties. In fact more than the conclusion of Lemma 1.12(b) holds: besides $V_0(\mathcal{F})$ being the whole $\mathrm{Pic}^0 X$, $\mathcal{F}$ is also

continuously globally generated. A survey on M-regularity and its applications is [Pareschi and Popa 2008]. A more recent development where the concept of M-regularity is relevant is the result of [Barja et al. 2012] on the bicanonical map of irregular varieties.

2. Generic vanishing theorems for the canonical sheaf and its higher direct images

Kollár's theorems on higher direct images of canonical sheaves. The following theorems will be of ubiquitous use in what follows.

Theorem 2.1 [Kollár 1986a, Theorem 2.1; 1986b, Theorem 3.1]. *Let X and Y be complex projective varieties of dimension d and $d - k$, with X smooth, and let $f : X \to Y$ a surjective map. Then:*

(a) $R^i f_* \omega_X$ *is torsion-free for all $i \geq 0$.*

(b) $R^i f_* \omega_X = 0$ *if $i > k$.*

(c) *Let L be an ample line bundle on Y. Then*

$$H^j(L \otimes R^i f_* \omega_X) = 0 \quad \text{for all } i \geq 0 \text{ and } j > 0;$$

(d) *in the derived category of Y,*

$$\mathbf{R} f_* \omega_X \cong \bigoplus_{i=0}^{k} R^i f_* \omega_X[-i].$$

In the next section, Theorem 2.1 will be used in the following variant, which is also a very particular case of more general formulations in [Kollár 1986b, §3].

Variant 2.2. *In the hypothesis and notation of Theorem 2.1, ω_X can be replaced by $\omega_X \otimes \beta$, where $[\beta]$ is a torsion point of $\mathrm{Pic}^0 X$.*

Generic vanishing theorems: Green–Lazarsfeld and Hacon. According to the previous terminology, a *generic vanishing theorem* is the statement that a certain sheaf is a GV_{-k}-sheaf. Within such terminology, the Green–Lazarsfeld generic vanishing theorem [Green and Lazarsfeld 1987], arisen independently of the theory of Fourier–Mukai transforms, can be stated as follows

Theorem 2.3 [Green and Lazarsfeld 1987, Theorem 1; Ein and Lazarsfeld 1997, Remark 1.6]. *Let $a : X \to A$ be a morphism from X to an abelian variety A, and let $k = \dim X - \dim a(X)$. Then ω_X is a GV_{-k}-sheaf (with respect to a). In particular, if a is generically finite onto its image then ω_X is a GV-sheaf.*

In fact Theorem 2.3 is sharp, as shown by the next proposition.

Proposition 2.4 [Barja et al. 2012, Proposition 2.7] (a similar result appears in [Lazarsfeld and Popa 2010, Proposition 1.2 and Remark 1.4]). *With the hypothesis and notation of Theorem 2.3, ω_X is a GV_{-k}-sheaf and it is not a $GV_{-(k-1)}$-sheaf.*

Proof. By the Green–Lazarsfeld generic vanishing theorem, ω_X is a GV_{-k}-sheaf. The fact that ω_X is not a $GV_{-(k-1)}$-sheaf means that there is a $j \geq 0$ (in fact a $j \geq k$) such that

$$\operatorname{codim} V^j(\omega_X) = j - k. \tag{9}$$

By Theorem 2.1(b,d), $H^d(\omega_X) = \oplus_{i=0}^k H^{d-i}(R^i a_*\omega_X)$. Since $H^d(\omega_X) \neq 0$, it follows that $H^{d-k}(R^k a_*\omega_X) \neq 0$. Hence, by Corollary 1.9, there is a $\bar{i} \geq 0$ such that

$$\operatorname{codim} V^{\bar{i}}(R^k a_*\omega_X) \leq \bar{i}. \tag{10}$$

Again by Theorem 2.1(b),(d), and projection formula

$$H^i(X, \omega_X \otimes a^*\alpha) = \bigoplus_{h=0}^{min\{i,k\}} H^{i-h}(A, R^h a_*\omega_X \otimes \alpha) \tag{11}$$

Therefore

$$V_a^{\bar{i}+k}(X, \omega_X) \supseteq V^{\bar{i}}(A, R^k a_*\omega_X).$$

Hence (10) yields codim $V_a^{\bar{i}+k}(\omega_X) \leq \bar{i} = (\bar{i} + k) - k$. In fact equality holds, since ω_X is a GV_{-k}-sheaf. Therefore (9) is proved. $\square$

In the argument of [Green and Lazarsfeld 1987] the GV_{-k} condition is verified by proving condition (a) of Theorem 1.4, i.e., the bound on the codimension of the cohomological support loci $V_a^i(\omega_X)$. This is achieved via an infinitesimal argument, based on Hodge theory. In fact, the Green–Lazarsfeld theorem holds, more generally, in the realm of compact Kähler varieties. Using the theory of Fourier–Mukai transforms, Hacon extended Theorem 2.3 to higher direct images of dualizing sheaves (in the case of smooth projective varieties). Hacon's result can be stated in several slightly different variants. A simple one, which is enough for the application of the present paper, is this:

Theorem 2.5 [Hacon 2004, Corollary 4.2]. *Let X be a smooth projective variety and let $a : X \to A$ be a morphism to an abelian variety. Then $R^i a_*\omega_X$ is a GV-sheaf on A for all $i \geq 0$.*

In fact, the Green–Lazarsfeld theorem follows from Hacon's via Kollár's theorems:

Proof that Theorem 2.5 implies Theorem 2.3. It follows from (11) that

$$V_a^i(X, \omega_X) = \bigcup_{h=0}^{\min\{i,k\}} V^{i-h}(A, R^h a_* \omega_X). \tag{12}$$

By Theorem 2.5, codim $V^{i-h}(R^h a_* \omega_X) \geq i - h \geq i - k$. □

Hacon's generic vanishing theorem will also be used in a variant form:

Variant 2.6 [Hacon and Pardini 2005, Theorem 2.2]. *Theorem 2.5 (hence also Theorem 2.3) still holds if ω_X is replaced by $\omega_X \otimes \beta$, where $[\beta]$ is a torsion point of $\mathrm{Pic}^0 X$.*

Let us describe briefly the proof of Theorem 2.5, which is completely different from the arguments of Green and Lazarsfeld. Hacon's approach consists in reducing a generic vanishing theorem to a vanishing theorem of Kodaira–Nakano type. Interestingly enough this is done by verifying directly condition (b) of Theorem 1.4, rather than condition (a), the dimensional bound for the cohomological support loci. The argument is as follows. Let L be an ample line bundle on $\mathrm{Pic}^0 A$ and consider, for a positive n, the locally free sheaf on A obtained as the "converse" Fourier–Mukai transform of L^n:

$$\mathbf{R}\Psi_{\mathscr{P}} L^n = R^0 \Psi_{\mathscr{P}} L^n$$

(see the notation above concerning the Fourier–Mukai transform), where the above equality follows from Kodaira vanishing. With an argument similar to the proof of Grauert–Riemenschneider vanishing, Hacon shows that, given a sheaf $\mathscr{F}$ on A, the condition (b) of Theorem 1.4 (namely $R^i \Phi_{\mathscr{P}}(R\Delta\mathscr{F}) = 0$ for $0 \neq \dim X$) is equivalent to the fact that there exist an n_0 such that, for all $n \geq n_0$,

$$H^i(A, (R\Delta\mathscr{F}) \otimes R^0 \Psi_{\mathscr{P}} L^n) = 0 \quad \text{for all } i < q, \qquad \text{where} q = \dim A.$$

By Serre duality, this is equivalent to

$$H^i(\mathscr{F} \otimes (R^0 \Psi_{\mathscr{P}} L^n)^*) = 0 \quad \text{for all } i > 0. \tag{13}$$

On the other hand, it is well known that, up to an isogeny, $R^0 \Psi_{\mathscr{P}} L^n$ is the direct sum of copies of negative line bundles. More precisely, let $\phi_{L^n} : \mathrm{Pic}^0 A \to A$ be the *polarization* associated to L^n. Then (see [Mukai 1981, Proposition 3.11(1)], for example)

$$\phi_{L^n}^*(R^0 \Psi_{\mathscr{P}} L^n) \cong H^0(\mathrm{Pic}^0 A, L^n) \otimes L^{-n} \tag{14}$$

Therefore, putting together (13) and (14) it turns out that, to prove that $\mathscr{F}$ is a GV-sheaf, it is enough to prove that, for n big enough,

$$H^i(\mathrm{Pic}^0 A, \phi_{L^n}^*(\mathscr{F}) \otimes L^n) = 0 \quad \text{for all } i > 0. \tag{15}$$

Condition (15) is certainly satisfied by the sheaves $\mathcal{F}$ enjoying the following property: *for each isogeny* $\pi : B \to A$, *and for each ample line bundle* N *on* B, $H^i(B, \pi^*\mathcal{F} \otimes N) = 0$ *for all* $i > 0$. Such property is satisfied by a higher direct image of a canonical sheaf since, via étale base extension, its pullback via an étale cover is still a higher direct image of a canonical sheaf, so that Kollár's Theorem 2.1(c) applies.

Notes 2.7. In [Pareschi and Popa 2011a] it is shown that Hacon's approach works in greater generality, and does not need the ambient variety to be an abelian variety. This yields other "generic vanishing theorems". For example, Green–Lazarsfeld's result (Theorem 2.3) works also for line bundles of the form $\omega_X \otimes L$, with L nef (see *loc. cit.* Corollary 5.2 and Theorem 5.8 for a better statement). In *loc. cit.* (Theorem A) is also shown that a part of Hacon's approach works in the setting of arbitrary integral transforms.

The subtorus theorems of Green, Lazarsfeld and Simpson. The geometry of the loci $V^i_a(\omega_X)$ is described by the Green–Lazarsfeld subtorus theorem, with an important addition due to Simpson.

Theorem 2.8 [Green and Lazarsfeld 1991, Theorem 0.1; Ein and Lazarsfeld 1997, Proof of Theorem 3, p. 249; Simpson 1993]. *Let* X *be a compact Kähler manifold, and* W *a component of* $V^i(\omega_X)$ *for some* i. *Then*
(a) *There exists a torsion point* $[\beta]$ *and a subtorus* B *of* $\mathrm{Pic}^0 X$ *such that* $W = [\beta] + B$.
(b) *Let* $g := \pi \circ \mathrm{alb} : X \to \mathrm{Pic}^0 B$, *where* $\pi : \mathrm{Alb}\, X \to \mathrm{Pic}^0 B$ *is the dual map of the embedding* $B \hookrightarrow \mathrm{Pic}^0 X$. *Then* $\dim g(X) \leq \dim X - i$.

Simpson's result (conjectured by Beauville and Catanese) is that $[\beta]$ is a torsion point. In [Simpson 1993] there are also, among other things, other different proofs of part (a) of the theorem. It is worth mentioning that, admitting part (a), the dimensional bound (b) is a direct consequence of the generic vanishing theorem:

Proof of (b). By Theorem 2.8(a) it follows that $V^i_g(\omega_X \otimes \beta) = \mathrm{Pic}^0(\mathrm{Pic}^0 B) = B$. Therefore, by definition, $\omega_X \otimes \beta$ is a GV$_{-h}$-sheaf with $h \geq i$, with respect to g. Hence, by Variant 2.6, $\dim X - \dim g(X) \geq i$. $\qquad\square$

Notes 2.9. (a) One defines the loci $V^i_m(\mathcal{F}) = \{[\alpha] \in \mathrm{Pic}^0 A \mid h^i(X, \mathcal{F} \otimes \alpha) \geq m\}$ and the Green–Lazarsfeld–Simpson result (Theorem 2.8) holds more generally for these loci as well. As noted in [Hacon and Pardini 2005, Theorem 2.2(b)], this implies that Theorem 2.8(a) holds replacing ω_X with higher direct images of $R^i f_*\omega_X$, where f is a morphism $f : X \to Y$, where Y is a smooth irregular variety (for example, an abelian variety).

(b) An explicit description and classification of all possible (positive-dimensional) components of the loci $V^i(\omega_X)$ is known only for $i = \dim X - 1$, by [Beauville 1992, Corollary 2.3]. Note that in this case, by Theorem 2.3(b), the image $g(X)$ is a curve.

3. Some results of Ein and Lazarsfeld

The content of this section is composed of four basic results of L. Ein and R. Lazarsfeld. We provide their proofs with the tools described here, both for sake of self-containedness, and also because they are good examples of application of the general principles of the previous section. The first two of them will be basic steps in the proof of the Chen–Hacon theorem appearing in the next section (they appear also in the original argument), while the last two will be used in the characterization of desingularizations of theta divisors (Section 5).

A theorem of Kawamata [1981], conjectured in [Ueno 1975], asserts that the Albanese map of a complex projective variety X of Kodaira dimension is zero is surjective and has connected fibers. As a consequence, one has Kawamata's characterization of abelian varieties as varieties of Kodaira dimension zero such that $q(X) = \dim X$. Subsequently Kollár [1986a; 1993; 1995] addressed the problem of giving effective versions of such results, replacing the hypothesis on the Kodaira dimension with the knowledge of finitely many plurigenera. In fact, he proved that the surjectivity conclusion of Kawamata's theorem and the characterization of abelian varieties held under the weaker assumption that $p_m(X) := h^0(\omega_X^m) = 1$ for some $m \geq 3$, and conjectured that $m = 2$ would suffice. The surjectivity part of Kawamata's theorem was settled by Ein and Lazarsfeld in the result quoted below as Theorem 3.1(b), while the characterization of abelian varieties is the content of the Chen–Hacon theorem (see the next section).[5]

Theorem 3.1 [Ein and Lazarsfeld 1997, Theorem 4]. *Let X be a smooth projective variety such that $p_1(X) = p_2(X) = 1$.*

(a) (Kollár) *There is no positive-dimensional subvariety Z of $\mathrm{Pic}^0 X$ such that both Z and $-Z$ are contained in $V^0(\omega_X)$.*

(b) *The Albanese map of X is surjective.*

Proof. (a) (This is as in [Kollár 1993, Theorem 17.10]; the proof is included here for self-containedness). Assume that there is a positive-dimensional subvariety Z of $\mathrm{Pic}^0 X$ as in the statement. The images of the multiplication maps of global sections

$$H^0(\omega_X \otimes \gamma) \otimes H^0(\omega_X \otimes \gamma^{-1}) \to H^0(\omega_X^2)$$

---

[5]Concerning an effective version of the connectedness part of Kawamata's theorem, recently Zhi Jiang has proved that if $p_1(X) = p_2(X) = 1$ then the Albanese morphism has connected fibers [Jiang 2011, Theorem 1.3]. The proof uses also the theorem of Chen and Hacon.

are nonzero for all $\gamma \in T$. Therefore $h^0(\omega_X^2) > 1$, since otherwise the only effective divisor in $|\omega_X^2|$ would have infinitely many components.

(b) By hypothesis, the identity point $\hat{0}$ of $\operatorname{Pic}^0 X$ belongs to $V^0(\omega_X)$. It must be an isolated point, since otherwise, by Theorem 2.8, a positive-dimensional component Z of $V^0(\omega_X)$ containing $\hat{0}$ would be a subtorus, thus contradicting (a). Therefore $\hat{0}$ is also an isolated point of $V^0(\operatorname{Alb} X, \operatorname{alb}_* \omega_X)$. Since $\operatorname{alb}_* \omega_X$ is a GV-sheaf on $\operatorname{Alb} X$ (Theorem 2.5), $\hat{0}$ is also a isolated point of $V^{q(X)}(\operatorname{Alb} X, \operatorname{alb}_* \omega_X)$ (Lemma 1.8). In particular $H^{q(X)}(\operatorname{Alb} X, \operatorname{alb}_* \omega_X)$ is nonzero. Therefore alb, the Albanese map of X, is surjective. $\square$

Next, we provide a different proof of the following characterization of abelian varieties. The same type of argument will be applied in the characterization of theta divisors of Section 5.

Theorem 3.2 (Ein and Lazarsfeld; see [Chen and Hacon 2001a, Theorem 1.8]). *Let X be a smooth projective variety of maximal Albanese dimension such that* $\dim V^0(\omega_X) = 0$. *Then X is birational to an abelian variety.*

Proof. By Theorem 2.3, ω_X is a GV-sheaf (with respect to the Albanese morphism). By Lemma 1.8, the hypothesis yields that $\dim X = q(X)$ and $V^0(\omega_X) = \{\hat{0}\}$. Using Proposition 1.14, $\mathbb{C}(\hat{0}) = R^q \Phi_P(\omega_X)$. Therefore, by Proposition 1.6, $\mathscr{E}xt^q(\widehat{\mathcal{O}_X}, \mathcal{O}_{\operatorname{Pic}^0 X}) = \mathbb{C}(\hat{0})$. Moreover, since $\widehat{\mathcal{O}_X}$ is supported at a finite set, $\mathscr{E}xt^i(\widehat{\mathcal{O}_X}, \mathcal{O}_{\operatorname{Pic}^0 X}) = 0$ for $i < q$. Summarizing: $R\Delta(\widehat{\mathcal{O}_X}) = \mathbb{C}(\hat{0})[-q]$. Since the functor $R\Delta$ is an involution, $\widehat{\mathcal{O}_X} = \mathbb{C}(\hat{0})$. In conclusion

$$\mathbf{R}\Phi_P(\mathcal{O}_X) = \mathbb{C}(\hat{0})[-q].$$

By Proposition 1.13 this means that $\mathbf{R}\Phi_{\widehat{\mathscr{P}}}(\mathbf{R}\operatorname{alb}_* \mathcal{O}_X) = \mathbb{C}(\hat{0})[-q]$. Then, by Mukai's inversion theorem (Theorem 1.11), $\mathbf{R}\operatorname{alb}_* \mathcal{O}_X = \mathcal{O}_{\operatorname{Alb} X}$. In particular $\operatorname{alb}_* \mathcal{O}_X = \mathcal{O}_{\operatorname{Alb} X}$. Since alb is assumed to be generically finite, this means that it is birational onto $\operatorname{Alb} X$. $\square$

The next result concerns varieties for which the Albanese dimension is maximal and $\chi(\omega_X) = 0$. Similarly to Theorem 3.1, here the proof is only partially different from the original argument of [Ein and Lazarsfeld 1997].

Theorem 3.3 [Ein and Lazarsfeld 1997, Theorem 3]. *Let X be a smooth projective variety of maximal Albanese dimension such that $\chi(\omega_X) = 0$. Then the image of the Albanese map of X is fibered by translates of abelian subvarieties of $\operatorname{Alb} X$.*

Proof. Since ω_X is GV, the condition $\chi(\omega_X) = 0$ is equivalent to the fact that $V^0(\omega_X)$ is a proper subvariety of $\operatorname{Pic}^0 X$ (Proposition 1.6(a)). Hence, by Lemma 1.8, there is a positive k such that $V^k(\omega_X)$ has a component W of codimension k. At this point the proof is that of Ein and Lazarsfeld: the subtorus theorem (Theorem 1.10) says that $W = [\beta] + T$, where T is a subtorus of $\mathrm{Pic}^0 X$, and provides the diagram

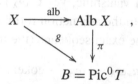

where $\dim g(X) \le \dim X - k$, π is surjective, and the fibers of π are translates of k-dimensional subtori of $\mathrm{Alb} X$. Since alb is generically finite, it follows that $\dim g(X) = \dim X - k$ and that a generic fiber of g surjects onto a generic fiber of π. □

Notes 3.4. (a) We recall that Theorem 3.3 settled a conjecture of Kollár, asserting that a variety X of general type and maximal Albanese dimension should have $\chi(\omega_X) > 0$. Ein and Lazarsfeld in [Ein and Lazarsfeld 1997] disproved the conjecture, producing a threefold X of general type, maximal Albanese dimension and $\chi(\omega_X) = 0$. But, at the same time, with Theorem 3.3, they showed that if $\chi(\omega_X) = 0$ then (a desingularization of) the Albanese image of X can't be of general type. However, the structure of varieties of general type and maximal Albanese dimension X with $\chi(\omega_X) = 0$ still remain mysterious. Results in this direction are due to Chen and Hacon [2001b; 2004].

(b) Corollary 5.1 of [Pareschi and Popa 2009] extends Theorem 3.3 to varieties with low $\chi(\omega_X)$ as follows: let X be a variety of maximal Albanese dimension. Then the image of the Albanese map of X is fibered by h-codimensional subvarieties of subtori of $AlbX$, with $h \le \chi(\omega_X)$ (see *loc. cit.* for a more precise statement). The proof uses k-syzygy sheaves and the Evans–Griffith syzygy theorem.

(c) Theorems 3.2 and 3.3, as well as the extension mentioned in (b) above, work also in the compact Kähler setting. Moreover, the present proof of Theorem 3.2 is algebraic, so that the results holds over any algebraically closed field.

We conclude with Ein–Lazarsfeld's result on the singularities of theta divisors. Here the difference with the original argument is that adjoint ideals are not invoked.

Theorem 3.5 [Ein and Lazarsfeld 1997, Theorem 1]. *Let Θ be an irreducible theta divisor of a principally polarized abelian variety A. Then Θ is normal with rational singularities.*

Proof. Let $a : X \to \Theta$ be a resolution of singularities of X. It is well known that, under our hypotheses, the fact that Θ is normal with rational singularities is equivalent to the fact that the trace map $t : a_*\omega_X \to \omega_\Theta$ is an isomorphism. We have that $a_* \omega_X$ is a GV-sheaf on A (this follows by Hacon's generic vanishing or, more simply, by the Green–Lazarsfeld generic vanishing and the fact that, by Grauert–Riemenschneider vanishing, $V^i(X, \omega_X) = V^i(A, a_* \omega_X)$). Moreover, an immediate calculation with the adjunction formula shows that $V^i(\omega_\Theta) = \{\hat{0}\}$ for all $i > 0$. We consider the exact sequence (the trace map t is injective)

$$0 \to a_* \omega_X \xrightarrow{t} \omega_\Theta \to \operatorname{coker} t \to 0. \qquad (16)$$

Tensoring with $\alpha \in \operatorname{Pic}^0 A$ and taking cohomology it follows that

$$\operatorname{codim} V^i(A, \operatorname{coker} t) > i \quad \text{for all } i > 0.$$

By Lemma 1.12(b), it follows that if $\operatorname{coker} t \neq 0$ then $V^0(A, \operatorname{coker} t) = \operatorname{Pic}^0 A$, or again, by Proposition 1.6(a), that $\chi(\operatorname{coker} t) > 0$. Then, since $\chi(\omega_\Theta) = 1$, from (16) it follows that $\chi(a_* \omega_X) = 0$, that is, $\chi(\omega_X) = 0$. At this point one concludes as in [Ein and Lazarsfeld 1997]. In fact, by Theorem 3.3, Θ would be fibered by subtori of A, which is not the case since Θ is ample and irreducible. $\square$

In [Ein and Lazarsfeld 1997] there are also results on the singularities of pluri-theta divisors, extending previous seminal results of Kollár in [Kollár 1993]. These, together with Theorem 3.5, have been extended to other polarizations of low degree, especially in the case of simple abelian varieties, by Debarre and Hacon [2007].

4. The Chen–Hacon birational characterization of abelian varieties

The goal of this section is to supply a new proof of the Chen–Hacon characterization of abelian varieties. We refer to the previous section for a short history and motivation.

Theorem 4.1 [Chen and Hacon 2001a]. *Let X be smooth complex projective variety. Then X is birational to an abelian variety if and only if $q(X) = \dim X$, and $h^0(\omega_X) = h^0(\omega_X^2) = 1$.*

Via the approach of Kollár and Ein–Lazarsfeld, the Chen–Hacon theorem will be a consequence of the following:

Lemma 4.2. *Let X be a projective variety of maximal Albanese dimension. If $\dim V^0(\omega_X) > 0$ then there exists a positive-dimensional subvariety Z of $\operatorname{Pic}^0 X$ such that both Z and $-Z$ are contained in $V^0(\omega_X)$.*

Proof of Theorem 4.1. Let X be a smooth projective variety such that $p_1(X) = p_2(X) = 1$ and $q(X) = \dim X$. By Theorem 3.1(b) the Albanese map of X is surjective, hence generically finite. By Lemma 4.2, combined with Theorem 3.1(a), it follows that $\dim V^0(\omega_X) = 0$. Therefore, thanks to the characterization provided by Theorem 3.2, X must be birational to an abelian variety. $\square$

Proof of Lemma 4.2. Let W be a positive-dimensional component of $V^0(\omega_X)$. If W contains the identity point then it is a subtorus by Theorem 2.8(a), and the conclusion of the Lemma is obviously satisfied. If W does not contain the identity point then, again by Theorem 2.8(a), $W = [\beta] + T$ where $[\beta]$ is a torsion point of $\mathrm{Pic}^0 X$ and T is a subtorus of $\mathrm{Pic}^0 X$ not containing $[\beta]$. To prove Lemma 4.2 it is enough to show then that there is a positive-dimensional subvariety Z of $[\beta^{-1}] + B$ which is contained in $V^0(\omega_X)$.

Let $d = \dim X$, $q = q(X)$, and $k = \mathrm{codim}_{\mathrm{Pic}^0 X} W$. We have the diagram

$$
\begin{array}{ccc}
X & \xrightarrow{\text{alb}} & \mathrm{Alb}\, X \\
& \searrow{\scriptstyle g} & \downarrow{\scriptstyle \pi} \\
& & B = \mathrm{Pic}^0 T
\end{array}
$$

and, as in the proof of Theorem 3.3,

$$\dim g(X) = d - k. \tag{17}$$

Next, we claim that

$$R^k g_*(\omega_X \otimes \beta) \neq 0. \tag{18}$$

Indeed, by Kollár splitting (Variant 2.2(d)) and the projection formula, we have (replacing $g(X)$ with B) that for all $\alpha \in T = \mathrm{Pic}^0 B$,

$$H^k(X, \omega_X \otimes \beta \otimes g^* \alpha) = \bigoplus_{i=0}^{k} H^{k-i}(B, R^i g_*(\omega_X \otimes \beta) \otimes \alpha) \tag{19}$$

We know that $H^k(X, \omega_X \otimes \beta \otimes g^* \alpha) > 0$ for all $\alpha \in \mathrm{Pic}^0 B$ (in other words: $[\beta] + g^* \mathrm{Pic}^0 B$ is contained in $V^k(X, \omega_X)$). By (19), this means that

$$\mathrm{Pic}^0 B = \bigcup_{i=0}^{k} V^{k-i}(B, R^i g_*(\omega_X \otimes \beta)).$$

But, by Hacon's generic vanishing theorem (as in Variant 2.6), all sheaves $R^i g_*(\omega_X \otimes \beta)$ are GV-sheaves (on B). In particular their $V^{k-i}()$ are *proper* subvarieties of $\mathrm{Pic}^0 B$ for $k - i > 0$. Therefore $T = \mathrm{Pic}^0 B$ must be equal to $V^0(B, R^k g_*(\omega_X \otimes \beta))$. This implies (18). By Kollár's torsion-freeness result (Variant 2.2(a)), $R^k g_*(\omega_X \otimes \beta)$ is torsion-free on $g(X)$.

Let $X \xrightarrow{f} Y \xrightarrow{a} g(X)$ be the Stein factorization of the morphism g. It follows from (18) that $R^k f_*(\omega_X \otimes \beta) \neq 0$. Therefore, denoting F a general fiber of f, $H^k(\omega_F \otimes \beta) > 0$. Since, by (17), the dimension of a general fiber F of f is k, $[\beta]$ must belong to the kernel of the restriction map $\mathrm{Pic}^0 X \to \mathrm{Pic}^0 F$. Hence so does $[\beta^{-1}]$. Therefore $R^k f_*(\omega_X \otimes \beta^{-1})$ is nonzero (in fact, again by Variant 2.2, it is torsion-free on Y). Hence

$$R^k g_*(\omega_X \otimes \beta^{-1}) \neq 0.$$

Finally, we claim that

$$\dim V^0(B, R^k g_*(\omega_X \otimes \beta^{-1})) > 0 \qquad (20)$$

(in fact it turns out that $V^0(B, R^k g_*(\omega_X \otimes \beta^{-1}))$ has no isolated point). Granting (20) for the time being, we conclude the proof. By (19) with β^{-1} instead of β, the positive-dimensional subvariety $V^0(B, R^k g_*(\omega_X \otimes \beta^{-1}))$ induces a positive-dimensional subvariety, say Z, of $[\beta^{-1}] + B$, which is contained in $V^k(X, \omega_X)^6$. By base change (Lemma 1.7), Z is contained in $V^0(X, \omega_X)$. This proves Lemma 4.2.

It remains to prove (20). Again by Hacon's generic vanishing (Variant 2.6), $R^k g_*(\omega_X \otimes \beta^{-1})$ is a GV-sheaf on B. Therefore, by the nonvanishing result of Lemma 1.12, the variety $V^0(B, R^k g_*(\omega_X \otimes \beta^{-1}))$ is nonempty. If $V^0(B, R^k g_*(\omega_X \otimes \beta^{-1}))$ had an isolated point, say $[\bar{\alpha}]$, then $[\bar{\alpha}]$ would belong also to $V^{q-k}(A, R^k g_*(\omega_X \otimes \beta^{-1}))$ (Lemma 1.8, recalling that $\dim B = q - k$). It would follow that $d - k = q - k$, and so $d = q$. Hence

$$H^{d-k}(A, R^k g_*(\omega_X \otimes \beta^{-1}) \otimes \bar{\alpha}) \neq 0.$$

Once again, by Kollár splitting as in (19), it would follow that

$$H^d(X, \omega_X \otimes \beta^{-1} \otimes g^* \bar{\alpha}) > 0,$$

implying that the line bundle $\beta^{-1} \otimes g^* \bar{\alpha}$ is trivial. But this is impossible since β^{-1} does not belong to $B = g^* \text{Pic}^0 A$. $\qquad\square$

It is perhaps worth mentioning that slightly more has been proved:

Scholium 4.3. *Let X be a variety of maximal Albanese dimension such that* $\dim V^0(\omega_X) > 0$. *Given a positive-dimensional component $[\beta] + B$ of $V^0(\omega_X)$, where $[\beta]$ is of order $n > 1$ and B is a subtorus of $\text{Pic}^0 X$, then, for all $k = 1, \ldots, n - 1$ coprime with n, there is a positive-dimensional subtorus C_k of B such that $[\beta^k] + C_k$ is contained in $V^0(\omega_X)$.*

5. On Hacon's characterization of theta divisors

Let Θ be an irreducible theta divisor in a principally polarized abelian variety A, and let $X \to \Theta$ be a desingularization. Thanks to the fact that Θ has rational singularities (Theorem 3.5), $V^i(X, \omega_X) = V^i(\Theta, \omega_\Theta)$. Hence the following conditions hold:

---

^{6}In fact Z is a translate of a subtorus; see Note 2.9(a).

(a) $V^i(X, \omega_X) = \{\hat{0}\}$ for all $i > 0$.

(b) $h^0(\omega_X \otimes \alpha) = 1$ for all $[\alpha] \in \mathrm{Pic}^0 X$ such that $[\alpha] \neq \hat{0}$.

In particular, it follows that $\chi(\omega_X) = 1$ and codim $V^i(\omega_X) > i + 1$ for all i such that $0 < i < \dim X$.

The following refinement of a theorem of Hacon's shows that desingularizations of theta divisors can be characterized — among varieties such that $\dim X < q(X)$ — by conditions (a) and (b). The proof illustrates in a simple case a principle — already mentioned before Lemma 1.8 — often appearing in arguments based on generic vanishing theorems and Fourier–Mukai transform: the interplay between the size of the cohomological support loci $V^i(\mathcal{F})$, where $\mathcal{F}$ is a GV-sheaf, and the sheaf-theoretic properties of transform $R\widehat{\Delta\mathcal{F}}$. The statement and the argument provided here are modeled on Proposition 3.1 of [Barja et al. 2012].

Theorem 5.1. *Let X be a smooth projective variety such that:*

(a) $\chi(\omega_X) = 1$;

(b) codim $V^i(\omega_X) > i + 1$ *for all i such that $0 < i < \dim X$;*

(c) $\dim X < q(X)$.

Then X is birational to a theta divisor.

Proof. Let us denote, as usual, $d = \dim X$ and $q = q(X)$. Conditions (b) and (c) imply that ω_X is a GV-sheaf. Therefore the Albanese map of X is generically finite (Proposition 2.4). Not only: (b) and (c) imply that codim $V^i(\omega_X) > i$ for all $i > 0$. Therefore, by Theorem 1.10, the sheaf $\widehat{\mathcal{O}_X}$ is torsion-free. Since its generic rank is $\chi(\omega_X) = 1$, it has to be an ideal sheaf twisted by a line bundle on $\mathrm{Pic}^0 X$:

$$\widehat{\mathcal{O}_X} = \mathcal{I}_Z \otimes L.$$

Next, we claim that, for each (non-embedded) component W of Z

$$\mathrm{codim}_{\mathrm{Pic}^0 X} W = d + 1 \qquad (21)$$

To prove this we note that, for $i > 1$,

$$\mathcal{E}xt^i(\widehat{\mathcal{O}_X}, \mathcal{O}_{\mathrm{Pic}^0 X}) - \mathcal{E}xt^i(\mathcal{I}_Z \otimes L, \mathcal{O}_{\mathrm{Pic}^0 X}) \cong \mathcal{E}xt^{i+1}(L \otimes \mathcal{O}_Z, \mathcal{O}_{\mathrm{Pic}^0 X}) \qquad (22)$$

Let W be one such component of Z, and let $j + 1$ be its codimension. We have that the support of $\mathcal{E}xt^{j+1}(L \otimes \mathcal{O}_Z, \mathcal{O}_{\mathrm{Pic}^0 X})$ contains W. Hence, combining (22), Grothendieck duality (Proposition 1.6(b)), and the Auslander–Buchsbaum–Serre formula, it follows that $R^j \Phi_P(\omega_X)$ is supported in codimension $j + 1$. This implies, by base-change, that codim $V^j(\omega_X) \leq j + 1$. Because of hypothesis (b), it must be that $j = \dim X$. This proves (21). Next, we claim that

$$R\mathcal{H}om(\widehat{\mathcal{O}_X}, \mathcal{O}_{\mathrm{Pic}^0 X}) = R\mathcal{H}om(\mathcal{I}_Z \otimes L, \mathcal{O}_{\mathrm{Pic}^0 X}) = \mathbb{C}(\hat{0})[-d] \quad \text{and} \quad d = q - 1 \qquad (23)$$

Indeed, arguing as in the proof of (21), from (21) it follows that $\mathcal{E}xt^i(\widehat{\mathbb{O}_X}, \mathbb{O}_{\mathrm{Pic}^0 X})$ is zero for $i < \mathrm{codim}_{\mathrm{Pic}^0 X} Z = d + 1$. By Proposition 1.6(b) again, this means that $R^i \Phi_P(\omega_X)$ is zero for $i < d + 1 = \mathrm{codim}_{\mathrm{Pic}^0 X} Z$. Since we know from Proposition 1.14 that $R^d \Phi_P(\omega_X) \cong \mathbb{C}(\hat{0})$, we get the first part of (23). Since the dualization functor is an involution, it follows that Z itself is the reduced point $\hat{0}$ and that $d = q - 1$, completing the proof of (23). At this point the proof is exactly as in [Barja et al. 2012, Proposition 3.1]. We report it here for the reader's convenience. By Proposition 1.13,

$$R\Phi_P(\mathbb{O}_X) = R\Phi_{\mathcal{P}}(R \text{ alb}_* \mathbb{O}_X) = \mathcal{I}_{\hat{0}} \otimes L[-q + 1].$$

Therefore, by Mukai's inversion theorem (Theorem 1.11),

$$R\Psi_{\mathcal{P}}(\mathcal{I}_{\hat{0}} \otimes L) = (-1)^*_{\mathrm{Pic}^0 X} R \text{ alb}_* \mathbb{O}_X[-1]. \tag{24}$$

In particular,

$$R^0 \Psi_{\mathcal{P}}(\mathcal{I}_{\hat{0}} \otimes L) = 0 \quad \text{and} \quad R^1 \Phi_{\mathcal{P}}(\mathcal{I}_{\hat{0}} \otimes L) \cong \text{alb}_* \mathbb{O}_X. \tag{25}$$

Applying $\Psi_{\mathcal{P}}$ to the standard exact sequence

$$0 \to \mathcal{I}_{\hat{0}} \otimes L \to L \to \mathbb{O}_{\hat{0}} \otimes L \to 0, \tag{26}$$

and using (25) we get

$$0 \to R^0 \Psi_{\mathcal{P}}(L) \to \mathbb{O}_{\mathrm{Alb} X} \to \text{alb}_* \mathbb{O}_X, \tag{27}$$

whence $R^0 \Psi_{\mathcal{P}}(L)$ is supported everywhere (since $\text{alb}_* \mathbb{O}_X$ is supported on a divisor). It is well known that this implies that L is *ample*. Therefore $R^i \Psi_{\mathcal{P}}(L) = 0$ for $i > 0$. Hence, by sequence (26), $R^i \Psi_{\mathcal{P}}(\mathcal{I}_{\hat{0}} \otimes L) = 0$ for $i > 1$. By (24) and (25), this implies that $R^i \text{alb}_*(\mathbb{O}_X) = 0$ for $i > 0$. Furthermore, (27) implies easily that $h^0(\mathrm{Pic}^0 X, L) = 1$; that is, L is a *principal* polarization on $\mathrm{Pic}^0 X$. Therefore, via the identification $\mathrm{Alb}(X) \cong \mathrm{Pic}^0(X)$ provided by L, we have $R^0 \Psi_{\mathcal{P}}(L) \cong L^{-1}$ (see [Mukai 1981, Proposition 3.11(1)]). Since the arrow on the right in (27) is onto, it follows that $\text{alb}_* \mathbb{O}_X = \mathbb{O}_\Theta$, where Θ is the only effective divisor in the linear series $|L|$. As we already know that alb is generically finite, this implies that alb is a birational morphism onto Θ. $\qquad \square$

Notes 5.2. (1) The cohomological characterization of theta divisors is due to Hacon [2000], who proved it under some extra hypotheses, subsequently refined in [Hacon and Pardini 2002]. A further refinement was proved in [Barja et al. 2012, Proposition 3.1] and [Lazarsfeld and Popa 2010, Proposition 3.8(ii)]. The present approach is the one in [Barja et al. 2012].

(2) Concerning the significance of the hypothesis of the above theorem, note that, removing the hypothesis $\dim X < q$ there are varieties nonbirational to theta

divisors satisfying (a) and (b) (e.g., sticking to varieties of maximal Albanese dimension, the double cover of a p.p.a.v. ramified on a smooth divisor $D \in |2\Theta|$). Moreover products of (desingularized) theta divisors are examples of varieties satisfying conditions (a) and (c), but not (b).

(3) Theorem 5.1 and its proof hold assuming, more generally, that X is compact Kähler. The argument works also for projective varieties over any algebraically closed field, except for the fact that Proposition 2.4 is used to ensure the maximal Albanese dimension. Therefore, up to adding the hypothesis that X is of maximal Albanese dimension and replacing condition (b) with the condition $\dim X < \dim \text{Alb } X$, Theorem 5.1 holds in any characteristic.

(4) With the same argument one can prove the following characterization of abelian varieties, valid in any characteristic: *Assume that X is a smooth projective variety of maximal Albanese dimension such that*

(a) $\chi(\omega_X) = 0$ *and*

(b) $\text{codim } V^i(\omega_X) > i$ *for all i such that $0 < i < \dim X$.*

Then X is birational to an abelian variety.

References

[Arbarello et al. 1985] E. Arbarello, M. Cornalba, P. A. Griffiths, and J. Harris, *Geometry of algebraic curves*, Grundlehren der Mathematischen Wissenschaften **267**, Springer, New York, 1985.

[Barja et al. 2012] M. A. Barja, M. Lahoz, J. C. Naranjo, and G. Pareschi, "On the bicanonical map of irregular varieties", *J. Alg. Geom.* (2012).

[Beauville 1992] A. Beauville, "Annulation du H^1 pour les fibrés en droites plats", pp. 1–15 in *Complex algebraic varieties* (Bayreuth, 1990), Lecture Notes in Math. **1507**, Springer, Berlin, 1992.

[Birkenhake and Lange 2004] C. Birkenhake and H. Lange, *Complex abelian varieties*, 2nd ed., Grundlehren der Mathematischen Wissenschaften **302**, Springer, Berlin, 2004.

[Chen and Hacon 2001a] J. A. Chen and C. D. Hacon, "Characterization of abelian varieties", *Invent. Math.* **143**:2 (2001), 435–447.

[Chen and Hacon 2001b] J. A. Chen and C. D. Hacon, "Pluricanonical maps of varieties of maximal Albanese dimension", *Math. Ann.* **320**:2 (2001), 367–380.

[Chen and Hacon 2002] J. A. Chen and C. D. Hacon, "On algebraic fiber spaces over varieties of maximal Albanese dimension", *Duke Math. J.* **111**:1 (2002), 159–175.

[Chen and Hacon 2004] J. A. Chen and C. D. Hacon, "On the irregularity of the image of the Iitaka fibration", *Comm. Algebra* **32**:1 (2004), 203–215.

[Debarre and Hacon 2007] O. Debarre and C. D. Hacon, "Singularities of divisors of low degree on abelian varieties", *Manuscripta Math.* **122**:2 (2007), 217–228.

[EGA 1963] A. Grothendieck, "Éléments de géométrie algébrique, III: étude cohomologique des faisceaux cohérents", *Inst. Hautes Études Sci. Publ. Math.* **17** (1963), 1–90. Zbl 0153.22301

[Ein and Lazarsfeld 1997] L. Ein and R. Lazarsfeld, "Singularities of theta divisors and the birational geometry of irregular varieties", *J. Amer. Math. Soc.* **10**:1 (1997), 243–258.

[Green and Lazarsfeld 1987] M. Green and R. Lazarsfeld, "Deformation theory, generic vanishing theorems, and some conjectures of Enriques, Catanese and Beauville", *Invent. Math.* **90**:2 (1987), 389–407.

[Green and Lazarsfeld 1991] M. Green and R. Lazarsfeld, "Higher obstructions to deforming cohomology groups of line bundles", *J. Amer. Math. Soc.* **4**:1 (1991), 87–103.

[Griffiths and Harris 1978] P. Griffiths and J. Harris, *Principles of algebraic geometry*, Wiley, New York, 1978.

[Hacon 2000] C. D. Hacon, "Fourier transforms, generic vanishing theorems and polarizations of abelian varieties", *Math. Z.* **235**:4 (2000), 717–726.

[Hacon 2004] C. D. Hacon, "A derived category approach to generic vanishing", *J. Reine Angew. Math.* **575** (2004), 173–187.

[Hacon and Pardini 2002] C. D. Hacon and R. Pardini, "On the birational geometry of varieties of maximal Albanese dimension", *J. Reine Angew. Math.* **546** (2002), 177–199.

[Hacon and Pardini 2005] C. D. Hacon and R. Pardini, "Birational characterization of products of curves of genus 2", *Math. Res. Lett.* **12**:1 (2005), 129–140.

[Huybrechts 2006] D. Huybrechts, *Fourier–Mukai transforms in algebraic geometry*, The Clarendon Press Oxford University Press, Oxford, 2006.

[Jiang 2011] Z. Jiang, "An effective version of a theorem of Kawamata on the Albanese map", *Commun. Contemp. Math.* **13**:3 (2011), 509–532.

[Kawamata 1981] Y. Kawamata, "Characterization of abelian varieties", *Compositio Math.* **43**:2 (1981), 253–276.

[Kempf 1991] G. R. Kempf, *Complex abelian varieties and theta functions*, Springer, Berlin, 1991.

[Kollár 1986a] J. Kollár, "Higher direct images of dualizing sheaves, I", *Ann. of Math.* (2) **123**:1 (1986), 11–42.

[Kollár 1986b] J. Kollár, "Higher direct images of dualizing sheaves, II", *Ann. of Math.* (2) **124**:1 (1986), 171–202.

[Kollár 1993] J. Kollár, "Shafarevich maps and plurigenera of algebraic varieties", *Invent. Math.* **113**:1 (1993), 177–215.

[Kollár 1995] J. Kollár, *Shafarevich maps and automorphic forms*, Princeton University Press, Princeton, NJ, 1995.

[Lazarsfeld and Popa 2010] R. Lazarsfeld and M. Popa, "Derivative complex, BGG correspondence, and numerical inequalities for compact Kähler manifolds", *Invent. Math.* **182**:3 (2010), 605–633.

[Mukai 1981] S. Mukai, "Duality between $D(X)$ and $D(\hat{X})$ with its application to Picard sheaves", *Nagoya Math. J.* **81** (1981), 153–175.

[Mumford 1970] D. Mumford, *Abelian varieties*, Tata Studies in Mathematics **5**, Tata Institute of Fundamental Research, Bombay, 1970.

[Okonek et al. 1980] C. Okonek, M. Schneider, and H. Spindler, *Vector bundles on complex projective spaces*, Progress in Mathematics **3**, Birkhäuser, Boston, 1980.

[Pareschi and Popa 2008] G. Pareschi and M. Popa, "M-regularity and the Fourier–Mukai transform", *Pure Appl. Math. Q.* **4**:3, part 2 (2008), 587–611.

[Pareschi and Popa 2009] G. Pareschi and M. Popa, "Strong generic vanishing and a higher-dimensional Castelnuovo–de Franchis inequality", *Duke Math. J.* **150**:2 (2009), 269–285.

[Pareschi and Popa 2011a] G. Pareschi and M. Popa, "GV-sheaves, Fourier–Mukai transform, and generic vanishing", *Amer. J. Math.* **133**:1 (2011), 235–271.

[Pareschi and Popa 2011b] G. Pareschi and M. Popa, "Regularity on abelian varieties, III: relationship with generic vanishing and applications", pp. 141–167 in *Grassmannians, moduli spaces and vector bundles*, edited by D. A. Ellwood et al., Clay Math. Proc. **14**, Amer. Math. Soc., Providence, RI, 2011.

[Popa 2009] M. Popa, "Generic vanishing filtrations and perverse objects in derived categories of coherent sheaves", preprint, 2009. To appear in the proceedings of the GCOE conference "Derived categories 2011 Tokyo". arXiv 0911.3648

[Simpson 1993] C. Simpson, "Subspaces of moduli spaces of rank one local systems", *Ann. Sci. École Norm. Sup.* (4) **26**:3 (1993), 361–401.

[Ueno 1975] K. Ueno, *Classification theory of algebraic varieties and compact complex spaces*, Lecture Notes in Mathematics **439**, Springer, Berlin, 1975.

pareschi@mat.uniroma2.it   *Dipartimento di Matematica, Università di Roma "Tor Vergata", Viale della Ricerca Scientifica, I-00133 Roma, Italy*

Current Developments in Algebraic Geometry
MSRI Publications
Volume **59**, 2011

Algebraic surfaces and hyperbolic geometry

BURT TOTARO

We describe the Kawamata–Morrison cone conjecture on the structure of Calabi–Yau varieties and more generally klt Calabi–Yau pairs. The conjecture is true in dimension 2. We also show that the automorphism group of a K3 surface need not be commensurable with an arithmetic group, which answers a question by Mazur.

1. Introduction

Many properties of a projective algebraic variety can be encoded by convex cones, such as the ample cone and the cone of curves. This is especially useful when these cones have only finitely many edges, as happens for Fano varieties. For a broader class of varieties which includes Calabi–Yau varieties and many rationally connected varieties, the Kawamata–Morrison cone conjecture predicts the structure of these cones. I like to think of this conjecture as what comes after the abundance conjecture. Roughly speaking, the cone theorem of Mori, Kawamata, Shokurov, Kollár, and Reid describes the structure of the curves on a projective variety X on which the canonical bundle K_X has negative degree; the abundance conjecture would give strong information about the curves on which K_X has degree zero; and the cone conjecture fully describes the structure of the curves on which K_X has degree zero.

We give a gentle summary of the proof of the cone conjecture for algebraic surfaces, with plenty of examples [Totaro 2010]. For algebraic surfaces, these cones are naturally described using hyperbolic geometry, and the proof can also be formulated in those terms.

Example 7.3 shows that the automorphism group of a K3 surface need not be commensurable with an arithmetic group. This answers a question by Barry Mazur [1993, Section 7].

Thanks to John Christian Ottem, Artie Prendergast-Smith, and Marcus Zibrowius for their comments.

2. The main trichotomy

Let X be a smooth complex projective variety. There are three main types of varieties. (Not every variety is of one of these three types, but minimal model theory relates every variety to one of these extreme types.)

Fano. This means that $-K_X$ is ample. (We recall the definition of ampleness in Section 3.)

Calabi–Yau. We define this to mean that K_X is numerically trivial.

ample canonical bundle. This means that K_X is ample; it implies that X is of "general type."

Here, for X of complex dimension n, the *canonical bundle* K_X is the line bundle Ω_X^n of n-forms. We write $-K_X$ for the dual line bundle K_X^*, the determinant of the tangent bundle.

Example 2.1. Let X be a curve, meaning that X has complex dimension 1. Then X is Fano if it has genus zero, or equivalently if X is isomorphic to the complex projective line $\mathbb{P}^1$; as a topological space, this is the 2-sphere. Next, X is Calabi–Yau if X is an elliptic curve, meaning that X has genus 1. Finally, X has ample canonical bundle if it has genus at least 2.

Example 2.2. Let X be a smooth surface in $\mathbb{P}^3$. Then X is Fano if it has degree at most 3. Next, X is Calabi–Yau if it has degree 4; this is one class of *K3 surfaces.* Finally, X has ample canonical bundle if it has degree at least 5.

Belonging to one of these three classes of varieties is equivalent to the existence of a Kähler metric with Ricci curvature of a given sign [Yau 1978]. Precisely, a smooth projective variety is Fano if and only if it has a Kähler metric with positive Ricci curvature; it is Calabi–Yau if and only if it has a Ricci-flat Kähler metric; and it has ample canonical bundle if and only if it has a Kähler metric with negative Ricci curvature.

We think of Fano varieties as the most special class of varieties, with projective space as a basic example. Strong support for this idea is provided by Kollár, Miyaoka, and Mori's theorem that smooth Fano varieties of dimension n form a bounded family [Kollár et al. 1992]. In particular, there are only finitely many diffeomorphism types of smooth Fano varieties of a given dimension.

Example 2.3. Every smooth Fano surface is isomorphic to $\mathbb{P}^1 \times \mathbb{P}^1$ or to a blow-up of $\mathbb{P}^2$ at at most 8 points. The classification of smooth Fano 3-folds is also known, by Iskovskikh, Mori, and Mukai; there are 104 deformation classes [Iskovkikh and Prokhorov 1999].

By contrast, varieties with ample canonical bundle form a vast and uncontrollable class. Even in dimension 1, there are infinitely many topological types of varieties with ample canonical bundle (curves of genus at least 2). Calabi–Yau varieties are on the border in terms of complexity. It is a notorious open question whether there are only finitely many topological types of smooth Calabi–Yau varieties of a given dimension. This is true in dimension at most 2. In particular, a smooth Calabi–Yau surface is either an abelian surface, a K3 surface, or a quotient of one of these surfaces by a free action of a finite group (and only finitely many finite groups occur this way).

3. Ample line bundles and the cone theorem

After a quick review of ample line bundles, this section states the cone theorem and its application to Fano varieties. Lazarsfeld's book [2004] is an excellent reference on ample line bundles.

Definition 3.1. A line bundle L on a projective variety X is *ample* if some positive multiple nL (meaning the line bundle $L^{\otimes n}$) has enough global sections to give a projective embedding

$$X \hookrightarrow \mathbb{P}^N.$$

(Here $N = \dim_{\mathbb{C}} H^0(X, nL) - 1$.)

One reason to investigate which line bundles are ample is in order to classify algebraic varieties. For classification, it is essential to know how to describe a variety with given properties as a subvariety of a certain projective space defined by equations of certain degrees.

Example 3.2. For X a curve, L is ample on X if and only if it has positive degree. We write $L \cdot X = \deg(L|_X) \in \mathbb{Z}$.

An $\mathbb{R}$-*divisor* on a smooth projective variety X is a finite sum

$$D = \sum a_i D_i$$

with $a_i \in \mathbb{R}$ and each D_i an irreducible divisor (codimension-one subvariety) in X. Write $N^1(X)$ for the "Néron–Severi" real vector space of $\mathbb{R}$-divisors modulo *numerical equivalence*: $D_1 \equiv D_2$ if $D_1 \cdot C = D_2 \cdot C$ for all curves C in X. (For me, a *curve* is irreducible.)

We can also define $N^1(X)$ as the subspace of the cohomology $H^2(X, \mathbb{R})$ spanned by divisors. In particular, it is a finite-dimensional real vector space. The dual vector space $N_1(X)$ is the space of 1-cycles $\sum a_i C_i$ modulo numerical equivalence, where C_i are curves on X. We can identify $N_1(X)$ with the subspace of the homology $H_2(X, \mathbb{R})$ spanned by algebraic curves.

Definition 3.3. The *closed cone of curves* $\overline{\mathrm{Curv}}(X)$ is the closed convex cone in $N_1(X)$ spanned by curves on X.

Definition 3.4. An $\mathbb{R}$-divisor D is *nef* if $D \cdot C \geq 0$ for all curves C in X. Likewise, a line bundle L on X is *nef* if the class $[L]$ of L (also called the first Chern class $c_1(L)$) in $N^1(X)$ is nef. That is, L has nonnegative degree on all curves in X.

Thus $\mathrm{Nef}(X) \subset N^1(X)$ is a closed convex cone, the *dual cone* to $\overline{\mathrm{Curv}}(X) \subset N_1(X)$.

Theorem 3.5 (Kleiman). *A line bundle L is ample if and only if $[L]$ is in the interior of the nef cone in $N^1(X)$.*

This is a *numerical* characterization of ampleness. It shows that we know the ample cone $\mathrm{Amp}(X) \subset N^1(X)$ if we know the cone of curves $\overline{\mathrm{Curv}}(X) \subset N_1(X)$. The following theorem gives a good understanding of the "K-negative" half of the cone of curves [Kollár and Mori 1998, Theorem 3.7]. A *rational* curve means a curve that is birational to $\mathbb{P}^1$.

Theorem 3.6. *(Cone theorem: Mori, Shokurov, Kawamata, Reid, Kollár). Let X be a smooth projective variety. Write $K_X^{\leq 0} = \{u \in N_1(X) : K_X \cdot u < 0\}$. Then every extremal ray of $\overline{\mathrm{Curv}}(X) \cap K_X^{\leq 0}$ is isolated, spanned by a rational curve, and can be contracted.*

In particular, every extremal ray of $\overline{\mathrm{Curv}}(X) \cap K_X^{\leq 0}$ is rational (meaning that it is spanned by a $\mathbb{Q}$-linear combination of curves, not just an $\mathbb{R}$-linear combination), since it is spanned by a single curve. A *contraction* of a normal projective variety X means a surjection from X onto a normal projective variety Y with connected fibers. A contraction is determined by a face of the cone of curves $\overline{\mathrm{Curv}}(X)$, the set of elements of $\overline{\mathrm{Curv}}(X)$ whose image under the pushforward map $N_1(X) \to N_1(Y)$ is zero. The last statement in the cone theorem means that every extremal ray in the K-negative half-space corresponds to a contraction of X.

Corollary 3.7. *For a Fano variety X, the cone of curves $\overline{\mathrm{Curv}}(X)$ (and therefore the dual cone $\mathrm{Nef}(X)$) is rational polyhedral.*

A rational polyhedral cone means the closed convex cone spanned by finitely many rational points.

Proof. Since $-K_X$ is ample, K_X is negative on all of $\overline{\mathrm{Curv}}(X) - \{0\}$. So the cone theorem applies to all the extremal rays of $\overline{\mathrm{Curv}}(X)$. Since they are isolated and live in a compact space (the unit sphere), $\overline{\mathrm{Curv}}(X)$ has only finitely many extremal rays. The cone theorem also gives that these rays are rational. □

It follows, in particular, that a Fano variety has only finitely many different contractions. A simple example is the blow-up X of $\mathbb{P}^2$ at one point, which is Fano. In this case, $\overline{\mathrm{Curv}}(X)$ is a closed strongly convex cone in the two-dimensional real vector space $N_1(X)$, and so it has exactly two 1-dimensional faces. We can write down two contractions of X, $X \to \mathbb{P}^2$ (contracting a (-1)-curve) and $X \to \mathbb{P}^1$ (expressing X as a $\mathbb{P}^1$-bundle over $\mathbb{P}^1$). Each of these morphisms must contract one of the two 1-dimensional faces of $\overline{\mathrm{Curv}}(X)$. Because the cone has no other nontrivial faces, these are the only nontrivial contractions of X.

4. Beyond Fano varieties

"Just beyond" Fano varieties, the cone of curves and the nef cone need not be rational polyhedral. Lazarsfeld's book [2004] gives many examples of this type, as do other books on minimal model theory [Debarre 2001; Kollár and Mori 1998].

Example 4.1. Let X be the blow-up of $\mathbb{P}^2$ at n very general points. For $n \le 8$, X is Fano, and so $\overline{\mathrm{Curv}}(X)$ is rational polyhedral. In more detail, for $2 \le n \le 8$, $\overline{\mathrm{Curv}}(X)$ is the convex cone spanned by the finitely many (-1)-curves in X. (A (-1)-curve on a surface X means a curve C isomorphic to $\mathbb{P}^1$ with self-intersection number $C^2 = -1$.) For example, when $n = 6$, X can be identified with a cubic surface, and the (-1)-curves are the famous 27 lines on X.

But for $n \ge 9$, X is not Fano, since $(-K_X)^2 = 9-n$ (whereas a projective variety has positive degree with respect to any ample line bundle). For $p_1, \ldots, p_n$ very general points in $\mathbb{P}^2$, X contains infinitely many (-1)-curves; see [Hartshorne 1977, Exercise V.4.15]. Every curve C with $C^2 < 0$ on a surface spans an isolated extremal ray of $\overline{\mathrm{Curv}}(X)$, and so $\overline{\mathrm{Curv}}(X)$ is not rational polyhedral.

Notice that a (-1)-curve C has $K_X \cdot C = -1$, and so these infinitely many isolated extremal rays are on the "good" (K-negative) side of the cone of curves, in the sense of the cone theorem. The K-positive side is a mystery. It is conjectured (Harbourne–Hirschowitz) that the closed cone of curves of a very general blow-up of $\mathbb{P}^2$ at $n \ge 10$ points is the closed convex cone spanned by the (-1)-curves and the "round" positive cone $\{x \in N_1(X) : x^2 \ge 0 \text{ and } H \cdot x \ge 0\}$, where H is a fixed ample line bundle. This includes the famous Nagata conjecture [Lazarsfeld 2004, Remark 5.1.14] as a special case. By de Fernex, even if the Harbourne–Hirschowitz conjecture is correct, the intersection of $\overline{\mathrm{Curv}}(X)$ with the K-positive half-space, for X a very general blow-up of $\mathbb{P}^2$ at $n \ge 11$ points, is bigger than the intersection of the positive cone with the K-positive half-space, because the (-1)-curves stick out a lot from the positive cone [de Fernex 2010].

Example 4.2. Calabi–Yau varieties (varieties with $K_X \equiv 0$) are also "just beyond" Fano varieties ($-K_X$ ample). Again, the cone of curves of a Calabi–Yau variety need not be rational polyhedral.

For example, let X be an abelian surface, so $X \cong \mathbb{C}^2/\Lambda$ for some lattice $\Lambda \cong \mathbb{Z}^4$ such that X is projective. Then $\overline{\mathrm{Curv}}(X) = \mathrm{Nef}(X)$ is a round cone, the positive cone

$$\{x \in N^1(X) : x^2 \geq 0 \text{ and } H \cdot x \geq 0\},$$

where H is a fixed ample line bundle. (Divisors and 1-cycles are the same thing on a surface, and so the cones $\overline{\text{Curv}}(X)$ and $\text{Nef}(X)$ lie in the same vector space $N^1(X)$.) Thus the nef cone is not rational polyhedral if X has Picard number $\rho(X) := \dim_{\mathbb{R}} N^1(X)$ at least 3 (and sometimes when $\rho = 2$).

For a K3 surface, the closed cone of curves may be round, or may be the closed cone spanned by the (-2)-curves in X. (One of those two properties must hold, by [Kovács 1994].) There may be finitely or infinitely many (-2)-curves. See Section 5.1 for an example.

5. The cone conjecture

But there is a good substitute for the cone theorem for Calabi–Yau varieties, the *Morrison–Kawamata cone conjecture*. In dimension 2, this is a theorem of Sterk, Looijenga, and Namikawa [Sterk 1985; Namikawa 1985; Kawamata 1997]. We call this Sterk's theorem for convenience:

Theorem 5.1. *Let X be a smooth complex projective Calabi–Yau surface (meaning that K_X is numerically trivial). Then the action of the automorphism group $\text{Aut}(X)$ on the nef cone $\text{Nef}(X) \subset N^1(X)$ has a rational polyhedral fundamental domain.*

Remark 5.2. For any variety X, if $\text{Nef}(X)$ is rational polyhedral, then the group $\text{Aut}^*(X) := \text{im}(\text{Aut}(X) \to \text{GL}(N^1(X)))$ is finite. This is easy: the group $\text{Aut}^*(X)$ must permute the set consisting of the smallest integral point on each extremal ray of $\text{Nef}(X)$. Sterk's theorem implies the remarkable statement that the converse is also true for Calabi–Yau surfaces. That is, if the cone $\text{Nef}(X)$ is not rational polyhedral, then $\text{Aut}^*(X)$ must be infinite. Note that $\text{Aut}^*(X)$ coincides with the discrete part of the automorphism group of X up to finite groups, because $\ker(\text{Aut}(X) \to \text{GL}(N^1(X)))$ is an algebraic group and hence has only finitely many connected components.

Sterk's theorem should generalize to Calabi–Yau varieties of any dimension (the Morrison–Kawamata cone conjecture). But in dimension 2, we can visualize it better, using hyperbolic geometry.

Indeed, let X be any smooth projective surface. The intersection form on $N^1(X)$ always has signature $(1, n)$ for some n (the Hodge index theorem). So $\{x \in N^1(X) : x^2 > 0\}$ has two connected components, and the positive cone $\{x \in N^1(X) : x^2 > 0 \text{ and } H \cdot x > 0\}$ is the standard round cone. As a result, we can identify the quotient of the positive cone by $\mathbb{R}^{>0}$ with *hyperbolic n-space*.

One way to see this is that the negative of the Lorentzian metric on $N^1(X) = \mathbb{R}^{1,n}$ restricted to the quadric $\{x^2 = 1\}$ is a Riemannian metric with curvature -1.

For any projective surface X, $\mathrm{Aut}(X)$ preserves the intersection form on $N^1(X)$. So $\mathrm{Aut}^*(X)$ is always a group of isometries of hyperbolic n-space, where $n = \rho(X) - 1$.

By definition, two groups G_1 and G_2 are *commensurable*, written $G_1 \doteq G_2$, if some finite-index subgroup of G_1 is isomorphic to a finite-index subgroup of G_2. A group *virtually* has a given property if some subgroup of finite index has the property. Since the groups we consider are all virtually torsion-free, we are free to replace a group G by G/N for a finite normal subgroup N (that is, G and G/N are commensurable).

5.1. *Examples.*

For an abelian surface X with Picard number at least 3, the cone $\mathrm{Nef}(X)$ is round, and so $\mathrm{Aut}^*(X)$ must be infinite by Sterk's theorem. (For abelian surfaces, the possible automorphism groups were known long before; see [Mumford 1970, Section 21].)

For example, let $X = E \times E$ with E an elliptic curve (not having complex multiplication). Then $\rho(X) = 3$, with $N^1(X)$ spanned by the curves $E \times 0$, $0 \times E$, and the diagonal Δ_E in $E \times E$. So $\mathrm{Aut}^*(X)$ must be infinite. In fact,

$$\mathrm{Aut}^*(X) \cong \mathrm{PGL}(2, \mathbb{Z}).$$

Here $\mathrm{GL}(2, \mathbb{Z})$ acts on $E \times E$ as on the direct sum of any abelian group with itself. This agrees with Sterk's theorem, which says that $\mathrm{Aut}^*(X)$ acts on the hyperbolic plane with a rational polyhedral fundamental domain; a fundamental domain for $\mathrm{PGL}(2, \mathbb{Z})$ acting on the hyperbolic plane (not preserving orientation) is given by any of the triangles in the figure.

For a K3 surface, the cone $\mathrm{Nef}(X)$ may or may not be the whole positive cone. For any projective surface, the nef cone modulo scalars is a convex subset of hyperbolic space. A finite polytope in hyperbolic space (even if some vertices are at infinity) has finite volume. So Sterk's theorem implies that, for a Calabi–Yau surface, $\mathrm{Aut}^*(X)$ acts with *finite covolume* on the convex set $\mathrm{Nef}(X)/\mathbb{R}^{>0}$ in hyperbolic space.

For example, let X be a K3 surface such that $\mathrm{Pic}(X)$ is isomorphic to $\mathbb{Z}^3$ with intersection form

$$\begin{pmatrix} 0 & 1 & 1 \\ 1 & -2 & 0 \\ 1 & 0 & -2 \end{pmatrix}.$$

P

Such a surface exists, since Nikulin showed that every even lattice of rank at most 10 with signature $(1, *)$ is the Picard lattice of some complex projective K3 surface [Nikulin 1979, Section 1.12]. Using the ideas of Section 6, one computes that the nef cone of X modulo scalars is the convex subset of the hyperbolic plane shown in the figure. The surface X has a unique elliptic fibration $X \to \mathbb{P}^1$, given by a nef line bundle P with $\langle P, P \rangle = 0$. The line bundle P appears in the figure as the point where $\mathrm{Nef}(X)/\mathbb{R}^{>0}$ meets the circle at infinity. And X contains infinitely many (-2)-curves, whose orthogonal complements are the codimension-1 faces of the nef cone. Sterk's theorem says that $\mathrm{Aut}(X)$ must act on the nef cone with rational polyhedral fundamental domain. In this example, one computes that $\mathrm{Aut}(X)$ is commensurable with the Mordell–Weil group of the elliptic fibration (Pic^0 of the generic fiber of $X \to \mathbb{P}^1$), which is isomorphic to $\mathbb{Z}$. One also finds that all the (-2)-curves in X are sections of the elliptic fibration. The Mordell–Weil group moves one section to any other section, and so it divides the nef cone into rational polyhedral cones as in the figure.

6. Outline of the proof of Sterk's theorem

We discuss the proof of Sterk's theorem for K3 surfaces. The proof for abelian surfaces is the same, but simpler (since an abelian surface contains no (-2)-curves), and these cases imply the case of quotients of K3 surfaces or abelian surfaces by a finite group. For details, see [Kawamata 1997], based on the earlier [Sterk 1985; Namikawa 1985].

The proof of Sterk's theorem for K3 surfaces relies on the Torelli theorem of Piatetski-Shapiro and Shafarevich. That is, any isomorphism of Hodge structures between two K3s is realized by an isomorphism of K3s if it maps the nef cone into the nef cone. In particular, this lets us construct automorphisms of a K3 surface X: up to finite index, every element of the integral orthogonal group $O(\mathrm{Pic}(X))$ that preserves the cone $\mathrm{Nef}(X)$ is realized by an automorphism of X. (Here $\mathrm{Pic}(X) \cong \mathbb{Z}^\rho$, and the intersection form has signature $(1, \rho(X) - 1)$ on $\mathrm{Pic}(X)$.)

Moreover, $\mathrm{Nef}(X)/\mathbb{R}^{>0}$ is a very special convex set in hyperbolic space $H_{\rho-1}$: it is the closure of a Weyl chamber for a discrete reflection group W acting on $H_{\rho-1}$. We can define W as the group generated by all reflections in vectors $x \in \mathrm{Pic}(X)$ with $x^2 = -2$, or (what turns out to be the same) the group generated by reflections in all (-2)-curves in X. By the first description, W is a *normal* subgroup of $O(\mathrm{Pic}(X))$. In fact, up to finite groups, $O(\mathrm{Pic}(X))$ is the semidirect product group

$$O(\mathrm{Pic}(X)) \doteq \mathrm{Aut}(X) \ltimes W.$$

By general results on arithmetic groups going back to Minkowski, $O(\mathrm{Pic}(X))$ acts on the positive cone in $N^1(X)$ with a rational polyhedral fundamental domain D. (This fundamental domain is not at all unique.) And the reflection group W acts on the positive cone with fundamental domain the nef cone of X. Therefore, after we arrange for D to be contained in the nef cone, $\mathrm{Aut}(X)$ must act on the nef cone with the same rational polyhedral fundamental domain D, up to finite index. Sterk's theorem is proved.

7. Nonarithmetic automorphism groups

In this section, we show for the first time that the discrete part of the automorphism group of a smooth projective variety need not be commensurable with an arithmetic group. (Section 5 defines commensurability.) This answers a question raised by Mazur [1993, Section 7]. Corollary 7.2 applies to a large class of K3 surfaces.

An *arithmetic group* is a subgroup of the group of $\mathbb{Q}$-points of some $\mathbb{Q}$-algebraic group $H_{\mathbb{Q}}$ which is commensurable with $H(\mathbb{Z})$ for some integral structure on $H_{\mathbb{Q}}$; this condition is independent of the integral structure [Serre 1979]. We view arithmetic groups as abstract groups, not as subgroups of a fixed Lie group.

Borcherds gave an example of a K3 surface whose automorphism group is not isomorphic to an arithmetic group [Borcherds 1998, Example 5.8]. But, as he says, the automorphism group in his example has a nonabelian free subgroup of finite index, and so it is commensurable with the arithmetic group $\mathrm{SL}(2, \mathbb{Z})$. Examples of K3 surfaces with explicit generators of the automorphism group have been given by Keum, Kondo, Vinberg, and others; see [Dolgachev 2008, Section 5] for a survey.

Although they need not be commensurable with arithmetic groups, the automorphism groups G of K3 surfaces are very well-behaved in terms of geometric group theory. More generally this is true for the discrete part G of the automorphism group of a surface X which can be given the structure of a klt Calabi–Yau pair, as defined in Section 8. Namely, G acts cocompactly on a CAT(0) space (a precise notion of a metric space with nonpositive curvature). Indeed, the nef cone modulo scalars is a closed convex subset of hyperbolic space, and thus a CAT(-1) space [Bridson and Haefliger 1999, Example II.1.15]. Removing a G-invariant set of disjoint open horoballs gives a CAT(0) space on which G acts properly and cocompactly, by the proof of [Bridson and Haefliger 1999, Theorem II.11.27]. This implies all the finiteness properties one could want, even though G need not be arithmetic. In particular: G is finitely presented, a finite-index subgroup of G has a finite CW complex as classifying space, and

G has only finitely many conjugacy classes of finite subgroups [Bridson and Haefliger 1999, Theorem III.Γ.1.1].

For smooth projective varieties in general, very little is known. For example, is the discrete part G of the automorphism group always finitely generated? The question is open even for smooth projective rational surfaces. About the only thing one can say for an arbitrary smooth projective variety X is that G modulo a finite group injects into $\mathrm{GL}(\rho(X), \mathbb{Z})$, by the comments in Section 5.

In Theorem 7.1, a *lattice* means a finitely generated free abelian group with a symmetric bilinear form that is nondegenerate $\otimes \mathbb{Q}$.

Theorem 7.1. *Let M be a lattice of signature $(1, n)$ for $n \geq 3$. Let G be a subgroup of infinite index in $O(M)$. Suppose that G contains $\mathbb{Z}^{n-1}$ as a subgroup of infinite index. Then G is not commensurable with an arithmetic group.*

Corollary 7.2. *Let X be a K3 surface over any field, with Picard number at least 4. Suppose that X has an elliptic fibration with no reducible fibers and a second elliptic fibration with Mordell–Weil rank positive. (For example, the latter property holds if the second fibration also has no reducible fibers.) Suppose also that X contains a (-2)-curve. Then the automorphism group of X is a discrete group that is not commensurable with an arithmetic group.*

Example 7.3. Let X be the double cover of $\mathbb{P}^1 \times \mathbb{P}^1 = \{([x, y], [u, v])\}$ ramified along the following curve of degree $(4, 4)$:

$$0 = 16x^4u^4 + xy^3u^4 + y^4u^3v - 40x^4u^2v^2 - x^3yu^2v^2 - x^2y^2uv^3$$
$$+ 33x^4v^4 - 10x^2y^2v^4 + y^4v^4.$$

Then X is a K3 surface whose automorphism group (over $\mathbb{Q}$, or over $\overline{\mathbb{Q}}$) is not commensurable with an arithmetic group.

Proof of Theorem 7.1. We can view $O(M)$ as a discrete group of isometries of hyperbolic n-space. Every solvable subgroup of $O(M)$ is virtually abelian [Bridson and Haefliger 1999, Corollary II.11.28 and Theorem III.Γ.1.1]. By the classification of isometries of hyperbolic space as elliptic, parabolic, or hyperbolic [Alekseevskij et al. 1993], the centralizer of any subgroup $\mathbb{Z} \subset O(M)$ is either commensurable with $\mathbb{Z}$ (if a generator g of $\mathbb{Z}$ is hyperbolic) or commensurable with $\mathbb{Z}^a$ for some $a \leq n - 1$ (if g is parabolic). These properties pass to the subgroup G of $O(M)$. Also, G is not virtually abelian, because it contains $\mathbb{Z}^{n-1}$ as a subgroup of infinite index, and $\mathbb{Z}^{n-1}$ is the largest abelian subgroup of $O(M)$ up to finite index. Finally, G acts properly and not cocompactly on hyperbolic n-space, and so G has virtual cohomological dimension at most $n - 1$ [Brown 1982, Proposition VIII.8.1].

Suppose that G is commensurable with some arithmetic group Γ. Thus Γ is a subgroup of the group of $\mathbb{Q}$-points of some $\mathbb{Q}$-algebraic group $H_{\mathbb{Q}}$, and Γ is commensurable with $H(\mathbb{Z})$ for some integral structure on $H_{\mathbb{Q}}$. We freely change Γ by finite groups in what follows. So we can assume that $H_{\mathbb{Q}}$ is connected. After replacing $H_{\mathbb{Q}}$ by the kernel of some homomorphism to a product of copies of the multiplicative group G_m over $\mathbb{Q}$, we can assume that Γ is a *lattice* in the real group $H(\mathbb{R})$ (meaning that $\mathrm{vol}(H(\mathbb{R})/\Gamma) < \infty$), by Borel and Harish-Chandra [Borel and Harish-Chandra 1962, Theorem 9.4].

Every connected $\mathbb{Q}$-algebraic group $H_{\mathbb{Q}}$ is a semidirect product $R_{\mathbb{Q}} \ltimes U_{\mathbb{Q}}$ where $R_{\mathbb{Q}}$ is reductive and $U_{\mathbb{Q}}$ is unipotent [Borel and Serre 1964, Theorem 5.1]. By the independence of the choice of integral structure, we can assume that $\Gamma = R(\mathbb{Z}) \ltimes U(\mathbb{Z})$ for some arithmetic subgroups $R(\mathbb{Z})$ of $R_{\mathbb{Q}}$ and $U(\mathbb{Z})$ of $U_{\mathbb{Q}}$. Since every solvable subgroup of G is virtually abelian, $U_{\mathbb{Q}}$ is abelian, and $U(\mathbb{Z}) \cong \mathbb{Z}^a$ for some a. The conjugation action of $R_{\mathbb{Q}}$ on $U_{\mathbb{Q}}$ must be trivial; otherwise Γ would contain a solvable group of the form $\mathbb{Z} \ltimes \mathbb{Z}^a$ which is not virtually abelian. Thus $\Gamma = R(\mathbb{Z}) \times \mathbb{Z}^a$. But the properties of centralizers in G imply that any product group of the form $W \times \mathbb{Z}$ contained in G must be virtually abelian. Therefore, $a = 0$ and $H_{\mathbb{Q}}$ is reductive.

Modulo finite groups, the reductive group $H_{\mathbb{Q}}$ is a product of $\mathbb{Q}$-simple groups and tori, and Γ is a corresponding product modulo finite groups. Since any product group of the form $W \times \mathbb{Z}$ contained in G is virtually abelian, $H_{\mathbb{Q}}$ must be $\mathbb{Q}$-simple. Since the lattice Γ in $H(\mathbb{R})$ is isomorphic to the discrete subgroup G of $O(M) \subset O(n, 1)$ (after passing to finite-index subgroups), Prasad showed that $\dim(H(\mathbb{R})/K_H) \le \dim(O(n, 1)/O(n)) = n$, where K_H is a maximal compact subgroup of $H(\mathbb{R})$. Moreover, since G has infinite index in $O(M)$ and hence infinite covolume in $O(n, 1)$, Prasad showed that either $\dim(H(\mathbb{R})/K_H) \le n - 1$ or else $\dim(H(\mathbb{R})/K_H) = n$ and there is a homomorphism from $H(\mathbb{R})$ onto $\mathrm{PSL}(2, \mathbb{R})$ [Prasad 1976, Theorem B].

Suppose that $\dim(H(\mathbb{R})/K_H) \le n - 1$. We know that Γ acts properly on $H(\mathbb{R})/K_H$ and that Γ contains $\mathbb{Z}^{n-1}$. The quotient $\mathbb{Z}^{n-1} \backslash H(\mathbb{R})/K_H$ is a manifold of dimension $n - 1$ with the homotopy type of the $(n - 1)$-torus (in particular, with nonzero cohomology in dimension $n - 1$), and so it must be compact. So $\mathbb{Z}^{n-1}$ has finite index in Γ, contradicting our assumption.

So $\dim(H(\mathbb{R})/K_H) = n$ and $H(\mathbb{R})$ maps onto $\mathrm{PSL}(2, \mathbb{R})$. We can assume that $H_{\mathbb{Q}}$ is simply connected. Since H is $\mathbb{Q}$-simple, H is equal to the restriction of scalars $R_{K/\mathbb{Q}}L$ for some number field K and some absolutely simple and simply connected group L over K [Tits 1966, Section 3.1]. Since $H(\mathbb{R})$ maps onto $\mathrm{PSL}(2, \mathbb{R})$, L must be a form of $\mathrm{SL}(2)$. We showed that $G \cong \Gamma$ has virtual cohomological dimension at most $n - 1$, and so Γ must be a noncocompact subgroup of $H(\mathbb{R})$. Equivalently, H has $\mathbb{Q}$-rank greater than zero [Borel and Harish-Chandra 1962, Lemma 11.4, Theorem 11.6], and so $\mathrm{rank}_K(L) = \mathrm{rank}_{\mathbb{Q}}(H)$ is greater than zero. Therefore, L is isomorphic to $\mathrm{SL}(2)$ over K.

It follows that Γ is commensurable with $\mathrm{SL}(2, o_K)$, where o_K is the ring of integers of K. So we can assume that Γ contains the semidirect product

$$o_K^* \ltimes o_K = \left\{ \begin{pmatrix} a & b \\ 0 & 1/a \end{pmatrix} \right\} \subset \mathrm{SL}(2, o_K).$$

If the group of units o_K^* has positive rank, then $o_K^* \ltimes o_K$ is a solvable group which is not virtually abelian. So the group of units is finite, which means that K is either $\mathbb{Q}$ or an imaginary quadratic field, by Dirichlet. If K is imaginary quadratic, then $H_{\mathbb{Q}} = R_{K/\mathbb{Q}} \mathrm{SL}(2)$ and $H(\mathbb{R}) = \mathrm{SL}(2, \mathbb{C})$, which does not map onto $\mathrm{PSL}(2, \mathbb{R})$. Therefore $K = \mathbb{Q}$ and $H_{\mathbb{Q}} = \mathrm{SL}(2)$. It follows that Γ is commensurable with $\mathrm{SL}(2, \mathbb{Z})$. So Γ is commensurable with a free group. This contradicts that $G \cong \Gamma$ contains $\mathbb{Z}^{n-1}$ with $n \geq 3$. $\square$

Proof of Corollary 7.2. Let M be the Picard lattice of X, that is, $M = \mathrm{Pic}(X)$ with the intersection form. Then M has signature $(1, n)$ by the Hodge index theorem, where $n \geq 3$ since X has Picard number at least 4.

For an elliptic fibration $X \to \mathbb{P}^1$ with no reducible fibers, the Mordell–Weil group of the fibration has rank $\rho(X) - 2 = n - 1$ by the Shioda–Tate formula [Shioda 1972, Corollary 1.5], which is easy to check in this case. So the first elliptic fibration of X gives an inclusion of $\mathbb{Z}^{n-1}$ into $G = \mathrm{Aut}^*(X)$. The second elliptic fibration gives an inclusion of $\mathbb{Z}^a$ into G for some $a > 0$. In the action of G on hyperbolic n-space, the Mordell–Weil group of each elliptic fibration is a group of parabolic transformations fixing the point at infinity that corresponds to the class $e \in M$ of a fiber (which has $\langle e, e \rangle = 0$). Since a parabolic transformation fixes only one point of the sphere at infinity, the subgroups $\mathbb{Z}^{n-1}$ and $\mathbb{Z}^a$ in G intersect only in the identity. It follows that the subgroup $\mathbb{Z}^{n-1}$ has infinite index in G.

We are given that X contains a (-2)-curve C. I claim that C has infinitely many translates under the Mordell–Weil group $\mathbb{Z}^{n-1}$. Indeed, any curve with finitely many orbits under $\mathbb{Z}^{n-1}$ must be contained in a fiber of $X \to \mathbb{P}^1$. Since all fibers are irreducible, the fibers have self-intersection 0, not -2. Thus X contains infinitely many (-2)-curves. Therefore the group

$$W \subset O(M)$$

generated by reflections in (-2)-vectors is infinite. Here W acts simply transitively on the Weyl chambers of the positive cone (separated by hyperplanes $v^\perp$ with v a (-2)-vector), whereas $G = \mathrm{Aut}^*(X)$ preserves one Weyl chamber, the ample cone of X. So G and W intersect only in the identity. Since W is infinite, G has infinite index in $O(M)$. By Theorem 7.1, G is not commensurable with an arithmetic group. $\square$

Proof of Example 7.3. The given curve C in the linear system $|O(4, 4)| = |-2K_{\mathbb{P}^1 \times \mathbb{P}^1}|$ is smooth. One can check this with Macaulay 2, for example. Therefore the double cover X of $\mathbb{P}^1 \times \mathbb{P}^1$ ramified along C is a smooth K3 surface. The two projections from X to $\mathbb{P}^1$ are elliptic fibrations. Typically, such a double cover $\pi : X \to \mathbb{P}^1 \times \mathbb{P}^1$ would have Picard number 2, but the curve C has been chosen to be tangent at 4 points to each of two curves of degree $(1, 1)$, $D_1 = \{xv = yu\}$ and $D_2 = \{xv = -yu\}$. (These points are $[x, y] = [u, v]$ equal to $[1, 1]$, $[1, 2]$, $[1, -1]$, $[1, -2]$ on D_1 and $[x, y] = [u, -v]$ equal to $[1, 1]$, $[1, 2]$, $[1, -1]$, $[1, -2]$ on D_2.) It follows that the double covering is trivial over D_1 and D_2, outside the ramification curve C: the inverse image in X of each curve D_i is a union of two curves, $\pi^{-1}(D_i) = E_i \cup F_i$, meeting transversely at 4 points. The smooth rational curves E_1, F_1, E_2, F_2 on X are (-2)-curves, since X is a K3 surface.

The curves D_1 and D_2 meet transversely at the two points $[x, y] = [u, v]$ equal to $[1, 0]$ or $[0, 1]$. Let us compute that the double covering

$$\pi : X \to \mathbb{P}^1 \times \mathbb{P}^1$$

is trivial over the union of D_1 and D_2 (outside the ramification curve C). Indeed, if we write X as $w^2 = f(x, y, z, w)$ where f is the given polynomial of degree $(4, 4)$, then a section of π over $D_1 \cup D_2$ is given by

$$w = 4x^2 u^2 - 5x^2 v^2 + y^2 v^2.$$

We can name the curves E_i, F_i so that the image of this section is $E_1 \cup E_2$ and the image of the section $w = -(4x^2 u^2 - 5x^2 v^2 + y^2 v^2)$ is $F_1 \cup F_2$. Then E_1 and F_2 are disjoint. So the intersection form among the divisors $\pi^* O(1, 0), \pi^* O(0, 1), E_1, F_2$ on X is given by

$$\begin{pmatrix} 0 & 2 & 1 & 1 \\ 2 & 0 & 1 & 1 \\ 1 & 1 & -2 & 0 \\ 1 & 1 & 0 & -2 \end{pmatrix}$$

Since this matrix has determinant -32, not zero, X has Picard number at least 4.

Finally, we compute that the two projections from $C \subset \mathbb{P}^1 \times \mathbb{P}^1$ to $\mathbb{P}^1$ are each ramified over 24 distinct points in $\mathbb{P}^1$. It follows that all fibers of our two elliptic fibrations $X \to \mathbb{P}^1$ are irreducible. By Corollary 7.2, the automorphism group of X (over $\mathbb{C}$, or equivalently over $\overline{\mathbb{Q}}$) is not commensurable with an arithmetic group. Our calculations have all worked over $\mathbb{Q}$, and so Corollary 7.2 also gives that $\mathrm{Aut}(X_{\mathbb{Q}})$ is not commensurable with an arithmetic group. $\square$

8. Klt pairs

We will see that the previous results can be generalized from Calabi–Yau varieties to a broader class of varieties using the language of pairs. For the rest of the paper, we work over the complex numbers.

A normal variety X is $\mathbb{Q}$-*factorial* if for every point p and every codimension-one subvariety S through p, there is a regular function on some neighborhood of p that vanishes exactly on S (to some positive order).

Definition 8.1. A pair (X, Δ) is a $\mathbb{Q}$-factorial projective variety X with an effective $\mathbb{R}$-divisor Δ on X.

Notice that Δ is an actual $\mathbb{R}$-divisor $\Delta = \sum a_i \Delta_i$, not a numerical equivalence class of divisors. We think of $K_X + \Delta$ as the canonical bundle of the pair (X, Δ). The following definition picks out an important class of "mildly singular" pairs.

Definition 8.2. A pair (X, Δ) is *klt* (Kawamata log terminal) if the following holds. Let $\pi : \widetilde{X} \to X$ be a resolution of singularities. Suppose that the union of the exceptional set of π (the subset of $\widetilde{X}$ where π is not an isomorphism) with $\pi^{-1}(\Delta)$ is a divisor with simple normal crossings. Define a divisor $\widetilde{\Delta}$ on $\widetilde{X}$ by

$$K_{\widetilde{X}} + \widetilde{\Delta} = \pi^*(K_X + \Delta).$$

We say that (X, Δ) is klt if all coefficients of $\widetilde{\Delta}$ are less than 1. This property is independent of the choice of resolution.

Example 8.3. A surface $X = (X, 0)$ is klt if and only if X has only quotient singularities [Kollár and Mori 1998, Proposition 4.18].

Example 8.4. For a smooth variety X and Δ a divisor with simple normal crossings (and some coefficients), the pair (X, Δ) is klt if and only if Δ has coefficients less than 1.

All the main results of minimal model theory, such as the cone theorem, generalize from smooth varieties to klt pairs. For example, the Fano case of the cone theorem becomes [Kollár and Mori 1998, Theorem 3.7]:

Theorem 8.5. *Let* (X, Δ) *be a klt Fano pair, meaning that* $-(K_X + \Delta)$ *is ample. Then* $\overline{\mathrm{Curv}}(X)$ *(and hence the dual cone* $\mathrm{Nef}(X)$*) is rational polyhedral.*

Notice that the conclusion does not involve the divisor Δ. This shows the power of the language of pairs. A variety X may not be Fano, but if we can find an $\mathbb{R}$-divisor Δ that makes (X, Δ) a klt Fano pair, then we get the same conclusion (that the cone of curves and the nef cone are rational polyhedral) as if X were Fano.

Example 8.6. Let X be the blow-up of $\mathbb{P}^2$ at any number of points on a smooth conic. As an exercise, the reader can write down an $\mathbb{R}$-divisor Δ such that (X, Δ) is a klt Fano pair. This proves that the nef cone of X is rational polyhedral, as proved by other methods in [Galindo and Monserrat 2005, Corollary 3.3; Mukai 2005; Castravet and Tevelev 2006]. These surfaces are definitely not Fano if we blow up 6 or more points. Their Betti numbers are unbounded, in contrast to the smooth Fano surfaces.

More generally, Testa, Várilly-Alvarado, and Velasco proved that every smooth projective rational surface X with $-K_X$ big has finitely generated Cox ring [Testa et al. 2009]. Finite generation of the Cox ring (the ring of all sections of all line bundles) is stronger than the nef cone being rational polyhedral, by the analysis of [Hu and Keel 2000]. Chenyang Xu showed that a rational surface with $-K_X$ big need not have any divisor Δ with (X, Δ) a klt Fano pair [Testa et al. 2009]. I do not know whether the blow-ups of higher-dimensional projective spaces considered by in [Mukai 2005] and [Castravet and Tevelev 2006] have a divisor Δ with (X, Δ) a klt Fano pair.

It is therefore natural to extend the Morrison–Kawamata cone conjecture from Calabi–Yau varieties to *Calabi–Yau pairs* (X, Δ), meaning that $K_X + \Delta \equiv 0$. The conjecture is reasonable, since we can prove it in dimension 2 [Totaro 2010].

Theorem 8.7. *Let* (X, Δ) *be a klt Calabi–Yau pair of dimension 2. Then* $\mathrm{Aut}(X, \Delta)$ *(and also* $\mathrm{Aut}(X)$*) acts with a rational polyhedral fundamental domain on the cone* $\mathrm{Nef}(X) \subset N^1(X)$.

Here is a more concrete consequence of Theorem 8.7:

Corollary 8.8 [Totaro 2010]. *Let* (X, Δ) *be a klt Calabi–Yau pair of dimension 2. Then there are only finitely many contractions of X up to automorphisms of X. Related to that*: $\mathrm{Aut}(X)$ *has finitely many orbits on the set of curves in X with negative self-intersection.*

This was shown in one class of examples [Dolgachev and Zhang 2001]. These results are false for surfaces in general, even for some smooth rational surfaces:

Example 8.9. Let X be the blow-up of $\mathbb{P}^2$ at 9 very general points. Then $\mathrm{Nef}(X)$ is not rational polyhedral, since X contains infinitely many (-1)-curves. But $\mathrm{Aut}(X) = 1$ [Gizatullin 1980, Proposition 8], and so the conclusion fails for X.

Moreover, let Δ be the unique smooth cubic curve in $\mathbb{P}^2$ through the 9 points, with coefficient 1. Then $-K_X \equiv \Delta$, and so (X, Δ) is a *log-canonical* (and even canonical) Calabi–Yau pair. The theorem therefore fails for such pairs.

We now give a classical example (besides the case $\Delta = 0$ of Calabi–Yau surfaces) where Theorem 8.7 applies.

Example 8.10. Let X be the blow-up of $\mathbb{P}^2$ at 9 points $p_1, \ldots, p_9$ which are the intersection of two cubic curves. Then taking linear combinations of the two cubics gives a $\mathbb{P}^1$-family of elliptic curves through the 9 points. These curves become disjoint on the blow-up X, and so we have an elliptic fibration $X \to \mathbb{P}^1$. This morphism is given by the linear system $|-K_X|$. Using that, we see that the (-1)-curves on X are exactly the sections of the elliptic fibration $X \to \mathbb{P}^1$.

In most cases, the Mordell–Weil group of $X \to \mathbb{P}^1$ is $\doteq \mathbb{Z}^8$. So X contains infinitely many (-1)-curves, and so the cone Nef(X) is not rational polyhedral. But Aut(X) acts transitively on the set of (-1)-curves, by translations using the group structure on the fibers of $X \to \mathbb{P}^1$. That leads to the proof of Theorem 8.7 in this example. (The theorem applies, in the sense that there is an $\mathbb{R}$-divisor Δ with (X, Δ) klt Calabi–Yau: let Δ be the sum of two smooth fibers of $X \to \mathbb{P}^1$ with coefficients $\frac{1}{2}$, for example.)

9. The cone conjecture in dimension greater than 2

In higher dimensions, the cone conjecture also predicts that a klt Calabi–Yau pair (X, Δ) has only finitely many small $\mathbb{Q}$-factorial modifications $X \dashrightarrow X_1$ *up to pseudo-automorphisms of X*. (See [Kawamata 1997; Totaro 2010] for the full statement of the cone conjecture in higher dimensions.) A pseudo-automorphism means a birational automorphism which is an isomorphism in codimension 1.

More generally, the conjecture implies that X has only finitely many birational contractions $X \dashrightarrow Y$ modulo pseudo-automorphisms of X, where a birational contraction means a dominant rational map that extracts no divisors. There can be infinitely many small modifications if we do not divide out by the group PsAut(X) of pseudo-automorphisms of X.

Kawamata proved a relative version of the cone conjecture for a 3-fold X with a K3 fibration or elliptic fibration $X \to S$ [Kawamata 1997]. Here X can have infinitely many minimal models (or small modifications) over S, but it has only finitely many modulo PsAut(X/S).

This is related to other finiteness problems in minimal model theory. We know that a klt pair (X, Δ) has only finitely many minimal models if Δ is big [Birkar et al. 2010, Corollary 1.1.5]. It follows that a variety of general type has a finite set of minimal models. A variety of nonmaximal Kodaira dimension can have infinitely many minimal models [Reid 1983, Section 6.8; 1997]. But it is conjectured that every variety X has only finitely many minimal models *up to isomorphism*, meaning that we ignore the birational identification with X. Kawamata's results on Calabi–Yau fiber spaces imply at least that 3-folds of positive Kodaira dimension have only finitely many minimal models up to isomorphism [Kawamata 1997, Theorem 4.5]. If the abundance conjecture

[Kollár and Mori 1998, Corollary 3.12] holds (as it does in dimension 3), then every nonuniruled variety has an Iitaka fibration where the fibers are Calabi–Yau. The cone conjecture for Calabi–Yau fiber spaces (plus abundance) implies finiteness of minimal models up to isomorphism for arbitrary varieties.

The cone conjecture is wide open for Calabi–Yau 3-folds, despite significant results by Oguiso and Peternell [1998], Szendrői [1999], Uehara [2004], and Wilson [1994]. Hassett and Tschinkel recently [2010] checked the conjecture for a class of holomorphic symplectic 4-folds.

10. Outline of the proof of Theorem 8.7

The proof of Theorem 8.7 gives a good picture of the Calabi–Yau pairs of dimension 2. We summarize the proof from [Totaro 2010]. In most cases, if (X, Δ) is a Calabi–Yau pair, then X turns out to be rational. It is striking that the most interesting case of the theorem is proved by reducing properties of certain rational surfaces to the Torelli theorem for K3 surfaces.

Let (X, Δ) be a klt Calabi–Yau pair of dimension 2. That is, $K_X + \Delta \equiv 0$, or equivalently

$$-K_X \equiv \Delta,$$

where Δ is effective. We can reduce to the case where X is smooth by taking a suitable resolution of (X, Δ).

If $\Delta = 0$, then X is a smooth Calabi–Yau surface, and the result is known by Sterk, using the Torelli theorem for K3 surfaces. So assume that $\Delta \neq 0$. Then X has Kodaira dimension

$$\kappa(X) := \kappa(X, K_X)$$

equal to $-\infty$. With one easy exception, Nikulin showed that our assumptions imply that X is *rational* [Alexeev and Mori 2004, Lemma 1.4]. So assume that X is rational from now on.

We have three main cases for the proof, depending on whether the Iitaka dimension $\kappa(X, -K_X)$ is 0, 1, or 2. (It is nonnegative because $-K_X \sim_{\mathbb{R}} \Delta \geq 0$.) By definition, the Iitaka dimension $\kappa(X, L)$ of a line bundle L is $-\infty$ if $h^0(X, mL) = 0$ for all positive integers m. Otherwise, $\kappa(X, L)$ is the natural number r such that there are positive integers a, b and a positive integer m_0 with $am^r \leq h^0(X, mL) \leq bm^r$ for all positive multiples m of m_0 [Lazarsfeld 2004, Corollary 2.1.38].

10.1. Case where $\kappa(X, -K_X) = 2$. That is, $-K_X$ is big. In this case, there is an $\mathbb{R}$-divisor Γ such that (X, Γ) is klt Fano. Therefore $\mathrm{Nef}(X)$ is rational polyhedral by the cone theorem, and hence the group $\mathrm{Aut}^*(X)$ is finite. So Theorem 8.7 is true in a simple way. More generally, for (X, Γ) klt Fano of any dimension, the

Cox ring of X is finitely generated, by Birkar, Cascini, Hacon, and McKernan [2010].

This proof illustrates an interesting aspect of working with pairs: rather than Fano being a different case from Calabi–Yau, Fano becomes a special case of Calabi–Yau. That is, if (X, Γ) is a klt Fano pair, then there is another effective $\mathbb{R}$-divisor Δ with (X, Δ) a klt Calabi–Yau pair.

10.2. Case where $\kappa(X, -K_X) = 1$. In this case, some positive multiple of $-K_X$ gives an elliptic fibration $X \to \mathbb{P}^1$, not necessarily minimal. Here Aut$^*(X)$ equals the Mordell–Weil group of $X \to \mathbb{P}^1$ up to finite index, and so Aut$^*(X) \doteq \mathbb{Z}^n$ for some n. This generalizes the example of $\mathbb{P}^2$ blown up at the intersection of two cubic curves.

The (-1)-curves in X are multisections of $X \to \mathbb{P}^1$ of a certain fixed degree. One shows that Aut(X) has only finitely many orbits on the set of (-1)-curves in X. This leads to the statement of Theorem 8.7 in terms of cones.

10.3. Case where $\kappa(X, -K_X) = 0$. This is the hardest case. Here Aut$^*(X)$ can be a fairly general group acting on hyperbolic space; in particular, it can be highly nonabelian.

Here $-K_X \equiv \Delta$ where the intersection pairing on the curves in Δ is negative definite. We can contract all the curves in Δ, yielding a singular surface Y with $-K_Y \equiv 0$. Note that Y is klt and hence has quotient singularities, but it must have worse than ADE singularities, because it is a singular Calabi–Yau surface that is rational.

Let I be the "global index" of Y, the least positive integer with $I K_Y$ Cartier and linearly equivalent to zero. Then

$$Y = M/(\mathbb{Z}/I)$$

for some Calabi–Yau surface M with ADE singularities. The minimal resolution of M is a smooth Calabi–Yau surface. Using the Torelli theorem for K3 surfaces, this leads to the proof of the theorem for M and then for Y, by Oguiso and Sakurai [2001, Corollary 1.9].

Finally, we have to go from Y to its resolution of singularities, the smooth rational surface X. Here Nef(X) is more complex than Nef(Y): X typically contains infinitely many (-1)-curves, whereas Y has none (because $K_Y \equiv 0$). Nonetheless, since we know "how big" Aut(Y) is (up to finite index), we can show that the group

$$\text{Aut}(X, \Delta) = \text{Aut}(Y)$$

has finitely many orbits on the set of (-1)-curves. This leads to the proof of Theorem 8.7 for (X, Δ). QED

11. Example

Here is an example of a smooth rational surface with a highly nonabelian (discrete) automorphism group, considered by Zhang [1991, Theorem 4.1, p. 438], Blache [1995, Theorem C(b)(2)], and [2010, Section 2]. This is an example of the last case in the proof of Theorem 8.7, where $\kappa(X, -K_X) = 0$. We will also see a singular rational surface whose nef cone is round, of dimension 4.

Let X be the blow-up of $\mathbb{P}^2$ at the 12 points: $[1, \zeta^i, \zeta^j]$ for $i, j \in \mathbb{Z}/3$, $[1, 0, 0]$, $[0, 1, 0]$, $[0, 0, 1]$. Here ζ is a cube root of 1. (This is the dual of the "Hesse configuration" [Dolgachev 2004, Section 4.6]. There are 9 lines $L_1, \ldots, L_9$ through quadruples of the 12 points in $\mathbb{P}^2$.)

On $\mathbb{P}^2$, we have

$$-K_{\mathbb{P}^2} \equiv 3H \equiv \sum_{i=1}^{9} \tfrac{1}{3} L_i.$$

On the blow-up X, we have

$$-K_X \equiv \sum_{i=1}^{9} \tfrac{1}{3} L_i,$$

where $L_1, \ldots, L_9$ are the proper transforms of the 9 lines, which are now disjoint and have self-intersection number -3. Thus $\left(X, \sum_{i=1}^{9} \tfrac{1}{3} L_i\right)$ is a klt Calabi–Yau pair.

Section 10.3 shows how to analyze X: contract the 9 (-3)-curves L_i on X. This gives a rational surface Y with 9 singular points (of type $\tfrac{1}{3}(1, 1)$) and $\rho(Y) = 4$. We have $-K_Y \equiv 0$, so Y is a klt Calabi–Yau surface which is rational. We have $3K_Y \sim 0$, and so $Y \cong M/(\mathbb{Z}/3)$ with M a Calabi–Yau surface with ADE singularities. It turns out that M is smooth, $M \cong E \times E$ where E is the Fermat cubic curve

$$E \cong \mathbb{C}/\mathbb{Z}[\zeta] \cong \{[x, y, z] \in \mathbb{P}^2 : x^3 + y^3 = z^3\},$$

and $\mathbb{Z}/3$ acts on $E \times E$ as multiplication by (ζ, ζ) [Totaro 2010, Section 2].

Since E has endomorphism ring $\mathbb{Z}[\zeta]$, the group $GL(2, \mathbb{Z}[\zeta])$ acts on the abelian surface $M = E \times E$. This passes to an action on the quotient variety $Y = M/(\mathbb{Z}/3)$ and hence on its minimal resolution X (which is the blow-up of $\mathbb{P}^2$ at 12 points we started with). Thus the infinite, highly nonabelian discrete group $GL(2, \mathbb{Z}[\zeta])$ acts on the smooth rational surface X. This is the whole automorphism group of X up to finite groups (*loc. cit.*).

Here $\text{Nef}(Y) = \text{Nef}(M)$ is a round cone in $\mathbb{R}^4$, and so Theorem 8.7 says that $PGL(2, \mathbb{Z}[\zeta])$ acts with finite covolume on hyperbolic 3-space. In fact, the quotient of hyperbolic 3-space by an index-24 subgroup of $PGL(2, \mathbb{Z}[\zeta])$ is familiar to topologists as the complement of the figure-eight knot [Maclachlan and Reid 2003, 1.4.3, 4.7.1].

References

[Alekseevskij et al. 1993] D. V. Alekseevskij, È. B. Vinberg, and A. S. Solodovnikov, "Geometry of spaces of constant curvature", pp. 1–138 in *Geometry*, vol. II, Encyclopaedia Math. Sci. **29**, Springer, Berlin, 1993.

[Alexeev and Mori 2004] V. Alexeev and S. Mori, "Bounding singular surfaces of general type", pp. 143–174 in *Algebra, arithmetic and geometry with applications* (West Lafayette, IN, 2000), Springer, Berlin, 2004.

[Birkar et al. 2010] C. Birkar, P. Cascini, C. D. Hacon, and J. McKernan, "Existence of minimal models for varieties of log general type", *J. Amer. Math. Soc.* **23**:2 (2010), 405–468.

[Blache 1995] R. Blache, "The structure of l.c. surfaces of Kodaira dimension zero. I", *J. Algebraic Geom.* **4**:1 (1995), 137–179.

[Borcherds 1998] R. E. Borcherds, "Coxeter groups, Lorentzian lattices, and *K*3 surfaces", *Internat. Math. Res. Notices* **1998**:19 (1998), 1011–1031.

[Borel and Harish-Chandra 1962] A. Borel and Harish-Chandra, "Arithmetic subgroups of algebraic groups", *Ann. of Math.* (2) **75** (1962), 485–535.

[Borel and Serre 1964] A. Borel and J.-P. Serre, "Théorèmes de finitude en cohomologie galoisienne", *Comment. Math. Helv.* **39** (1964), 111–164.

[Bridson and Haefliger 1999] M. R. Bridson and A. Haefliger, *Metric spaces of non-positive curvature*, Grundlehren der Mathematischen Wissenschaften **319**, Springer, Berlin, 1999.

[Brown 1982] K. S. Brown, *Cohomology of groups*, Graduate Texts in Mathematics **87**, Springer, New York, 1982.

[Castravet and Tevelev 2006] A.-M. Castravet and J. Tevelev, "Hilbert's 14th problem and Cox rings", *Compos. Math.* **142**:6 (2006), 1479–1498.

[Debarre 2001] O. Debarre, *Higher-dimensional algebraic geometry*, Springer, New York, 2001.

[Dolgachev 2004] I. V. Dolgachev, "Abstract configurations in algebraic geometry", pp. 423–462 in *The Fano Conference*, Univ. Torino, Turin, 2004.

[Dolgachev 2008] I. V. Dolgachev, "Reflection groups in algebraic geometry", *Bull. Amer. Math. Soc.* (*N.S.*) **45**:1 (2008), 1–60.

[Dolgachev and Zhang 2001] I. V. Dolgachev and D.-Q. Zhang, "Coble rational surfaces", *Amer. J. Math.* **123**:1 (2001), 79–114.

[de Fernex 2010] T. de Fernex, "The Mori cone of blow-ups of the plane", preprint, 2010. arXiv 1001.5243

[Galindo and Monserrat 2005] C. Galindo and F. Monserrat, "The total coordinate ring of a smooth projective surface", *J. Algebra* **284**:1 (2005), 91–101.

[Gizatullin 1980] M. H. Gizatullin, "Rational *G*-surfaces", *Izv. Akad. Nauk SSSR Ser. Mat.* **44**:1 (1980), 110–144. In Russian; translated in *Math. USSR Izv.* **16** (1981), 103–134.

[Hartshorne 1977] R. Hartshorne, *Algebraic geometry*, Graduate Texts in Mathematics **52**, Springer, New York, 1977.

[Hassett and Tschinkel 2010] B. Hassett and Y. Tschinkel, "Flops on holomorphic symplectic fourfolds and determinantal cubic hypersurfaces", *J. Inst. Math. Jussieu* **9**:1 (2010), 125–153.

[Hu and Keel 2000] Y. Hu and S. Keel, "Mori dream spaces and GIT", *Michigan Math. J.* **48** (2000), 331–348.

[Iskovkikh and Prokhorov 1999] V. Iskovkikh and Y. Prokhorov, "Fano varieties", pp. 1–247 in *Algebraic geometry*, vol. V, Encyclopaedia of Mathematical Sciences **47**, Springer, Berlin, 1999.

[Kawamata 1997] Y. Kawamata, "On the cone of divisors of Calabi–Yau fiber spaces", *Internat. J. Math.* **8**:5 (1997), 665–687.

[Kollár and Mori 1998] J. Kollár and S. Mori, *Birational geometry of algebraic varieties*, Cambridge Tracts in Mathematics **134**, Cambridge University Press, Cambridge, 1998.

[Kollár et al. 1992] J. Kollár, Y. Miyaoka, and S. Mori, "Rational connectedness and boundedness of Fano manifolds", *J. Differential Geom.* **36**:3 (1992), 765–779.

[Kovács 1994] S. J. Kovács, "The cone of curves of a $K3$ surface", *Math. Ann.* **300**:4 (1994), 681–691.

[Lazarsfeld 2004] R. Lazarsfeld, *Positivity in algebraic geometry, I: classical setting: line bundles and linear series*, Ergebnisse der Math. (3) **48**, Springer, Berlin, 2004.

[Maclachlan and Reid 2003] C. Maclachlan and A. W. Reid, *The arithmetic of hyperbolic 3-manifolds*, Graduate Texts in Mathematics **219**, Springer, New York, 2003.

[Mazur 1993] B. Mazur, "On the passage from local to global in number theory", *Bull. Amer. Math. Soc. (N.S.)* **29**:1 (1993), 14–50.

[Mukai 2005] S. Mukai, "Finite generation of the Nagata invariant rings in A-D-E cases", preprint 1502, RIMS, 2005.

[Mumford 1970] D. Mumford, *Abelian varieties*, Tata Studies in Mathematics **5**, Tata Institute of Fundamental Research, Bombay, 1970.

[Namikawa 1985] Y. Namikawa, "Periods of Enriques surfaces", *Math. Ann.* **270**:2 (1985), 201–222.

[Nikulin 1979] V. V. Nikulin, "Integer symmetric bilinear forms and some of their geometric applications", *Izv. Akad. Nauk SSSR Ser. Mat.* **43**:1 (1979), 111–177. In Russian; translated in *Math. USSR Izv.* **14** (1979), 103–167.

[Oguiso and Peternell 1998] K. Oguiso and T. Peternell, "Calabi–Yau threefolds with positive second Chern class", *Comm. Anal. Geom.* **6**:1 (1998), 153–172.

[Oguiso and Sakurai 2001] K. Oguiso and J. Sakurai, "Calabi–Yau threefolds of quotient type", *Asian J. Math.* **5**:1 (2001), 43–77.

[Prasad 1976] G. Prasad, "Discrete subgroups isomorphic to lattices in Lie groups", *Amer. J. Math.* **98**:4 (1976), 853–863.

[Reid 1983] M. Reid, "Minimal models of canonical 3-folds", pp. 131–180 in *Algebraic varieties and analytic varieties* (Tokyo, 1981), Adv. Stud. Pure Math. **1**, North-Holland, Amsterdam, 1983.

[Serre 1979] J.-P. Serre, "Arithmetic groups", pp. 105–136 in *Homological group theory* (Durham, 1977), London Math. Soc. Lecture Note Ser. **36**, Cambridge Univ. Press, Cambridge, 1979. Reprinted in his *Œuvres*, v. 3, Springer (1986), 503–534.

[Shioda 1972] T. Shioda, "On elliptic modular surfaces", *J. Math. Soc. Japan* **24** (1972), 20–59.

[Sterk 1985] H. Sterk, "Finiteness results for algebraic $K3$ surfaces", *Math. Z.* **189**:4 (1985), 507–513.

[Szendrői 1999] B. Szendrői, "Some finiteness results for Calabi–Yau threefolds", *J. London Math. Soc.* (2) **60**:3 (1999), 689–699.

[Testa et al. 2009] D. Testa, A. Várilly-Alvarado, and M. Velasco, "Big rational surfaces", preprint, 2009. To appear in *Math. Ann.* arXiv 0901.1094

[Tits 1966] J. Tits, "Classification of algebraic semisimple groups", pp. 33–62 in *Algebraic groups and discontinuous subgroups* (Boulder, CO, 1965), Amer. Math. Soc., Providence, RI, 1966, 1966.

[Totaro 2010] B. Totaro, "The cone conjecture for Calabi–Yau pairs in dimension 2", *Duke Math. J.* **154**:2 (2010), 241–263.

[Uehara 2004] H. Uehara, "Calabi–Yau threefolds with infinitely many divisorial contractions", *J. Math. Kyoto Univ.* **44**:1 (2004), 99–118.

[Wilson 1994] P. M. H. Wilson, "Minimal models of Calabi–Yau threefolds", pp. 403–410 in *Classification of algebraic varieties* (L'Aquila, 1992), Contemp. Math. **162**, Amer. Math. Soc., Providence, RI, 1994.

[Yau 1978] S. T. Yau, "On the Ricci curvature of a compact Kähler manifold and the complex Monge–Ampère equation, I", *Comm. Pure Appl. Math.* **31**:3 (1978), 339–411.

[Zhang 1991] D.-Q. Zhang, "Logarithmic Enriques surfaces", *J. Math. Kyoto Univ.* **31**:2 (1991), 419–466.

b.totaro@dpmms.cam.ac.uk   *Department of Pure Mathematics and Mathematical Statistics, Cambridge University, Wilberforce Road, Cambridge CB3 0WB, England*

Printed in the United States
By Bookmasters